Advances in Biochemical and Molecular Mechanisms of Plant–Pathogen Interaction

Online at: https://doi.org/10.1088/978-0-7503-5673-2

Advances in Biochemical and Molecular Mechanisms of Plant–Pathogen Interaction

Edited by
Hitendra K Patel
CSIR-Centre for Cellular and Molecular Biology, Hyderabad, Telangana, 500007, India

and

Academy of Scientific and Innovative Research (AcSIR), Ghaziabad, Uttar Pradesh, 201002, India

Anirudh Kumar
Department of Botany, Indira Gandhi National Tribal University, Amarkantak, India

and

Department of Botany, Central Tribal University of Andhra Pradesh, Vizianagaram, Andhra Pradesh, 535003, India

IOP Publishing, Bristol, UK

ISBN 978-0-7503-5673-2 (ebook)
ISBN 978-0-7503-5671-8 (print)
ISBN 978-0-7503-5674-9 (myPrint)
ISBN 978-0-7503-5672-5 (mobi)

DOI 10.1088/978-0-7503-5673-2

Version: 20250601

IOP ebooks

British Library Cataloguing-in-Publication Data: A catalogue record for this book is available from the British Library.

Published by IOP Publishing, wholly owned by The Institute of Physics, London

IOP Publishing, No.2 The Distillery, Glassfields, Avon Street, Bristol, BS2 0GR, UK

US Office: IOP Publishing, Inc., 190 North Independence Mall West, Suite 601, Philadelphia, PA 19106, USA

This book is dedicated to all the farmers, whose tireless efforts feed the world's 8.2 billion people, and to the authors of the chapters, in recognition of their hard work and unwavering dedication.

Contents

Preface

Plants' sessile nature makes them vulnerable to a variety of biotic stresses. Pests and pathogens such as viruses, nematodes, bacteria, fungus, and herbivorous insects are biotic stressors in plants. Crop productivity is severely harmed as a result of these biotic stressors. The great Bengal famine of 1943, the potato blight in Ireland, and coffee rot in Brazil are only a few examples of historical biotic stresses. Furthermore, climate change has the potential to alter the biotic stress paradigm. Changes in global temperature, for example, can cause insect pests to spread across more territory, produce more generations per year, and increase the incidence of insect-transmitted plant diseases. These changes result in significant crop productivity losses, posing a serious danger to world food security.

Plants have evolved to protect themselves from invading pests and pathogens by responding to these biotic stresses through a variety of morphological, biochemical, and molecular processes by detecting pathogen attacks and stop them before they can cause significant damage to the host tissues. Understanding the important defence systems, like pathogen triggered immunity (PTI), effectors triggered immunity (ETI), pathogenesis-related proteins (PR-proteins) etc, may help to protect crops from various biotic constraints, resulting in increased crop output. The emergencein recent years of omics methodologies, such as next-generation sequencing (NGS), genomics, transcriptomics, proteomics, and metabolomics has pushed agriculture research toward the development of climate-smart crops that are resistant to pathogens and insect pests. In light of the above, the book *Advances in Biochemical and Molecular Mechanisms of Plant–Pathogen Interaction* provides a complete and cohesive assessment of plant defence mechanisms in response to various biotic stressors. It also discusses the use of genetic engineering and molecular breeding to generate pathogen-resistant crops, as well as several case studies in this area. In addition, this book offers a path forward for dealing with impending issues, which will pique the readers' interest.

The number of research articles currently being published on the issue demonstrates a major increase in interest in the study of the biochemistry, physiology, and molecular events of host–pathogen interactions in recent years. The increased attention is most likely due to evidence that efficient control of many plant diseases is largely dependent on a thorough understanding of host–pathogen interactions. This increased study effort necessitates the publication of more books like this one, which is designed to collect, compile, analyse and evaluate widely dispersed pieces of knowledge on the subject. The study of host–pathogen interactions is concerned with the ongoing battle between plants and pathogens throughout their co-evolution and such interactions are generally complex. Plants have evolved both generic and highly specific defence systems to combat the bulk of microbial infections they encounter. Specific interactions between pathogen avr (avirulence) gene loci and alleles of the corresponding plant disease resistance (R) loci control highly specialised defence. Microbes from the same species can function differently in their co-evolution with the specific host plant, which has similarly evolved its reaction to external threats, making these defences incredibly dynamic. Because of newly developed methodologies and the availability of genomic, proteomic and metabolomic information, there have been significant advancements in the field of plant–pathogen interactions in recent years. Molecular and biochemical plant–pathogen interactions explore novel discoveries with an emphasis on the mechanisms that control plant disease resistance, cross-talk across the pathways involved, and pathogen strategies for overturning these defences. This book will be valuable for researchers and students in plant pathology and plant biology-related fields since it documents developments in plant defences, pathogen counter-defence, and their interactions.

It also sheds light on pathogen classes, the arms race that occurs between host and pathogen, effector molecules like pathogen associated molecular patterns (PAMPs), receptor molecules like pattern recognition receptors (PRRs) and nucleotide-binding site and leucine-rich repeats (NBS-LRR) proteins, signalling components like mitogen-activated protein kinases (MAPKs), regulatory molecules like phytohormones and microRNA (miRNA), transcription factors like WRKY, defence-related proteins like pathogenesis-related proteins (PR-proteins), and defensive metabolites like secondary metabolites. In addition, it provides a case study of crop plants concerning how genetic engineering, epigenetic modifications, and mutation breeding strategies have been used for enhancing productivity, resilience and food security. The updated information makes this book a valuable source of knowledge for all those working in crop development since it provides a comprehensive picture of plant–pathogen interaction.

Briefly, the topics covered in this book include:

- Providing both fundamental and cutting-edge information on the molecular and biochemical mechanisms underlying plant–pathogen interactions.
- Providing a comprehensive picture of the plant–pathogen interaction, from the microbe's sensing to the molecular and biochemical changes that occur in plants to attain tolerance/resistance.
- Based on a thorough latest review of the literature, discusses all areas of pathogen recognition, signal transduction and immune development.
- This book also provides the way forward approach to addressing the upcoming challenges which will be the interest of readers.

The book provides a comprehensive overview of plant–pathogen interactions at molecular level. It goes over every step of the process, from the microbe–host–cell contact at the surface to the plant's recognition of the bacterium to the activation of a plant's defence response, which includes biochemical and molecular changes to attain tolerance/resistance. It also discusses the arms race between host and pathogens, as well as effector molecules like PAMPs, receptor molecules like PRRs and NBS-LRR proteins, signalling components like MAPKs, regulatory molecules like phytohormones and miRNA, transcription factors like WRKY, defence-related proteins like PR-proteins, and defensive metabolites like secondary metabolites. The role of epigenetic changes, genome editing and mutation breeding in developing disease resistance will be discussed. We also look at how post-genomics, high-throughput technology (transcriptomics and proteomics) might be used to investigate disease outbreaks that cause crop losses in a variety of plants. The updated information in this book will be beneficial for all those working in crop development since it provides a comprehensive picture of plant–pathogen interaction.

Acknowledgements

Professor T V Kattimani

Dr Vinay K Nandicoori

We wish to express our heartfelt gratitude to **Professor T V Kattimani**, Vice-Chancellor, Central Tribal University of Andhra Pradesh, and **Dr Vinay K Nandicoori**, Director, CSIR-Centre for Cellular and Molecular Biology whose wisdom, love, patience, and passion for knowledge have profoundly shaped our professional and personal journeys. Their unwavering support and enduring inspiration have been a constant source of motivation throughout this endeavor.

This book, *Advances in Biochemical and Molecular Mechanisms of Plant–Pathogen Interaction*, is a humble reflection of the many invaluable lessons we have learned under their guidance. Their belief in our potential and steadfast encouragement have been the cornerstone of this work.

Thank you, Prof T V Kattimani and Dr Vinay K Nandicoori Sir, for igniting in us the flame of hope and confidence—truly the most powerful and effective vector of transformation.

With deep respect and admiration.

Hitendra K Patel
Anirudh Kumar

Editor biographies

Hitendra Kumar Patel

Dr Hitendra Kumar Patel completed his master's degree in biotechnology from Guru Ghasidas Central University, Bilaspur, Chhattisgarh, India and PhD in Molecular Life Sciences from International Centre for Genetic Engineering and Biotechnology (ICGEB), Trieste, Italy. He is currently working as a Senior Principal Scientist at the CSIR-Centre for Cellular and Molecular Biology (CSIR-CCMB), Hyderabad, Ministry of Science and Technology, Government of India. He is also affiliated as Associate Professor with the Academy of Scientific and Innovative Research (AcSIR), Ghaziabad, India. He has been involved in the field of plant–pathogen interactions for more than 20 years and has made significant contributions to the molecular understanding of Rice–*Xanthomonas oryzae* pv. *oryzae* interactions. His group has also developed several climate-resilient and high-yielding rice varieties. He is serving as an editorial board member of *Journal of Genetics* and recognizing his contributions, he has been selected as Associate Fellow of Telangana, India.

Anirudh Kumar

Dr Anirudh Kumar is an Associate Professor in the Department of Botany, Central Tribal University of Andhra Pradesh (CTUAP), India. He has research experience of more than 10 years in the area of plant molecular biology and plant pathology. He has received his MSc and PhD degrees from the University of Hyderabad (India's Institution of Eminence), and Postdoc from CSIR-Centre for Cellular and Molecular Biology (CSIR-CCMB), Hyderabad, India and Agricultural Research Organization (ARO), Volcani Institute, Israel. His current research interests span from antioxidants studies of medicinal plants to plant pathology. He is author and co-author of several high impact papers on different aspects of plant biology and has been the editor of two Springer books related to omics technologies for sustainable agriculture and global food security and two Institute of Physics (IOP) books related to genetic engineering and stress responses in plants. He also teaches courses for BSc, MSc and PhD degrees.

List of contributors

Rekha Balodi
ICAR-National Research Centre for Integrated Pest Management, Rajpur Khurd, Mehrauli, New Delhi 110068, India

Anjan Barman
Department of Biotechnology, Pandu College, Maligaon, Guwahati-781012, Assam, India

Niranjan Gattu
CSIR-Centre for Cellular and Molecular Biology, Hyderabad 500007, India

C G Gokulan
CSIR-Centre for Cellular and Molecular Biology, Hyderabad 500007, India

Gargi Jauhari
Department of Biotechnology, School of Life Sciences and Biotechnology, Chhatrapati Shahu Ji Maharaj University, Kanpur 208024, Uttar Pradesh, India

A S Kotasthane
Department of Plant Pathology, College of Agriculture, Indira Gandhi Krishi Vishwavidyalaya, Raipur, 492001, Chhattisgarh, India

Anirudh Kumar
Central Tribal University of A.P., Vizianagaram 535003, A.P., India

Apoorva Masade
CSIR-Centre for Cellular and Molecular Biology, Hyderabad 500007, India

Monika Naik
Department of Plant Pathology, College of Agriculture, Indira Gandhi Krishi Vishwavidyalaya, Raipur, 492001, Chhattisgarh, India

Praveen Kumar Nayak
CSIR-Centre for Cellular and Molecular Biology, Hyderabad 500007, India

Ranjith Pamirelli
College of Horticulture, Odisha University of Agriculture and Technology, Odisha, India

Alok Pandey
Department of Biotechnology, School of Life Sciences and Biotechnology, Chhatrapati Shahu Ji Maharaj University, Kanpur 208024, Uttar Pradesh, India

Hitendra K Patel
CSIR-Centre for Cellular and Molecular Biology, Hyderabad 500007, India

Ravinayak Patlavath
Department of Botany, Faculty of Science, The M S University of Baroda, Vadodara 390 002, India

Shakuntala E Pillai
Faculty of Pharmacy, The M S University of Baroda, Vadodara 390 002, India

Priyanshi Porwal
Department of Biotechnology, School of Life Sciences and Biotechnology, Chhatrapati Shahu Ji Maharaj University, Kanpur 208024, Uttar Pradesh, India

S J S Rama Devi
School of Life Sciences, Sambalpur University, Sambalpur, Odisha, India

Ashish Ranjan
Department of Plant Pathology, University of Minnesota—Twin Cities, St. Paul, MN 55108, USA

Anjana Sharma
CSIR-Centre for Cellular and Molecular Biology, Hyderabad 500007, India

Astha Singh
Department of Biotechnology, School of Life Sciences and Biotechnology, Chhatrapati Shahu Ji Maharaj University, Kanpur 208024, Uttar Pradesh, India

Lavanya Tayi
Department of Biochemistry, St. Ann's College for Women, Mehdipatnam, Hyderabad 500028, India

Kriti Tyagi
Department of Plant Pathology, University of Minnesota—Twin Cities, St. Paul, MN 55108, USA

Somya Yadav
Department of Biotechnology, School of Life Sciences and Biotechnology, Chhatrapati Shahu Ji Maharaj University, Kanpur 208024, Uttar Pradesh, India

IOP Publishing

Advances in Biochemical and Molecular Mechanisms of Plant–Pathogen Interaction

Hitendra K Patel and Anirudh Kumar

Chapter 1

The arms race of plants and pathogens: an introduction

Kriti Tyagi and Ashish Ranjan

A range of pathogens constantly challenges plants. Plants have evolved active and passive defence responses to recognize them and defend themselves. Similarly, pathogens have also developed different pathogenic and survival strategies. These counter-defence tactics are continuously evolving in the plant–microbe arms race. The focus of this chapter is to provide a molecular perspective on the strategies deployed by pathogens and plants to cstablish dominance over each other. We introduce and briefly discuss various aspects of plant immunity, including passive and active defences and their signalling components. Passive defence includes physical and chemical barriers, while active defence includes PAMP/MAMP/DAMP-triggered immunity (PTI/MTI/DTI) and effector-triggered immunity (ETI). Pathogens use countermeasures to detoxify the toxic compounds, evade recognition, and suppress these defence responses using a plethora of effectors. We briefly discuss techniques to modulate these interactions for generating disease resistance in economically important crops. However, the details of these interactions will be covered in greater detail in the following chapters of the book.

1.1 Introduction

Plants being sessile, are prone to continuous threats from various pathogens, including fungi, viruses, nematodes, bacteria and insect pests. These organisms have adopted different strategies to attack plants and obtain nutrition from them. On the basis of the feeding model and pathogenic lifestyle, microbes have been grouped into three classes: necrotrophs, biotrophs, and hemibiotrophs [1]. Biotrophs such as *Cladosporium fulvus*, and *Ralstonia solanacearum*, feed on living plant tissues while necrtrophs, such as *Sclerotinia sclerotiorum*, *Pectobacterium carotovorum*, kill the plants and feed on dead tissues. Hemibiotrophs such as *Magnaporthe oryzae*,

doi:10.1088/978-0-7503-5673-2ch1

Pseudomonas syringae initially adopt a biotrophic phase followed by a necrotrophic phase [2, 3]. Both plants and pathogens are engaged in continuous combat to dominate each other. On one hand, plants have evolved to fortify their defence system, while pathogens have developed various mechanisms to overcome plant immune responses and colonize them.

Primarily, there are four phases of plant defence responses, also called the zig-zag model of plant defence, which depend on the mode of pathogen attack [4]. In the first phase, plants utilize pattern-recognition receptors (PRRs) to recognize pathogen/microbe-associated molecular patterns (PAMPs/MAMPs, such as flagellin) or host-derived damage-associated molecular patterns (DAMPs, such as plant cell wall degradation products). This recognition triggers PAMP/MAMP-triggered immunity (PTI/MTI) or DAMP-triggered immunity (DTI) that prevent host colonization [5]. PTI includes the induction of cell wall defences (e.g., callose deposition or lignification), a burst of reactive oxygen species (ROS) within the host, and the secretion of hydrolytic enzymes like chitinases and proteases to disrupt the pathogen cell wall. However, pathogens can deploy effector molecules to overcome PTI, leading to effector-triggered susceptibility (ETS). In the next phase, plants recognize these effectors employing specialized resistance (R) proteins or NB-LRR (nucleotide-binding leucine-rich repeat) proteins, which induce effector-triggered immunity (ETI). R genes are typically effective against biotrophic pathogens but not necrotrophic ones [6]. ETI is specific to pathogens producing specific effectors and is often associated with cell death or a hypersensitive response (HR) at the infection site. It also induces the production of various signalling molecules, including phytohormones such as ethylene (ET), salicylic acid (SA), and jasmonic acid (JA) as well as antimicrobial compounds such as phytoalexins, in the infected tissues [7]. These signalling molecules spread to uninfected areas leading to transcriptional reprogramming and activation of induced systemic response (ISR) or systemic acquired resistance (SAR) within the host [8]. To suppress ETI, pathogens modulate their effector proteins, exerting selection pressure that may eventually lead to the evolution of new R genes in plants. In the ongoing arms race, some pathogens have evolved to manipulate host cell machinery to promote susceptibility. The host genes involved in enhancing disease susceptibility are called susceptibility genes (S genes) [9]. The hijacking of these susceptibility genes may further induce plant defences.

Researchers have made considerable efforts to study the molecular and evolutionary basis of these compatible or non-compatible plant–pathogen interactions. Unravelling the dynamics between plants and pathogens in terms of establishing dominance over each other can help researchers deploy various crop protection strategies. In this chapter, we have provided a brief insight into the critical events that occur during plant–pathogen encounters, including pathogen strategies to overcome plant defence, plant structural and biochemical barriers, plant cell wall based immunity, various components, including proteins and signalling molecules involved in pathogen and effector-triggered immunity. We also briefly discuss various techniques utilized to generate disease resistant plants. These topics will be elaborated in detail in different chapters of the book.

1.2 Passive defence responses in host

To infect plants and obtain nutrition, pathogens, especially fungi, and insects, need to access the plant cells. These pathogens adhere to a plant surface and penetrate into the plant cell with the help of various hydrolytic enzymes. However, plants prevent this invasion using their physical as well as biochemical barriers [153].

1.3 Physical barriers

Before penetrating the cell wall, phytopathogens must overcome the plant's physical barriers, mainly trichomes, cuticles, and stomata (figure 1.1(A)).

1.3.1 Trichomes

Trichomes are unicellular or multicellular hair-like structures protruding from the epidermal cells of plants. They play a key role as a protective barrier to pathogens and herbivore attacks. These trichomes can be secretory (glandular) and non-secretory (non-glandular). The non-secreting trichomes entrap bacteria, insects, and fungal spores by fortifying the plant cell wall. This fortification may sometimes occur through the acetylation of cell wall polymers mediated through various genes present within the trichomes [11]. Glandular secreting trichomes produce leaf exudates consisting mainly of terpenoids and phenylpropanoids that may have antimicrobial activities. These secreted compounds may vary depending on different

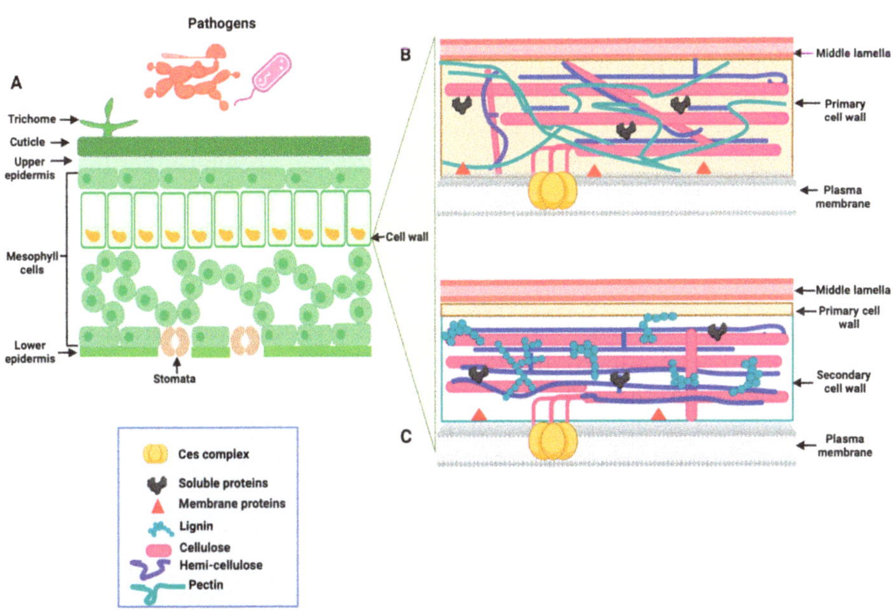

Figure 1.1. Physical barriers and cell wall mediated immunity in plants. (A) Cross-section of leaf indicating various physical barriers which interfere with pathogen colonization. Magnified cross-section of (B) primary cell wall and (C) secondary cell wall, depicting various complexes involved in cell wall immunity.

plant species and are involved in providing resistance against insect and microbial attack. Glandular trichomes in tomatoes induce JA-mediated defence signalling upon contact with herbivores [12]. A previous report suggested that mutation of two trichome birefringence-like (tbl) proteins (ostbl1 and ostbl2) were involved in xylan O-acetylation leading to susceptibility in rice to leaf blight disease [13].

1.3.2 Cuticle

The outermost epidermal layer of the aerial part of the plant consists of a waxy cuticle, one of the foremost physical barriers to an invading pathogen encounter. During penetration into plant cells, the fungal pathogens secrete cutinases to degrade the leaf cuticle and facilitate attachment on the leaf surface. The action of cutinase further leads to the production of cutin monomers, such as hexadecanediol in rice, which induces spore germination and appressorium differentiation in fungi, including *Botrytis cinerea* and *Magnaporthe grisea* [14, 154]. The release of cutin monomers can also act as DAMP elicitors to activate DTI in plants. The pathogen invasion also leads to remodelling of cuticles and secretion of antimicrobial compounds. Cuticle biosynthesis upregulates in tomatoes during attachment and appressorium formation of the fungal pathogen *Colletotrichum gloeosporioides* [15]. In addition, infection of *Colletotrichum acutatum*, in the *Citrus sinensis* leads to enhanced lipid synthesis in cuticles [10].

1.3.3 Stomata

Plants have small pores on their leaf surface, known as stomata, for gaseous exchange from the environment [16]. Various phytopathogens exploit these stomata as an entry point for efficient colonization, e.g., bacterial pathogen *Xanthomonas campestris* pv. *campestris (Xcc)* causes infection in the members of the *Brassicaceae* family through stomata [17]. Pathogens induce stomatal opening either by targeting them through type III secretion system (T3SS) proteins, also called effectors such as AvrB and HopZ1, or by using phytotoxins such as coronatine. Coronatine is a phytotoxin secreted by *P. syringae pv. Tomato (Pst) strain* DC3000 [18]. Coronatine perturbs hormonal signalling by mimicking JA-Ile (Jasmonic acid-Isoleucine) and directly engages with JA signalling pathway genes [19]. A previous study has shown that the coronatine produced by *P. syringae* facilitates stomatal opening and effective colonization through modulation of JA/SA homeostasis under high humidity conditions in *Arabidopsis* [20]. Pathogens not only favour the opening of the stomata to colonize the apoplast, but they also induce stomatal closure to enhance their multiplication. Some of the effector proteins, AvrE and HopM1 from *P. syringae* induced (ABA)-mediate stomatal closure in *Arabidopsis* providing assistance in its multiplication [21, 22]. Besides bacteria, many pathogenic fungi, including basidiomycete rust fungus, invade plants through stomatal openings [17, 23]. However, the perception of PAMPs such as flagellin or fungal chitosan promotes stomatal closure in plants [18].

1.4 Biochemical barriers

In addition to the structural barriers, plant's biochemical barriers also play a crucial role in controlling pathogens. These biochemical barrier repertoires include the production of secondary metabolites, saponins, phytoalexins, and pathogenesis-related proteins (PRs) such as chitinases, β-1,3-glucanases and proteases, polyamines, carotenoids and melatonin [24, 25]. Plants also produce phenolic metabolites such as flavanones, flavones, chalcones, isoflavones, and anthocyanins. These compounds show antimicrobial activity against various pathogens, e.g., sakuranetin, a flavone produced by rice, shows inhibitory activity against sheath blight pathogen *Rhizoctonia solani* [26]. Moreover, the metabolic studies conducted for a mutant variety of upland cotton (S156) compared to the control cultivar (S78) show a higher accumulation of flavonoids such as myricetin, quercetin, and naringenin. These flavonoids show increased resistance against *Verticillium dahliae* and *B. cinerea* infection [27, 28]. Saponins are another set of glycosylated secondary metabolites which are usually preformed or are released from constitutively stored precursors in plants. These mainly consist of three groups; triterpenoids, steroids, and steroidal glycoalkaloids [29]. α-Tomatine, a steroidal glycoalkaloid saponin present in tomato leaves, provides resistance against various fungal pathogens through ROS-mediated cell death response [30]. However, certain fungi, including *B. cinerea*, *Fusarium oxysporum*, *Verticillium albo-atrum*, and *Alternaria solani*, have evolved the ability to enzymatically degrade α-tomatine by use of enzymes tomatinases [31, 32]. Another example of triterpene glycosides saponin is avenacins, which are present in the roots of oats (*Avena* spp.) and exhibit antimicrobial activity against various soil-borne pathogens, including *Fusarium avenaceum* [33]. Moreover, recent studies have shown great metabolic variety and abundance of antimicrobial components in root exudates of plants [34]. The metabolic analysis of *Fusarium graminearum* infected barley shows induction of five antifungal phenylpropanoids, i.e., syringic, cinnamic, vanillic, ferulic, and *p*-coumaric acids which inhibit the fungal spore germination [35]. Phenylpropanoid pathway intermediates, such as ferulic acid, cinnamic acid, and 4-hydroxybenzoate were found to be upregulated in the resistant line compared to the susceptible soybean line following the *S. sclerotiorum* challenge [36]. Momilacton A, an antimicrobial diterpene, is secreted in higher amounts from roots of resistant rice cultivars upon infection with the blast fungus. In addition, exogenous application of Momilacton A on rice leaves reduced the disease symptoms and fungal growth [37]. Few of the previous studies have shown that volatile organic compounds (VOCs) having antimicrobial properties, such as monoterpene 1,8-cineole, are released from *Arabidopsis* roots in response to pathogens, *P. syringae* strain DC3000 and *Alternaria brassicola* [38]. In *Zea mays*, another class of defence molecules, benzoxazinoids are released during developmental stages or when the plant is more susceptible to prevent pathogen invasion [39]. Another set of compounds that are involved in primary defence is glucosinolates. These are the secondary metabolites, produced to a great extent by the members of the Brassicaceae family. Glucosinolates comprise a thioglucose group, a sulfonated oxime group and an amino acid side chain [40]. Glucosinolates have been effective against bacteria such as *P. syringae* and *X. campestris*, and fungi including *Fusarium* spp., *Alternaria brassicae*, *Phytophthora* spp. and *S. sclerotiorum* [41]. However, some pathogens, such as *A. brassicicola*, can avoid glucosinolates mainly by modulating the expression of plant genes [42]. In addition to this, plants also secrete hydrolases, such as chitinases, to degrade chitin (a homopolymer of β-(1 → 4)-linked N-acetylglucosamine; GlcNAc units), the major building block of the fungal cell wall.

1.4.1 Cell wall immunity

The plant cell wall serves as a intricate bridge between passive and active immunity in the ongoing evolutionary arms race between plants and pathogens. Plant cell walls comprise complex polysaccharides such as cellulose, hemicellulose, lignin, and pectin (figure 1.1(B)). Pathogens secrete various cell wall degrading enzymes (CWDEs), including glycoside hydrolases, cellulases, pectinases, and polygalactouranses, to degrade the plant cell wall. Even bacterial pathogens deliver their effector proteins within host cells through various secretion systems that penetrate the cell wall. Thus, cell wall integrity (CWI) is essential for mounting a robust defence response, and any attack on the CWI triggers cell wall–associated defence responses, including host ROS (reactive oxygen species) burst, callose deposition, lignin formation, impairment, or remodelling of the proteins involved in cell wall biosynthesis.

1.4.1.1 Pectin

After the pathogen breaches the outer cutin layer of the plant's surface, pectin is the next barrier involved in suppressing invasion. Pectin is a complex polysaccharide mainly consisting of α-1,4-galacturonic acid. Methyl esterification of pectin plays a critical role in plant immunity, as elevated levels of pectin methyl esterification correlate with heightened tolerance to pathogens. Pectin methyl esterification is mediated by pectin methylesterases (PMEs), which are in turn regulated by pectin methylesterase inhibitors (PMEIs) [43]. In *Arabidopsis* and wheat, overexpression of *PMEI1* reduces susceptibility towards bacteria and fungi such as *P. carotovorum*, *B. cinerea*, and *F. graminearum* [44, 45]. Besides, to avoid degradation of pectin from the activity of enzymes like polygalacturonase, plants deploy polygalacturonase-inhibiting proteins (PGIPs). In *Arabidopsis*, increased expression of PGIP1 and PGIP2 enhances resistance against *B. cinerea*. Moreover, *pmr5* and *pmr6 mutants* of *Arabidopsis* have increased pectin content that reduces growth of powdery mildew fungus *Erysiphe cichoracearum*. Overexpression of *Brassica napus BnPGIP2* inhibits the polygalacturonases SsPG1 and SsPG6 of *S. sclerotiorum*. *BnPGIP2* overexpression also induces defence gene expression and decreases rapeseed ROS accumulation, providing enhanced resistance against *S. sclerotiorum* [46].

1.4.1.2 Cellulose

The other component that contributes to cell wall integrity is cellulose. Cellulose is a polymer of unbranched β-1,4-linked glucan chains synthesized by cellulose-synthase

(CESA) complexes in the plasma memebrane. Mutations in *CESA* often lead to the elevation of the immune response in plants. Mutation in *Arabidopsis CESA4/7/8* enhances disease resistance against the necrotrophic fungal pathogens *Plectosphaerella cucumerina* and *B. cinerea* and bacterial pathogen *R. solanacearum* [47, 48]. Impairment of cellulose synthesis has also been linked to the activation of phytohormones mediated defence. In *Arabidopsis CESA3* mutants have elevated levels of jasmonic acid (JA) and ethylene (ET) that, in turn, induce defence responses against various fungal pathogens [49]. Similar to cellulose, hemicellulose is also a crucial integral constituent of the plant cell wall, and its manipulation can trigger plant cell wall immunity. Hemicellulose is composed of xylan and xyloglucan as its main components. Modification of the xyloglucan in the *Arabidopsis xyl1-2* mutant enhances tolerance to the necrotroph *P. cucumerina* [50] while overexpressing the glycosyltransferase family gene in barley enhances heteroxylan accumulation and provides resistance against powdery mildew fungus [51].

1.4.1.3 Lignin

Lignin is a complex strucutral component of the secondary cell wall of plants. It primarily consists of three monolignols, coniferyl, *p*-coumaryl, and sinapyl alcohol. The first enzyme in the lignin biosynthesis pathway is phenylalanine ammonia-lyase (PAL) [52, 53]. Accumulation and reduction in lignin synthesis are both known to affect the host resistance level against various pathogens. Enhancement in the expression of *PAL* gene and lignin content were observed in tomato varieties showing *Ve* gene mediated tolerance to *V. dahliae* [54]. Overexpressing *cinnamoyl-CoA Reductase 2*, a lignin biosynthesis gene in *B. napus*, provides tolerance against *S. sclerotiorum* [55]. Mutants of phenylalanine ammonia-lyase (*PAL*) [56], caffeic acid O-methyltransferase (*COMT*) [57], and cinnamyl alcohol dehydrogenase (*CAD*) in *Arabidopsis* were highly susceptible to *P. syringae* [58], and the fungal pathogens, *A. brassicicola*, *Blumeria graminis* and *B. cinerea*, respectively [56, 59]. These mutants lacked monolignols, an essential component of lignin. In contrast, inhibition of lignin biosynthesis achieved through suppression of hydroxycinnamoyl CoA: shikimate hydroxycinnamoyl transferase (*HCT*) in Alfalfa (*Medicago sativa*) enhances defence gene expression (PR1 and PR5) and levels of salicylic, jasmonic and abscisic acids, thereby reducing susceptibility towards fungus *Colletotrichum trifoli* [60].

1.4.1.4 Callose

Callose is a polymer of β-1,3-glucan monomers. Like lignin, the deposition of callose, especially in papillae, is also regarded as one of the cell wall reinforcement strategies during pathogen attack [61]. Callose synthesis is mainly regulated by callose synthase (*CALSs*) genes. *CALS12/GSL5* (*GLUCAN SYNTHASE-LIKE 5*) is involved in papillary callose deposition in *Arabidopsis* during wounding and pathogen attack. Silencing of glucan synthase-like 6 (HvGSL6) in barley, a homologue of *Arabidopsis* GSL5, reduces accumulation of callose and increases susceptibility to powdery mildew pathogen *B. graminis* f. sp. *hordei* [62, 63]. The *powdery mildew resistant 4* (*pmr4*) *Arabidopsis* mutants show reduced expression of the callose synthase gene, thereby low callose deposition and enhanced SA-mediated defence response towards powdery mildew fungus [64]. Moreover, the bacterial pathogens deficient in T3SS also affect papillary callose accumulation [65]. The T3SS-deficient *P. syringae* pv *tomato* DC3000 *hrcC* strain induces callose deposition in *Arabidopsis*. Expressing *P. syringae* effector protein AvrPto in *Arabidopsis* compromised callose accumulation in the plant cell wall and increased the number of T3SS-deficient strain of the bacteria [21]. Many viral pathogens such as *Potato virus X* (PVX), *Tobacco mosaic virus* (TMV), and *Maize dwarf mosaic virus* (MDMV), also enhance callose accumulation in the plasmodesmata of host plants [66, 67].

1.5 Active immune responses in the host

Plants induce cellular responses once the pathogens overcome structural barriers. These induced defences are known as active immune responses, and they usually get activated at the onset of pathogen or microbe recognition by receptor–ligand binding interactions (figure 1.2).

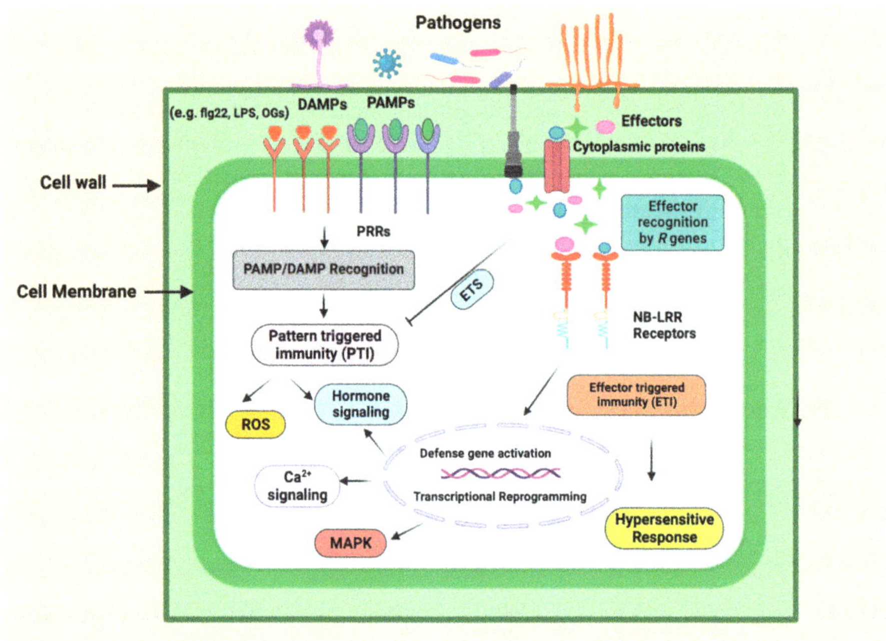

Figure 1.2. Overview of the PTI and ETI-mediated signalling in plants. PTI involves recognition of PAMPs/DAMPs by the host PRRs, whereas ETI involves recognition of pathogen effector molecules by host R proteins. Recognition of pathogens in both cases induces various defence responses including activation of ROS (Reactive Oxygen Species), Ca2+ signalling, MAPK (Mitogen-Activated Protein Kinase cascade, phytohormone signalling, and defence gene activationresponses triggering plant immunity. ETS—Effector-triggered Susceptibility, R gene—Resistance gene, NB-LRR receptors—Nucleotide-binding leucine-rich repeat receptors, flg22—flagellin 22 amino acid peptide, LPS—Lipopolysaccharides, and OGs—Oligogalacturonides.

1.5.1 PAMP/MAMP-triggered immunity

Once the pathogen breaches the host cell wall, the microbes come into contact with the plasma membrane, where they encounter extracellular surface receptors called PRRs. These PRRs recognize pathogen/microbe-associated molecular patterns (PAMPs/MAMPs), initiating the PAMP/MAMP-triggered immunity (PTI or MTI).

1.5.2 Pathogen/microbe-associated molecular patterns

PAMPs/MAMPs are typically highly conserved molecules originating from pathogens/microbes [68]. Amongst the bacterial PAMPs highly conserved N-terminal domain of flagellin (flg22) is the one that is recognized by most plant species. Lipid A is a type of lipopolysaccharide (LPS), along with core oligosaccharide and O-antigen that acts as PAMPs [69]. Another group of bacterial PAMPs that the plants recognize are the peptidoglycans (PGNs) which are components of gram-positive bacterial cell wall. Apart from extracellular PAMPs, intracellular components, including the translation elongation factor Tu (EF-Tu) and bacterial cold shock protein (CSP) are also recognized as PAMPs, especially by the members of *solanaceae* and *Brassicaceae* [70] plant family. Pep13 is a 13-amino acid epitope derived from the oomycete *Phytophthora sojae*. It is also found on the surface of transglutaminase and cellulose-binding elicitor lectin from *Phytophthora parasitica* var. *nicotianae*. The host recognizes it as PAMPs [71]. Other bacterial and fungal PAMPs include necrosis-inducing proteins, fungal chitin, ergosterol cerebrosides A, C, and β-glucans [72]. Apart from PAMPs, plants can also recognize damage-associated molecular patterns (DAMPs) and induce DAMP-triggered immunity (DTI). These DAMPs (e.g. oligo-α-galacturonides, OGs) are released by the action of fungal/bacterial hydrolytic enzymes on plant cell walls, cello-oligomers generated from the breakdown of cellulose, and cutin monomers formed by the action of cutinases [73].

1.5.3 Pattern recognition receptors

The pathogen induced PAMPs and DAMPs are recognized by PRRs localized on the plasma membrane of plants and lead to PAMP and DAMP-triggered immunity (PTI and DTI), respectively. PRRs are mainly of two types; receptor-like kinases (RLKs) and receptor-like proteins (RLPs) families. Both RLKs and RLPs comprise an extracellular domain such as leucine-rich repeats (LRRs), lysine motifs (LysMs), lectin motifs, and epidermal growth factor (EGF)-like domains that determine the specificity of the receptors to a particular ligand and a transmembrane domain [74]. RLPs lack the intracellular kinase domain found in RLKs. The *Arabidopsis* LRR-RK flagellin-sensing 2 (FLS2) senses bacterial flg22, while tomato LRR-RLK/FLS3 perceives a different epitope of flagellin known as flgII-28 [75, 76]. In Brassicaceae, the EF-Tu receptor (EFR) recognizes Elf18, while the receptors PEPR1/PEPR2 recognize AtPEPs. Treatment with AtPep peptides induces the expression of defence genes against the fungal root pathogen *Pythium irregulare* [77]. *Arabidopsis* RLK7 detects DAMPs, PIP1, and PIP2 (PAMP-induced secreted peptides), triggering defence responses against both *P. syringae* DC3000 (Pst DC3000) and the fungal pathogen *F. oxysporum* [78]. Rice XA21 binds to tyrosine-sulfated protein RaxX from *Xanthomonas oryzae pv. oryzae* and confers resistance against different *Xoo* strains [79]. Another class of kinases called the wall-associated receptor kinases (WAKs) have also been associated with DTI in plants. *WAK1* in *Arabidopsis* and tobacco recognizes oligogalacturonides (OGs) and activates defence against pathogens such as *B. cinerea* [80]. RLPs also induce PTI by recognizing various fungal and bacterial PAMPs. Potato plants expressing RLP23 sense nlp20 and are resistant to *Phytophthora infestans* and *S. sclerotiorum* [81]. In *Arabidopsis*, RLP30 (receptor-like protein 30) detects *S. sclerotiorum* SCFE1 (sclerotinia culture filtrate elicitor 1), eliciting immune responses against *S. sclerotiorum* [82]. Moreover, tobacco LRR-RLP CSP receptor (NbCSPR) binds to bacterial cold shock protein (Csp22) and triggers defence responses against bacterial diseases [83]. LysMdomain containing PRRs mainly recognizes fungal chitin elicitors. Rice LysM-RLK chitin elicitor receptor kinase 1 (OsCERK1) and rice chitin elicitor-binding protein (OsCEBiP) complex binds to chitin oligosaccharides and induces defence signalling in rice, while a mutation in *Arabidopsis cerk1* enhances susceptibility towards fungi *E. cichoracearum* and *A. brassicicola* [84, 85]. In addition, soybean glucan-binding proteins (GBPs) perceive branched 1,6–1,3-β-glucans and elicit a defence response against *Phytophthora* [86]. Sometimes these receptors do not work alone but need co-receptors to function efficiently. BRI1-associated receptor kinase 1 (BAK1) and other somatic embryogenesis act as a co-receptor of FLS2, EFR, and PEPR1. Increased expression of BAK1 in *Arabidopsis* has been observed to decrease infection by *A. brassicicola*, whereas plants with silenced BAK1 are more susceptible to the bacterial pathogen *Pseudomonas* and the oomycete *Hyaloperonospora parasitica* [87].

1.5.4 Effector triggered immunity

Plant pathogens such as fungi, oomycetes, and bacteria secrete or deliver effectors into plant cells to suppress PTI and induce ETS, thereby causing disease. The bacterial effectors are delivered into the plant cell mainly by three distinct secretion systems (types III, IV and VI), while fungi make use of haustoria to deliver effector molecules directly into the apoplast or cytoplasm of the host [88, 89]. However, resistant plants have evolved and developed resistance (R) genes to recognize these effector molecules and initiate ETI. R proteins primarily consist of nucleotide-binding domain leucine-rich repeat (NLR) proteins, which feature conserved nucleotide-binding (NB) and leucine-rich repeat (LRR) domains. Effectors for which the corresponding R-protein is present in the host are called avirulence (Avr) proteins. A pathogen that possesses the Avr protein is classified as avirulent, while virulent pathogens lack Avr and can successfully infect plants [90, 91]. Significant progress has been made in identifying various fungal and bacterial effectors and revealing the mechanisms of R-protein activation that lead to ETI in host plants. A few of these interactions are discussed below.

1.5.5 Fungal effector-R gene interaction

To date, 13 Avr genes (Avr2, Avr4, Avr4E, Avr5, Avr9, Ecp1, Ecp2–1, Ecp4, Ecp5, Ecp6, and CfTom1) have been identified in *C. fulvum*, the pathogen responsible for tomato leaf mold. However, the functions of only a few of these genes are currently understood. Most of these AVR genes encode small cysteine-rich proteins that are secreted during infection [92]. Amongst these, Avr4 is known to bind with chitin and protect the fungal cell wall against the action of plant-secreted chitinases [93]. Silencing *Avr4* has been demonstrated to impair the pathogenicity of *C. fulvum* in tomatoes [94]. These Avr proteins are recognized by tomato PRR-like receptor-like proteins (RLPs) known as Cf (such as Cf-2, Cf-4, Cf-5, Cf-9, Cf-10, Cf-12, Cf-16, and Cf-19). Numerous *Cf* genes have been identified in tomatoes, and introducing these genes into various tomato varieties confers resistance against *C. fulvum* [95]. The most studied Avr-R gene interactions in tomato include Cf-2/Avr2 and Cf-9/Avr9, and Avr4/Cf-4n [96]. These interactions elicit hypersensitive response (HR), ROS generation, and suppression of disease [97]. In necrotrophic fungus *S. sclerotiorum*, a group of researchers have reported 70 putative effector genes, including (SsNEP1, SsNEP2, Ss-Cafl, SsSSVP1, SsNE1–6 and the SsCP1) [98] that are primarily involved in promoting host cell death. These effectors have been used as targets for disease resistance, as host-induced gene silencing of *Ss-caF1* (a putative Ca+ binding protein) has been shown to provide tolerance against the fungus in *Arabidopsis* [99]. Sspg1d, an endo-polygalacturonase (PG) encoding effector of *S. sclerotiorum*, targets canola IPG-1 (having a Ca2+ regulatory domain) to prevent downstream calcium signalling and induce PCD (programmed cell death), facilitating the fungal infection [100]. *V. dahliae*, which causes vascular wilt in plants, also secrete effector proteins to manipulate host defence. Most effectors are small and cysteine-rich, including Ave1, VdCBM1, and VdSCP7. In tomato, two leucine-rich repeat receptor-like proteins, Ve1 and Ve2 provide race-specific resistance against *Verticillium* [101]. *Ve1* recognizes Ave1 effector and induces ROS, JA, and phenylpropanoid signalling mediated immune response against *V. dahliae*. *Gbve1* is the orthologue of tomato *Ve1* identified in cotton. Silencing of *GbVe1* in cotton enhances susceptibility, while its over-expression in *Arabidopsis* and cotton provides resistance against *V. dahliae* [102]. Moreover, *V. dahliae* effector Av2 is reported to induce race-specific resistance in tomatoes upon interaction with *R* gene *V2* [103]. Various avirulence effectors have been identified from oomycetes fungus *Phytophthora* including Avr1b from *P. sojae* (infects soybean); Avr3a, Avr4, Avr-blb1, and AVRblb2 from *P. infestans* (infects potato and tomato) [89]. *P. infestans*, effectors contain RXLR and EER motifs and a nuclear localization signal that helps in transcriptional regulation in the host [104]. AVR3a; the RXLR effector targets a host U box containing E3 ligase, CMPG1, that triggers infestin-1 mediated cell death in the host. However, AVR3a is recognized by an LRR domain-containing R protein, R3a, in resistant potato cultivars inducing ETI. A *P. infestans* cytoplasmic effector Avr-blb1 (also known as IPiO1) binds to LecRK-I.9 (Lectin Receptor Kinase) and disrupts the integrity of cell wall and plasma membrane in the host. This IPiO1 is sensed by Rpi-blb1 or RB (a CC-NB-LRR containing R-protein) identified from *S. bulbocastanum* and induces an HR mediated disease resistance [105]. AVRblb2, an effector, blocks the focal secretion of the host papain-like cysteine protease C14, which normally degrades fungal haustoria during colonization, thereby increasing host plant susceptibility to *P. infestans*. However, AVRblb2 also interacts with Rpi-blb2, an R-protein containing CC-NBS-LRR domains, triggering a defence response in *Nicotiana benthamiana* [106].

1.5.6 Bacterial effector-R gene interaction

P. syringae infects a wide range of plants and delivers numerous effector proteins directly into plant cells using the type III secretion system (T3SS) [107]. Several effectors secreted by *P. syringae* have been identified, including cysteine proteases: AvrPphB, AvrRpt2, and HopZ1; E3 ubiquitin ligase: AvrPtoB; protein tyrosine phosphatase: HopAO1 and kinase inhibitor: AvrPto [108]. Effector A*vrPto* induces necrosis in the tomato plants infected with *P. syringae* pv. *tomato* strain T1 [109]. AvrPto and AvrPtoB trigger ETI in tomato plants having NB-LRR, *R* gene, Prf, whereas effectors AvrB, AvrRpml, AvrRpt2, and AvrPphB are recognized by *R* genes RPM1, RPS2, and RPS5, resepctively, in *Arabidopsis*. AvrRpt2 and AvrRpml interact with RPM1-INTERACTING PROTEIN4 (RIN4), leading to its phosphorylation and resulting in suppression of basal defence [110]. AvrRpml triggers an immune response in *Nicotiana species* via interaction with *Rpal* [111]. Besides *Pseudomonas*, pathogenic bacterial species *Xanthomonas* causes a variety of diseases, e.g. *X. oryzae* pv. (pathovar) *oryzae* causes bacterial blight of rice and *X. axonopodis* pv. *Citri* causes citrus canker. It encompasses most of the known protein transport systems found in gram-negative bacteria, including the Sec and TAT pathways, as well as type I to VI secretion systems [112]. *Xanthomonas* encodes several effector proteins, including *AvrBs2*, *AvrBs3*, *AvrXa27*, *Ecf XopX*, *XopN*, and *XopQ* etc [65]. XopD is a cysteine protease/isopeptidase transported to the nucleus, where it targets plant SUMO (small ubiquitin-like modifiers) isopeptidases [113]. AvrBs3 belongs to a protein family also known as TAL (transcription activator-like) effectors. These effectors bind to the promoters of host genes and regulate their transcription. AvrBs3 regulates the expression of UPA20 (upregulated by AvrBs3) and promotes cell hypertrophy in pepper. *X. oryzae* pv. *oryzae* secretes AvrBs3-like effectors AvrXa7 and PthXo1, which enhance bacterial growth and accelerate disease progression in rice [112, 114]. Many resistance (*R*) genes have been identified that recognize Xanthomonas effectors and induce host resistance. These include *Xa1*, *Xa4*, *Xa3/Xa26*, *Xa10xa5*, *Xa21*, *Xa23*, *Xa27*, *xa13*, *xa25*, and *xa41*. *lRR-RLK* gene *Xa21* is the most widely studied and is known to provide broad-spectrum resistance to *X. oryzae* pv. *oryzae* [115]. *XA21* recognizes RaxX, a sulfated microbial peptide secreted by *Xoo*, and elicits an immune response in rice [116]. Another *R* gene *Xa4* that encodes a cell wall-associated kinase strengthens the cell wall integrity and prevents *Xoo* invasion in rice [117, 118]. Some *R* genes are race-specific, e.g., xa13 provides resistance against *Xoo* race 6 in rice cultivars BJ1 and Juma [119]. Besides these, some necrotrophic pathogens, such as *Stagonospora nodorum* causal agent of *S. nodorum* blotch (SNB), produce host specific toxins (HSTs) like ToxA, which acts as effectors and target host *R* genes and induce programmed cell death [120].

1.6 PTI and ETI mediated immune signalling

Although different receptors trigger PTI and ETI, eventually, they induce similar downstream immune responses such as induction of mitogen-activated protein kinase (MAPK) cascades, calcium signalling, reactive oxygen species (ROS) burst, and phytohormone signalling during induction of plant immunity [121].

1.6.1 MAPK signalling

MAPK cascades are evolutionarily conserved pathways that swiftly activate a diverse array of downstream genes crucial for plant immunity. In *Arabidopsis*, MPK4 phosphorylates ASR3 (ARABIDOPSIS SH4-RELATED3) the transcriptional repressor, negatively regulating genes-induced by flg22 [122]. MAPKs also play an important role in ETI. It has been shown that recognition of *P. syringae* effector AvrRpt2 by CNL RPS2 activates MPK3 and MPK6 and further induces SA-mediated immune response in *Arabidopsis thaliana* [123].

1.6.2 ROS signalling

ROS act as a key defence signalling molecule during PTI as well as ETI. In *Arabidopsis*, phosphorylation of respiratory burst oxidase homologue D (RbohD) by protein kinases such as BIK1 or CPKs upon activation by flg22 enhances ROS production. Phosphorylation of RbohD at Ser343 and Ser347 residues enhances immunity mediated by RPS2 and RPM1 [124]. However, sometimes pathogens modulate host ROS production to promote pathogenesis. *S. sclerotiorum* modulates respiratory burst oxidase homologue, *GmRBOH-VI*, to suppress ROS production and enhance host susceptibility in soybean [154]. Similarly, *R. solani* upregulates tomato and rice alternative NADH-ubiquinone oxidoreductase *(NUOR)* to enhance ROS-mediated necrosis and promote pathogenesis [125]. These genes serve as targets to enhance disease resistance in plants.

1.6.3 Calcium signalling

Calcium acts as a secondary messenger in both PTI and ETI-mediated signalling processes. CPK5 and CPK6, calcium-dependent protein kinases, act as positive regulators of ETI. Mutants lacking *cpk5* and *cpk6* in *Arabidopsis* are impaired in NLRs (nucleotide-binding leucine-rich repeat), RPS2 and RPM1 mediated resistance against *P. syringae* [126]. Moreover, OsCIPK14 and OsCIPK15, the calcineurin B-like interacting protein kinases, induce ROS production and HR in rice upon infection and trigger PTI in rice [121]. The flg22 induced calmodulin (CaM)-binding transcription factor, CBP60g, reduces the growth of *P. syringae* pv *maculicola ES4326* and enhances SA-mediated defence in *Arabidopsis* [127]. AtCNGC11 and AtCNGC12 calcium channels contribute to the ETI-mediated defence response against *H. parasitica* in *Arabidopsis* [128].

1.6.4 Hormone signalling

Several studies demonstrate that plant hormones, including salicylic acid (SA), jasmonic acid (JA), and ethylene (ET) are key signalling molecules playing a crucial role in plant defence against pathogens. SA is primarily associated with resistance against biotrophic pathogens, whereas JA and ET are typically linked to defence against necrotrophic pathogens, although there are exceptions to these generalizations. These hormones mainly act antagonistic to each other, and pathogens exploit this to suppress plant defences. For example, *P. syringae* pv. *tomato* DC3000 produces coronatine, a phytotoxin analogous to JA-isoleucine (JA-Ile), and mimics most of its functions. This suppresses the SA-mediated defence response and enhances the susceptibility of the host toward the bacteria [129]. Some fungi, like *M. grisea*, secrete antibiotic biosynthetic monooxygenase (Abm) that can metabolize JA into 12-OH-JA to suppress JA-mediated defence in hosts and promotes its colonization [130]. Salicylic acid also induces flg22 based defence response against *Pto* DC3000 as SA-deficient *Arabidopsis* mutants are partially compromised in their ability to activate flg22-triggered immunity [123]. SA provides resistance against a variety of pathogens, including *Verticillium* spp. The phenylpropanoid pathway is crucial for SA biosynthesis in plants [131]. In moderately resistant potato cultivars compared to susceptible ones, expression levels of key SA signalling genes such as *phenylalanine ammonia-lyase 1* (*PAL1*) and *phenylalanine ammonia-lyase 2* (*PAL2*), which are part of the phenylpropanoid pathway, as well as pathogenesis-related genes PR1, PR2, and PR5, were significantly upregulated following infection with *V. dahliae*. These genes are major constituents of SA signalling [131, 132]. SA also contributes to immunity against the necrotrophic pathogen *R. solani* as the exogenous treatment of SA decreases the fungal biomass in rice and *Brachypodium distachyon* [133]. The jasmonate and ethylene hormones generally work together in defence against necrotrophs as well as in generating ISR (induced systemic response). Plants overexpressing *CARBOXYL METHYLTRANSFERASE* (*JMT*) gene show upregulation of the JA-responsive genes such as *VEGETATIVE STORAGE PROTEIN 2* (*VSP2*) and *PLANT DEFENSIN 1.2* (*PDF1.2*) and are tolerant towards the necrotrophic fungus *B. cinerea* [134]. Exogenous application of MeJA upregulates the expression of defence genes *PR4*, *PR5*, and *PEROXIDASE (PEROX)* in wheat and induces defence against *F. graminearum* [135]. Elevated expression of the *OsbHLH034* gene, a positive regulator of JA signalling in rice, led to heightened lignin production and improved resistance against bacterial blight disease [136]. Ethylene is shown to be involved in the production of phytoalexins and regulating defence-related genes transcriptionally during pathogen attacks [137, 138]. Induced expression of ethylene responsive transcription factor *AtERF1* or heterologous expression of its tomato homologue *Pti4* upregulates PR-genes' expression in *Arabidopsis*. Disruption of ethylene signalling in the *Arabidopsis* mutant *ein2* increases susceptibility to the fungal pathogen *B. cinerea* and the bacterium *Erwinia carotovora* [139]. Phytohormone abscisic acid (ABA), which is majorly involved in physiological processes, also influences plant defence. Priming with ABA induces callose deposition and boosts resistance against necrotrophic pathogens in *Arabidopsis* [140].

1.7 Transcriptional control of plant defence

The perception of signal pathways leads to transcriptional reprogramming by master regulators or transcription factors such as W-box containing WRKY, basic leucine zipper (bZIP), MYB, APETALA2/ETHYLENE-RESPONSE ELEMENT BINDING FACTOR (AP2/ERF), MADS box, NAC and basic-helix-loop-helix (bHLH), [141]. An instance of this transcriptional reprogramming is the phosphorylation of WRKY33 by MPK3/MPK6 in *Arabidopsis*, which is crucial for activating the camalexin biosynthetic gene cluster and promoting camalexin production in response to *B. cinerea* infection [142].

1.8 Plant disease management strategies

RNA interference (RNAi) is a technique employed to selectively target and silence any pathogen or plant gene either at the transcriptional or post-transcriptional level [143]. Most important players of RNAi are small interfering (siRNAs), and microRNA (miRNAs). RNAi-mediated silencing was used to target mycotoxin producing genes in *F. graminearum* that reduced its virulence in wheat [144]. A recent study shows *A. thaliana* secrete small RNAs encapsulated in vesicles that are transferred to fungal pathogen *B. cinerea* upon infection. These siRNAs silence the genes critical for pathogenicity of *B. cinerea* [145]. Moreover, RNAi has also been used to target host susceptibility *(S)* genes, one such example is silencing of a *Vitis vinifera* susceptibility gene (*VviLBDIf7*) provides tolerance against downy mildew [146].

Over the past decade, several genome editing techniques such as transcription activator-like effector nucleases (TALENs), zinc finger nucleases (ZFNs), and clustered regularly interspaced short palindromic repeats (CRISPR/Cas) have emerged, enhancing the efficiency of gene engineering in plants [147]. CRISPR/Cas9 mediated knockout of *OsSWEET13* an *S* gene in rice, which is modulated by TALE effector PthXo2, enhanced resistance to *Xanthomona oryzae* [148]. These techniques have also enabled the targeting of multiple genes simultaneously. Editing of three *Mlo* (mildew resistance locus o) orthologues of wheat (*TaMlo-A1*, *B1* and *D1*) using CRISPR/Cas9 and TALEN based genome editing provides tolerance towards the powdery mildew pathogen *B. graminis* f. sp. *tritici* [149].

Advances in molecular and mutation breeding strategies have enabled scientists to increase selection efficiency and enhance plant disease resistance. *R* genes from the resistant wild varities are being widely identified using genome-wide association studies and QTL mapping tools [143]. A combination of mutation and modern breeding techniques is being used for NLR genes in plants e.g., MutRenSeq, a technique combining RenSeq (resistance gene enrichment sequencing) and EMS mutagenesis, identified *Sr22* and *Sr45* as stem rust resistance genes [150]. Some other examples of genes introgressed for disease resistance are; *CcRpp1*, isolated from a wild pigeon pea which confers resistance to Asian soybean rust in soybean, and *Fhb7*, which confers resistance to *Fusarium* head blight was retrieved from wild wheat [151, 152].

1.9 Conclusions

Plants and pathogens continuously evolve and change in terms of their infection process and the ability to recognize pathogenic invasion. The understanding of microbes–host interactions helps us to know the real impact of plant diseases on food security and devise strategies to control them. Widespread use of pesticides has been one of the widely adopted control measures for economically important crop diseases. However, with the advent of new techniques such as NGS (next-generation sequencing), transcriptomics, metabolomics, and proteomics, it has become easier to identify the putative host and pathogen targets for disease resistance. Newer transgene-free genome editing techniques are also being widely used to edit genes involved in plant–pathogen interactions. Since plant–pathogen coevolution is a dynamic field, it always poses a constant challenge for researchers. Thus, having a mechanistic insight and developing new approaches to study plant immunity and pathogen attack strategies would prepare us to better control current and future crop disease problems.

References

[1] Liao C-J *et al* 2022 Pathogenic strategies and immune mechanisms to necrotrophs: differences and similarities to biotrophs and hemibiotrophs *Curr. Opin. Plant Biol.* **69** 102291

[2] Faris J D and Friesen T L 2020 Plant genes hijacked by necrotrophic fungal pathogens *Curr. Opin. Plant Biol.* **56** 74–80

[3] Glazebrook J 2005 Contrasting mechanisms of defense against biotrophic and necrotrophic pathogens *Annu. Rev. Phytopathol.* **43** 205–27

[4] Dangl J L and Jones J D 2001 Plant pathogens and integrated defence responses to infection *Nature* **411** 826–33

[5] Anderson J P *et al* 2010 Plants versus pathogens: an evolutionary arms race *Funct. Plant Biol.: FPB* **37** 499–512

[6] Zhang R *et al* 2019 Evolution of disease defense genes and their regulators in plants *Int. J. Mol. Sci.* **20** 335

[7] Koornneef A and Pieterse C M J 2008 Cross talk in defense signaling *Plant Physiol.* **146** 839–44

[8] Heil M and Bostock R M 2002 Induced systemic resistance (ISR) against pathogens in the context of induced plant defences *Ann. Bot.* **89** 503–12

[9] van Schie C C N and Takken F L W 2014 Susceptibility genes 101: how to be a good host *Annu. Rev. Phytopathol.* **52** 551–81

[10] Marques J P R *et al* 2016 Ultrastructural changes in the epidermis of petals of the sweet orange infected by *Colletotrichum acutatum Protoplasma* **253** 1233–42

[11] Shepherd R W and Wagner G J 2007 Phylloplane proteins: emerging defenses at the aerial frontline? *Trends Plant Sci.* **12** 51–6

[12] Peiffer M *et al* 2009 Plants on early alert: glandular trichomes as sensors for insect herbivores *New Phytol.* **184** 644–56

[13] Gao Y *et al* 2017 Two trichome birefringence-like proteins mediate xylan acetylation, which is essential for leaf blight resistance in rice *Plant Physiol.* **173** 470–81

[14] Serrano M *et al* 2014 The cuticle and plant defense to pathogens *Front. Plant Sci.* **5** 274

[15] Alkan N *et al* 2015 Simultaneous transcriptome analysis of *Colletotrichum gloeosporioides* and tomato fruit pathosystem reveals novel fungal pathogenicity and fruit defense strategies *New Phytol.* **205** 801–15

[16] Melotto M *et al* 2006 Plant stomata function in innate immunity against bacterial invasion *Cell* **126** 969–80

[17] Gudesblat G E, Torres P S and Vojnov A A 2009 Stomata and pathogens: warfare at the gates *Plant Signal Behav.* **4** 1114–6

[18] Wu J and Liu Y 2022 Stomata–pathogen interactions: over a century of research *Trends Plant Sci.* **27** 964–7

[19] Geng X *et al* 2014 The phytotoxin coronatine is a multifunctional component of the virulence armament of *Pseudomonas syringae Planta* **240** 1149–65
[20] Panchal S and Melotto M 2017 Stomate-based defense and environmental cues *Plant Signal. Behav.* **12** e1362517
[21] Hauck P, Thilmony R and He S Y 2003 A *Pseudomonas syringae* type III effector suppresses cell wall-based extracellular defense in susceptible *Arabidopsis* plants *PNAS* **100** 8577–82
[22] Melotto M, Underwood W and He S Y 2008 Role of stomata in plant innate immunity and foliar bacterial diseases *Annu. Rev. Phytopathol.* **46** 101–22
[23] Schulze-Lefert P and Robatzek S 2006 Plant pathogens trick guard cells into opening the gates *Cell* (Amsterdam: Elsevier) pp 831–4
[24] Jashni M K *et al* 2015 The battle in the apoplast: further insights into the roles of proteases and their inhibitors in plant–pathogen interactions *Front. Plant Sci.* **6** 584
[25] Kaur S *et al* 2022 How do plants defend themselves against pathogens—biochemical mechanisms and genetic interventions *Physiol. Mol. Biol. Plants* **28** 485–504
[26] Park H L *et al* 2014 Antimicrobial activity of UV-induced phenylamides from rice leaves *Molecules* **19** 18139–51
[27] Laouane H and Lazrek H B 2011 Synthesis and toxicity evaluation of cinnamyl acetate: a new phytotoxin produced by a strain of *Verticillium dahliae* pathogenic on olive tree *Int. J. Agric. Biol.* **13** http://webagris.inra.org.ma/doc/sedra201104.pdf
[28] Long L *et al* 2019 Flavonoid accumulation in spontaneous cotton mutant results in red coloration and enhanced disease resistance *Plant Physiol. Biochem.* **143** 40–9
[29] Zaynab M *et al* 2021 Saponin toxicity as key player in plant defense against pathogens *Toxicon* **193** 21–7
[30] Ito S-I *et al* 2007 alpha-Tomatine, the major saponin in tomato, induces programmed cell death mediated by reactive oxygen species in the fungal pathogen *Fusarium oxysporum FEBS Lett.* **581** 3217–22
[31] Mert-Türk F 2006 Saponins versus plant fungal pathogens *J. Cell Mol. Biol.* **5** 13–17
[32] Ruiz-Rubio M *et al* 2001 Metabolism of the tomato saponin α-tomatine by phytopathogenic fungi ed Atta-ur-Rahman *Studies in Natural Products Chemistry* (Amsterdam: Elsevier) pp 293–326
[33] Inagaki Y-S *et al* 2013 Infection–inhibition activity of avenacin saponins against the fungal pathogens *Blumeria graminis* f. sp. *Hordei*, *Bipolaris oryzae*, and *Magnaporthe oryzae J. Gen. Plant Pathol.* **79** 69–73
[34] Baetz U and Martinoia E 2014 Root exudates: the hidden part of plant defense *Trends Plant Sci.* **19** 90–8
[35] Lanoue A *et al* 2010 De novo biosynthesis of defense root exudates in response to Fusarium attack in barley *New Phytol.* **185** 577–88
[36] Ranjan A *et al* 2019 Resistance against *Sclerotinia sclerotiorum* in soybean involves a reprogramming of the phenylpropanoid pathway and up-regulation of antifungal activity targeting ergosterol biosynthesis *Plant Biotechnol. J.* **17** 1567–81
[37] Hasegawa M *et al* 2010 Phytoalexin accumulation in the interaction between rice and the blast fungus *Mol. Plant-Microbe. Interact.* **23** 1000–11
[38] Chen F *et al* 2004 Characterization of a root-specific Arabidopsis terpene synthase responsible for the formation of the volatile monoterpene 1,8-cineole *Plant Physiol.* **135** 1956–66

[39] De-la-Pena C *et al* 2010 Root secretion of defense-related proteins is development-dependent and correlated with flowering time *J. Biol. Chem.* **285** 30654–65

[40] Mitreiter S and Gigolashvili T 2021 Regulation of glucosinolate biosynthesis *J. Exp. Bot.* **72** 70–91

[41] Sotelo T *et al* 2015 *In vitro* activity of glucosinolates and their degradation products against brassica-pathogenic bacteria and fungi *Appl. Environ. Microbiol.* **81** 432–40

[42] Eugui D *et al* 2022 Glucosinolates as an effective tool in plant–parasitic nematodes control: exploiting natural plant defenses *Appl. Soil Ecol.* **176** 104497

[43] Lionetti V *et al* 2017 Three pectin methylesterase inhibitors protect cell wall integrity for arabidopsis immunity to botrytis *Plant Physiol.* **173** 1844–63

[44] Lionetti V *et al* 2007 Overexpression of pectin methylesterase inhibitors in Arabidopsis restricts fungal infection by *Botrytis cinerea Plant Physiol.* **143** 1871–80

[45] Volpi C *et al* 2011 The ectopic expression of a pectin methyl esterase inhibitor increases pectin methyl esterification and limits fungal diseases in wheat *Mol. Plant-Microbe Interact.* **24** 1012–9

[46] Wang Z *et al* 2021 Interaction between *Brassica napus* polygalacturonase inhibition proteins and *Sclerotinia sclerotiorum* polygalacturonase: implications for rapeseed resistance to fungal infection *Planta* **253** 34

[47] Hernández-Blanco C *et al* 2007 Impairment of cellulose synthases required for Arabidopsis secondary cell wall formation enhances disease resistance *Plant Cell* **19** 890–903

[48] Ramírez V *et al* 2011 MYB46 modulates disease susceptibility to *Botrytis cinerea* in Arabidopsis *Plant Physiol.* **155** 1920–35

[49] Ellis C *et al* 2002 The Arabidopsis mutant cev1 links cell wall signaling to jasmonate and ethylene responses *Plant Cell* **14** 1557–66

[50] Delgado-Cerezo M *et al* 2012 Arabidopsis heterotrimeric G-protein regulates cell wall defense and resistance to necrotrophic fungi *Mol Plant* **5** 98–114

[51] Chowdhury J *et al* 2017 Altered expression of genes implicated in xylan biosynthesis affects penetration resistance against powdery mildew *Front. Plant Sci.* **8** 445

[52] Bacete L *et al* 2018 Plant cell wall-mediated immunity: cell wall changes trigger disease resistance responses *Plant J.: Cell Mol. Biol.* **93** 614–36

[53] Miedes E *et al* 2014 The role of the secondary cell wall in plant resistance to pathogens *Front. Plant Sci.* **5** 358

[54] Gayoso C *et al* 2010 The Ve-mediated resistance response of the tomato to *Verticillium dahliae* involves H_2O_2, peroxidase and lignins and drives *PAL*gene expression *BMC Plant Biol.* **10** 232

[55] Wang D *et al* 2023 Pectin, lignin and disease resistance in *Brassica napus* L.: an update *Horticulturae* **9** 112

[56] Lee M-H *et al* 2019 Lignin-based barrier restricts pathogens to the infection site and confers resistance in plants *EMBO J.* **38** e101948

[57] Quentin M *et al* 2009 Imbalanced lignin biosynthesis promotes the sexual reproduction of homothallic oomycete pathogens *PLoS Pathog.* **5** e1000264

[58] Tronchet M *et al* 2010 Cinnamyl alcohol dehydrogenases-C and D, key enzymes in lignin biosynthesis, play an essential role in disease resistance in Arabidopsis *Mol. Plant Pathol.* **11** 83–92

[59] Bhuiyan N H *et al* 2009 Gene expression profiling and silencing reveal that monolignol biosynthesis plays a critical role in penetration defence in wheat against powdery mildew invasion *J. Exp. Bot.* **60** 509–21

[60] Gallego-Giraldo L *et al* 2011 Selective lignin downregulation leads to constitutive defense response expression in alfalfa (*Medicago sativa* L.) *New Phytol.* **190** 627–39
[61] Ellinger D and Voigt C A 2014 Callose biosynthesis in Arabidopsis with a focus on pathogen response: what we have learned within the last decade *Ann. Bot.* **114** 1349–58
[62] Chowdhury J *et al* 2016 Down-regulation of the glucan synthase-like 6 gene (HvGsl6) in barley leads to decreased callose accumulation and increased cell wall penetration by *Blumeria graminis* f. sp. Hordei *New Phytol.* **212** 434–43
[63] Wang Y *et al* 2021 Regulation and function of defense-related callose deposition in plants *Int. J. Mol. Sci.* **22**
[64] Nishimura M T *et al* 2003 Loss of a callose synthase results in salicylic acid-dependent disease resistance *Science* **301** 969–72
[65] Alfano J R and Collmer A 2004 Type III secretion system effector proteins: double agents in bacterial disease and plant defense *Annu. Rev. Phytopathol.* **42** 385–414
[66] Iglesias V A and Meins F Jr 2000 Movement of plant viruses is delayed in a beta−1,3-glucanase-deficient mutant showing a reduced plasmodesmatal size exclusion limit and enhanced callose deposition *Plant J.* **21** 157–66
[67] Lee J-Y and Lu H 2011 Plasmodesmata: the battleground against intruders *Trends Plant Sci.* **16** 201–10
[68] Zipfel C 2009 Early molecular events in PAMP-triggered immunity *Curr. Opin. Plant Biol.* **12** 414–20
[69] Dow M, Newman M-A and von Roepenack E 2000 The induction and modulation of plant defense responses by bacterial lipopolysaccharides *Annu. Rev. Phytopathol.* **38** 241–61
[70] Newman M-A *et al* 2007 Invited review: priming, induction and modulation of plant defence responses by bacterial lipopolysaccharides *J. Endotoxin Res.* **13** 69–84
[71] Schwessinger B and Zipfel C 2008 News from the frontline: recent insights into PAMP-triggered immunity in plants *Curr. Opin. Plant Biol.* **11** 389–95
[72] Nürnberger T *et al* 2004 Innate immunity in plants and animals: striking similarities and obvious differences *Immunol. Rev.* **198** 249–66
[73] Hou S *et al* 2019 Damage-associated molecular pattern-triggered immunity in plants *Front. Plant Sci.* **10** 646
[74] Yu X *et al* 2017 From chaos to harmony: responses and signaling upon microbial pattern recognition *Annu. Rev. Phytopathol.* **55** 109–37
[75] Chinchilla D *et al* 2006 The Arabidopsis receptor kinase FLS2 binds flg22 and determines the specificity of flagellin perception *Plant Cell* **18** 465–76
[76] Hind S R *et al* 2016 Tomato receptor FLAGELLIN-SENSING 3 binds flgII-28 and activates the plant immune system *Nat. Plants* **2** 16128
[77] Yamaguchi Y, Pearce G and Ryan C A 2006 The cell surface leucine-rich repeat receptor for AtPep1, an endogenous peptide elicitor in Arabidopsis, is functional in transgenic tobacco cells *PNAS* **103** 10104–9
[78] Hou S *et al* 2014 The secreted peptide PIP1 amplifies immunity through receptor-like kinase 7 *PLoS Pathog.* **10** e1004331
[79] Song W Y *et al* 1995 A receptor kinase-like protein encoded by the rice disease resistance gene, Xa21 *Science* **270** 1804–6
[80] Brutus A *et al* 2010 A domain swap approach reveals a role of the plant wall-associated kinase 1 (WAK1) as a receptor of oligogalacturonides *PNAS* **107** 9452–7

[81] Albert I *et al* 2015 An RLP23–SOBIR1–BAK1 complex mediates NLP-triggered immunity *Nat. Plants* **1** 1–9

[82] Wang Z *et al* 2019 Recent advances in mechanisms of plant defense to sclerotinia sclerotiorum *Front. Plant Sci.* **10** 1314

[83] Saur I M L *et al* 2016 NbCSPR underlies age-dependent immune responses to bacterial cold shock protein in *Nicotiana benthamiana PNAS* **113** 3389–94

[84] Shimizu T *et al* 2010 Two LysM receptor molecules, CEBiP and OsCERK1, cooperatively regulate chitin elicitor signaling in rice *Plant J.* **64** 204–14

[85] Wan J *et al* 2008 A LysM receptor-like kinase plays a critical role in chitin signaling and fungal resistance in Arabidopsis *Plant Cell* **20** 471–81

[86] Zipfel C 2014 Plant pattern-recognition receptors *Trends Immunol.* **35** 345–51

[87] Kemmerling B *et al* 2007 The BRI1-associated kinase 1, BAK1, has a brassinolide-independent role in plant cell-death control *Curr. Biol.: CB* **17** 1116–22

[88] Cui H, Xiang T and Zhou J-M 2009 Plant immunity: a lesson from pathogenic bacterial effector proteins *Cell. Microbiol.* **11** 1453–61

[89] Pradhan A *et al* 2021 Fungal effectors, the double edge sword of phytopathogens *Curr. Genet.* **67** 27–40

[90] Li X, Kapos P and Zhang Y 2015 NLRs in plants *Curr. Opin. Immunol.* **32** 114–21

[91] Wu L *et al* 2014 Go in for the kill: how plants deploy effector-triggered immunity to combat pathogens *Virulence* **5** 710–21

[92] Mesarich C H *et al* 2018 Specific hypersensitive response–associated recognition of new apoplastic effectors from *Cladosporium fulvum* in wild tomato *Mol. Plant-Microbe Interact.* **31** 145–62

[93] van Esse H P *et al* 2007 The chitin-binding *Cladosporium fulvum* effector protein Avr4 is a virulence factor *Mol. Plant-Microbe Interact.* **20** 1092–101

[94] Ellis J G *et al* 2009 Recent progress in discovery and functional analysis of effector proteins of fungal and oomycete plant pathogens *Curr. Opin. Plant Biol.* **12** 399–405

[95] Kanyuka K and Rudd J J 2019 Cell surface immune receptors: the guardians of the plant's extracellular spaces *Curr. Opin. Plant Biol.* **50** 1–8

[96] Zhao T *et al* 2019 Transcriptome profiling reveals the response process of tomato carrying Cf-19 and *Cladosporium fulvum* interaction *BMC Plant Biol.* **19** 572

[97] Rivas S and Thomas C M 2005 Molecular interactions between tomato and the leaf mold pathogen *Cladosporium fulvum Annu. Rev. Phytopathol.* **43** 395–436

[98] Derbyshire M and Denton-Giles M 2017 The complete genome sequence of the phytopathogenic fungus *Sclerotinia sclerotiorum* reveals insights into the genome architecture of broad host range *Genome Biol.* **9** 593–618

[99] Maximiano M R *et al* 2022 Host induced gene silencing of *Sclerotinia sclerotiorum* effector genes for the control of white mold *Biocatal. Agric. Biotechnol.* **40** 102302

[100] Wang X *et al* 2009 Characterization of a canola C2 domain gene that interacts with PG, an effector of the necrotrophic fungus *Sclerotinia sclerotiorum J. Exp. Bot.* **60** 2613–20

[101] Wang D *et al* 2020 Functional analyses of small secreted cysteine-rich proteins identified candidate effectors in *Verticillium dahliae Mol. Plant Pathol.* **21** 667–85

[102] Song R *et al* 2020 An overview of the molecular genetics of plant resistance to the verticillium wilt pathogen *Verticillium dahliae Int. J. Mol. Sci.* **21**

[103] Chavarro-Carrero E A *et al* 2021 Comparative genomics reveals the in planta-secreted *Verticillium dahliae* Av2 effector protein recognized in tomato plants that carry the V2 resistance locus *Environ. Microbiol.* **23** 1941–58

[104] Wang W and Jiao F 2019 Effectors of Phytophthora pathogens are powerful weapons for manipulating host immunity *Planta* **250** 413–25

[105] Chen Y, Liu Z and Halterman D A 2012 Molecular determinants of resistance activation and suppression by *Phytophthora infestans* effector IPI-O *PLoS Pathog.* **8** e1002595

[106] Oh S-K, Kim H and Choi D 2014 Rpi-blb2-mediated late blight resistance in *Nicotiana benthamiana* requires SGT1 and salicylic acid-mediated signaling but not RAR1 or HSP90 *FEBS Lett.* **588** 1109–15

[107] Cunnac S, Lindeberg M and Collmer A 2009 *Pseudomonas syringae* type III secretion system effectors: repertoires in search of functions *Curr. Opin. Microbiol.* **12** 53–60

[108] Block A and Alfano J R 2011 Plant targets for *Pseudomonas syringae* type III effectors: virulence targets or guarded decoys? *Curr. Opin. Microbiol.* **14** 39–46

[109] Chang J H *et al* 2000 avrPto enhances growth and necrosis caused by *Pseudomonas syringae* pv. tomato in tomato lines lacking either Pto or Prf *Mol. Plant-Microbe Interact.* **13** 568–71

[110] Bretz J R *et al* 2003 A translocated protein tyrosine phosphatase of *Pseudomonas syringae* pv. tomato DC3000 modulates plant defence response to infection *Mol. Microbiol.* **49** 389–400

[111] Yoon M and Rikkerink E H A 2020 Rpa1 mediates an immune response to avrRpm1Psa and confers resistance against *Pseudomonas syringae* pv. actinidiae *Plant J.: Cell Mol. Biol.* **102** 688–702

[112] Kay S and Bonas U 2009 How xanthomonas type III effectors manipulate the host plant *Curr. Opin. Microbiol.* **12** 37–43

[113] Hotson A *et al* 2003 Xanthomonas type III effector XopD targets SUMO-conjugated proteins in planta *Mol. Microbiol.* **50** 377–89

[114] Römer P *et al* 2007 Plant pathogen recognition mediated by promoter activation of the pepper Bs3 resistance gene *Science* **318** 645–8

[115] Jiang N *et al* 2020 Resistance genes and their interactions with bacterial blight/leaf streak pathogens (*Xanthomonas oryzae*) in rice (*Oryza sativa* L.)—an updated review *Rice* **13** 1–12

[116] Ercoli M F *et al* 2022 Plant immunity: rice XA21-mediated resistance to bacterial infection *PNAS* **119** e2121568119

[117] Hu K *et al* 2017 Improvement of multiple agronomic traits by a disease resistance gene via cell wall reinforcement *Nat. Plants* **3** 17009

[118] Sun X *et al* 2003 Identification of a 47-kb DNA fragment containing Xa4, a locus for bacterial blight resistance in rice *Theor. Appl. Genet.* **106** 683–7

[119] Yang B, Sugio A and White F F 2006 Os8N3 is a host disease-susceptibility gene for bacterial blight of rice *PNAS* **103** 10503–8

[120] Friesen T L *et al* 2008 Host-specific toxins: effectors of necrotrophic pathogenicity *Cell. Microbiol.* **10** 1421–8

[121] Peng Y, van Wersch R and Zhang Y 2018 Convergent and divergent signaling in PAMP-triggered immunity and effector-triggered immunity *Mol. Plant-Microbe Interact.* **31** 403–9

[122] Li B *et al* 2015 Phosphorylation of trihelix transcriptional repressor ASR3 by MAP KINASE4 negatively regulates Arabidopsis immunity *Plant Cell* **27** 839–56

[123] Tsuda K *et al* 2013 Dual regulation of gene expression mediated by extended MAPK activation and salicylic acid contributes to robust innate immunity in *Arabidopsis thaliana* *PLoS Genet.* **9** e1004015

[124] Kadota Y *et al* 2019 Quantitative phosphoproteomic analysis reveals common regulatory mechanisms between effector-and PAMP-triggered immunity in plants *New Phytol.* **221** 2160–75
[125] Kant R *et al* 2019 Host alternative NADH: ubiquinone oxidoreductase serves as a susceptibility factor to promote pathogenesis of *Rhizoctonia solani* in plants *Phytopathology* **109** 1741–50
[126] Gao X *et al* 2013 Bifurcation of arabidopsis NLR immune signaling via Ca^{2+}-dependent protein kinases *PLoS Pathog.* **9** e1003127
[127] Truman W *et al* 2013 The CALMODULIN-BINDING PROTEIN60 family includes both negative and positive regulators of plant immunity *Plant Physiol.* **163** 1741–51
[128] Yoshioka K *et al* 2006 The chimeric Arabidopsis CYCLIC NUCLEOTIDE-GATED ION CHANNEL11/12 activates multiple pathogen resistance responses *Plant Cell* **18** 747–63
[129] López M A, Bannenberg G and Castresana C 2008 Controlling hormone signaling is a plant and pathogen challenge for growth and survival *Curr. Opin. Plant Biol.* **11** 420–7
[130] Yang J *et al* 2019 The crosstalks between jasmonic acid and other plant hormone signaling highlight the involvement of jasmonic acid as a core component in plant response to biotic and abiotic stresses *Front. Plant Sci.* **10** 1349
[131] Dhar N *et al* 2020 Hormone signaling and its interplay with development and defense responses in verticillium–plant interactions *Front. Plant Sci.* **11** 584997
[132] Derksen H *et al* 2013 Differential expression of potato defence genes associated with the salicylic acid defence signalling pathway in response to weakly and highly aggressive isolates of *Verticillium dahliae J. Phytopathol.* **161** 142–53
[133] Kouzai Y *et al* 2018 Salicylic acid-dependent immunity contributes to resistance against *Rhizoctonia solani*, a necrotrophic fungal agent of sheath blight, in rice and *Brachypodium distachyon New Phytol.* **217** 771–83
[134] Pieterse C M J *et al* 2012 Hormonal modulation of plant immunity *Annu. Rev. Cell Dev. Biol.* **28** 489–521
[135] Wang Y *et al* 2021 Function and mechanism of jasmonic acid in plant responses to abiotic and biotic stresses *Int. J. Mol. Sci.* **22** 8568
[136] Onohata T and Gomi K 2020 Overexpression of jasmonate-responsive OsbHLH034 in rice results in the induction of bacterial blight resistance via an increase in lignin biosynthesis *Plant Cell Rep.* **39** 1175–84
[137] Adie B *et al* 2007 Modulation of plant defenses by ethylene *J. Plant Growth Regul.* **26** 160–77
[138] Broekaert W F *et al* 2006 The role of ethylene in host–pathogen interactions *Annu. Rev. Phytopathol.* **44** 393–416
[139] Lorenzo O *et al* 2003 Ethylene response factor1 integrates signals from ethylene and jasmonate pathways in plant defense *Plant Cell* **15** 165–78
[140] Derksen H, Rampitsch C and Daayf F 2013 Signaling cross-talk in plant disease resistance *Plant Sci.* **207** 79–87
[141] Tsuda K and Somssich I E 2015 Transcriptional networks in plant immunity *New Phytol.* **206** 932–47
[142] Buscaill P and Rivas S 2014 Transcriptional control of plant defence responses *Curr. Opin. Plant Biol.* **20** 35–46
[143] Rani M, Tyagi K and Jha G 2020 Advancements in plant disease control strategies *Advancement in Crop Improvement Techniques* ed N Tuteja *et al* (Cambridge: Woodhead Publishing) ch 10 pp 141–57

[144] Kuo Y-W and Falk B W 2020 RNA interference approaches for plant disease control *BioTechniques* **69** 469–77
[145] Cai Q *et al* 2018 Plants send small RNAs in extracellular vesicles to fungal pathogen to silence virulence genes *Science* **360** 1126–9
[146] Marcianò D *et al* 2021 RNAi of a putative grapevine susceptibility gene as a possible downy mildew control strategy *Front. Plant Sci.* **12** 667319
[147] Yin K and Qiu J-L 2019 Genome editing for plant disease resistance: applications and perspectives *Philos. Trans. R. Soc. London, Ser.* B **374** 20180322
[148] Zhou J *et al* 2015 Gene targeting by the TAL effector PthXo2 reveals cryptic resistance gene for bacterial blight of rice *Plant J.: Cell Mol. Biol.* **82** 632–43
[149] Wang Y *et al* 2014 Simultaneous editing of three homoeoalleles in hexaploid bread wheat confers heritable resistance to powdery mildew *Nat. Biotechnol.* **32** 947–51
[150] Steuernagel B *et al* 2016 Rapid cloning of disease-resistance genes in plants using mutagenesis and sequence capture *Nat. Biotechnol.* **34** 652–5
[151] Deng Y *et al* 2020 Molecular basis of disease resistance and perspectives on breeding strategies for resistance improvement in crops *Mol. Plant* **13** 1402–19
[152] Wang H *et al* 2020 Horizontal gene transfer of Fhb7 from fungus underlies Fusarium head blight resistance in wheat *Science* **368** 6493
[153] Martin J T 1964 Role of cuticle in the defense against plant disease *Annu. Rev. Phytopathol.* **2** 81–100
[154] Ranjan A *et al* 2018 The pathogenic development of *Sclerotinia sclerotiorum* in soybean requires specific host NADPH oxidases *Mol. Plant Pathol.* **19** 700–14
[155] Ziv C *et al* 2018 Multifunctional roles of plant cuticle during plant–pathogen interactions *Front. Plant Sci.* **9** 1088

IOP Publishing

Chapter 2

Surveillance at the surface: pattern- and damage-triggered immunity in plants

Anjan Barman

Plant surfaces act as the frontline of resistance to any incoming biotic stressor. Appositely, both constitutive and inducible sectors of plant defences ardently contribute to this exigency. This chapter commences with a brief insight on the constitutive sector of plant defence and is followed by a rich summary of the crucial events ensuing during pattern-triggered immunity (PTI) cascades prompted by varied exogenous and endogenous elicitors in plants. Additionally, a discussion on regulation of PTI signalling episodes along with updates on utility of PTI machinery for betterment of plant disease resistance phenotype have been included towards the end.

2.1 Introduction

On planet earth, environmental ingredients comprising both abiotic as well as biotic components play decisive roles in the success of a plant's life course. These components can afford indispensable survival benefits to the plant [1, 2], and at times under varying situations, may alter a plant's life processes critically [3–5]. Inability to move hence pose substantial hindrance for plants to escape harsh environmental challenges. In recent years, a wealth of literature has discussed variable impacts of abiotic and biotic components on plants either separately or by integrating both of them [3–5].

Keeping aside those biotic components involved in beneficial interactions, which notably are essential to a plant's very existence (e.g. pollination [2], nitrogen fixations [6], mineral acquisitions, growth-promotions [1] etc), there are several others which accrue considerable negative impact on a plant's growth, development as well as survivability [7]. Plant pathogenic viruses [8], bacteria [9], fungi [10], nematodes [11] especially represent the negatively interacting microbial biotic

doi:10.1088/978-0-7503-5673-2ch2

communities. In addition, many avid heterotrophs (especially metazoans including both invertebrate and vertebrate herbivores) comprising macroscopic biotic component do threaten plant health to a great extent [12, 13]. A number of herbivorous insect pests can further act as carriers of plant pathogenic microbial agents (e.g. bacteria, viruses) thereby posing an indirect threat to a plant's fitness [14, 15]. Some of the notable phytopathogenic bacteria like *Xylella fastidiosa*, *Spiroplasma* spp., *Liberibacter* spp., and *Candidatus* Phytoplasma spp. have been known to be transmitted by different insect pests to plants while feeding on their hosts [15]. The fascinating aspect is, despite persistence of colossal opposing biotic arbitrations, plants have acquired remarkable abilities to defend themselves and have managed to live and evolve through the ages. Possibly, an advantage is that, in the absence of movement, accumulating unexpended energy reserves could also be rerouted towards a plant's defence functions quite remarkably.

It is difficult to ascertain exactly when and where the realizations of plant host–pathogen/pest interactions were seriously perceived initially. In nature nevertheless, plant–pathogen or plant and pest interactions have persisted ever since such evolutionary alliances began [16]. On the same note, evolution of compatible associations between beneficial or commensal microbial populations with host plants too draws critical attention [17, 18]. In fact, now, it is being perceived that simultaneous evolution of interacting microbial communities has also contributed toward shaping of plant genomes immensely [16, 19]. Pathogens and pests in nature can be of varied kinds and the degree of afflictions brought about by them to different plant species/varieties can be variable as well [20]. Nonetheless, every plant species possesses distinctive layers of defence ploys to limit damage caused by these biotic interferences [21–23]. Likewise, pathogens/pests are also bestowed with unique strategies to subvert and overcome host defences on many occasions [24, 25]. In a way, this more or less resembles a 'tug-of-war' episode between pathogens/pests and respective plant hosts [26], whose outcome may lead to triumph of either of the opponents over the other.

2.2 Plant defence at the surface: generalized and specialized arms

In respect of the different layers of restrictions laid down by plant systems against biotic intrusions, usually two broad and distinctive mechanisms can persist: the first that is active at the surface of the plant cell, tissue, organ and the second that remains vigilant within the cell interior. The active protection at the surface is again rendered by both generalized and specialized classes of arsenals encoded by plant genetic factors.

2.2.1 The generalized arm

The very first layer of rather generalized defence that prevails on the surface of the plant cell, tissue or organ can be further attributed to various factors [27]. To begin with, the 'epidermis', constituting the outermost coating of plant surfaces, apparently becomes the first canonical barrier to biotic interferences, although it is further indispensable for various other key functions (like growth and development, exchanges of materials and gaseous substances etc), necessary for cell survival [28]. A number of components can be constitutively present on the shoot and root surfaces of the plant and these might serve as a non-specific barrier to most of the biotic invasions [29]. In shoots, these may comprise certain structural modifications of epidermal tissues (like trichomes, cuticles) and secretions thereof (like cutin, waxes etc), varied antimicrobial secretory products (e.g. camalexin, sulphoraphane, 3-deoxyanthocyanidins, napthoquinones), cell wall reorganizing factors (e.g. extensins), secondary metabolites etc [21, 28, 30]. Prevalence of cuticles, trichomes, spines as well as rigid, gluey or softer surfaces have been implicated in deterring pathogen/pest from settling down successfully which thereby check damage initiation on hosts [31]. Occasionally, the presence of simple non-secretory trichomes or sticky hairs for instance, could also be assisting in obstructing mobility of damaging insect herbivores considerably [27].

Likewise, roots of the plant located below ground may have characteristic surface constituents which confer abilities to restrict numerous phytopathogenic intrusions across the rhizopsphere. Normally, root endodermis comprising suberin as well as lignin-based casparian strips, is observed to provide immunity to plants against pathogens like root-nematodes, cysts etc [32]. Again, in the root region, the presence of cell wall extensins, root border cells, root border-like cell types have been implicated in the defence against various phytopathogens [30, 33]. Root border cells for instance, release a number of antimicrobial compounds (viz., arabinogalactan proteins, phytoalexins, histones, reactive oxygen species, enzymes etc), and also secrete DNA into the extracellular matrix that helps to trap microbial pathogens thereby disabling subsequent pathogenic afflictions significantly [33, 34]. Root border cells and root border-like cells are generally derived from root-cap cells which along with mucilages are said to constitute the first line of defence for root-cap and apical meristems against various microbial agents in the rhizosphere [34]. In fact, a characteristic root extracellular trap (RET), a mechanism akin to the neutrophil extracellular trap (NET) phenomenon prevalent in mammalian defence strategy has been correlated to root-cap-derived cells and mucilage-mediated immunity functions against phytopathogens in the root [34]. For interested readers some notable literature is available in this particular domain [35, 36].

Below the epidermal barrier, at an individual cellular level, the cell wall forms the next rigid enclosure before most phytopathogens [37]. Generally, plant cell walls are comprised of associations of cellulose micro-fibrils interconnected to hemicellulose heteropolymers [37, 38]. These associated structures can be again surrounded by another class of complex heteropolysaccharide molecules called pectin and they all constitute the primary cell wall [39]. Cellulose, hemicelluloses and pectins together provide essential protection against the majority of incoming biotic aggressions [37]. Further, defects in any one of these constituents, e.g. errors during cellulose synthesis in primary cell walls, have been found to confer significant resistance towards some of the devastating phytopathogens like *Plectosphaerella cucumerina* (a fungus) and *Ralstonia solanacearum* (bacterial wilt causal agent) [40] via induction of alternative mechanisms like enhancement of lignifications at the defective site [41]. Interestingly, absence of the intact plant cell wall can still serve essential needs through induction of surrogate defensive responses whenever required. In the case of the secondary cell wall, however, pectin composition is low as compared to the primary wall and instead the phenolic polymers called lignin become a crucial ingredient [42]. In vascular plants lignin (of the secondary wall) is known to play an important role in restricting various phytopathogens' entry at the cell surface in a somewhat non-specific manner [37, 43]. For example, in the absence of proper lignin precursor biosynthesis, enhanced susceptibility of diploid wheat leaves towards powdery mildew fungus (*Blumeria graminis* f. sp. *tritici*) invasions clearly suggest significance of the lignification process in cell wall immunity [44].

Several plants biosynthesize and store varied defence-related compounds such as resins, mucilages, gum etc in their stem and leaves that may be secreted upon pathogenic ambush. These compounds may comprise secondary metabolites like oleoresin (as in *Abies grandis*), latex (as in some species of *Asclepias* spp.) etc capable of dissuading pathogens/pests considerably [45]. Notably, some of the constituents of these secondary metabolites include diterpenoid acids, sesquiterpenes, monoterpenes, cardiac glucosides and cardenolides [29] bearing essential defensive attributes. Peculiar cell types called 'Laticifers' found in several plant species like *Hevea brasiliensis*, *Papaver somniferum*, *Carica papaya*, *Asclepias* spp., *Cannabis sativa* etc generally pile up latex which may comprise a variety of organic substances like carbohydrates, organic acids, fats, proteins, mucilages, sterols, rubbers, essential oils etc [46, 47]. Usually stored under pressure, these latex substances may ooze out of laticifers due to impending physical injuries and spread on to the surface. The secreted stuff can then trap herbivorous insect pests by immobilizing their mouthparts as well as entire body parts and could further cover up wounded sites in the ruptured tissues [47, 48].

Additionally, in instances when plants perceive pathogenic intrusions, they are able to pile up callose enriched cell wall around the locations where pathogenic attack caused injuries. This may further induce mounting up of phenolic substances and a range of toxic compounds in the cell wall, which may be followed by biosynthesis of lignin-like substances for repairing the damaged wall [49]. Several phytopathogenic microbes release proteases/enzymes to loosen/disrupt plant cell wall and render infection initiation in the host. Cell wall degrading enzymes such as polygalacturonases, pectolyases, xylanases, pectinases, cellulases, membrane destabilizing lipases, are some of the notable examples of proteases that are released by different phytopathogens such as *Erwinia carotovora*, *R. solanacearum*, *Cochliobolus carbonum*, *Verticillium dahliae*, *Stagonospora* (*Septoria*) *nodorum*, *Phytophtora infestans* etc [50]. In the same manner, many plant species are able to secrete a number of inhibitory factors such as protease inhibitors to impede these enzymes/proteases released by phytopathogens [50]. Generally, plant protease inhibitors can be synthesized constitutively or through induction mediated by pathogenic/pest elicitations [51]. Considering persistence of diverse groups of proteases across varied types of phytopathogens [52, 53], plant protease inhibitors of a multitude of types have been reported [50, 54]. Previously, seeds of soy, bean and potato tubers containing trypsin and chymotrypsin inhibitors were observed to curtail proteinases released by phytopathogenic fungus *Fusarium solani* [55]. Likewise, a distinct protease inhibitor from barley seeds (*Hordeum vulgare* L.) was found to curb proteinase actions of several microbes' viz. *Aspergillus oryzae*, *Bacillus subtilis*, *Streptomyces griseus*, and *Alternaria tennuissima* [50] suggesting broad deterrence potential of specific plant protease inhibitors against different pathogen species.

2.2.2 The specialized arm

The specialized arm further constitutes two branches: one active at the external surface of the plasma membrane and the other branch effectively functioning at the intracellular side. The first dedicated branch of elements which remain anchored to plasma membranes are somewhat more specialized in recognizing conserved patterns of pathogenic (as well as non-pathogenic) microbial origin [56] or damaged products of their own cellular constituents [57] at the surface, following which they relay a battery of defence signalling events running down the cell interior [58]. These plasma membrane localized entities are formed of proteins with variable domain constitution and might contain both extracellular and intracellular components [59]. Through recognition of conserved molecular patterns at plasma membrane and inducing appropriate cellular defences, these entities provide efficient protection towards numerous pathogenic invaders [60]. A detailed account on this specialized category of arsenal is going to be the prime focus of the ongoing chapter.

Next is the second layer of committed protective barrier that usually lies intracellularly and is lacking on the cell surfaces [61]. A section of these intracellular defenders can, however, modulate amplitudes of the defensive responses due to the pattern recognizing elements anchored at plasma membrane quite significantly [62–65]. Here in this case, participation of specialized proteins, also dubbed as resistance factors, help to recognize pathogenic secretory products introduced into the cell and induce more rapid and amplified immune responses to thwart pathogenic invasions [66]. Plenty of rich literature elaborates on the intracellular mode of immunity in plants which may be referred to for more details [61, 67, 68] and here discussion on this part would be limited to their essential connection with the main theme of this chapter.

Relying on perception of conserved molecular signatures, the precise and more dexterous shielding functions conferred by both categories of arsenals, i.e. one that remains exterior to plasma membrane and the other which serves as intracellular guards, generally constitute machinery of the 'innate immune system' in plants [69]. Pertinently, the plant innate immune system turns out to be the next formidable barrier for those pathogens who had earlier successfully overcome the restrictions laid down by the generalized defensive arm described above [70].

2.3 Plant innate immune system: distinct receptors engages in key roles

Immune systems persisting in both the animal and plant kingdom, are highly evolved molecular machinery that primarily distinguishes self from non-self and simultaneously operates to fend off negative biotic interferences dynamically [71]. Nevertheless, there are both semblances as well as differences between animal and plant immune mechanisms [61, 72]. For instance, absence of any dedicated circulatory immune cells/factors in plants (like in animals) clearly segregates them from animals in terms of immunity functions. In plants, the predominant immune functions are conferred by innate immunity mechanisms, whereas, animals can have both innate and acquired (or adaptive) immunity routes [73]. At the same time, some relatable convergences between two branches of plant innate immunity and innate as well as adaptive immunity functions in animals need important considerations [74]. The complete absence of dedicated defence cell/organ types necessitates alternative strategies that can serve efficiently in a plant's defence against pathogens or pests. By now, it is already perceived that every plant cell can act as defence units and are capable of initiating immune responses as and when pathogenic incursion takes place [75]. Although defence against biotic interventions is a central theme of the current discussion, it also needs mentioning that immunity against abiotic factor-mediated stresses is also of immense importance. However, the components involved in defence against abiotic and biotic impinges seems to be substantially varying in plants [76].

Propitiously, in order to serve as defence units, all plant cells are afforded with essential repertoires on their surfaces as well as internally which could be deployed for specifically recognizing and restricting pathogens [75]. This recognition process also works effectively to distinguish beneficial microbial members from harmful ones [77]. Notably, what turns out to be an obvious aspect now is that a range of beneficial microbes remain associated with plants and these aid in growth and survival of the latter quite significantly [78, 79].

Consistent efforts to comprehend the molecular entities involved in plant innate immunity mechanisms have resulted in illumination of numerous aspects relating to this sphere [80]. For instance, dedicated participation of some special types of receptor proteins viz. receptor kinases (RK), receptor-like proteins (RLPs) and intracellular receptors, in the crucial events of plant innate immunity function was uncovered a few decades ago that in turn unlocked the routes to unleash fundamental events of defence signalling downstream [81–83]. Pertinently, RKs are a distinct category of protein kinases (PKs). PKs generally regulate function of their target protein substrates via characteristic phosphorylation steps at specific amino acid residues such as serine/threonine, tyrosine, histidine etc [84, 85]. PKs are indispensable for critical signalling processes of plant growth, differentiation, maturation, immunity etc [84, 85]. RKs are again the largest and diverse sub-group of PKs [86] quintessential for development, reproduction as well as defence functions in plants [87, 88]. RKs are popularly known as receptor-like kinases (RLKs) [89] and the majority of them are localized in the plasma membrane of plant cells [90, 91]. RLKs are unique in the sense that they are capable of perceiving specific signals (or ligands) at one end and also carry an intrinsic kinase function at the other end. In fact, the structure of the first plant RLK elucidated almost three decades ago, revealed three essential components: an extracellular ligand binding domain (ECD), a transmembrane (TM) domain representing a hydrophobic region, and an internal kinase catalytic domain at their carboxyl terminal [89, 92] that further resembles the representative members of the receptor tyrosine kinase (RTK) family in animals [89]. The RLPs, that form another important arsenal involved in surface immunity, are also residents of plant cell plasma membranes and they too possess an extracellular ECD domain along with a TM hydrophobic region [93]. What distinguishes RLPs from RLKs is that they lack the catalytic kinase domain towards cytoplasmic side [94] unlike the latter. Notably, RLKs and RLPs both form the basic framework of Pattern Recognition Receptors (PRRs) [95] which are predominantly involved in guarding cell surfaces through their immaculate perception of various biological patterns as danger signals followed by subsequent induction of appropriate cellular defence. The third category of receptors that generally reside in the cytoplasmic side, constitute the intracellular receptors (IRs) [69]. The IR class may include three distinct categories which can have variable structural constitutions [69]. The largest and diverse group of IRs presently known constitute the nucleotide-binding leucine-rich repeat (NLR) family proteins [96] and they embrace two of the three distinct IR categories within [69]. NLRs are also referred to as nucleotide-binding oligomerization domain-like receptors (Nod-like receptors) which generally recognize pathogen delivered effector molecules in the cytoplasmic side of plant cells [97]. They are in fact central to generation of a more robust and stronger defence cascade designated as ETI in plant cells [66]. The focus on ETI branch of innate immunity would be limited to the context of their convergence with the other branch i.e. PTI that is being elaborated in the following sections.

Besides those stated above, there are other essential proteins in the transmembrane regions of plasma membrane and towards the cytoplasmic side of the plant cells which play a pivotal part in the generation and regulations of signal transduction processes involved in plant innate immune responses [98]. In fact, these proteins cooperate with primary plant immune receptors (i.e. PRRs) and thus assist significantly in the downstream defence signalling cascades [99, 100]. Notable among them are some of the well studied co-receptors (e.g. SERK, SOBIR etc) [100, 101], receptor-like cytoplasmic kinases (like BIK1, BSK1, PBLs, PCRKs etc) [102, 103] and other plasma membranes as well as cytosol localized proteins (e.g. RBOHD, MAPKs, CDPKs etc) [103, 104]. These aspects are considered in the subsequent discussions below.

2.4 Pattern recognition receptors in surface immunity

2.4.1 Pattern recognition receptors and associated receptor kinases

Located at plant cell plasma membranes, PRRs recognize various biological patterns at the cell surfaces. The PRRs are generally composed of a specific class of RLK or RLP proteins [58, 105]. However, it needs mentioning that there are many RLKs/RLPs that are also transmembrane residents but not all of them comprise PRRs [106]. In fact, PRRs are a distinct class of surface receptors and they are specialized for detecting various patterns which subsequently elicit signalling for cellular defence downstream [95]. Pertinently, the host of biological patterns recognized by PRRs can be both of exogenous as well as endogenous origin [107].

In most instances, RLKs harbouring intracellular serine/threonine kinase domains are common across plant PRRs [108] suggesting that substrate targets downstream to PRR signalling could be frequently serine or threonine residues.

Nevertheless, phosphorylation events at definite tyrosine (Y) residues have been further observed to ensure activation or functionality of these kinases to great extents [109]. PKs coupled to PRRs and all others in general transfer phosphoryl residues to hydroxyl moieties present on different amino acids of signalling proteins taking part in the downstream signal relay. However, a certain number of PKs have also been observed to execute autophosphorylations for their self-activation [110]. Importantly, PKs attached to plant PRRs may not necessitate autophosphorylation events and most often they belong to the family constituting known non-RD kinases [111]. Non-RD kinases are usually characterized by lack of an arginine (R) residue ahead of a conserved aspartate (D) moiety in the catalytic pockets of this family of PKs [24, 111].

In regard to ECDs found in the RLKs/RLPs of PRRs, distinct diversities in their constitution as well as sizes are apparent and these are thought to determine crucial specificities during recognition of peculiar/conserved molecular patterns persisting across varied sources of biological origin [58]. For example, ECDs with variable size and number of motifs of LRR (leucine-rich repeat), Lys-M (lysin motif), Lectin (C, G or L type), and other classes of domains such as malectin, cysteine-rich domains of anonymous function (DUF26), and thaumatin domains have been observed for both RLKs as well as RLPs [94, 106, 112]. In general, there can be as many as 44 sub-categories of RLKs, classified based on their ectodomain compositions [108, 113]. However, not all types of ectodomains prevalent in RLK/RLPs might be represented in PRRs. Other ectodomains in RLK/RLPs are known or anticipated to play a pivotal role in recognition of various ligands necessary for signalling events associated with varied physiological as well as developmental pathways in a plant's life processes [91, 113]. Meanwhile, with respect to ECD type harboured, PRRs can be again categorized as LRR, lysin motif (LysM), lectin, wall-associated kinase (WAK) and other sub-categories [112, 114]. Unique PRR classes (RLK based) with their ectodomains having epidermal growth factor (EGF)-like domain [115] and S-domains [116] have also been identified. EGF-domain containing PRRs forms the WAK category which senses damage-associated molecular patterns (DAMPs) [115]. Interestingly, some members of S-domain RLKs were also implicated in determination of 'self-incompatibility' of their own pollen grains in the *Brassicaceae* family [117]. Figure 2.1 shows examples of a few of the PRRs along with their distinct extra-celluar domains.

Among various kinds of PRR categories identified so far, LRR sub-family of PRRs (comprising both RLK/RLPs) are observed to be predominant [59, 91, 118]. Again, among RLK and RLP containing PRR types, RLK-bearing PRRs have been investigated in more numbers than the latter [59]. Examples of some of the frequently studied PRRs include FLS2 [Flagellin-Sensitive 2], EFR [Elongation Factor Tu (EF-Tu) Receptor], Xa21 [rice receptor for *Xanthomonas oryzae* pv. *oryzae* (*Xoo*) PAMP], RLP 23 (Receptor-like Protein 23), PEPR1/2 [Pep Receptor 1/2], WAK1 [Wall-Associated Kinase 1], Cf-9 [Receptor for *Cladosporium fulvum* effector], CEBiP [Chitin Elicitor-Binding Protein] etc [90, 91, 119]. The first ever plant PRR to be characterized was an RLK member from rice plant which is referred to as Xa21 [82]. Xa21 is in fact a LRR-type PRR [82] that recognizes an

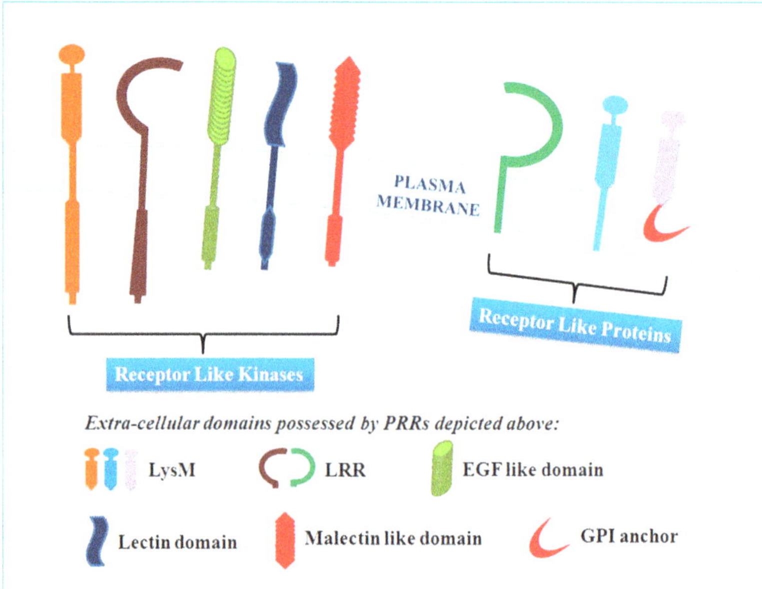

Figure 2.1. Schematic depicting representative PRR protein (RLK/RLP) kinds located on the plant cell membrane. The PRRs can be categorized into various types on the basis of characteristic ECD harboured by them and ECDs are instrumental in recognizing immunogenic patterns. Presented here are some primary ECD types such as LRR, LysM, EGF, Lectin, Malectin-like domain etc possessed by specific PRR types [507]. Examples of each ECD type are cited in the text. Importantly, RLPs lack intracellular kinase domain unlike in RLKs. Certain RLPs may not possess transmembrane domain and may remain anchored to the membrane through chemical moieties like GPI [507], as shown in the diagram.

ABC transporter secreted tyrosine sulphated peptide (i.e. RaxX) from the phytopathogenic bacterium *X. oryzae* pv. *oryzae (Xoo)* [120]. However, chronologically FLS2 (another LRR RLK PRR member) was the first PRR to be characterized with its corresponding flg22 ligand (bacterial PAMP) simultaneously [95]. Since then the number of identified PRR members in plants have continued to rise [95, 118]. Notably, most of the reported PRR members have been identified only in a few of the model plant/crop species representing a handful of plant families [90, 91]. Diversity of PRR members' observable even within these studied plant species may suggest that other plant members from various plant families are also likely to possess related or novel surface receptors for recognizing biological patterns. However, investigation at this end extended beyond these model plant species/families would be quintessential considering the necessity of sustainable agronomic practices.

Plant PRRs are able to detect varied kinds of pathogenic patterns from bacterial origin (e.g. fragments of flagellin protein like flg22 and flg28, elongation factor Tu or EF-Tu, lipopolysacccharides), fungal origin (chitin, fungal xylanase), and also certain secreted pathogenic effectors [121–123]. PRRs can further sense imminent signs of threats to plants in the form of oligogalacturonide (OG) molecules (liberated as a result of damage/injury to plant cell wall components), ATP molecules in the cell exterior as well as certain signalling peptides released by plants themselves [123]. It is noteworthy that OGs are small-sized hydrolysed products of 'homogalacturonans' which form key constituents of the pectin component of the plant cell wall [124]. During biotic ingressions, enzymes called polygalacturonases (PGs) secreted by various phytopathogenic microbial agents can generate OGs. Moreover, plants have endogenous PGs which can further release OGs upon experiencing physical wounds and distress [124]. These fragmented products liberated from host cells due to mechanical wounds or biotic ingressions can act as DAMPs [125] and are sensed by specific surface PRRs (called WAKs) [126]. As in the case of endogenous carbohydrate patterns, endogenous proteinaceous ligands released from the plant cell wall too can be recognized by specific PRRs to initiate appropriate defence responses. For instance, PRR pair PEPR1 and PEPR2 (both LRR RLK members) can discern proteinaceous ligands called plant elicitor protein 1 (Pep1) liberated from a precursor protein called PROPEP as a consequence of plant cell wall injury [127, 128]. Importantly, like OGs, Pep1 also behaves as DAMP. It would be noteworthy that the PRRs and their complementary pattern (PAMPs/MAMPs or DAMPs) pairs elucidated so far may not indicate their universal presence across all families of plant species. The well established PRR-PAMP pair EFR-elf18 exemplifies such a distinction. EFR (an LRR RLK) has been found to be confined to the *Brassicaceae* family that includes *Arabidopsis* [73] suggesting EFR-elf18-mediated signalling to be mostly restricted to this family. Interestingly, the angiosperm family *Solanaceae* that embraces several economically important crop species viz. tomato, potato, chilli, egg-plants, banana etc lack EFRs and thus elf18-induced signalling [73]. On the other hand, two important representatives of the *Solanaceae* family, i.e. *Solanum lycopersicum* and *Nicotiana benthamiana* have been examined to carry a unique LRR RLK PRR called CORE (cold shock protein (CSP) receptor) that detects csp22 (a fragment of CSPs from *Pseudomonas syringae* pv. tomato DC3000) as PAMP [129].

Additionally, PRRs responsible for recognition of a specific PAMP or MAMP from certain microbial species may not necessarily detect similar patterns in other microbes. For example, the *Arabidopsis* lectin S-domain receptor kinase *At*LORE was revealed to perceive lipopolysaccharide (LPS) fractions from *Pseudomonas* and *Xanthomonas* spp. but not from *Escherichia coli* or *Burkholderia* spp. [122]. On the other hand, rice PRR called *Os*CERK1 (a Lys-M RLK) was shown to sense LPS fractions from *E. coli*, *Salmonella minnesota*, *P. aeruginosa* as well as *X. oryzae* pv. *oryzae* [130]. Distinct receptors for LPS pattern perception apparent in *Arabidopsis* and rice implies, persistence of separate sensory modules for bacterial LPS across plant families *Brassicaceae* and *Poaceae*, as seen in this case [130].

Apart from insights gained from numerous bacterial, fungal ligand and a plant's endogenous molecular pattern detection, PRR members engaged in recognizing ligands from viruses, herbivores (including insects) and nematodes have also been under critical investigation. Until recently, recognition of molecular patterns from viruses by plant PRRs rarely had much evidence. In fact, plant defence responses towards pathogenic viruses were mostly considered to be based on intracellular resistance factors leading to 'ETI' [131]. Nevertheless, earlier attempts toward understanding of PTI machinery participation in the plant anti-viral defence mechanisms had revealed contributions of some RLKs (such as BAK1, BKK1, NIK1 etc), although testimony of canonical PTI events initiating at distinct surface-localized PRRs were inadequate [131]. Lately, involvement of an LRR RLP PRR called *Os*RLP1 (in rice) in recognition of rice black-streaked dwarf (RBSD) virus has been validated [132]. However, in this case, appropriate RBSD PAMP for *Os*RLP1 has yet not been isolated [132]. In regard to detection of HAMPs (herbivore associated molecular patterns) associated with oral and salivary secretions of herbivorous organisms (that include many plant damaging insects), several PRRs have been implicated although corresponding ligands to these have not been determined so far [133]. Interestingly, a recent study has identified a LRR RLP PRR called *Vu*INR (cowpea, *Vigna unguiculata* inceptin receptor) along with its corresponding HAMP ligand 'inceptin' (from beet armyworm *Spodoptera exigua* oral secretions) [134]. Conversely, a clear demonstration of PRR mediated recognition of distinct parasitic nematode molecular pattern (also dubbed as nematode associated molecular pattern, NAMP) hasn't been possible until recently. Of late, an LRR RLK PRR referred as NILR1 (nematode-induced LRR-RLK 1) was shown in *Arabidopsis* to detect some components in 'NemaWater' and induce immune signalling [135]. 'NemaWater' is a preparation obtained by incubating juvenile stages of parasitic nematodes viz. *Heterodera schachtii* and *Meloidogyne incognita* in an aqueous solution [135]. The corresponding NAMP from NemaWater is, however, yet to be determined. Very recently, a report unveiled recognition of ascaroside #18 (Ascr18) ligands by NILR1 that could set off immune signalling cascades with hallmark PTI responses in *Arabidopsis* [136]. Ascr18 and related ascarosides are established nematode derived elicitors (or NAMPs) which were already revealed to induce immune signalling in *Arabidopsis*, maize, rice, soybean, tomato, potato and barley plants [137, 138]. Importantly, Ascr18 is yet not validated to be a constituent of 'NemaWater' [136] and it is possible that NILR1 might be able to perceive both Ascr18 and NemaWater in distinct manners leading to induction of appropriate defence responses.

Plant PRR members have been further observed to recognize parasitic plants' invasion patterns. For illustration, some current reports have suggested involvement of LRR ectodomain carrying PRR members CuRe1 (RLP PRR in tomato) and HaOr07 (RLK PRR in sunflower) in detecting molecular patterns from parasitic plants *Cuscuta* and *Orobanche* species, respectively [139], thereby initiating defence responses downstream.

It is further worth mentioning that unlike most PRRs which generally detect exogenous or endogenous molecular patterns (PAMPs or DAMPs) [112], there are instances when despite being surface-localized, certain PRRs have been divulged to detect secreted effector molecules (or Avrs) of pathogenic microbes [140]. One of the suitable examples of such a receptor type is that of a known PRR cluster (i.e. Cf-2, Cf-4 and Cf-9) from tomato that recognizes secreted pathogenic effectors (Avr2, Avr4, and Avr9) from the fungal pathogen *C. fulvum*, and they are basically comprised of RLPs [141, 142]. *Cf* gene cluster is in fact the first plant RLP cluster to be reported in tomato. It detects secreted effectors of *C. fulvum* at the leaf apoplastic region [143]. Notably, the Cf group of surface receptors belongs to the LRR class of PRR. Likewise, another PRR member in tomato encoded by *I-3* gene was reported to recognize Avr-3 effector secreted by the fungal pathogen *Fusarium oxysporum* f. sp. *lycopersici* (*Fol*) [116]. *I-3* encoded receptor is said to possess a unique ectodomain referred to as S-domain and is rather known as SRLK or S-receptor-like kinase [116]. Importantly, this PRR also possesses an intracellular serine/threonine kinase domain and other features like most of the reported RLK PRRs [116].

Generally, PRRs formed of RLPs are devoid of their own intracellular kinase domains and these may bear at their carboxyl terminal a short cytoplasmic tail [93, 94] or occasionally they may be anchored to the plasma membrane through glycosylphosphatidylinositol (GPI) moieties [144]. All RLP PRRs indispensably depend on partner RLKs for their kinase function in the signalling processes downstream. Often, the partner RLKs assisting them (i.e. RLP PRRs) may be one or more in number [145–147]. Regularly, these assisting RLKs have been referred to as co-receptors. For instance, *Arabidopsis* RLP PRR member RLP23 requires support of two RLKs namely SOBIR1 (Suppressor Of BAK1-Interacting Receptor like Kinase1 (BIR1)-1) and BAK1 (BRI1-Associated Receptor Kinase1) for transducing defence signal after perception of a microbial PAMP called 'nlp20'derived from Nep1-like proteins (NLPs) [147]. In fact, RLP PRRs essentially form heteromers during signal transduction processes downstream [123].

As in the case of RLP PRRs, most often RLK PRRs also require assistance of other RLKs which are called co-receptors [98]. Usually, co-receptors are distinguished from their corresponding PRR partners (that recognize specific molecular patterns) through possession of much shorter ectodomains in them as against the latter [148]. Further, from the viewpoint of owning kinase domain type, co-receptors reported till now are belonging to the RD-kinase category as against most PRR RLKs which are often allied to the non-RD kinase group (baring exceptions) [98, 149]. Certain common co-receptors like BAK1 can aid both RLP PRRs as well as RLK PRRs in the signal relay events after perception of distinct categories of ligands [99]. However, selective association of SOBIR1 to LRR RLP PRR complexes but not LRR RLK PRRs has been well perceived [99, 150, 151]. Interestingly, association of BAK1 and SOBIR1 to respective PRRs has been suggested to be essentially determined by the presence of LRR containing ectodomain in specific PRRs [101]. However, exceptions may still persist. For example, *Arabidopsis* LRR-RK PRR XPS1 (that detects XUP25 peptide, derived from *P. syringae* [152]) has yet not been found to associate with any co-receptor for signal transduction [140]. Nevertheless, there are examples of other PRR categories as well which don't necessitate any co-receptor association for transducing immune signalling. Lectin ectodomain bearing RLK PRRs, namely LORE (lipooligosaccharide-specific reduced elicitation) and DORN1 (Does not Respond to Nucleotides1), for instance doesn't require co-receptors for their immune signalling events downstream [99].

An important aspect in regard to co-receptor BAK1 which is extensively recruited by LRR RLK/RLP PRR members in immunogenic ligand perception and defence signalling [99], is that it also forms a crucial partner of other non-immunogenic (e.g. growth and development) signalling modules [153, 154]. For instance, perception of brassinosteroid hormone and its subsequent signalling for plant growth and development requires LRR RLK brassinosteroid-insensitive 1 (BRI1) and BAK1 involvement [153]. LRR RLKs BAK1 along with SERK1, and BAK1-like Kinase 1 (BKK1, also referred as SERK4) cumulatively belong to the somatic embryogenesis receptor kinase (SERK) family [155]. Notably, the SERKs were initially established as BRI1 interacting proteins with an essential role in BRI1-mediated signalling [154, 155]. Apart from SERKs, another important co-receptor belonging to the Lys-motif-containing RLK category is CERK1 (chitin elicitor receptor kinase 1) which plays a pivotal role in fungal chitin-triggered signalling via Lys-M containing PRRs viz. LYK5 (Lysin motif containing receptor-like kinase5), LYM1, LYM3 (Lysin motif containing Receptor protein), CEBiP (chitin elicitor-binding protein) etc [145].

Apart from co-receptor RLKs, PRRs further rely on another distinct RLK cluster for their signalling events during plants' defence responses. These are known as receptor-like cytoplasmic kinases (RLCKs) which play an indispensable role in the signalling processes passing downstream of RLK/RLP PRR and co-receptor complexes [156]. Notably, RLCKs are crucial for signal transduction in several other physiological and developmental processes as well [156]. RLCKs generally don't possess any transmembrane segment and they are observed to interact intimately with PRR members in the plasma membrane [156]. There are seventeen sub-categories of RLCKs with respect to the extent of sequence homology manifested by these and like most plant RLKs they too predominantly harbour serine/threonine kinase domain [156]. Within the RLCK sub-family itself, among others, group VII members have been found to actively cooperate in PRR-triggered defence signalling, although their functions are not just restricted to the plant immunity realm alone [157]. BIK1 (Botrytis-Induced Kinase 1), PBL1 (PBS1-Like1), PBL19 (PBS1-Like19), RIPK (RPM1-Induced Protein Kinase), TRK1 (TPK1b Related Kinase1) etc are some examples of RLCK members which connect the pattern-elicited triggers from PRR-immune complexes and subsequently transduce varied signalling modules situated downstream [156–159].

2.4.2 PRRs and varied patterns they detect

Plant cell surface-localized PRRs are devised to recognize unique molecular patterns of varied biological origin through their respective ectodomains. In fact, PRRs are the first defence arsenals to detect the intruding danger signals in the form of peculiar patterns. In respect of particular molecular patterns perceived by PRRs, the chemical origin of the pattern and the category of PRR defined by their ectodomain constitution (i.e. array of peptide motifs/domains borne) seem to play critical roles. For example, the PRR class belonging to the LRR-RLK type have been known to recognize mostly the proteinaceous ligands which could be both MAMPs/PAMPs as well as DAMPs [160]. The protienaceous ligands can be again of varied kinds.

Some of them like flagellin [161], elongation factor [73], necrosis and ethylene (ET)-inducing peptide 1-like proteins (NLPs) [147], CSPs [129], elicitins [150], ET-inducing xylanase [Eix;121] etc can act as MAMPs/PAMPs, whereas fragments derived from plant elicitor propeptides (Pep) generally serve as DAMPs [162]. Essentially, they are sensed by specific LRR type PRRs at the plant cell surface. Additionally, a newly identified LRR RLK PRR called PERU (Pep-13 receptor unit) from *Solanum tuberosum* that recognizes proteinous MAMP 'Pep13' conserved across pathogenic oomycete *Phytophthora* species was brought to light very recently [163]. Nevertheless, NILR1 (an *Arabidopsis* LRR RLK family member) was recently elucidated to sense a modified carbohydrate elicitor from parasitic nematodes referred to as ascaroside#18 (Ascr18) [136] which demonstrate ligands for LRR RLK recognition may not be just restricted to proteinous origin. Table 2.1 enlists some of the recently divulged PRR receptors with corresponding immunogenic patterns from biotic agents.

Lysine-motif containing PRRs on the other hand, are largely implicated in detecting carbohydrate ligands e.g., chitin, peptidoglycan (PGN) etc (having a common constituent N-acetylglucosamine, GlcNAc) as PAMPs derived from fungi and bacteria, respectively [164]. Notably, Lys-M PRR members used for chitin and PGN detection can be distinct in most occasions [164]. However, the existence of

Table 2.1. Some of the recently elucidated PRR receptors with corresponding patterns from varied biotic agents.

PRR nomenclature	Host plant	PRR type	Molecular pattern/organism recognized	References
RXEG1	*N. benthamiana*	LRR-RLP	XEG1 from *Phytophthora sojae*	[537, 538]
RLP42	*Arabidopsis*	LRR-RLP	pg9 (At) epitope of fungal endopolygalacturonases	[539]
RLP32	*Arabidopsis*	LRR-RLP	IF1 from proteobacteria	[540]
RE02	*N. benthamiana*	LRR-RLP	SCP from *Sclerotinia sclerotiorum*	[541]
Unknown?	Soybean	Predicted RLP	PsRLK6 from *Phytophthora sojae*	[542]
NILR1	*Arabidopsis*	LRR-RLK	Ascr18 from nematodes	[136]
PERU	Potato	LRR-RLK	Pep-13 from *Phytophthora* spp.	[163]
RCAP1	*N. benthamiana*	LRR-RLP	PsCAP1 from *Phytophthora sojae*	[543]
REL	*N. benthamiana*	LRR-RLP	Clade 1 Elicitins from *Phytophthora*	[193]
OsRLP1	Rice	LRR-RLP	RBSD (rice black-streaked dwarf) viral PAMP	[132]
VuINR	*Vigna unguiculata*	LRR-RLP	Inceptin	[134]
HaOr7	Sunflower	LRR-RLK	*Orobanche cumana* (sunflower broomrape)	[544]

common Lys-M PRR members utilized for sensing both chitin and PGN PAMPs (e.g., *Os*LYP4 and *Os*LYP6 in rice) have also been described [165]. Apart from LRR and Lys-M types, a unique EGF motif containing PRR called WAK1 is known to recognize host-derived small carbohydrate (e.g., OGs) residues generally considered as DAMPS [126]. Likewise, two lectin-type RLKs designated as DORN1 and LecRK-I.8 were shown to involve in detection of extracellular nucleotides (like ATP, NAD $^{+}$) which are also reported to act as DAMPs in plants [166, 167].

In the same manner, inquiry towards perception of lipid constituting MAMPs/PAMPs has also revealed at least two PRRs members lately. One of these two PRRs is a member of lectin-type RLK known as LORE (lipooligosaccharide-specific reduced elicitation) which was shown to detect free medium-chain 3-hydroxy fatty acid (mc-3-OH-FAs) residues in *Arabidopsis* [168]. In fact, mc-3-OH-FAs form building blocks of bacterial LPS and RLK LORE's perception of LPS was confirmed to be only due to these precursors [168]. Studies have indicated conservation of PRR member LORE to be restricted to only the *Brassicaceae* family [169], thereby suggesting probable prevalence of other PRR types across different plant families for detection of LPS or derivatives of LPS [170]. The other PRR member for sensing LPS was reported in rice and is a lysine-motif-carrying RLK called *Os*CERK1 [130]. Importantly, the ability of *Os*CERK1to sense varied LPS preparations from both pathogenic as well as non-pathogenic bacterial members has also been stated [170]. These revelations may suggest that at least some PRR members with lectin and lysine-motif ectodomains seem to have the definite ability to recognize PAMPs of carbohydrate (e.g. PGN, chitin) and lipid (e.g. LPS or its derivatives) origin. These PRR members are also likely to be represented in varied plant species and might function in a similar manner. In addition, recently *Arabidopsis* PRR LORE has also been shown to detect a precursor of pathogenic bacterial rhamnolipid called (R)-3-hydroxyalkanoate (HAAs) as MAMP that could eventually elicit immune responses in the host plant [171].

It is not overemphasis to assert that the hallmarks of successful plant innate immunity induction processes are governed by immunogenic biological patterns and the precision with which these patterns are recognized by plant PRRs. Considerable detail on plant PRR members has been explored in previous sections of this chapter. Also, a sizeable number of biological patterns have been mentioned in connection with respective PRR members in earlier segments. However, in the context of plant PRR and pattern interactions in nature, inclusion of some additional aspects on the biological patterns would be relevant.

Notably, critical insights into a wide range of biological patterns sensed by PRR members have resulted in realization of their different origins. Importantly, these patterns can primarily be derived from two sources. One class of patterns can arise from diverse microbial communities (which can be pathogenic, non-pathogenic or beneficial), pathogenic nematodes, herbivore insect pests etc and therefore are considered exogenous [151, 172]. Patterns from these exogenous sources have been termed as PAMPs [173], MAMPs [174] or HAMPs [134] according to their origins. The terms MAMP and PAMP are alternatively used to describe conserved molecular patterns which serve vital functions across a wide range of microbes.

However, MAMP is a much broader term that embraces conserved patterns from both pathogenic as well as non-pathogenic microbes [73]. In contrast, PAMP would mean preserved molecular patterns largely from a pathogenic group of microbes [22]. Some of the MAMP/PAMP components from bacteria, fungi and their mechanism of immune response induction in different plant hosts have been revealed to a great extent. Considering association of immensely diverse microbial populations with plants under natural conditions, many more MAMP/PAMP candidates are anticipated to be revealed in near future. Among all, some of the established MAMP/PAMP candidates from pathogenic bacteria include Flagellin (flg22, flgII-28), EF-Tu, peptidoglycan, lipopolysaccharide, Nep1-like protein (nlp20), CSP etc [118]. Likewise, a few examples of notable MAMP/PAMP representatives elucidated from pathogenic fungal/oomycetes species are NLPs, elicitins, pep-13, cellulose-binding elicitor lectin (CBEL), chitin, glycoside hydrolase, β-glucans etc [56]. Given below is a succinct account on some of these immunogenic patterns.

2.4.2.1 Flagella

Flagella form an essential armoury of bacteria that aid the latter in their movement and during adherence to varied surfaces. In pathogenic bacteria, flagella act as an important virulence factor as well. Protein flagellin predominantly serves as the structural unit of flagellum. Initially, a conserved peptide domain of 22 amino acids length located at the N-terminus end of flagellin protein designated as 'flg22' was found to act as PAMP/MAMP in different plants (including tomato and *Arabidopsis*) and elicits immune responses [175]. Importantly, another variant peptide domain of bacterial flagellin referred to as flgII-28 (28 amino acids long) was confirmed to be detected as PAMP by some plant species of *Solanaceae* family [176] suggesting occurrence of distinct PAMP components within flagellin protein capable of eliciting immune responses across varied plant species.

2.4.2.2 Elongation factor thermal unstable Tu (EF-Tu)

EF-Tu is a G-protein that occurs abundantly in bacteria and serves a multitude of functions including a crucial one during protein synthesis [177]. EF-Tu executes aminoacyl-tRNA allocation to acceptor (A) site in ribosome. Owing to several other functions (including virulence functions in pathogens) performed by EF-Tu, it is also considered a moonlighting protein in bacteria [177]. Bacterial EF-Tu sequence is highly conserved and several plant pathogenic bacteria secrete out EF-Tu that in turn can serve as PAMP/MAMP to many plant species of *Brassicaceae* family (including *Arabidopsis*) [177, 178]. In fact, a peptide segment comprising first 18 conserved amino acid sequences in the N-terminal end of EF-Tu was confirmed to specifically elicit immune responses in *Arabidopsis* [178]. This peptide PAMP/MAMP was designated as 'elf18' [178].

2.4.2.3 Peptidoglycan (PGN)

It forms a crucial and distinctive element of the outer cell envelope across virtually all bacteria (baring exceptions) that render bacterial cell shape and firmness [179].

PGN being a heteropolysaccharide comprises repeating units of N-acetylglucosamine (GlcNAc) and N-acetyl-muramic acid (MurNAc) residues linked via β-1–4 bonds and these units are subsequently interconnected through smaller peptides. Importantly, for the first time, the carbohydrate backbone of PGN from *Staphylococcus aureus* was shown to elicit immune responses in *Arabidopsis* and eventually PGN was included in the catalogue of bacterial PAMPs [179]. Notably, *S. aureus* is a human pathogen and yetthe PGN constituent of this Gram-positive bacterium could elicit defence responses in plants. In a later study, PGN constituents from Gram-negative plant pathogens, namely *Xanthomonas campestris* pv. *campestris* and *Agrobacterium tumefaciens*, was shown to serve as a PAMP for *Arabidopsis*, eliciting characteristic immune responses in this host [180]. In the subsequent studies that ensued, PGN as an active PAMP elicitor was corroborated for rice and tobacco plants as well [164].

2.4.2.4 Lipopolysaccharides

Lipopolysaccharides (LPS) are chemically of glycolipidin origin, and form an important constituent of Gram-negative bacterial outer membrane [122]. LPS offer vital benefits to Gram-negative bacteria (irrespective of pathogens or non-pathogens) to sustain under unfavourable conditions by conferring protection against varied cytotoxic compounds such as antibiotics, detergents etc [181]. LPS embraces three crucial segments, namely a lipid-loving region called 'lipid A', a core oligosaccharide region and the O-antigen (or O-polysaccharide) region [182]. LPS ultrastructure varies considerably across different bacterial species and plausible occurrence of altered LPS compositions within outer membrane of single bacterium has also been confirmed [183]. LPS fractions from plant pathogens *Pseudomonas* spp., *X. campestris* were found to elicit immune responses in several plant members of *Brassicaceae* family (that embrace *Arabidopsis*) [169]. Interestingly, all the three segments of LPS can be individually detected as immune elicitors by plants, although via different yet undetermined mechanisms [169]. A key ingredient of LPS called medium-chain 3-hydroxy fatty acid (mc-3-OH-FA) has been recently validated to be an essential precursor of LPS fractions which can serve as a MAMP and trigger immune responses in plants [168].

2.4.2.5 Nep1-like proteins

Necrosis-and-ET-inducing peptide 1 (Nep1) like proteins (NLPs) constitute a cytotoxic family of proteins synthesized by bacteria, fungi and oomycetes microbes [147]. In dicot plants, NLPs have been shown to cause leaf necrosis and also induce typical immunity-related responses [147]. Nonetheless, non-cytotoxic NLPs from certain fungi and oomycetes members have also been reported recently [147]. Analyses of large numbers of NLPs from bacteria, fungi and oomycetes have indicated the presence of a conserved protein motif (referred to as 'nlp20') in NLPs that was adequate enough to trigger immune responses in *Arabidopsis* and other members of *Brassicaceae* [147, 184]. Importantly, this protein motif, i.e. nlp20, serves as a MAMP for plant hosts irrespective of the former's association with cytotoxic or non-cytotoxic NLPs [147]. In addition, NLPs are also believed to play a significant role in liberation of cellular DAMP particles via their cytotoxic effect on plant cells and thereby induce DAMP-triggered immunity in the plant host in an indirect manner [184]. Moreover, significant resemblances observed amidst inductions of similar defence-related genes triggered by cytotoxic NLPs and flg22 in *Arabidopsis* indicate considerable convergences in the immune signalling mechanisms due to both the patterns [185].

2.4.2.6 Cold shock proteins

CSPs form a small family of conserved proteins in all bacteria [186]. Bacterial CSPs are said to be expressed in response to cold stress caused by sharp decline in temperature and also induced by other kind of stresses [186]. These small proteins are characterized by the presence of single-stranded nucleic acid binding motifs RNP1 and RNP2 and have essential functions in bacteria [186]. Importantly, this small family of proteins was found to elicit immune responses in plant species restricted to the *Solanaceae* family only [129, 186]. More precisely, peptide fragments containing 22 and 15 amino acid sequences (denoted as csp22 and csp15, respectively) corresponding to a conserved segment of bacterial CSP protein were shown to act as MAMPs in the tomato plant [186, 187]. The corresponding PRR for csp22 is an LRR receptor kinase designated as CORE (CSP receptor) that was identified in tomato and *N. benthamiana* plants almost seven years ago [129].

2.4.2.7 Pep-13

Pep-13 is the nomenclature given to a 13-amino acids length peptide fragment that forms an intrinsic part of a cell wall glycoprotein (GP42) possessed by oomycete phytopathogen *Phytophthora sojae* [188]. The GP42 protein is in fact a calcium-dependent transglutaminase whose appropriate function in this oomycete is unknown. However, pep-13 fragment of GP42 from *P. sojae* was earlier revealed to elicit PTI responses in parsley (*Petroselinum crispum*) and potato (*S. tuberosum*), likely aided by some yet to be determined surface receptors [188, 189]. Importantly, pep-13 motif was identified as a conserved feature of GP42 in most of the *Phytophthora* species that include devastating phytopathogens like *P. infestans*, *P. capsici* etc [188]. Recently, Pep-25, another longer fragment within GP42 protein was also confirmed to possess the ability to elicit immune response in various potato cultivars and hence qualified as an important MAMP of *Phytophthora* species identical to pep-13 [189].

2.4.2.8 Elicitins

Elicitins are extracellular proteins secreted by several members of phytopathogenic *Phytophthora* and *Pythium* species [190]. Elicitins generally constitute a group of small-sized (~10 kDa) proteins bearing a conserved cysteine-rich elicitin domain within, and these have the ability to trigger immune responses in several *Solanaceae* species (including *Nicotiana* spp.) and also in selected species of *Brassicaceae*, largely characterized by hypersensitive responses and cell death [190, 191]. A major elicitin type secreted by *P. infestans* is referred to as INF1 [190]. INF1 acts as a bona fide MAMP-triggering immune responses across several species of Solanaceae [190, 191].

Recently, two phylogenetically unrelated LRR-RP PRRs namely ELR (in *Solanum microdontum*) [192] and REL (in *N. benthamiana*) [193] were divulged to detect INF1 elicitin and initiate PTI in respective hosts.

2.4.2.9 Cellulose binding elicitor lectin

Cellulose Binding Elicitor Lectin (CBEL) is a conserved glycoprotein located in the cell walls of oomycete phytopathogens like *P. infestans*, *P. sojae*, *P. ramorum* and *P. parasitica* [194]. Initially identified in *P. parasitica* var *nicotianae*, CBEL can bind to cellulose components of the plant cell wall and are suggested to have adhesion function [195]. Two important domains known as cellulose-binding domains (CBDs) which form an essential component of CBEL proteins are thought to aid in the cellulose-binding process. CBDs are held to be a close kin of the Carbohydrate Binding Module1 protein family and are rather restricted to only fungi [194]. CBEL lacks any enzymatic activity on cellulose but serves as a MAMP that induces characteristic PTI in tobacco as well as *Arabidopsis* [195, 196].

2.4.2.10 Chitin

Chitin, an abundant linear homopolysaccharide, is composed of N-acetyl-D-glucosamine (GlcNAc) repeating units linked via β-(1,4) glycosidic bonds. Chitin forms a vital and a highly conserved ingredient in cell walls of fungi, diatoms and also in the exoskeletons of insects, crustaceans as well as nematodes [170, 197]. During plant–fungal interactions, smaller fragments of chitin chains from fungal cell walls may be liberated by plant-secreted or fungal-derived chitinolytic enzymes [170]. These small fragments are referred to as chitin oligomers (COs) [170]. COs with variable numbers of monomeric constituents have been shown to induce plant cellular defence responses differentially, and these are detected as MAMPs [165, 198]. In fact, COs possessing seven or more monomers (e.g. heptamers, octamers) were revealed to trigger maximal immune responses in plant cells, whereas COs with five or fewer monomeric constituents could rarely induce any immune signalling [165, 198]. Apart from this, the acetylation state of the monomeric constituents in chitin has also been divulged to dictate MAMP-triggering potential [170]. For instance, the completely deacetylated state of chitin, generally referred to as 'chitosan' was shown to evoke feeble-to-diminished immune responses in *Arabidopsis* [199].

2.4.2.11 Glycoside hydrolases

Glycoside hydrolases (GHs) are one of the six broad classes of carbohydrate-active enzymes (CAZymes) that can hydrolyze complex carbohydrate compounds into smaller constituents [200]. Phytopathogenic bacteria and fungi deploy a large class of GHs generally referred to as cell wall degrading enzymes (CWDEs) for successful infection and colonization of plant tissues [201, 202]. CWDEs may comprise a diverse group of enzymes e.g. pectinases, polygalacturonases, glucanases, cellulases, xyloglucanases etc, and these are primarily utilized for dismantling composite carbohydrate polymers of the plant cell wall [203]. Importantly, several GHs secreted by phytopathogenic fungi are revealed to be detected as MAMPs as these can elicit defence responses in host plants [202, 204, 205]. In certain cases, the detection of GHs by host plants might be independent of the former's catalytic activities [204–206]. Some of the reported fungal GHs which are also established as important MAMPs/PAMPs eliciting immune responses in host plants include Fg05851, Fg11037 (*Fusarium graminearum*), FoEG1 (*F. oxysporum*), PsXEG1, EG1 (*Rhizoctonia solani*), BcXyl1 (*Botrytis cinerea*), etc [121, 202, 204, 205, 207, 208].

2.4.2.12 β-Glucans

Glucans are polymeric composites of glucose monomers abundantly found across cell walls of plants, algae as well as fungi [170, 209]. Usually, fungal cell walls may comprise two distinct kinds of glucans: α-glucans and β-glucans. Most of the α-glucan components in the fungal cell wall are connected via α-1,3-linkages and in some instances α-1,4-linkages may persist. On the other hand, the majority of β-glucan fractions persist as β-1,3 glucans which may again be connected to β-1,6-glucan components to generate branched polysaccharides in the fungal cell wall [209]. Apparently missing from plant cell walls, β-1,6-glucan residues form a unique constituent of cell walls in fungi as well as oomycetes organisms and hence form a strong MAMP candidate [209]. In fact, β-glucan fractions (comprising β-1,3 and β-1,6-glucan linked heptaglucosides) from *P. sojae* (an oomycete phytopathogen) were first shown to elicit immune responses in potato and several leguminous plant species more than four decades ago [170]. Later on, some of the other derivatives of cell wall β-glucans e.g. tetraglucosyl glucitols (from *Magnaporthe oryzae*), laminarin (from *Laminaria digitata*) etc were found to serve as MAMPs in rice and grapevine plants, respectively [170]. In recent findings, straight and unbranched β-1,3-glucan segments from cell walls of pathogenic fungi, oomycetes have also been observed to trigger immune responses in *Arabidopsis* [210]. It has also been suggested that length of cell wall β-1,3-glucan segments from pathogens as well as the type of plant species interacting with the former may determine efficient MAMP detection and subsequent immune response activation in the host [170]. A new variant of β-1,3-glucan termed as β -1,3–1,4-glucans has also been revealed recently to evoke defence responses in *Arabidopsis* and rice [170] indicating that many more β-glucan derivatives might be divulged in the near future.

Apart from the PAMP/MAMP candidates briefly described above, there are several other such candidates which can elicit defence responses in respective host plants. For instance, exopolysaccharides (from pathogenic bacteria as well as fungi) [211, 212], arachidonic acid (from oomycete pathogen) [213], SNP22 (from *Ustilaginoidea virens*) [114], ergosterol (found in fungal cell membranes) [214] etc are notable immune response elicitors. Importantly, many of the known and newly added PAMP/MAMP members are yet to be associated with their corresponding PRRs in respective plant species.

2.4.2.13 NAMPs

Nematodes have global distributions and some of them inflict considerable damage to plant health [215]. At present, the list of plainly characterized nematode associated molecular patterns (NAMPs) recognized by plant hosts, constitute just one member and this group of ligands has been found to induce characteristic pattern-triggered defence responses in respective hosts [137, 215]. An archetypal glycolipid pheromone compound found across parasitic nematodes are referred to as 'ascarosides' and these form the only confirmed NAMP representative of parasitic nematodes to date. A distinct member of ascarosides, Ascr18 was found to elicit typical PTI responses in different plant species including *Arabidopsis*, tomato, barley, soybean, potato, wheat, maize etc and qualify as a NAMP [136–138]. Further, an aqueous solution called 'NemaWater' prepared by incubating juvenile infective stages of parasitic nematodes was shown to demonstrate an immunomodulatory property as NAMPs [135]. The exact composition of NemaWater has not been revealed yet, however, a proteinous constituent in the concoction is thought to function as a NAMP [135].

2.4.2.14 HAMPs

HAMPs are specific immunomodulatory substances present in oral secretions (OS) of most insect herbivores that have been implicated in stimulation of plant defence responses [216, 217]. HAMPs could be released along with OS while insects chew/feed on plant tissues. The cognate plant cell surface receptors for many of the known herbivore associated elicitor patterns have yet not been elucidated, although these elicitors induce defence signalling in plants [218]. Volicitin (chemically 17-OH-C18:3-Gln), inceptin peptides and phospholipase Cfrom fall armyworm (*Spodoptera frugiperda*), fatty acid-amino acid conjugates (FACs) like linoleic acid-Glu (C18:3-Glu) from *Manduca* sp., volatile sex pheromone constituent like *E,S*-conophthorin spiroacetal found in goldenrod gall fly (*Eurosta solidaginis*) etc are a few examples of HAMPs triggering plant defence [217, 218]. Besides, β-glucosidases (from the cabbage white butterfly as well as the eastern tiger swallowtail), caeliferins (from grasshoppers), bruchins (from pea and cowpea weevils) etc have also been identified in OS of insect herbivores which elicits distinct defence responses in afflicted plant hosts [216]. Additionally, herbivore insect excrement, endosymbionts associated with herbivores can further act as HAMPs [218]. Recently, two new HAMP members from oral secretions (OS) of brown plant hopper (BPH, *Nilaparvata lugens*) namely NlDNAJB9 and NlG14, were described [219, 220] after confirming immune signalling responses mediated by both in *N. benthamiana* plants. It may, however, be noted that the defence output induced by HAMPs against herbivores may not exactly resemble that of pathogen elicited defence responses in plants [221, 222].

The second group of patterns generally include those which are basically derived from the constituents of the host cell itself and hence these are considered to be of endogenous origin [223]. The endogenous patterns generally emerge as a consequence of damage to plant cellular ultrastructures and could be released to extracellular milieu. The causes of cellular damage can be manifold—e.g., pathogen/herbivore attack, pathogen-induced plant pectin methyl esterase (PMEs) activity, mechanical injuries etc [223]. These endogenous patterns have been referred to as DAMPs. Notably, DAMPs are further categorized into two classes: (i) classical or constitutive DAMPs (cDAMPs) and (ii) inducible DAMPs (iDAMPs) [223]. The classical DAMPs are usually liberated by damaged plant cells in a latent manner, whereas inducible DAMPs (also known as phytocytokines) could be released by infected cells [224]. Importantly, some iDAMPs have been said to be evoked by PAMP triggered immune responses as well to additionally reinforce host defence outputs [223]. Another crucial function carried out by DAMPs is that these can serve as signalling molecules transmitting damage cues to far distances across the plant and elicit systemic wound responses [225]. Some of the notable DAMP candidates include OGs, esDNA, propeps, systemin, rapid alkalinization factors (RALFs), PAMP-induced secreted peptides (PIPs) etc [226–231] and a brief explanation of these candidates follows in the subsection below.

2.4.2.15 Oligogalacturonides

OGs are short segments of homogalacturonan (HG) polysaccharides usually found in the pectin component of plant cell walls [232]. HGs chemically comprise residues of α-1,4-linked galacturonic acids and are the predominant fraction among the three polysaccharide types constituting pectin [233]. OGs are known to be produced by partial degradation of HGs brought about by enzymatic action of PGs. PGs can be secreted by different microbes (including several pathogens) as well as by the plant itself upon induction via mechanical injuries [125]. OGs with 10–15 galacturonic acids residues, i.e. degree of polymerization between 10 and 15, were found to elicit characteristic immune responses in plants [227]. Subsequent investigations in regard to OGs have revealed a multidimensional impact on plant defence responses due to OGs [125]. In respect of OG generation from the plant cell wall, apart from the enzymatic action of pathogenic microbial PGs, plant cell apoplast-based PG-inhibiting proteins (PGIPs) also play an important role. In fact, it is thought that PGIPs regulate microbial PG activities in such a way that the OG fragments that result generally retain potent MAMP activities [125]. OGs generally belong to the classical DAMP category.

2.4.2.16 Extracellular self DNA

Release of nuclear/organellar DNA outside of the cells can serve as a signal for the latter's damaged or injured status. Cell damage can occur due to multifaceted causes like pathogen/pest ingression or mechanical injuries [234]. The liberated extracellular DNA rather referred to as extracellular self-DNA (esDNA) has been found to elicit immune responses in the same plant species from where it is discharged. Depending upon quantity of esDNA and relatedness of its biologic source, esDNA has been revealed to act as a classical DAMP triggering defence responses in respective plants [231, 235]. For instance, treatment of *Phaseolus vulgaris* (common bean) plants with esDNA obtained from conspecific individuals stimulated expression of characteristic defence-related genes, consequently making the plant resistant to a bacterial pathogen and herbivore infestations [231]. Similar kinds of esDNA-mediated defence responses have also been recorded in plant species such as *Arabidopsis* as well as tomato [235, 236].

2.4.2.17 Propeps

Propeps are a precursor of a small group of peptides (i.e. plant elicitor peptides) which serve as DAMPs and trigger conspicuous defence responses in plants [229].

Propeps can be induced by wounds (that can also be herbivore-mediated), other MAMP-mediated triggers, as well as through stimulation of Propep derived products [237, 238]. One of the notable elicitor peptides, *At*Pep1 was the first to be isolated from leaves of *Arabidopsis*. *At*Pep1 was determined to be a 23aa long peptide which is derived from C-terminal end of its 92aa long precursor *At*proPep1 [239]. *Arabidopsis* genome encodes eight Propeps in total and each of them is thought to release elicitor peptides, Peps, upon induction via different danger triggers [238]. Likewise, *At*Pep homologues of *Zea mays*, *Zm*Pep1 and *Zm*Pep3 were found to trigger defence responses against different insect herbivores [240, 241]. Peps are in fact an inducible category of DAMPs.

2.4.2.18 Systemin

Systemin, an inducible DAMP, is an 18 amino acids long peptide produced from its precursor protein called Prosystemin [242]. Systemin was first isolated from extracts of wounded tomato leaves and was confirmed to be induced by pathogens, herbivore rendered injuries in plant tissues [226]. Later, systemin was purified from other solanaceaous plant species as well [243]. Prosystemin is about 200 amino acids in length [226] and the former in its intact form can also trigger defence responses in plants against insect pests as well as pathogenic fungi [242]. Systemins, apart from stimulating proteinase inhibitor sysnthesis locally, can also act as long distance signal carriers throughout the plant and induce defence responses far from wounded sites [242].

2.4.2.19 Rapid alkalinization factors

RALFs are small plant based cysteine-rich peptides that are generated from their precursor proteins via proteolytic cleavages [225, 244]. RALFs were traditionally known to induce rapid alkalinization in the extracellular milieu of tobacco cells and are subsequently associated with developmental functions as well as regulation of defence responses in plants [228, 244]. They are also an inducible DAMP type and are detected by a dedicated class of plant PRRs called FERONIA. A RALF member, RALF23 had been shown to prevent the plant's PTI responses via its receptor FERONIA [245].

2.4.2.20 PAMP-induced secreted peptides

PIPs are a family of endogenous peptides liberated by plant cells when elicited by pathogenic PAMPs or sometimes also via abiotic stress triggers [230]. PIPs are generated from their precursors (prePIPs) which are thought to be secreted in the extracellular spaces of plant cells [230, 246]. As an inducible DAMP, PIPs can induce immune responses in host plants. For instance, *Arabidopsis* PIP members, PIP1 and PIP2 were earlier shown to induce immune responses in the host plant via interaction with RLK7 (receptor-like kinase 7) [230]. Recently, *St*PIP1 (a PIP member from *S. tuberosum*) was reported to be induced by potato virus Y (PVY) infection in potato and subsequently some important insights have been gained towards *St*PIP1-mediated anti-viral immunity in potato [247].

In addition to the above, there are other DAMP members which can trigger immune responses in plants as well, e.g., eATP [248], CAP-derived peptide (CAPE) 1 [249], IDA-like peptides [250], inceptin [251], phytosulfokines [252] etc. As in the case of MAMPs/PAMPs, corresponding PRRs for many of the DAMP candidates have yet not been identified.

Recent assessments have led to categorization of all these varied biological patterns i.e. PAMP, MAMPs, NAMPs, HAMPs, DAMPs etc into a common cluster called 'danger signals' [151, 172, 223]. It would be relevant to mention that there are various exogenous ligands which actually play essential roles in generation of plant endogenous patterns [253]. Both these pattern types, despite their variable sources of origin, generally serve to carry somewhat similar danger 'signals' to plant cells and accordingly the plant cellular defence responds to them [151, 172]. The PRR candidates that recognize exogenous patterns are, however, distinct from those that sense endogenous patterns and this could be an essential mechanism to distinguish self from non-self. In fact, a few PRRs with unique ECDs containing EGF, malectin-like domains have been found to perceive DAMPs specifically [223]. Nevertheless, in *Arabidopsis*, a common PRR (i.e. MIK2; MALE DISCOVERER 1-INTERACTING RECEPTOR-LIKE KINASE 2) for detection of a characteristic MAMP and a DAMP component have also been divulged recently [224]. MIK2 (a LRR RLK) PRR recognizes peptide fragments harbouring distinct SCOOP (SERINE RICH ENDOGENOUS PEPTIDE) signature motifs and interestingly the SCOOP family of peptides are encoded by *Arabidopsis* as well as certain fungal pathogens (e.g. *Fusarium*) [254] suggesting evolutionary conservation of SCOOP-like motifs across these organisms. SCOOP peptides (alternatively termed as a phytocytokine belonging to the iDAMP category) of *Arabidopsis* serve as a DAMP, whereas SCOOP-like peptides from *Fusarium* are detected as PAMP by MIK2, thereby triggering ligand-specific immune signalling cascades [224].

2.5 Pattern- and damage-triggered immunity: mechanisms

An important indication evident from earlier and recent findings is that irrespective of pattern types (either MAMP/PAMPs or DAMPs) detected by PRRs at the plant cell exterior, they generally trigger a similar kind of intracellular immune signalling networks [255]. In fact, plant cellular defence responses resulting from both PAMP- and DAMP-mediated elicitations have been portrayed as PTI [118]. This further drives curiosity as to why there persist two distinct perception strategies for endogenous (i.e. DAMPs) and exogenous (i.e. PAMPs/MAMPs/HAMPs etc) patterns, when they elicit similar intracellular defence signalling. Presently, an acceptable hypothesis forwarded in this regard enumerates that concomitant perception of MAMP and DAMP from microbes can render plant cells to discriminate pathogens from non-pathogens [118]. This is because MAMP is a common feature in most microbes, whereas DAMP is a restricted attribute of the pathogenic manoeuvre [118]. Moreover, in addition to PAMP/MAMP-elicited immunity, the ability of DAMPs to transmit danger signals to distances from the site of infection and induce similar kinds of immune responses elsewhere across the plant nevertheless afford immense survival benefits. In addition, studies attempting to investigate immune signalling mechanisms by recruiting PAMP as well as DAMP elicitors simultaneously have indicated further escalation of PTI responses in the host plant [223, 255].

In the immediate section below, a generalized view of PTI signalling corresponding to MAMP/PAMP as well as DAMP ligands has been portrayed. Afterwards, a concise account on secondary metabolites, defence-related compounds/gene products resulting from final events of PTI signalling have been illustrated. The canonical machinery for PTI signalling mediated by NAMP and HAMP ligands had not been fully elucidated until recently [134, 136]. For instance, the first fully validated PRR for corresponding NAMP detection has been described only recently [136]. Further, in regard to HAMP-induced PTI, key distinctions have been observed when compared to MAMP-triggered immunity signalling [221, 222]. Accordingly, PTI signalling mechanisms elicited by NAMP and HAMP elicitors have been considered in a separate section.

2.5.1 PTI generalised scheme

Upon cognition of specific patterns, PRRs along with co-receptor partner(s) initiate immune signals downstream via RLCKs. Mechanistically, PRRs and RLCKs are said to remain in close contact with each other even before a ligand (either PAMP or DAMP) is detected by PRR. Once PRR senses the ligand, next a co-receptor is recruited to the complex after which follows activation of the RLCKs via phosphorylations executed by PRRs [58]. Usually, PRRs and co-receptors remain disengaged prior to any ligand perception. However, binding of the ligand (or specific pattern) to the PRR ectodomain is thought to act as a mediator that could facilitate close proximity of PRRs with co-receptors, thereby giving rise to heteromeric immune complexes at the plasma membrane [114, 184]. Down the lane, activated RLCKs transduce the PRR relayed signals further downhill [103]. The resultant is an array of the plant's offensive counteractions categorized under the broad terminology PTI, manifested as swift release of reactive oxygen species (ROS), calcium entry to the cell interior, triggering of mitogen-activated protein kinase (MAPK) pathways, defence-related gene expressions, stimulation of calcium-dependent protein kinase (CDPK)-mediated activities, differential hormonal modulations etc [256]. Therefore, it appears, after perception of pattern triggers by PRR-immune complexes and subsequent relay of the triggers to RLCKs, the major events of signal dissemination to varied signal transduction modules downstream are largely catalyzed by RLCKs. Figure 2.2 illustrates a simplified model of MAMP-induced immune signalling initiated at the plant cell surface.

It has been corroborated that more than one RLCK member can aid PRR-immune complexes in the downstream signalling events [105, 157]. Possibilities of indirect association of certain RLCKs to PRR-immune complexes have also been envisaged [123]. Since a single RLCK may not suffice for emanation of so many quick and variable responses concurrently, engagement of multiple RLCKs in the proximity of activated PRR-immune complexes might be an effective tactic for plant

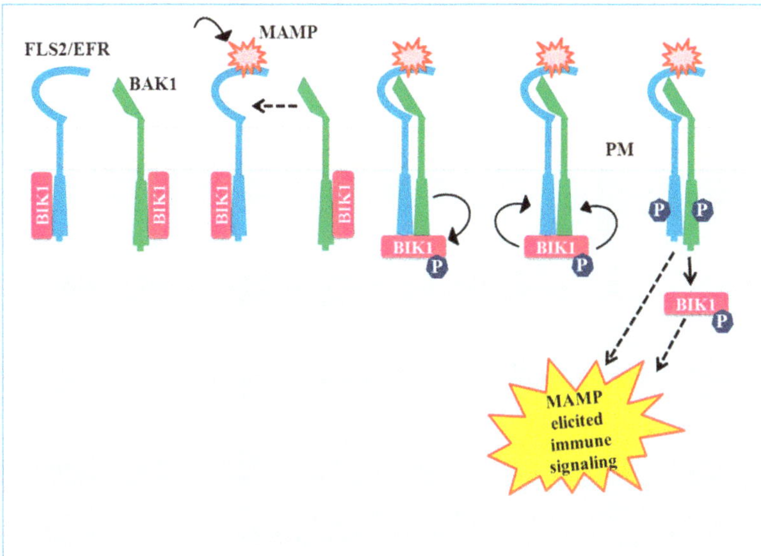

Figure 2.2. Simple illustration of MAMP (such as flg22, elf18 etc)-triggered immune signalling initiated at the plant cell surface [343, 531]. Shown here is a LRR RLK PRR (e.g., FLS2/EFR) that perceives conserved MAMP pattern (such as flg22 or elf18) and deploys a co-receptor (shown here is BAK1) to relay defence signalling via RLCK members (like BIK1 and others). Pep (an inducible DAMP pattern category) recognizing PEPR receptors (another LRR RLK PRR) also rely on BAK1 and RLCKs like BIK1 along with others for immune signalling downhill. Notably, phosphorylation and transphosphorylation steps are crucial for immunity signalling relay in the downstream of PRR-immune complexes.

cells to deploy them in multitasking. Several PRRs have been further explicated to recruit certain common RLCK members for signal transduction between them and signalling components downstream. BIK1 and PBL1 (both members of RLCK VII sub-family) for instance directly collaborate with FLS2, EFR, CERK1 and PEPR1 (Pep1 receptor kinase 1) to assist in the surge of pattern-elicited defence against microbial pathogens as well as insect pests [157]. Key events of PTI signalling triggered through PRRs such as FLS2, CEBiP, CERK1 and PEPR (1/2) are schematically represented in figures 2.3, 2.4 and 2.5, respectively.

Upon receiving a threshold trigger from PRR-immune complexes, RLCKs next activate varied signalling proteins downstream of them to generate appropriate immune responses. A number of such proteins that act as substrates of RLCKs have been explored and some have already been identified [158, 257]. These substrates include respiratory burst oxidase homologue (RBODH) enzymes, MAP kinases etc [258, 259]. RLCKs can further induce CDPKs indirectly [260, 261] or directly in a mutual phosphorylation dependent manner [471]. It is noteworthy that CDPKs are involved in calcium-dependent phosphorylation of RBODHs which in turn produces ROS as a defence response to danger signals. RBODHs are phosphorylated also by RLCKs, but in a calcium-independent manner. That implies RBODHs entail both calcium-dependent as well as calcium-independent phosphorylation events for their total activation and function [261]. CDPKs further regulate functions of

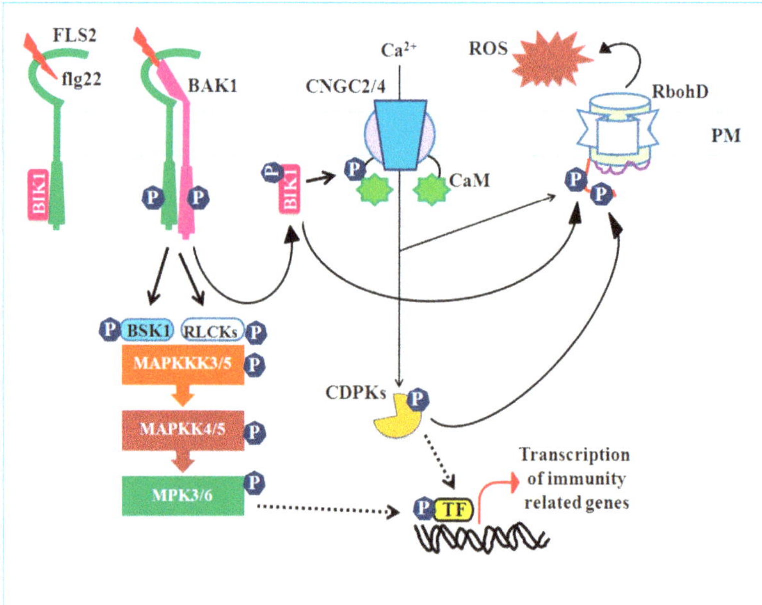

Figure 2.3. A simplified scheme of FLS2-flg22 pair mediated PTI signalling process adapted from Wang *et al* 2020 [532], with certain modifications incorporated. In *Arabidopsis*, after flg22 perception, LRR-RLK FLS2 (PRR) associates with BAK1 (LRR-RLK co-receptor) to set off events of PTI signalling. The beginning of immune trigger involves auto and transphosphorylation events depicted here as green spheres. Group VII RLCK members like BIK1, PBLs, BSK1 etc [260, 346] are immediate substrates of FLS2-BAK1 complex that takes part in the phosphorylation (auto and trans) steps. RLCKs are then liberated from the PRR-ligand complex. It was shown that BSK1 (but not BIK1) could trigger MAP kinase cascade (the MPK3/6 module) directly by phosphorylating MAP3K5 that in turn induced transcriptional reprogramming of defence-related genes. The appearances of canonical PTI markers e.g., calcium ion (Ca^{2+}) influx towards plant cell cytoplasm and generation of ROS are facilitated by BIK1-mediated phosphorylation and activation of membrane localized RbohD (respiratory burst oxidase homologue D) and CNGC2/4 (CYCLIC NUCLEOTIDE GATED CHANNEL 2/4) complexes, respectively. Elevated intracellular Ca^{2+} gradient could further trigger calcium-responsive proteins like CDPKs that in turn influences RbohD activity as well as transcription of PTI-related genes. Several other details relating to FLS2-flg22 immune signalling are provided in the text.

RLCKs, and MAPKs to fine tune defence outcomes downhill. They are also implicated in regulating transcriptional mechanisms leading to expression of specific immunity-related genes [261].

During PTI, activated RLCKs trigger MAPKs in their downstream that in turn can incite intracellular MAPK cascades subsequently leading to diverse responses [262]. It is, however, unknown whether all RLCKs directly phosphorylate MAPKs or there are other intermediary factors involved in the process [263]. Nevertheless, a few examples of RLCKs which activate MAP kinase pathway by phosphorylation have been evident. PBL27 (a RLCK member of the same BIK1 family) phosphorylates MAPKKK5 at its C-terminus, in response to chitin-mediated trigger [264]. Upon induced by chitin elicitors, rice RLCK member OsRLCK185 phosphorylates C-terminus of OsMAPKKKε to activate MAP kinase cascade downstream [265].

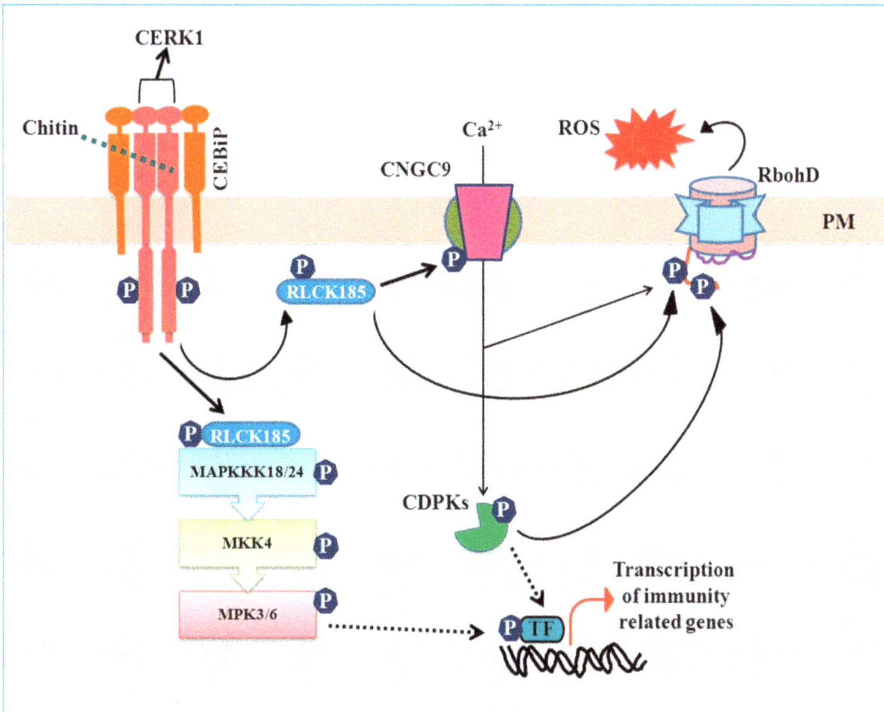

Figure 2.4. Illustration of chitin (fungal MAMP)-induced PTI signalling in rice through involvement of RLP (CEBiP) and RLK (CERK1) membrane localized proteins adapted from Wang *et al* [532] with some modifications. CEBiP being an RLP lacks intracellular kinase domain and remains attached to the plasma membrane via GPI-anchors. Both CEBiP and CERK1 have extracellular domains consisting of lysine motifs that aid in recognizing chitin. Importantly, upon chitin perception, CEBiP and CERK1 first undergo homodimerization and then heterodimerization events [146], which is unlike FLS2-BAK1 couple. CERK1 intracellular kinase activity is utilized by CEBiP to initiate chitin-induced downstream PTI signalling course. CERK1 is said to get autophosphorylated which may simultaneously phosphorylate and trigger RLCK185 [533]. In rice, activated RLCK185 subsequently induce ROS generation via RbohD and promote Ca^{2+} influx through CNGC9 channels. RLCK185 could further activate MPK3/6 kinase cascade by directly phosphorylating MAP3K18/24 kinase member to facilitate downstream immune signalling and transcription of defence-related genes. Calcium-responsive proteins (like CDPKs) can further function synergistically to promote canonical PTI responses as observable in the case of flg22 elicitation in *Arabidopsis*.

In some cases engagement of heterotrimeric G proteins during activation of certain MAPK members has also come into light [266]. MAPKs typically operate downstream of the PRR, co-receptor and RLCKs in the signalling pathway. A representative MAPK cascade may constitute at least three kinase members viz. a MAPK kinase kinase (MAPKKK or MEKK), a MAPK kinase (MKK) and a MAPK [262]. The phosphorylation triggers in this cascade usually descend vertically from MAPKKK toward MAPK via MAPKK. Sequential phosphorylations, which are basically reversible ones, lead to activation of MAPKs that subsequently regulate downstream players in the signalling trail again through a series of phosphorylations [262]. Importantly, there are several members belonging to each

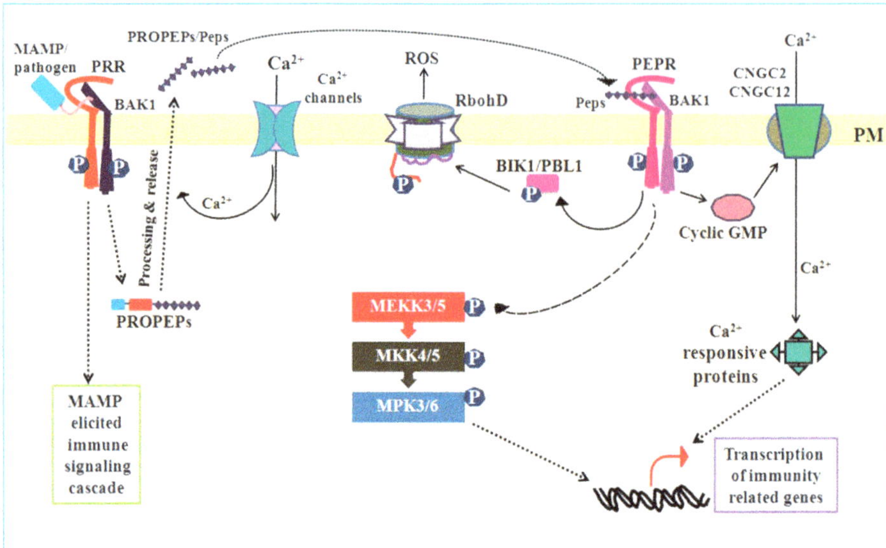

Figure 2.5. Schematic demonstrating generation of an inducible DAMP (Pep) that consequently triggers immune signalling in the plant cell [91, 528]. Synthesis of plant elicitor peptides (Peps) can be induced by various factors including wounds as well as MAMP stimulation [229, 239, 264]. Shown here is induction of Pep precursors (PROPEPs) by MAMP (such as flg22/elf18) stimulation (alongside MAMP-triggered immunity signalling). PROPEPs (barring exceptions) are processed in the cytoplasm by action of metacaspases (not shown here) that are activated by elevated cytosolic Ca^{2+} ion gradient. Peps, the end products of PROPEP processing, are then liberated towards the apoplastic region possibly via damaged sites in the cell membrane. Peps in the apoplastic region can next be sensed by membrane-localized PEPR (1/2) receptors which subsequently relay immune signalling towards the cell interior. PEPR (1/2) receptor associates with BAK1 co-receptor upon Pep perception and subsequently enables phosphorylations and release of group VII RLCK family members like BIK1, PBL1 to execute downstream signalling steps [162, 258]. Pep-PEPR induced signalling significantly resembles that of MAMP-elicited signalling, which is usually characterized by generation of ROS, Ca^{2+} ion influx to the cell interior, MAP kinase cascade activation and finally transcriptional reprogramming directed at expression of defence-related genes. Unlike the case of flg22 triggering, Pep evoked Ca^{2+} influx, however, is said to occur primarily through cGMP-stimulated CNGC2 and CNGC12 (CYCLIC NUCLEOTIDE GATED CATION CHANNEL 2 and 12) channels [353, 534, 535]. Importantly, cGMPs are generated by PEPR RLK through the latter's intrinsic gunylyl cyclase function [353, 395]. The elevated intracellular Ca^{2+} ion gradient can next activate certain calcium-responsive proteins that in turn regulate expression of immunity-related genes [353, 395, 536]. Other details of the Pep-PEPR complex signalling is discussed in the text.

type of MAPK cluster (i.e. MAPKKK, MAPKK and MAPK) which participate in distinct combinations in MAPK cascades and thus relay specific immune signals downstream [267]. Evidently, two distinct MAPK cascades are reported to be stimulated by currently known PRR members in their downstream [257]. One of these cascades includes the MAPK member MPK4 and the second cascade embraces two MAPKs namely MPK3 and MPK6 along with other participatory members upstream of them [257]. Initially, the MPK4 cascade was identified as a negative regulator of PTI responses [268]. Accumulating evidence, however, indicates both a positive and negative role of this cascade in plant PTI [269, 270]. The second cascade (comprising MPK3 and MPK6 kinases) is said to play a prominent role in conferring plant defence via induction of several immunity-related genes as well as various defence regulatory compounds (e.g., phytoalexin, ET etc) [257].

The MAPK cascades can induce several key substrates in their downstream. The host of substrates for MAPKs can comprise CDKCs (Cyclin-Dependent Kinase Cs), several transcription factors (e.g. WRKY, ERFs, BES1, ASR3 etc), certain transcriptional regulatory proteins (like VQ-motif-containing proteins) etc [271]. It would be relevant to mention here that unlike transcription factors (TFs), there are other transcriptional regulatory proteins that may not directly bind to DNA but can still modulate transcription via their interactions with TFs in response to various stimuli emanating out of biotic as well as abiotic interferences in the cell's external milieu [272]. A class of protein with short VQ (valine-glutamine) motif called VQ-proteins (VQPs) possessing similar transcriptional regulatory role has been revealed to be triggered by certain MAPK members to modulate defence gene expressions downstream [271, 273]. Notably, all the above-mentioned substrates (including VQPs) can be phosphorylated by MAPKs to regulate immune functions via defence hormone synthesis and signalling, elicitation of defence genes, production of pathogen inhibitory metabolites, hypersensitive responses (HRs) etc [267, 271]. A suitable example of a VQ-protein is MKS1 (*Arabidopsis* MAP kinase 4 substrate 1) which is phosphorylated by MAP kinase 4 (MPK4) through direct interaction with the former and enable *WRKY33*-mediated defence gene expression downstream [269, 270]. Another *Arabidopsis* VQ-protein which is phosphorylated by MPK3/MPK6 kinases after MAMP elicitations is MVQ1. MVQ1 was shown to interact with *WRKY33* and it regulates MAMP-induced defence responses negatively [273]. Interestingly, VQ-motif in these proteins has been demonstrated to be crucial for interacting with *WRKY* transcription factors but dispensable for associating with MAPKs [273].

Meanwhile, it has been revealed that MAPK cascade activation (beyond triggered PRR-immune complex) is not the only means for initiating transcription-aided defence responses downstream. Occasionally, involvement of certain associated components from ETI (effector-triggered immunity) branch has also been unveiled [274]. For instance, RLP23 (a LRR type PRR) triggered by nlp20 (MAMP) can activate bi-directional immune signalling pathways where one of the pathway may accompany MAPK cascade activation [147, 274] and the other route skips elicitation of MAPK module but recruits certain factors from ETI machinery (e.g., EDS1, PAD4, ADR1 etc) to evoke nuclear response [274]. Interestingly, in both the routes, transcriptional modulation of defence genes seems to be convergent, thereby indicating that either of the pathway can generate defensive responses even when the other one may be functionally inoperative [274]. In fact, currently surfacing evidence indicates engagement of ETI components with PTI machinery during stimulation of downstream nuclear responses rather facilitate synergism in the plant's resistance mechanisms [62, 63, 65, 274]. So far, both ETI as well as PTI components have been documented to independently elicit MAPK cascades and derive appropriate defence responses [64]. Moreover, it is unknown whether there is any common window through which both PTI and ETI components could activate MAPK networks [64].

The events of PTI generally begin with perception of distinct molecular patterns at the cell surface via PRRs that finally evoke a series of nuclear events culminating in transcriptional modulation of several defence genes [271]. A number of regulatory players have been implicated in these nuclear transcription events. These players largely constitute several transcription factors (TFs), co-regulators, mediators as well as chromatin modifiers etc [275]. A concise account on these transcriptional regulators is provided below.

2.5.1.1 Transcription factors

Among the known TFs, members of WRKY [276], MYB [277], NAC [278], bHLH [279], TCP [280], ERF [281], CAMTA [282] families have been shown to be involved in the regulation of nuclear transcriptional events associated with PTI responses. Transcriptional analyses of model plant hosts following PAMP-elicitations have indicated the earliest and most actively expressing gene categories in them to comprise varied TF members [277]. Importantly, many of these TFs are reported to be activated or regulated by different signalling components associated with PTI network lying in their upstream, e.g., MAPKs [267, 279], calcium-responsive proteins (like CDPKs, Calmodulin-binding proteins, CAMTA (calmodulin-binding transcription activators TFs)) [283] etc. For instance, members of *WRKY* (e.g., *WRKY33*), *MYB* (e.g., *MYB30*, *MYB55* and *MYB110*) [277], bHLH (e.g., Rac Immunity1, *RAI1*) [279] TFs were shown to be activated by MAPKs after specific PAMP perception by respective hosts. A particular type of TFs (e.g., RAI1) may further facilitate expression of other category of TFs (e.g., WRKY19 in rice) in response to specific PAMP stimulations [279]. In some instances, besides MAPKs some additional proteins (e.g., GTPases) may also engage in the activation process of TFs [279]. Like MAPKs, calcium-responsive proteins can also influence transcriptional dynamics of varied nuclear genes during PTI. Ca^{2+} influx toward cell cytoplasm, a fated event following PAMP induction of PRRs, subsequently stimulate various Ca^{2+} responsive proteins like CDPKs, calmodulin-binding proteins (e.g., CBP60g), CBLs (calcineurin B-like proteins), calmodulin-like proteins (CML) etc [284, 285]. Several of these calcium-responsive proteins can regulate transcription of varied defence genes by controlling activities of transcriptional regulators (e.g., TFs) during PTI [68, 284]. For instance, CAMTA3 (a member of the *Arabidopsis* CAMTA TF family) which is responsive to a rise in intracellular Ca^{2+} upon PAMP perception, regulates another calmodulin (a secondary messenger for Ca^{2+}) binding protein CBP60g to control salicylic acid (SA) biosynthesis and thereby influence subsequent defence responses [68, 286]. Moreover, another PAMP-responsive TF called SARD1 (SAR DEFICIENT1) is also known to regulate SA biosynthesis although independent of Ca^{2+} binding or responsive factors [286]. It is noteworthy that salicylic acid (SA) (emanating from PAMP elicitation) plays an essential role in regulation of several defence gene expressions downstream and thereby facilitates propagation of resistance phenotype in the host plant through mechanisms viz. SAR (systemic acquired resistance), LAR (local acquired resistance) etc [287].

2.5.1.2 Co-regulators

In addition to TFs, transcriptional co-regulators were implicated in the regulation of defence gene expression during MAMP/pathogen-triggered immune responses [288]. Transcriptional co-regulators are those factors that generally do not bind to DNA, but can influence functions of specific TFs by directly or indirectly interacting with them [287, 289, 290]. Notably, these co-regulators can be co-activators or co-repressors [290]. Crucial roles of co-regulators have been evident in hormone-induced defence gene transcription regulation. One of the suitable examples of such transcriptional co-regulators is NPR1 (non-expressor of PR gene 1) which also acts as a receptor for SA hormone. After binding to SA, NPR1 next joins to and stimulates a number of TGA family TFs (notably, TGA2, TGA5 and TGA6) thereby promoting expression of pathogenesis-related (PR) genes that contribute to SA-induced plant immunity [287, 289]. NPR1 in fact, acts as transcriptional co-activator for SA-induced PR gene expression [287]. Recently, EDS1 (enhanced disease susceptibility 1) that forms an essential component of ETI pathway, was shown to act as co-activator of SA-induced defence gene expression [291]. Notably, EDS1 and NPR1 both act synergistically to facilitate PR gene expression by recruiting 'mediator complexes' for activating appropriate TFs [291]. Two paralogs of NPR1, namely NPR3 and NPR4, on the other hand, were identified as redundantly functioning co-repressors of SA-induced transcription processes [289]. Interestingly, SA binding to NPR3 and NPR4 subdued NPR3/NPR4 driven transcription repression which is in marked contrast to the fact that NPR1 induced transcriptional activation is heightened upon SA binding to NPR1 [289]. Another example of transcriptional co-repressor is TOPLESS/TOPLESS RELATED (TPL/TPR) proteins in *Arabidopsis*, which interact with specific transcriptional repressors involved in suppression of Auxin and jasmonic acid (JA)-induced transcription mechanisms downstream during plant development and defence [290, 291].

2.5.1.3 Mediators

Mediators are another important category of transcriptional co-regulators which form connecting links amidst RNA Polymerase II enzyme and TFs (generally bound to DNA). They also contribute immensely towards varied facets of transcription processes [293]. Pertinently, these multi-protein conjugates also aid in the transcriptional regulatory mechanisms essential for appropriate plant defence responses [293]. In plants, several mediator sub-units e.g., MED8, MED14, MED15, MED16, MED18, MED19, MED21, MED25, CDK8 etc were associated with immunity responses [294]. During SA-induced immune signalling, NPR1 and EDS1 co-activator couple cooperatively stimulate CDK8 (cyclin-dependent kinase 8) a subunit of the mediator complex to initiate transcription of SA-responsive gene *PR1* [291]. In a recent study, after elf18 MAMP-triggered elicitations in *Arabidopsis*, SA-responsive *PR1* gene expression was strongly correlated with involvement of mediator subunit MED19a. It was subsequently unveiled that for efficient activation of *PR1* transcription, MED19a actually relied on a long non-coding RNA molecule ELENA1 (ELF18-INDUCED LONG-NONCODING RNA1) [288]. Long non-coding RNA regulating a MAMP-triggered terminal immune response outcome undoubtedly introduces a new dimension for exploration in the plant immunity research arena.

2.5.1.4 Chromatin modifiers

Apart from the genetic context, epigenetic control of immunity signalling has also emerged as an essential realm influencing plant–pathogen/elicitor interactions [295, 296]. Alterations in DNA methylation profile, tailoring of chromatin architecture, histone remodelling mechanisms etc are conspicuous epigenetic manoeuvres which have a prominent affect on pathogen/elicitor-induced plant immune responses downstream [297]. These epigenetic makeovers have also been envisaged to act as immunogenic memories facilitating survival benefits to individuals in subsequent generations [297]. The actors serving as crucial epigenetic modulators (also called chromatin modifiers) include histone methyl transferases (HMTs), histone deacetylases (HDACs), histone demethylases (HDMs) etc [295, 296, 298]. These modulators through their catalytic activities bring about key rearrangements in histone proteins and hence influence transcriptional outcomes. Rice histone deacetylase member HDT701 (histone deacetylase701) was found to act as a negative regulator of key PTI responses via repression of acetylation status in distinct histone monomers (specifically at H4K5 and H4K16 sites) after flg22 and chitin triggering [295]. Transgenic rice plants silenced in HDT701 function showed elevated levels of *FLS2*, *CEBiP* (PRR genes) transcripts irrespective of elicitor induction, suggesting an inhibitory role of HDT701 on PTI associated responses [295]. Histone demethylases serve to remove methylation marks in histone proteins and thus influence chromosomal transcription processes noticeably. Induced by flg22 elicitations, *Arabidopsis* histone demethylase ROS1 (Repressor of Silencing 1) eliminates methylation signatures occurring in the promoter sequences of defence-related genes (e.g., RMG1, RLP43) and thereby promote defence responses by permitting expression of those genes [296]. ROS1 was in fact found to expunge methylated spots generated by RNA-directed DNA methylase (RdDM) activities in the promoter regions of aforementioned genes which subsequently allowed transcription to proceed [296]. On the other hand, histone methyl transferases (HMTs) are known to introduce methyl groups in histone proteins and hence control gene expression patterns differentially. For instance, HMTs methylating histone proteins at specific lysine residues (e.g., H3 lysine 4, H3 lysine 36) could be linked to active gene expression events, whereas methylation at H3K27 targets rendered inhibition of gene expression markedly [298]. In *Arabidopsis*, flg22, Pep1 as well as effector stimulated defence signalling was shown to be positively regulated by a pair of histone methyl transfereases viz. SDG8 (SET DOMAIN GROUP8) and SDG25 (SET DOMAIN GROUP25) [298]. Notably, in response to elicitor trigger, SDG8 and SDG25 could induce expression of important immunity-related genes viz. *CAROTENOID ISOMERASE2* (*CCR2*) and *ECERIFERUM3* (*CER3*) via methylation of histone3 monomers at two lysine residues i.e. H3K4 and H3K36. Notably, *CCR2* is involved in the carotenoid biosynthesis pathway, and *CER3* contributes toward production of cuticular wax. Mutant analyses of *sdg* plants indicated impairment in carotenoid production, lipid build-up as well as abnormal cuticle development suggesting a conspicuous role of SDG8 and SDG25 during elicitor-induced plant immunity [298]. In addition, *CER3* and *CCR2* functions could be correlated to methylation status of histone lysine residues at specific loci as well as overall histone repertoire regulated via SDG8 and SDG25 [298]. A number of relevant reviews extensively elaborate upon epigenetic control of plant innate immunity which may be further referred to in [297, 299, 300].

2.5.2 Compounds/metabolites generated during PTI and their roles in plant defence

Numerous studies implicate that the pattern (i.e. MAMP/DAMP) or pathogen-induced trigger transmitted through a plethora of plant immune signalling components ultimately leads to genesis of varied defence-related compounds/metabolites in the downstream [240, 301–306]. These metabolites (e.g., defensins, phytoalexins (like camalexins), glucosinolates, chitinases etc) serve as an essential repertoire of plant PTI signalling against the elicitor/pathogen trigger, which in turn confer significant resistance responses against a large number of plant pathogens/pests. For instance, OGs can stimulate biosynthesis of antimicrobial substances such as phytoalexins, chitinases, β-1,3-glucanases etc and contribute to defence responses in *Arabidopsis* as well as grapevine toward *B. cinerea* and other fungal phytopathogens [301]. Likewise, soybean endogenous peptide iDAMPs (GmPep914 and GmPep890) were revealed to stimulate immune responses via rapid induction of *CYP93A1* gene (encoding a cytochrome P450 protein) required for phytoalexin biosynthesis along with other defence-associated proteins viz. chitinases, chalcone synthases etc [304]. In *Arabidopsis*, flg22 MAMP trigger leads to expression of pathogen-related 1 (PR1) proteins (regulated through SA-hormone signalling) which initiate an active defence against *P. syringae* pv. *tomato* DC3000 pathogen [306]. In addition, flg22 ligand was also found to stimulate biosynthesis of camalexin (an indole phytoalexin), indole glucosinolates (IGs), indole-carboxylic acids (ICAs) etc in *Arabidopsis* hosts [302]. These antimicrobial metabolites could significantly aid in plant defences towards varied pathogenic fungi and oomycete members like *B. cinerea, P. (=Phytophthora) brassicae, Colletotrichum gloeosporioides, P. cucumerina* etc [307–309]. In *Z. mays*, inducible peptide DAMP *Zm*Pep1 (an ortholog of *At*Pep1) was divulged to activate synthesis of defence-associated metabolites like endochitinase A, PR-4, PRms, SerPIN along with a camalexin compound referred to as benzoxazinoids [240]. These hosts of *Zm*Pep1 induced antimicrobials could render maize plant resistant towards *Cochliobolis heterostrophus* (agent for southern leaf blight) and *Colletotrichum graminicola* (cause of anthracnose stalk rot) fungal pathogens [240]. Importantly, in an earlier report [303], benzoxazinoid compounds have also been implicated in maize plant immune responses against fungal pathogen (*Setosphaeria turtica*) as well as insect herbivore (*Rhopalosiphum padi*, a cereal aphid)-mediated afflictions. Recently, chitooctaose (a chitin MAMP representative) induced elicitation was shown to activate camalexin defence compound synthesis in *Arabidopsis* that resembled immune responses triggered by pathogenic fungi *B. cinerea* via the cooperative effect of MAP kinase cascade, ET and JA hormonal pathways [305]. In *Arabidopsis*, two MAMP members namely nlp20 and PG3 were stated to induce enhanced synthesis of camalexin defence compounds as a part of distinct PTI response [310]. Notably, transgenic potato lines engineered with RLP23 PRRs were earlier shown to possess a resistance phenotype against an oomycete (*P. infestans*) and a fungal (*S. sclerotiorum*) pathogen, after nlp20 induced elicitations [147].

Induction of callose synthesis and deposition in plants is one of the striking effects triggered by several MAMP (e.g. flg22, elf18, chitin, chitosan etc) as well as DAMP (such as *At*Pep1, OGs, PIP1 etc) candidates [169, 230, 311]. Callose is a homopolysaccharide, made up of glucose monomers linked via β-1,3-glycosidic bonds [312]. Callose deposition is triggered at the site of pathogen/pest induced injury and serves as an essential protective impediment to restrict the entry of unwanted intruders towards the plant's interior [311, 312]. Callose accumulations at the sites of plant cell wall injuries are referred to as papillae which are said to be instrumental in aiding plant defence [311]. Build-up of callose in the wounded plant sites due to nematode and insect herbivore driven damage has been documented as well. For instance, the bona fide NAMP member Ascr18 [313] along with legitimate HAMP candidates NlDNAJB9 [220] and NlG14 [219] from OS of BPH (*N. lugens*), were revealed to stimulate callose deposition in tested host plant species. Callose accretion as a typical PTI response triggered by MAMP, DAMP, NAMP as well as HAMP members thus indicates a commonality persisting in the immune signalling cascades initiated by all these ligands.

Apart from MAMPs and DAMPs, nematode associated elicitors (i.e. NAMPs) have also been revealed to evoke synthesis of different immunity-related compounds/metabolites as part of plant defence responses [136, 137]. For instance, Ascr18 (a typical NAMP) induced immune signalling in *Arabidopsis* led to activation of SA and JA hormonal pathways characterized by appearance of *PR-1* and *PDF1.2* gene products, respectively [137]. Stimulation of SA and JA pathways through Ascr18 contributed towards local and systemic immunity in plants. Ascr18-induced defence responses were found to confer resistance phenotypes in tomato, barley and *Arabidopsis* plants against varied microbial phytopathogens (e.g., *P. infestans*, *B. graminis* f. sp. *hordei*, *P. syringae* pv. *tomato*) as well as parasitic nematodes (e.g., *H. schachtii*, *M. incognita*) [136, 137]. An unidentified NAMP constituent in NemaWater was recently revealed to activate JA/ET hormone responsive defence network in *Arabidopsis*, along with induction of several immunity-related gene products. Eventually, NemaWater-triggered PTI signalling could be correlated with *Arabidopsis* resistance responses towards parasitic nematodes (*H. schachtii* and *M. incognita*) and a bacterial phytopathogen (*P. syringae* pv. *tomato*) [135].

In the same manner, HAMPs have also been implicated in inducing defence-related metabolites in plants and thus influence defence responses in the latter. The newly added HAMP candidate from BPH (*Nilapar vatalugens*) designated as *Nl*VgN (the short N-terminal segment of Vg protein), was shown to stimulate biosynthesis of JA, JA-Ile hormones in rice plants. *Nl*VgN further rendered transcription of JA-responsive genes (*OsJAZ8, OsJAZ11, OsPR10a* etc) that positively regulate rice immunity. *In vivo* expression of *Nl*VgN could aid in negative regulation of BPH egg hatching rate in rice plants [314]. Interestingly, one of the JA-responsive genes viz. *OsPR10a*, cited above encodes a pathogenesis-related RNAse protein that was earlier demonstrated to confer adequate protections against *X. oryzae* pv. *oryzae* (*Xoo*) and *X. campestris* pv. *campestris* (*Xcc*) pathogens in rice as well as *Arabidopsis* plants upon overexpression of *OsPR10a* in them [315]. In a way, induction of *OsPR10a* gene expression could be a convergent point for HAMP as well as the pathogen-induced immune signalling network in rice plants. Another insect-related ligand called *Vu*-In (an inceptin HAMP ubiquitous in legume herbivore insect OS) when perceived by INR (RLP PRR) receptor proteins expressed in transgenic tobacco plants (*N. benthamiana* and *N. tabacum*) stimulated generation of ROS burst, ET hormone synthesis, peroxidase protein production along with synthesis of anti-herbivory-related immunity proteins like trypsin inhibitors, ascorbate oxidase etc. Interestingly, *Vu*-In-induced defence responses could be correlated with decreased growth rate (up to 37%) of *S. exigua* (beet armyworm) caterpillar stages when fed on those transgenic tobacco plant lines [134].

Leaving aside the direct HAMP-induced immune signalling routes, insect herbivores can further evoke indirect plant defences by triggering release of herbivore-induced plant volatiles (HIPVs) during herbivory. The chemical origin of HIPVs may be variable and can include monoterpenes (e.g. β-ocimene, linalool), homoterpenes, sesquiterpenes (such as α-farnesene, β-caryophyllene), GLVs (green leaf volatiles derived from fatty acids) etc [316]. HIPVs can render indirect defences to plants by attracting varied kinds of natural herbivore antagonists like predatory arthropods, entomopathogenic nematodes, birds etc [316]. HAMPs in the oral secretions of insect herbivores can trigger synthesis of HIPVs in host plants [317] and subsequent release of these volatiles can occur. The range of volatile substances released may vary in composition and quantity with respect to plant species as well as specific herbivore species triggering HIPV synthesis [318, 319]. For example, infestation of two Lepidopteran moth herbivores, namely *Heliothis virescens* (tobacco budworm) and *Helicoverpa zea* (cotton boll worm) in cotton plants led to release of plant volatiles differing significantly in composition. Importantly, the distinction persisting amidst these two released volatile types was prudently manipulated by Hymenopteran parasitic wasp *Cardiochiles nigriceps* to particularly target its victim *H. virescens* and not *H. zea* [319]. Sometimes, a less damaging herbivore insect pest infesting simultaneously with a more damaging one on the same plant host may promote plant defence against the latter (i.e. more damaging one) particularly as well as effectively. *Nicotiana attenuata* plants afflicted by *Manduca quinquemaculata* (tobacco hornworm) and *Tupiocoris notatus* (mirid bug) induces synthesis and release of plant volatiles that differ in certain composition [319]. Both released volatile types, however tempted, carnivorous *Geocoris pallens* (western big-eyed bug) to assault *M. quinquemaculata* larvae with more proclivities [319]. It needs mentioning that *T. notatus* is also a prey for *G. pallens* under natural settings, but in the presence of *Manduca* larvae, *G. pallens's* affinity shifts towards the latter quite significantly [319]. Newer aspects in respect of HIPV-mediated plant defence responses have began to emerge in recent times and readers may refer to these studies for more details [320–322].

2.5.3 NAMP-triggered immunity

Parasitic nematodes that threaten plant health immensely include two devastating sub-groups, namely root-knot and cyst nematodes, alongside others [11, 137]. Investigation into the nature of plant immune responses elicited by nematodes had unveiled a class of conserved glycolipid compounds called 'ascaroside' that can act as NAMPs. A particular ascaroside member termed 'ascaroside#18' (Ascr18) as NAMP was shown to elicit typical PTI responses in leaves as well as roots of *Arabidopsis*, tomato and barley [137]. Recently, the plant PRR receptor for this NAMP has been elucidated as NILR1 (NEMATODE-INDUCED LRR-RLK1) [136]. Like in other characterized LRR PRR-ligand pairings, Ascr18 binds to the LRR containing the ectodomain of NILR1 receptor [136]. Previous findings suggested Ascr18 triggered immune signalling in selected dicot and monocot plants could lead to: activation of MAP kinase cascade, synthesis of calcium-dependent protein kinase-related PHOSPHATE-INDUCED1(PHI1) proteins, induction of SA and JA hormonal pathways and expression of defence genes responsive to these (i.e. SA/JA) hormones [137]. In later experiments, a considerable degree of Ascr18-induced positive defence responses could be extended to other crop species as well, which include soybean, rice, potato, maize, wheat etc [138]. Interestingly, via a new report, Ascr18-induced expression of various *Arabidopsis* PTI marker genes such as *MPK3* and *MPK6* (MAP kinase pathway genes), *FRK1* (*Flg22-INDUCED RECEPTOR KINASE1*), *GSTF6* (*GLUTATHIONE S-TRANSFERASE*), *PR1* and *PR4* (*PATHOGENESIS-RELATED1/4*), *SAUR34* (*SMALL AUXIN UP RNA*), *PDF1.2* (*PLANT DEFENSIN1.2*) etc were found to be markedly reliant on NILR1 PRR function [136]. A yet-to-be-identified NAMP in 'NemaWater' was earlier found to be sensed by *Arabidopsis* NILR1 RLK that subsequently relayed immune responses with the involvement of BAK1 co-receptor [135]. Additionally, earlier studies had already divulged significance of BAK1 in the PTI responses against nematodes in hosts such as *Arabidopsis* and tomato [323, 324]. Unknown elicitors in NemaWater extracts from two parasitic nematode members (*H. schachtii* and *M. incognita*) were shown to trigger strong and steady generation of ROS in *Arabidopsis*. NemaWater elicitations further hindered *Arabidopsis* seedling growth indicating typical MAMP/DAMP-induced effect. Stimulated expression of genes engaged in biosynthesis and signalling routes of JA/ET hormones was recorded after NemaWater mediated triggering. NemaWater-driven induction also led to enhancements in the expression of typical PTI-related genes such as *RbohD*, *CYP81F2*, *CYP82C2*, *PROPEP1/2/3* and *NLH10* etc [135]. Importantly, emanation of all these immune responses necessitates a cooperative role of NILR1 and BAK1 receptors during recognition of unknown NAMP from NemaWater [135].

2.5.4 HAMP-triggered immunity

HAMPs originating from several insect herbivores have been cited as elicitors of PTI-like defence signalling in plants [133, 324, 325]. These insights have emerged from numerous studies documenting induction of a handful to many PTI marker genes, pathways or metabolites by predicted HAMP candidates in different insect–

plant host systems. Several plant cell surface receptors (or PRR) molecules have been associated with defence responses towards varied insect herbivores, e.g., *Na*RLK1, *Os*RLK1, *Os*RLK2, *Gm*HAK1/2, *Os*LecRK1/2/3, *Os*LecRK etc [133]. Moreover, lack of a completely characterized plant cell surface receptor (or PRR) recognizing particular HAMP elicitors had been posing sufficient hindrances in interlinking all the key components of the HAMP-mediated defence responses. Recently, discovery of an RLP PRR receptor (inceptin receptor, INR) capable of sensing inceptin peptides (HAMP) has augmented possibilities of exploring HAMP-triggered defence signalling in much detail [134]. The RLP PRR member INR from *V. unguiculata* (cowpea) was shown to recognize an inceptin peptide *Vu*-In as HAMP. *Vu*-In is about an eleven amino acid long peptide found principally in the OS of legume insect herbivores like *Anticarsia gemmatalis*. Validation of INR association with co-receptors such as *At*SERK, *At*SOBIR1 and *Vu*SOBIR1 through co-immunoprecipitation (co-IP) studies suggested analogous links seen amidst RLP PRRs and co-receptors while sensing MAMP/DAMP ligands. Expression of INR genes from *V. unguiculata* and *P. vulgaris* in transgenic *N. benthamiana* as well as *N. tabacum* plants and subsequent elicitation of these tobacco plants via *Vu*-In HAMP-induced synthesis of ET hormone and peroxidase proteins during defence response. In addition, *Vu*-In-mediated induction of two characteristic anti-herbivory defence gene products, namely a Kunitz trypsin inhibitor (KTI) and ascorbate oxidase (AscOx), was reported in both the transgenic tobacco plant species [134].

Among the growing list, some of the candidate HAMP members that were recently revealed to stimulate PTI-like responses include RP309 (an apoplastic effector produced by legume pest *Riptortus pedestris*) [327], OS of cotton ball weevil [329], vitellogenins from BPH [314] etc. Transient expression of *R. pedestris* effector RP309 in *N. benthamiana* plants could reveal induction of pattern-triggered defence marker genes such as *PTI5*, *Acre31*, and *WRKY8*. RP309 was again found to rely on *N. benthamiana* co-receptor proteins *Nb*BAK1 and *Nb*SOBIR1 for stimulating plant cell death phenotypes [327]. The OS from cotton boll weevil (*Anthonomus grandis*) as HAMP candidate was found to induce MAP kinase gene expressions in *Arabidopsis* and cotton plants with noticeable dominant expression apparent for *AtMAPK6* and *GhMAPK6* over *MAPK3* counterparts [329]. *A. grandis* OS could further trigger expression of a PTI marker gene *FRK1* (flg22-induced receptor kinase 1) suggesting HAMP-induced defence response could considerably resemble microbial pattern-triggered signalling. However, a distinct difference between (*A. grandis* OS) HAMP and (flg22/chitin/Pep1) MAMP-mediated immunity elicitations could be witnessed as MAP kinase activation induced by *A. grandis* OS in *Arabidopsis* didn't rely on functional co-receptors such as BAK1/SERK4, SOBIR1, CERK1, LYK4 and LYK5 [329]. This may point to a somewhat non-canonical route of PTI activation as against the canonical type that usually involves PRRs as well as co-receptors during sensing of specific immunogenic ligand. Nevertheless, persistence of other unfounded receptor machineries at the plant surface for HAMP-induced PTI activation cannot be undermined.

Vitellogenins (Vgs) represent a ubiquitous protein in egg-yolks and various organs/tissues of insects including BPH (a serious rice insect pest) [314].

Interestingly, *Nl*VgN forms a small subunit of Vgs (from plant hopper *N. lugens*) at its N-terminus, which was subsequently shown to act as HAMP [314]. In rice, *Nl*VgN elicitations resulted in marked enhancement in cytosolic Ca^{2+} influx, generation of H_2O_2 as seen in the case of MAMP/DAMP inductions. Exogenous application of *Nl*VgN resulted in stimulation of JA, JA-Ile hormones in tested rice plants. Notably, JA hormonal pathway was earlier stated to be instrumental in rice defence towards BPH [329]. In addition, upregulated expression of JA-responsive genes namely *OsJAZ8, OsJAZ11 and OsPR10a* along with *OsWRKY26*, after *Nl*VgN triggering contributed to rice defence responses. Moreover, in rice transgenic lines, heterologous expression of *Nl*VgN could restrict BPH egg hatching incidences quite significantly [314].

Studies indicate HAMP-induced defence signalling in plants has crucial reliance on JA hormonal pathway [330]. OS from herbivore insect pest *Manduca sexta* that comprises FAC (fatty acid-amino acid conjugate) HAMP as elicitor, was able to induce JA biosynthesis in *N. attenuata* plants through involvement of *Na*BAK1 co-receptor. *Na*BAK1 participation was, however, not necessary for the activation of MAP kinase cascade in *N. attenuata* during FAC triggering. Therefore, after FAC induction, *Na*BAK1 could facilitate JA synthesis in this plant independently of MAP kinase pathway [331]. Contrary to this, MAMP/DAMP-induced JA signalling through MAP kinase route generally relies on BAK1 input [100, 240, 332]. HAMP and MAMP/DAMP-induced immunity triggering thus differs at some crucial junctures quite distinctively. Earlier findings had indicated FACs in *M. sexta* OS are able to trigger SA-induced protein kinase (SIPK) and wound-induced protein kinase (WIPK) activities in *N. attenuata* leaves as responses to wound and herbivory elicitations [325]. Meanwhile, SIPK and WIPK are known to be homologues of *At*MPK6 and *At*MPK3 kinases, respectively, from *Arabidopsis*. Therefore, resemblance in MAP kinase cascade activation by HAMP like MAMP or DAMP is conceivable. Moreover, SIPK- and WIPK-induced JA biosynthesis was found to be crucial for *N. attenuata* defence responses against *M. sexta* inflicted wounding and herbivory, further underlining the significance of MAP kinase cascade in JA-mediated immune signalling [325, 326]. In *N. attenuata, M. sexta*-mediated herbivory was shown to activate another MAP kinase MPK4 that represses the JA biosynthesis pathway significantly. MPK4 was further revealed to dampen *N. attenuata* defence against *M. sexta*, but that inhibitory mechanism was independent of JA-signalling components [333].

2.6 Redundant role of DAMP, NAMP and HAMP in mediating PTI

Damage-associated molecular patterns are endogenous signalling ligands which can be generated by varied kinds of stimulants [229, 238, 239, 264]. Insect herbivores (including defoliators, chewing–piercing insects) during their feeding induce considerable wounds/injuries in plant organs. The wounds caused consequently can release plant cell wall fragments such as OGs, intracellular molecules such as ATPs (eATP) to extracellular regions such as apoplast [334, 335]. Wounding can further stimulate biosynthesis of elicitor PROPEPs leading to formation and release of Pep ligands to the apoplastic region [238, 335]. Herbivore insects are generally known to carry candidate HAMP molecules in their OS and also in other forms. Therefore, during insect feeding on hosts, HAMPs as well as DAMPs both can be liberated and these may act as elicitors of PTI responses. For instance, rice Pep3 (*Os*Pep3) elicitors are induced by the BPH (*N. lugens*) that in turn stimulate conspicuous resistant responses against a fungal (*M. oryzae*) and a bacterial phytopathogen (*Xanthamonas oryzae pv. oryzae*) in the host [336]. The BPH was earlier shown to introduce a salivary elicitor protein NlSP1 into rice tissues during the former's feeding [337]. NlSP1 was shown to trigger typical PTI responses in rice [337]. This suggests BPH may simultaneously liberate DAMP as well as HAMP patterns which can trigger appropriate immune responses in rice in somewhat overlapping or distinct manners [336, 337]. Generation of AtPep1 [241] for instance had been shown to be elicited by insect herbivory too. Another illustration that demonstrates significant redundancy is the inceptin peptides released as HAMP by legume herbivore insect (e.g., *A. gemmatalis*) [134, 334]. Inceptin peptides are derived from host chloroplastic ATP synthase γ-subunit (cATPC) components through proteolytic processing and are released in insect OS [134, 334]. In a sense, although inceptin is considered a HAMP, its qualification as DAMP (as endogenous host constituent) cannot be undermined as well. Therefore, inceptin may be redundantly triggering PTI responses both as HAMP and DAMP.

In the same manner, plant parasitic nematodes can cause release of damage-associated patterns from plant hosts during their advances through plat root tissues and beyond. For instance, the invasion process of cyst nematode *H. schachtii* through *Arabidopsis* root could facilitate emergence of OGs via induction of PGIPs [338]. Subsequently, synthesis of camalexin and activation of indole glucosinolate pathways could be correlated to PGIP induction and also OG mediated defence triggering that eventually conferred resistance to *H. schachtii* infection [338]. Earlier, an unknown proteinaceous NAMP in the 'NemaWater' extract derived from *H. schachtii* was shown to elicit typical PTI-like responses depending on *Arabidopsis* PRR RLK NILR1 [135]. Therefore, it is possible that during *H. schachtii* invasion of plant roots, both DAMP (i.e. OGs) and NAMP (unknown 'NemaWater' pattern) ligands might induce PTI responses in the host. Additionally, 'Ascr#18' (a significantly conserved NAMP identified in parasitic nematodes) homologues in *H. schachtii* may contribute to canonical PTI responses similar to what was reported in a recent study [136].

2.7 Representative MAMP-PRR and DAMP-PRR systems

2.7.1 FLS2-flg22 system

One of the prominent PRR-MAMP conjugate systems in the plant immunity study domain would be the FLS2-flg22 system. The trigger initiated at FLS2 (PRR) through flg22 epitope (MAMP) detection and subsequent transmission of immunity signalling cascades within plant cell is notably one of the best characterized schemes so far. Nonetheless, new updates and findings in this regard have continued to pour in.

FLAGELLIN SENSING2 (FLS2) is a PRR belonging to the LRR RLK category which is primarily conserved across dicot plant species [339]. FLS2 recognize flg22—the 22 amino acids long peptide fragment (from conserved N-terminus of bacterial flagellin protein) and can transmit flg22-triggered immune signalling intracellularly [83]. Figure 2.3 depicts some of the important events of FLS2-flg22-induced PTI signalling process in the plant cell. Notably, tomato plants have two flagellin sensing PRRs namely FLS2 and FLS3 and these detect two different epitopes from flagellin protein i.e. flg22 and flgII-28. Tomato FLS2 sense flg22, whereas FLS3 recognizes flgII-28 fragments and induce defence responses downhill [340]. In *Arabidopsis*, the contact between FLS2 and flg22 was shown to be a direct one involving hydrophobic as well as hydrogen bonding interactions and the region spanning LRR3 to the LRR16 in the ectodomain of FLS2 was revealed to involve in the flg22 binding process [114, 341, 342]. In the paucity of elicitor (i.e. flg22), FLS2 generally remain bound to BIK1 (a RLCK sub-family VII member) and allied PBL proteins [343, 344]. Once flg22 comes in contact with FLS2, BAK1 (co-receptor) and BKK1 (another name for SERK4) then associate with FLS2 [341, 344]. In fact, flg22 binding renders heterodimerization of FLS2 with BAK1 [114]. Triggered by flg22 binding, FLS2 deploys BAK1 to phosphorylate BIK1 and renders BIK1 release from FLS2 [341, 343, 344]. Meanwhile, BAK1 phosphorylates FLS2 too [345]. In the meantime, activated BIK1 in turn phosphorylates FLS2 as well as BAK1 [343]. Importantly, sites in FLS2 for phosphoryl group additions carried out by BAK1 and BIK1 are different [345]. Apart from BIK1, Brassinosteroid signalling kinase 1 (BSK1) [a RLCK (sub-family-XII) member] is also known to directly associate with FLS2 and trigger downstream immune signalling when elicited by flg22 [346]. Moreover, BAK1 is also inducted to phosphorylate BSK1 as soon as flg22 is perceived by FLS2 [346]. BSK1 in turn is able to initiate relay of downstream immune signalling via induction of another key signalling module known as MAP kinase cascade [347]. Recently, two new RLCK members from brassinosteroid-signalling kinase family, namely BSK7 and BSK8 were shown to be associated with FLS2 and involved in the immune signalling triggered by flg22-FLS2 complex [348]. In tomato, a new RLCK member Fir1 (FLS2/FLS3-interacting receptor-like cytoplasmic kinase 1) was recently shown to play a vital role in flagellin-triggered immunity through interaction with both FLS2 and FLS3 PRR receptors [349].

Meanwhile, as soon as flg22 is perceived by FLS2 and receptor heteromerization occurs, a rapid and strong Ca^{2+} influx takes place which radically enhances cytosolic Ca^{2+} ion concentration [256, 350]. This abrupt rise in Ca^{2+} entry into the cell has been associated with induction of varied plasma membrane localized Ca^{2+} penetrable channels such as Ionotropic glutamate receptor (iGluR)-like channels, cyclic nucleotide-gated channels (CNGCs), reduced hyperosmolality induced Ca^{2+} increase channels (OSCAs) etc [351, 352]. Contribution of intracellular Ca^{2+} release had also been cited [353]. Besides Ca^{2+}, other ion influxes do occur towards cytosolic side rendering the cell's external environment alkaline and subsequent membrane depolarizations [354]. Generation of ROS following Ca^{2+} influx into the cell has been a recognized event and this is said to occur within minutes of flg22-FLS2 triggering [260]. Calcium ion concentration is stated as one of the essential prerequisites for ROS production [355]. The enzyme complex responsible for ROS generation in plant cells is a plasma membrane-localized NADPH oxidase called RBOHD [356]. RBOHD-mediated ROS generation was found to be stimulated by direct contact of calcium to RBOHD and also via phosphorylations carried out by calcium-dependent protein kinase 5 (CPK5) [356, 357]. RBOHD is further induced by BIK1-mediated phosporylation events though in a manner independent of calcium signalling [355]. In fact, RBOHD phosphorylation sites used by BIK1 and CPK5 were found to be distinct for both [355]. Notably, RBOHD Serine residues at the 39th and 343rd positions were predicted targets for BIK1, whereas Serine residue at the 148th position was thought to be phosphorylated by CPK5 [355, 357]. This implies ROS generation by RBOHD during flg22 triggering can be both dependent as well as independent of calcium signalling [355].

Calcium signalling/responsive proteins like calcium-dependent protein kinases (CDPKs or CPKs), calcineurin B-family proteins (CBLs) and calmodulins (CaMs) have also been divulged to get induced by enhanced calcium ion influx into cell cytosol [357]. In response to flg22 elicitations, activated CDPKs, CBL-CIPKs (CBL–CBL-interacting protein kinases) and CaMs were found to influence immune signalling networks downstream [284, 358–360]. CaMs were already reported to regulate calcium channels (like CNGCs), Ca^{2+} pumps during MAMP triggering [358]. In addition, CML9 (a CaM relative, also termed as CaM like protein 9), was shown to be rapidly activated by flg22 stimulations in *Arabidopsis* and was observed to negatively regulate late PTI responses [284]. Likewise, flg22-induced *Arabidopsis* CBL-interacting protein kinases viz. CIPK6 and CIPK14, were found to regulate overall PTI responses negatively [359, 360]. Certain calcium-responsive proteins may impede PTI responses by modifying essential components of the PTI armoury. For example, BIK1 was demonstrated to be negatively regulated by CPK28 (calcium-dependent protein kinase 28) mediated phosphorylations in response to flg22 induction [361]. CPK5 (calcium-dependent protein kinase 5), a member of the CDPK family in plants, was earlier shown to be phosphorylated and activated in response to flg22 elicitation in *Arabidopsis* [357]. Subsequently, CPK5 was revealed to induce RBOHD-mediated ROS generation and also bolster PTI responses in *Arabidopsis* [357]. Flg22-FLS2 induced activation of Ca^{2+} signalling network relayed via calcium-responsive/binding proteins therefore plays a significant role in PTI outcomes. A number of defence-related genes (e.g. *CYP81F2*, *PEN2*, *CYP71A12*, *CYP71A13*, *PHI1*, *NHL10*, *PER62*, *PER4*, *PR1* etc) [284, 362], TFs like MYB51, MYB122, ANAC042, WRKY7, CAMTA3 [363] are regulated via calcium-responsive proteins upon flg22 elicitations. These sets of TFs in turn modulate the transcriptional fate of several genes downstream and contribute to flg22-elicited immune responses in the respective host remarkably.

Another vital signalling module that executes downstream of PRR-ligand immune complexes is the mitogen-activated protein kinase cascades [262]. Subsequent to flg22 detection via FLS2 receptor and immunity signal relay, rapid activation of MAP kinase cascades has been well documented [350]. MAP kinase networks may be triggered by varied factors which may generally act upstream to them. Some of these known factors can include OXI1 protein kinases, calcium signalling pathways, RLCKs, hormonal signalling routes etc [262, 364]. OXI1 (OXIDATIVE SIGNAL-INDUCIBLE1) protein kinases were shown to activate two MAP kinase cascade members MPK3 and MPK6, respectively [262, 364]. Notably, ahead of this, the serine/threonine kinase OXI1 can be induced by ROS generated via RBOHD in response to various MAMP elicitations including flg22 as well as other factors [364, 365]. BSK1, an RLCK member that associates with FLS2 and plays an important role during flg22 triggering, was subsequently revealed to phosphorylate MAPKKK5 (at N-terminal Ser-289 position) and regulate its flg22 evoked downstream immunity signalling activities remarkably [347]. Notably, MAPKKK5 also connects with other MAP kinase members viz. MKK1, MKK2, MKK4, MKK5, and MKK6, thereby influencing the latter's signalling activities downhill [347]. It is further relevant to note that in response to chitin elicitations too, two RLCK members namely PBL27 (in *Arabidopsis*) and *Os*RLCK185 (in rice) were described to activate MAP kinases MAPKKK5 and *Os*MAPKKKε, respectively, by phosphorylating them at their corresponding C-termini [264, 265], and hence, pointing towards RLCK's pivotal role in downstream signal relays via MAP kinase cascades.

Seminal surveys have documented activation of two primary MAP kinase cascade modules in *Arabidopsis*, upon stimulated by flg22 [366, 367]. One module constitutes MAPKKK3/MAPKKK5–MKK4/MKK5–MPK3/MPK6 and the second one being MEKK1–MKK1/MKK2–MPK4 [366, 367]. Sequential relay of phosphorylation events from upstream components (i.e. MAPKKK3/MAPKKK5 or MEKK1) to elements at the base (MPK3/MPK6 or MPK4), consequently leads to emergence of diverse immunity-associated gene products in the downstream [366, 367].

The distinct correlations ascertained so far between PTI resistance response and MAPKKK3/MAPKKK5–MKK4/MKK5–MPK3/MPK6 signalling cascade indicates the MPK3/MPK6 module to have a boosting effect on PTI responses triggered by flg22 as well other elicitors [368]. MAPKKK3/MAPKKK5–MKK4/MKK5–MPK3/MPK6 cascade, for instance is strongly linked with regulated production of important defence-related substrates viz. phytoalexins, ET, indole glucosinolates etc [369]. A recent report has suggested that the proximal MEK kinase MAPKKK5 may be dispensable for activating MPK3/MPK6 in response to flg22 and elf18 but might be required for other MAMP signalling [264]. In a study that followed, in response to flg22 and other MAMPs or DAMP triggers, the identical role of both MAPKKK3 and MAPKKK5 ahead of MKK4/MKK5–MPK3/MPK6 cascade was revealed that contributed significantly to defence against phytopathogens [368]. This implies, for stimulation of MPK3/MPK6 cluster, plants may deploy MAPKKK3 or MAPKKK5 depending upon the nature of elicitors triggering immune responses or some other factors that are yet to be disclosed. Mutations in both MAPKKK3 and MAPKKK5 (i.e. in *mapkkk3 mapkkk5* background) but not the single genes, could be correlated with decreased phosphorylation status of MPK3/MPK6 after flg22 elicitations that eventually hindered MAP kinase signalling and apparently reduced *FRK1* and *WRKY29* gene expression downhill [368]. Lack of complete absence of MPK3/MPK6 phosphorylations, however, suggests there are other MAPKKK member(s) as well in the cascade [370]. Initially, during early PTI responses triggered by flg22, *FRK1* (*FLG22-INDUCED RECEPTOR-LIKE KINASE 1*) and *WRKY29* (a TF containing a conserved WRKY DNA-binding domain) genes were shown to be actively expressed via MPK3/MPK6 route [366]. Interestingly, CIPK6 [Calcineurin B-like protein (CBL)-interacting protein kinase 6] on the other hand was, however, found to down-regulate expression of these genes (i.e. *FRK1* and *WRKY29*) in response to flg22 induction [359], suggesting negative control of MAP kinase cascade by CIPK6 during PTI. This was further confirmed by detection of faster and increased phosphorylation status of MPK3/MPK6 cluster under *cipk6* mutant plant background than in a wild type of plant [359]. Under such opposing circumstances, how the MAP kinase pathway still operates a signalling relay to its downstream and facilitates defence responses is really intriguing. Importantly, ROS generation was stated to be unaltered in *mapkkk3 mapkkk5* plants when treated with flg22 [359] indicating no direct effect of MAPKKK3 or MAPKKK5 on RBOHD-mediated ROS generation during early events of PTI. The MKK4 and MKK5 kinases situated in the next tier downstream to MAPKKK3/MAPKKK5 were similarly found to act redundantly during flg22-triggered PTI responses [371]. Further, in *Arabidopsis mkk4/mkk5* double mutant background, flg22 treatment failed to stimulate MPK3/MPK6 downstream activities although flg22 mediated activation of MPK3/MPK6 was sustained under *mkk4* or *mkk5* single mutant backgrounds [371]. MPK3/MPK6 along with its downstream MKK4/MKK5 kinase pair were linked to effective apoplastic and stomatal immunity. Flg22 or pathogen stimulated MPK3/MPK6 was also responsible for stomatal closure which was found to be independent of abscisic acid (ABA) signalling [371]. It is noteworthy that ABA mediated stomatal aperture closure can be resultant of biotic as well as abiotic stresses in plants [372]. Attempts to elucidate plant immunity responses towards MAMPs or pathogens under *mpk3/mpk6* double mutant background had met with some fundamental challenges owing to the fact that *mpk3/mpk6* double mutant embryos could not survive, signifying yet another crucial role played by MPK3/MPK6 kinases in the plant embryogenesis processes [267, 371]. Following a modified chemical genetic approach incorporating an inducible system, elicitor-triggered immune response studies in *mpk3/mpk6* double mutant *Arabidopsis* lines were made possible. These studies further revealed redundant roles of MPK3 and MPK6 kinases in plant defence responses triggered by elicitor/pathogen [371]. In a recent report, convergent role of MPK3 and MPK6 kinases in conferring resistance to *P. syringae* was divulged during flg22 evoked PTI responses [373]. In fact, the same report demonstrated mutual dependence of *MYB44* (TF) and MPK3/MPK6 kinases for their expression and immune signalling downstream, highlighting a new TF mediated regulatory mechanism for MAP kinase activity during elicitor-induced plant defence [373]. Meanwhile, MAP kinases were already reported to trigger downstream TFs for subsequent regulation of immunity-related gene expressions [271, 374]. Apart from relevant TFs, both MPK3/MPK6 and MPK4 cascades can further phosphorylate certain other transcriptional regulators in their downstream when elicited by flg22 [271, 273]. VQ-motif-containing protein members namely MKS1 and MVQ1, for example are a type of transcriptional regulators which happen to be downstream substrates of both these MAP kinase cascades. Following flg22 elicitations, MPK3/MPK6 kinases phosphorylate MVQ1, whereas MKS1 is specifically phosphorylated by MPK4 thereby eliminating WRKY33 transcription repression by both these regulators [269, 270, 273]. It was subsequently concluded that both MKS1 and MVQ1 acts as negative regulators of MAMP-induced defence responses [270, 273].

Moreover, post flg22 elicitations, distinctive functional attributes of MPK3 and MPK6 have also been evident [376]. For instance, ROS build-up for extended duration, considerable increase in growth retardation etc are more prominent under *mpk3* background than that of *mpk6* after flg22 treatment [350]. On the other hand, MPK6 in particular, plays a vital role in activating ERF104 (an ET responsive TF) in response to flg22 elicitation and subsequent ET signalling where MPK3 is not involved [376]. MPK6 was divulged to activate BRASSINOSTEROID INSENSITIVE1-ETHYL METHANESULFONATE-SUPPRESSOR1 (BES1) TF through direct phosphorylation, when flg22-induced elicitation [377]. The BES1 function subsequently afforded resistance to *Pto* DC3000 infection in the host [377].

Further, MPK3 and MPK6 kinases influence plant basal immunity and elicitor-induced defences variably [378]. MPK3 disabled plants for example, displayed weak basal resistance to *B. cinerea*, but were able to hoist immune responses induced by elicitors such as flg22 (and OGs). In contrast, *mpk6* mutant plants were compromised in flg22- and/or OG-induced defence responses against *B. cinerea*, although basal immunity was unaffected. This also suggests, in comparison to MPK3, MPK6-mediated immune signalling (triggered by flg22 or OG elicitors) has a leading role in deciding plant resistance to *B. cinerea* [378].

The next vital MAP kinase cascade induced by flg22 is MEKK1–MKK1/MKK2–MPK4 [366, 367]. Early experiments had concluded that MPK4 cascade functions as a negative regulator of PTI signalling [379, 380]. The negative regulatory role of MPK4 cascade in flg22-elicited immune signalling was perceptible from scores of mutational studies conducted on constituent MAP kinase genes in the cascade [268, 367, 379, 380]. Flg22-treated *mpk4* mutant plants manifested dwarf stature, enhanced ROS level build-up, unprompted cell death, callose deposition, continuously expressed PR (pathogenesis-related) genes etc, all contributing towards autoimmune responses and resistance [379, 380]. Similar autoimmune responses were apparent in the case of *mekk1* (single mutant) and *mkk1 mkk2* (double mutant) plant lines in response to flg22 elicitations [380]. However, *mkk1* or *mkk2* single mutant plants did not display autoimmune phenotypes demonstrating the redundant role of MKK1 and MKK2 upstream of MPK4 [380]. As opposed to a canonical MAP kinase pathway, a study highlighted that MEKK1 kinase function was not necessary for flg22-triggered MAPK4 activation process [379]. However, the structural role of MEKK1 (without kinase activity) could be very significant in provisioning efficient interaction between MPK4 and MKK1/MKK2 [379] since complete lack of MEKK1 is identical to disabled/inactive MPK4 phenotype. Further, kinase function of MEKK1 in the cascade may be compensated by other MEKK member(s) present in the upstream [379]. It would be interesting to see which upstream MEKK member(s) perform this task and how and what factors induces them. Further inquest for causes of elicitor-evoked autoimmunity phenotype in MPK4 pathway mutant plants has unmasked the role of a nucleotide-binding leucine-rich repeat (NB-LRR) protein called SUMM2 [381] and another MAP kinase, MEKK2 (designated as SUMM1) [382]. SUMM2 is a member of the coiled-coil class of NB-LRR, which was ascertained to be activated by impairment of MEKK1–MKK1/MKK2–MPK4 cascade [381]. Activation of SUMM2 leads to sundry immune responses downstream akin to autoimmunity phenotype [381]. SUMM1 (or MEKK2) was shown to interact with MPK4 [382]. A later study explicated the role of MEKK2 in impeding MPK4 phosphorylation by upstream MAP kinases [383]. SUMM1 as a plausible interlink (with perhaps additional members) between MPK4 activation status and SUMM2 has been proposed [382]. Besides, SUMM1 was already correlated with induction of autoimmunity in *Arabidopsis* that critically relied upon SUMM2 activation and function [382], although direct interaction amidst SUMM1 and SUMM2 is yet to be demonstrated. Later, a third factor called SUMM3 (also referred to as CALMODULIN-BINDING RECEPTOR-LIKE CYTOPLASMIC KINASE 3; CRCK3) was uncovered which was subsequently validated to interact with MPK4 [384]. In wild-type plants, MPK4 phosphorylates SUMM3 (or CRCK3). Importantly, SUMM3 phosphorylation status was found to depend on intact MEKK1–MKK1/MKK2–MPK4 cascade. SUMM3 also interacts with SUMM2 and it is the phosphorylation status of SUMM3 that decide SUMM2 activation and subsequent autoimmune responses downstream [384]. A recent report stated interdependence and collective involvements of SUMM3 and MEKK2 in regulating SUMM2-triggered immune signalling downstream [385]. Now, it appears that MEKK1–MKK1/MKK2–MPK4 cascade actually facilitates basal defence to the plants when triggered by MAMPs (such as flg22) [381]. SUMM2, on the other hand, has evolved as a crucial surveillance system that keeps a close watch on functional status of MPK4 cascade signalling evoked by MAMPs [381–385]. The positive regulatory role of MPK4 cascade in basal defence is again apparent from the fact that despite the dearth of any elicitor/pathogen trigger, functionally disabled *mekk1*, *mkk1/mkk2*, and *mpk4* mutants can stimulate the launch of persistent immune reaction along with unprompted cellular death and build-up of ROS, SA hormones etc [268, 380] that in turn can suppress normal growth, stature of the plant [376]. Notably, MEKK1–MKK1/MKK2–MPK4 cascade also seems to have conspicuous defence signalling against biotrophic and necrotrophic pathogens. Earlier, elevated resistance phenotype of *mkk1 mkk2* and *mpk4* plants towards biotrophic phyto-pathogens like *Hyaloperonospora arabidopsidis* and *P. syringae* was delineated. In contrast, vulnerability to *B. cinerea* (in *mkk1 mkk2* mutants) and compromised immunity towards *Alternaria brassicicola* (in *mpk4* plants) indicated weak defence responses of *mpk4* cascade mutants to both these necrotrophic fungal pathogens [268, 332, 380].

The ultimate outcomes of flg22-induced PTI responses emerge through transcriptional modulation of varied TFs which may be regulated via calcium signalling pathway and/or MAP kinase routes, as cited above. A number of defence-related genes are expressed after flg22 induction. Biosynthesis of different secondary metabolites that act as defensive shields against numerous pathogens/pests are evoked by flg22 trigger. These metabolites include callose [302], camalexin [386], lignin [387], indole glucosinolate [302] compounds etc. Flagellin epitope (flg22) stimulated PTI responses in specific plant hosts could confer resistance against varied kinds of pathogens e.g., *B. cinerea* [378], *Pto* DC3000 [388], *P. infestans* [389], *Xanthomonas citri* ssp. *citri* (*Xcc*) [390], etc.

2.7.2 Pep-PEPR system

The Pep-PEPR system is one of the extensively investigated DAMP-PRR couples with respect to plant immunity signalling. Peps stands for plant elicitor peptides and these are also called phytocytokines or inducible DAMPs [224]. PEPRs refer to Pep receptors which are generally LRR RLK type PRRs localized on plant cell surfaces [391]. Peps are usually derived from PRECURSOR OF PEPTIDE (PROPEP) proteins. PROPEPs are synthesized in diverse plant species and their biosynthesis can be evoked by numerous stimulators such as MAMPs, phytopathogens, herbivores, external injuries, effectors, defence hormones, phytocytokines etc [229, 238, 249, 264]. The first ever characterized plant elicitor peptide was *At*Pep1 from *Arabidopsis thaliana* [239] and subsequent studies have outpoured several details in this regard. *At*Pep1 is a short peptide (about 23 amino acid residues) that originates from AtPROPEP1 precursor protein having about 92 amino acid residues. Importantly, AtPep1 could stimulate synthesis of its own precursor protein, i.e. AtPROPEP1, as is the case for other inducible DAMPs like PIP1, SCOOP12, and RGF7, respectively [224]. *At*Pep1 was revealed to be generated from C-terminus end of AtPROPEP1. Evidently, a mechanism for AtPep1 generation has been advanced in a recent study [392]. PROPEP1 synthesis can be induced by varied elicitors including MAMPs after recognition through specific surface based PRRs. The signalling components leading to PROPEP1 generation can comprise MAPKs, calcium-responsive proteins as well as certain TFs (like WRKY) [392]. As per findings, after generation, PROPEP1 may remain associated with tonoplast (vacuolar membrane) via the N-terminal end of the precursor [229]. Meanwhile, MAMP/elicitor induction concomitantly enhances Ca^{2+} influx towards the cytoplasmic side [353]. Ensuing cytoplasmic Ca^{2+} build-up then activate type-II metacaspases (MCs) like MC4 in the cell cytosol. Notably, MCs (including MC4) are Ca^{2+}-dependent proteases functioning in the plant cell cytoplasm. Type II MCs are again said to be activated by a self-tailoring process within. Triggered MCs next bring about slicing of PROPEP1, thereby releasing trimmed AtPep1 to cytosol. Subsequently, AtPep1 is translocated to the apoplastic region through a yet undefined mechanism [392]. Intriguingly, unlike other secreted peptides, Peps don't possess any N-terminal signal peptide for secretion [229, 239] and even then, they get discharged to apoplastic side. A plausible route for AtPep1 exit has, however, been predicted to be via collapsed membrane structures [127]. In the apoplast, AtPep1 can then bind to PEPR receptors to initiate a host of pattern-triggered defence responses signalling downstream of the latter [392]. In the *Arabidopsis* genome itself sequences encoding eight (8) PROPEPs have been identified. All the PROPEPs (1–8) have been reported to be conspicuous with respect to their protein sequences. Out of the eight, some of the PROPEPs may also occur in the cytosol unlike PROPEP1 and PROPEP6 that remain linked to tonoplast. It may be relevant to note that, like all other PROPEPs, PROPEP6 may not undergo trimming through type II MCs [392] and there may be other routes for its cleavage. Besides, beyond the plant immunity regime, other roles played by PROPEP derived AtPeps have also been divulged [229]. Further, in view of defence elicitation potential, Pep1-3 members have been investigated more frequently than others [128].

In *Arabidopsis,* there are two PEPR receptors for AtPep recognition, namely PEPR1 and PEPR2. PEPR1 and PEPR2 resemble a typical LRR RLK PRR (e.g., FLS2) member in architecture [391]. PEPRs pertain to LRR XI sub-family of the described LRR-RLK repertoires [119]. The number of LRR repeats in the ectodomain of PEPR1 was ascertained to be 27 [393]. Again, PEPR1 and PEPR2 shares sufficient homology at their amino acid composition level (64% identity and 76% similarity) [393]. In terms of ligand preferences, PEPR1 and PEPR2 have distinct variations. PEPR1 can be triggered by AtPep1-8, although PEPR2 showed selective affinity for AtPep1-2 induced elicitations [119, 229]. Pertinently, the mechanism of immune triggering occurring via AtPep1-PEPR1 ligand–receptor pair seems to resemble that of flg22-FLS2 and elf18-EFR couples [391]. AtPep1 contacts the PEPR1 lRR ectodomain via former's C-termini amino acid residues (more precisely Asn^{23}) [391]. AtPep1 attachment to PEPR1 leads to PEPR1 and BAK1 heterodimer formation and subsequently PEPR1 and BAK1 duo gets phosphorylated [394]. RLCK members—BIK1 and its relative PBL1 are thought to remain associated with both PEPRs i.e. PEPR1 and PEPR2 as integral components [162]. Moreover, the moment AtPep1 binds with PEPR1, BIK1 and PBL1 can be directly phosphorylated by PEPR1 (without requiring BAK1) [162] and consequently BIK1 would be released to induce downstream immune signalling [128]. AtPEPR1 (and also AtPEPR2) may then augment formation of cGMP (cyclic GMP) molecules by virtue of their intrinsic cytosolic guanylyl cyclase (GC) domain. Elevated cGMPs next stimulate plasma membrane localized CNGC2 (CYCLIC NUCLEOTIDE GATED CATION CHANNEL 2) calcium channels thereby inducing Ca^{2+} influx towards cytosol and subsequently incite Ca^{2+} signalling pathways [353, 395]. Generation of ROS and NO (nitric oxide) in the downstream of Pep1-PEPR1 immune triggering has been recorded [396, 397]. Meanwhile, production of ROS has been further correlated with calcium signalling routes involving crucial roles of CDPKs (Ca^{2+} -DEPENDENT PROTEIN KINASES) in it [362, 397]. ROS generation after AtPep1 induction was also found to associate with functional states of three RLCKs viz. BIK1, PBL1 and PCRK1 [162]. Next, activation of MAP kinase cascades consisting MPK3/6 as well as MPK4/11 kinases in *Arabidopsis* indicated semblance of AtPep1 elicitations with that of other established MAMPs [398]. AtPep1 was able to induce expression of MAP kinase as well as CDPK dependent genes (e.g., *NHL10*, *CYP82C2*, *PER4*, *PHI-1* etc) as part of the PTI response [362]. Expression of defence genes such as *PDF1.2* and *PR-1* that relied on the functional state of JA and SA hormonal pathways, respectively, could be induced by AtPep1 as well [237–239]. Besides, AtPep1 was also found to activate expression of other immunity-related genes such as *WRKY22, WRKY33, FRK1*, *BIR1* and *ANAC019* in *Arabidopsis* roots [399]. Further, upon treatment with AtPep1, *Arabidopsis* shoot and root regions manifested enhanced callose [162, 400, 401] and lignin deposition [401]. Like ROS generation, callose accumulation mediated by AtPep1 signalling was also revealed to depend on function of the RLCK members BIK1, PBL1 [162] and PCRK1 [400]. Dependence of AtPep1 on BIK1 and PBL1 for inducing lignin accumulation was also apparent [401]. Some of the key events of PEPR-Pep signalling module are being illustrated in figure 2.5.

It is relevant to note that apart from *Brassicaceae* (includes AtPeps), Pep-mediated signalling has been prevalent in other plant families (e.g., *Fabaceae*, *Poaceae*, *Rosaceae*, *Solanaceae*) as well [402]. For instance, ZmPep1 is the maize orthologue of AtPep1, which was determined to induce expression of varied PTI-related gene products including endochitinase A, PR-4, PRms, SerPIN and also a class of phytoalexin called benzoxazinoid [240]. Likewise, Pep orthologues from soybean, rice, broccoli and tomato plants have been divulged to be instrumental in conferring PTI mediated defences against varied pathogen/pests [402].

2.8 MAMP- and DAMP-triggered PTI: additional resemblances

The considerable resemblances existing amidst the initial and final events of immune triggering due to MAMP as well as DAMP patterns have been elaborated in the 'PTI generalized scheme' section of the ongoing chapter. Here are a few additional instances to the premise.

A similar mode of immune response activation by MAMP and DAMP, for instance, has been further accentuated by some recent findings corresponding to stability of a crucial plant immune kinase protein at the molecular level. BIK1 (BOTRYTIS-INDUCED KINASE 1) as a well-known plant RLCK member, participates in multitude of defence signalling steps that occur in the downstream of PRRs [58]. Evidently, BIK1 protein stability is an essential aspect that appreciably dictates downstream PTI signalling and responses in plants. CPK28 (CALCIUM-DEPENDENT PROTEIN KINASE28) was identified as a negative regulator of PTI as it promotes BIK1 decay via recruitment of two E3 ubiquitin ligase family proteins viz. PUB25 and PUB26. CPK28 phosphorylates PUB25 and PUB26, thereby inducing both the E3 ligases to polyubiquinate BIK1. Polyubiqutinated BIK1 is then destined to abolition through scheme of proteasome complexes [361, 471]. Interestingly, *At*Pep1 (DAMP) induced derepression of CPK28 mediated immunity suppression and thus activation of downstream PTI responses have been elucidated in a recent study [398]. The underlying mechanism behind ablation of CPK28 activity appears to involve an alternative splicing (AS) event that generates an intron-retained *CPK28* (*CPK28*-IR) mRNA intermediate. *CPK28*-IR mRNA forms a truncated protein (CPK28-IR) which is unable to activate E3 ligases. Generation of *CPK28*-IR variant was found to occur after *At*Pep1 elicitations that further correlated with dephosphorylated state of a novel RNA-binding protein IRR (IMMUNOREGULATORY RNA-BINDING

PROTEIN). Importantly phosphorylated state of IRR was necessary for generation of functional CPK28 proteins that was, however, reversed after *At*Pep1 elicitations [398]. Notably, appearance of AS rendered *CPK28*-isoform was also realized after flg22 (MAMP)-induced triggering [403] although, involvement of a similar IRR protein in this occasion has not been confirmed yet. Meanwhile, a couple of flg22-induced ubiquitin ligases, namely ATL31 and ATL6 (ATL = ARABIDOPSIS TÓXICOS EN LEVADURA), were recently revealed to ubiquitinate CPK28 kinase and thereby direct the latter toward proteasome-mediated decay [404]. Through mediation of CPK28 degradation, ATL31/6 ligases ensure BIK1 stability and hence, regulate BIK1-arbitrated defence signalling positively [404]. Presently, it is palpable that both flg22 and *At*Pep1 induced triggers can positively regulate BIK1 protein stability through impairment of CPK28 function. However, final events leading to regulation of CPK28 (after its inactivation) may differ in respect of the signalling triggered by either of these ligands. Moreover, incidence of CPK28-RI (truncated) protein interaction with ATL31 was recently communicated [405], although ATL31 mediated ubiquitination of CPK28-RI appeared to be almost negligible. It is possible that unlike ATL31/6-mediated decay of CPK28, some other ubiquitinases might involve in deciding CPK28-RI's fate.

Likewise, a recent report stating inhibition of inorganic phosphate (Pi) translocation across *Arabidopsis* root hairs being after triggered by flg22, elf18 and AtPep1 ligands further underscores distinct resemblances in the PTI signalling mechanisms rendered through these MAMP and DAMP elicitors [406]. Notably, immune signalling driven by these three ligands activates RLCK members, namely BIK1 and PBL1, which in turn represses *Arabidopsis* root phosphate transporter (PHT1;4) via phosphorylation and thus hinders root Pi uptake [406]. This suggests, MAMP- and DAMP-triggered immunity signalling can cause overlapping effects on plant physiological responses as well.

A number of studies have revealed that MAMPs and DAMPs can cooperatively contribute to enhanced PTI signalling responses in plants. For example, both flg22 and Pep1 stimulates synthesis of PEPRs and its ligands [230, 407]. In addition, the PEPR-Pep and PRR-MAMP pairs recruit similar machinery for PTI signalling indicating both the signalling routes to perform in synergy. In fact, DAMP-mediated elicitation has been said to intensify MAMP-triggered defence significantly. The two MAMP members namely flg22 and elf18 can also stimulate expression of another inducible DAMP called PIPs [230]. Further, PIPs (i.e., PIP1 and PIP2) like Pep1 were found to magnify flg22-triggered immune responses as well [230]. Host resistance towards *Pst*DC3000 was elevated upon flg22 and PIP1 application concurrently than when treated separately suggesting concerted effect of both [230].

2.9 Variations amidst MAMP- and DAMP-induced PTI signalling mechanisms

Besides numerous similarities, a range of conspicuous differences can be observed between MAMP- and DAMP-triggered immune signalling processes as well. This may be with respect to varied aspects relating to the immune signalling route impelled by both these pattern kinds, such as the PRR and associated signalling machinery composition, unique peptide domain constituent, PRR–co-receptor–RLCK interaction type, variable defence signalling in root/shoot regions of plants etc. A few among these differences are being discussed below.

2.9.1 On the basis of kinase type carried

MAMP- and DAMP-recognizing PRRs may have distinct kinase types and functions that can split them as well. For instance, AtPep (a DAMP) sensing receptors PEPR1 and PEPR2 are LRR RLK PRRs and are designated as RD-kinase types [408] corresponding to the presence of RD-motif in their catalytic kinase domain [111]. Another DAMP-sensing PRR called WAK (wall-associated kinase) that detects OGs [409] can, however, be categorized into RD-kinase group (WAK-RD) as well as non-RD kinase (WAK-non-RD) group. WAK-non-RD kinases were again shown to be limited to monocotyledonous plant species alone [411]. Contrary to this, the bulk of the known RLK PRRs (including FLS2, EFR) discerning MAMP ligands is considered as non-RD kinase types [411]. Earlier, the RD (Arginine-Aspartate)-motif in the kinase domain of RLKs had been correlated with autophosphorylation function [110, 111]. Accordingly, PEPRs, WAK-RDs are suggested to possess autophosphorylation capacity which would be wanting in WAK-nonRDs and MAMP binding PRRs like FLS2, EFR etc.

2.9.2 Presence/absence of guanylyl cyclase domain and Ca^{2+} signalling

DAMP receptors PEPR1/2 bear an inherent cyclic guanylyl cyclase (GC) domain which is wanting in MAMP recognizing PRRs such as FLS2/EFR etc [353, 395]. After AtPep1 perception, PEPR1/2 can stimulate CNGC2 channels to permit Ca^{2+} entry towards cytosolic side from extracellular spaces. PEPR1/2 GC domain mediated accumulation of cGMP molecules in turn activates CNGC2 channels [353, 395]. In contrast, flg22-triggered Ca^{2+} influx is reported not only to rely on CNGC2/CNGC4 but can occur via other Ca^{2+} channels (e.g. ionotropic glutamate receptor-like channels etc) as well [351, 352].

2.9.3 On the basis of co-receptor and RLCK interaction

MAMP- and DAMP-mediated immune triggering may further differ corresponding to the necessity of RLCK and co-receptor interaction. Subsequent to binding of respective MAMPs FLS2 and EFR receptors (for example) require BAK1 to phosphorylate BIK1. In contrast, when triggered by AtPep1, PEPR1 can directly phosphorylate BIK1 without needing BAK1 activity [162]. Nevertheless, AtPep1 induces union of PEPR1 and BAK1 in the same manner as observed in the case of flg22/elf18-mediated FLS2/EFR and BAK1 heterodimerization [394]. Disposability of BAK1 for AtPep1-triggered immune signalling has been further accentuated by a recent report that revealed BAK1 disruption could in fact lead to induction of a Pep-PEPR immune signalling pathway for restoration of host basal defence [264]. The study implicated BAK1 function as crucial for MAMP-elicited PTI as against DAMP such as AtPep1 [264].

2.9.4 Differential immune signalling in *Arabidopsis* root

Variations in the immune signalling induced by DAMP and MAMP stimulations have been clearly discernible through studies conducted on *Arabidopsis* roots after specific elicitor treatments [412, 413]. When equal concentrations of Pep1 and flg22 ligands were used as elicitors, *Arabidopsis* roots manifested enhanced root growth inhibition due to Pep1 rather than that of flg22 triggering [412]. An independent study further revealed Pep1-induced root growth inhibition to be even higher than what seemed to be caused by nlp20, apart from flg22 [413]. Such a strong inhibitory effect on *Arabidopsis* root growth due to Pep1 might involve distinct signalling routes varying from flg22 or nlp20 MAMPs. Pep1 treatment further causes quantitative enhancement in the expression of its cognate receptor (i.e. *PEPR1*) gene in *Arabidopsis* root as compared to flg22 and nlp20 MAMPs [413]. An earlier study had indicated that PEPR2 (one of the two PEPRs) occurrence in the roots, could play a decisive role in the Pep1 perception and subsequent defence signalling events [412]. Further, Pep1 triggered defence responses in roots accompanied elevated expression levels of varied immunity-related genes relative to flg22 or chi7 (MAMPs) treatments [412]. Pep1 was further revealed to induce SA, JA and ET hormonal pathways in *Arabidopsis* root and strongly displayed expression of hormone responsive defence gene repertoires like *HEL* (a tobacco *PR-4* gene homologue responsive to JA/ET), *PDF1.2* (responsive to JA), etc [412]. In fact, induction of *PDF1.2* (defensin gene) expression was sufficiently enhanced by Pep1 elicitation as compared to flg22 or elf18. In addition, Pep1 could uniquely activate expression of genes involved in SA and JA biosynthesis pathways e.g., *AOS* (in JA synthesis), *ICS1* (in SA synthesis). However, unlike pep1, flg22 or chi7 treatments failed to stimulate these marker genes involved in the JA/SA hormone synthesis routes [412]. In a separate study, Pep1 but not elf18 was shown to induce expression of an oxidative stress related gene *PER5* (*PEROXIDASE SUPERFAMILY PROTEIN 5*) in *Arabidopsis* root [414]. Notably, *PER5* encodes a protein affiliated to peroxidase superfamily and is also inducible by flg22 trigger [362, 414].

2.9.5 Activation of DAMP expression by MAMP treatment and not vice versa

Expression of certain DAMP elicitor peptide precursors can be induced by MAMP application. However, the opposite case is not feasible. For instance, flg22 elicitor can stimulate expression of PROPEP1, PROPEP2 and PROPEP3 precursor peptides for Pep1, Pep2 and Pep3, respectively, in *Arabidopsis* [407]. Further, inducible DAMPs such as Pep1, PIP1, SCOOP12 etc can stimulate their own expression as well [224]. On the other hand, DAMPs are incapable of inducing MAMP synthesis. Also, MAMPs themselves cannot stimulate their own synthesis.

2.9.6 Quantitative variation in immune response markers

Studies indicate, treatments with DAMP ligands such as Pep1, PIP1 and PIP2 generated remarkably reduced ROS and leaf callose build-up when compared to flg22 (MAMP) induction. Likewise, the flg22-induced plant defence responses against *Pst* DC3000 was found to be stronger as opposed to PIP-mediated elicitation [230].

2.9.7 Unique RLCK involvement

MAMP and DAMP signalling may involve a common as well as unique set of signalling components downstream to PRR-pattern complexes. For instance, FLS2-flg22 conjugate may employ unique RLCK members BSK7, BSK8 downstream to them, not known to be utilized by PEPR-Pep (a PRR-DAMP complex) systems in *Arabidopsis* [348]. PEPR1-Pep1 pair, in particular, relies on BIK1 and PBL1 (both RLCKs) functions for downstream signal relays. In contrast, MAMP-triggered immune signalling usually relies on several RLCK members from the BIK1 family [162]. Interestingly, another DAMP member PIP1, unlike Pep1, was revealed to relay downstream immune signalling avoiding BIK1 involvement [230].

2.10 Relieving PTI signalling

Continuation of uninterrupted PTI signalling through PRR machinery after elicitor triggering can have a deleterious effect on the plant host. This can be manifested in terms of tissue damage, compromised growth and anomalous developmental pathways in the plant [244]. Further, disengagement of PTI machinery becomes essential to kick-start a new set of immune signalling events that is about to follow after another immunogenic ligand/pattern is detected [24]. In this context, various strategies are deployed by host to alleviate PTI signalling continuum. Some of the approaches applied by plant hosts for PTI mitigation have been illustrated in the 'Regulation of PTI' section of the ongoing chapter. Besides, degradation/disarming of immunogenic patterns/ligands, elimination of PRR receptors etc are prevalent contrivances employed by hosts to attenuate PTI signalling [105]. For instance, desensitization of well characterized DAMP candidates namely OGs and cellodextrins (CDs) were shown to be catalyzed by enzymes of a specific plant oxidase family. These oxidases are members of a plant berberine bridge enzyme-like (BBE-like) protein family [415]. The BBE-like protein family enzymes viz. oligogalacturonide oxidase (OGOX) 1 and cellodextrin oxidase (CELLOX) could oxidize OGs and CDs, respectively, thereby converting these to inactive elicitor forms [416]. Again, unlike unmodified structures, the oxidized states of OGs and CDs cannot serve as carbon sources for microbes thereby restricting the latter's growth in apoplastic region [415]. It is anticipated that mechanisms for degradation/removal of other MAMP/DAMP members might be elucidated in the near future as well.

Effecting PRR receptor depletion is a crucial strategy deployed by plants to restrain PTI signalling. Accordingly, mechanisms for depletion of FLS2 receptor kinase (PRR for flg22 ligand) have been explored along with others and crucial details have emerged subsequently [355, 417]. Upon flg22 binding, stimulation of FLS2 PRRs occurs that subsequently leads to initial events of PRR triggering. Stimulation of FLS2 (with flg22 attachment) is said to simultaneously trigger endocytosis of this receptor towards the plant cell interior [418]. Meanwhile, flg22 attachment to FLS2 brings FLS2 and BAK1 in close proximity. BAK1 next phosphorylates PUB12 and PUB13 proteins, belonging to plant U-box E3 ubiquitin ligase family [419]. Phosphorylated PUB12/13 ligases next associate with the FLS2 complex. Once formation of FLS2—PUB12/13 composite occurs, PUB12/13 then ubiquitinates FLS2 at multiple sites. Polyubiquitinated FLS2 is thereby destined to degradation via 26S proteasome complex [419]. Interestingly, neither BAK1 nor BIK1 is targeted by PUB12/13 ligases [419]. It was recorded that flg22 elicitations could promote FLS2 endocytosis/depletion even within 30 min of ligand treatment that again indicated FLS2-flg22 interaction to be a crucial event for removal of the PRR receptor from the membrane [418, 419]. The endocytosis pathway for flg22 bound FLS2 receptors was unveiled to require a clathrin-coated vesicle route through assistance of self-polymerized clathrin chains [420, 421]. In the same manner, PRR-ligand congujate pairs for EFR, LYK5, PEPR1, Cf-4 etc have also been affirmed to follow a clathrin-mediated membrane internalization path [421–424]. The FLS2 internalized clathrin vesicles eventually proceed towards the late endocytic stage forming multi-vesicular bodies (MVBs) [420, 425]. MVBs are reported to undergo internal rearrangements within, consequently giving rise to intraluminal vesicles through assistance of ESCRT (Endosomal Sorting Complex Required for Transport) complexes [426]. FLS2 receptors are then segregated to these intraluminal vesicles (within MVBs) and are next propelled to central vacuole for their destined elimination [420, 424, 425]. In addition to FLS2, other PRR RLKs which are targeted to MVB/LE (late endosome) pathway for degradation depending on respective ligand perception including EFR, LYK5, PEPR1 etc [427].

Apart from elimination of activated FLS2 receptors, the mechanism for endocytic recycling of inactive FLS2 receptors has also been forwarded. Interestingly, this process does not involve clathrin coating for endocytosis to occur [424]. Accordingly, FLS2 repertoire that remains unbound by flg22 is subject to definite internalization via endosomal pathway [420]. The internalized FLS2 proteins may have two fates—the majority of these could be recycled back to the plasma membrane and a portion of the whole can be depleted via selective autophagy mechanism [417]. In the selective autophagy process of FLS2, involvement of two orosomucoid (ORM) receptor proteins was recently confirmed in *Arabidopsis* [417]. Findings indicate, both the endocytosis pathways, i.e. clathrin-mediated and constitutive, could be involved in assuring perpetuation of threshold FLS2 receptor abundance on the plasma membrane, thereby also regulating PTI signalling. Henceforth, endocytosis-mediated receptor degradation and recycling can distinctly regulate PRR arbitrated PTI triggering in plant [428]. In addition to this, transport of newly synthesized PRRs such as FLS2 to the plasma membrane is also under regulation of an E3 ubiquitin ligase called PUB22. PUB22 can ubiquitinate and thereby reroute Exo70B2 (a subunit of exocyst constituting secretory vesicle) for degradation [429]. Exo70B2 was indicated to be essential for newly synthesized FLS2 transit into the plasma membrane [430] and depletion of Exo70B2 could be correlated with diminished receptor movement to the membrane. In a way, PUB22 can further restrain FLS2 signalling significantly [428].

2.11 PTI towards non-pathogens

Non-pathogenic microbial colonizers of plants (as against pathogens), can be symbionts, endophytes or commensals which can provide immense survival benefits to their hosts through nutrient acquisitions and also confer remarkable defences against abiotic as well as biotic interferences [431, 432]. Nevertheless, both pathogenic and non-pathogenic microbial members may trigger similar kind of PTI responses in plants and also they may have comparable strategic manoeuvres to overcome host defence mechanisms [433, 434]. For instance, few of the innumerable countermeasures deployed by pathogens to win over host defences have been cited in the 'Regulation of PTI' section of this chapter. Likewise, non-pathogenic microbes do attempt to subvert host immunity. Preventing recognition through host immune machinery, for instance, is a rewarding ploy for many microbes that colonizes plants. Several bacterial colonizers may possess undetectable/immobile flagellin MAMP repertoires that could diminish emergence of FLS2-mediated immune triggering [435]. Certain bacterial species may possess minimal immunogenic MAMP versions than others that aid them in escaping intense host defence responses [436]. For instance, the bacterium *Sinorhizobium meliloti* (a legume symbiont) carries intriguingly varied forms of flagellin appendages, making it difficult to be detected by host PRRs [175]. Whereas, in *Lotus japonicus*, the PRR receptor for flagellin epitope i.e. *Lj*FLS2 failed to sense flagellin ligands from *Mesorhizobium loti* (the symbiotic bacterial partner of *L. japonicus*) [437]. Some rhizobial bacterial species may deploy type III effector molecules to suppress host defence and promote symbiotic nodulation [438, 439]. In contrast, certain endophytic bacterial species may recruit their type III effector molecules for triggering host immune responses [432]. Instead of suppressing immune signalling, the ability to induce host defence responses largely assured their threshold populations within hosts. For instance, recently, an endophytic *Burkholderia gladioli* strain was revealed to trigger PTI responses in tomato host plants that in turn conferred significant resistance in tomato towards two phytopathogenic microbes viz. *Rhizoctonia solani* and *R. solanacearum* (F1C1 strain; [440]), respectively [432]. It was further realized that *B. gladioli*-induced host defence was essential for perpetuation of the endophyte's threshold population in the tomato plant [432]. In addition to the above, several other mechanisms adopted by non-pathogenic microbes to pacify host plant defences have been discussed in detail by various authors and readers may refer to these [17, 441, 442].

In particular, two examples of non-pathogenic microbe groups that have evolved conspicuous strategies to establish symbiotic association with plant systems (and attenuate host PTI signalling) include nodulating rhizobial bacteria and arbuscular mycorrhiza (AM) members, need mention. Evidently, the symbiotic microbial members and their respective host plants have coevolved with essential amenities to coexist quite spectacularly. Notably, these host plants have mechanisms to segregate symbionts in one way and restrict pathogenic microbes through the innate immunity routes. Nodulating rhizobial bacterial species are known to secrete Nod factors (NFs) that are perceived by specific receptors in their hosts which then initiate nodulation signalling [433]. Nod factor receptors in *L. japonicus* for instance, include LjNFR1 and LjNFR5, which belong to the LysM class of receptors and these sense NFs from respective symbionts [433]. NFs have been determined to comprise lipo-chitooligosaccharides (LCOs) which resemble chitin oligosaccharides

(COs) with additional lipid modifications occurring in the former [77]. COs with a variable degree of polymerizations can act as MAMP, inducing PTI responses in different plant hosts. *Arabidopsis* LysM RLK AtCERK1 can detect COs to induce defence signalling as well [443]. Interestingly, LjNFR1 is a close relative of AtCERK1 and unlike later, LjNFR1 recognizes LCOs to trigger signalling for symbiotic association [433]. Similarly, LjNFR5 was indicated to be equivalent of AtLYK5 [433]. Apparently, LCOs were revealed to down-regulate PTI responses thereby enabling a symbiotic relationship to occur [433].

In the case of arbuscular mycorrhiza (AM), the signalling for symbiosis is also carried by LCOs along with short chain chitooligosaccharide (CO4/CO5) molecules (collectively constituting the Myc-factors) exuded by the former [444]. The rice RLK members viz. OsLYK2 and OsCERK1 (LysM receptors) are known to detect Myc-factors that subsequently induce signalling for symbiosis. Moreover, close relatives of OsCERK1 recognized in *L. japonicus* (LjLYS6) [445], pea plant (PsLYK9) [446] and *Medicago truncatula* (MtLYK9) [448] also take part in AM symbiosis signalling, apart from contributing to innate immunity triggering [444]. Apart from LCOs, CO4/CO5 molecules, contribution of peptidoglycans, cyclic β-glucans, lipopolysaccharides, exopolysaccharides etc were indicated to be crucial for appropriate plant–rhizobia symbiotic association to take place [77].

2.12 Regulation of pattern-triggered immunity

The host of biological pattern-elicited PTI signalling in plants further comes under regulation of diverse factors [103, 112]. These factors can be endogenous as well as exogenous in origin. Importantly, both endogenous and exogenous factors can influence PTI signalling in positive as well as negative manners. The endogenous factors regulating PTI cascades may include protein phosphatases, micro RNAs (miRNAs), PRRs, RLKs, RLCKs, MAP kinases, calcium-responsive proteins, components of ETI sector etc. The endogenous factor-mediated control of PTI triggering becomes indispensable for attenuating unregulated immunity responses as well as for maintaining a deliberate balance between defence and developmental goals of the plant systems. In contrast, exogenous factors attempt to diminish plant PTI signalling for making hosts more susceptible to phytopathogen and pest-driven afflictions. The exogenous factors may include varied effector components produced by pathogens/pests and these effectors may be remarkably exploited by the former to disable host PTI machinery in numerous ways. Some of the factors from both endogenous and exogenous origin are stated below.

2.12.1 Protein phosphatase mediated regulation

Phosphorylation and dephosphorylation steps are central to precise regulation of immunity signalling events in plants [24]. Plant protein phosphatases (PPs) usually remove phosphoryl groups (i.e. dephosphorylate) from varied types of immunity-related proteins, those could be phosphorylated by different protein kinases involved in the PTI signalling processes such as RLKs, RLCKs, MAPKs, CDPKs etc [448]. Dephosphorylation-mediated attenuation of immune triggering ensures minimum damage to host cells/tissues as opposed to incessantly activated immune signalling. Hence, PP functions have vital implications during defence signalling.

The significant roles played by co-receptors (e.g. BAK1) and RLCKs (e.g. BIK1) in association with specific PRRs (like FLS2, EFR, PEPR etc) during PTI triggering events have been already delineated [162, 342, 355]. Subsequently, PP-mediated regulation of co-receptors as well as RLCKs was found to be an effective strategy to check PTI signalling in plants [449, 450]. It has been revealed that BIK1 (RLCK) and BAK1 (co-receptor) generally remain constitutively bound to PP2C38 [450] and PP2A [449] phosphatases, respectively. Both these PPs maintain the dephosphorylated status of the respective immune kinases. PP2C38 and PP2A functions become critical for preventing uncontrolled PTI signalling due to BIK1 and BAK1 kinases in the absence of any elicitor [449, 450]. Alternatively, PPs regulating PTI responses by attenuating MAP kinase pathways have also been documented. For example, MAPK phosphatase AP2C1 negatively regulates MPK3, MPK4 and MPK6 functions after stimulation by MAMPs such as flg22/elf18 as well as pathogenic *Pto* DC3000 bacteria [451]. In a currently concluded study, the *S. tuberosum* type one protein phosphatase 6 (*St*TOPP6) was similarly shown to disable *St*MAPK3 kinase expression and ROS build-up, thereby restraining potato defence against *R. solanacearum* pathogen [452]. In an interesting recent report, a MAP kinase member recruiting an affiliate of protein phosphatase for preventing PTI responses in the host has come into the picture. Evidently, potato MAP kinase *St*MKK1 was found to directly activate *St*PTP1a (a protein tyrosine phosphatase in *S. tuberosum*) via phosphorylation steps, in response to *P. infestans* (pathogenic oomycete) induced trigger. Stimulated *St*PTP1a in turn dephosphorylates two MAP kinase members namely *St*MPK4 and *St*MPK7 in its downstream to finally attenuate SA-hormone-related defence responses regulated by these kinases (i.e. *St*MPK4/7) [453]. In another work, evidence in support of a PRR RLK deploying a PP member to dampen the former's PTI signalling events was provided. More precisely, the CERK1 surface receptor in *Arabidopsis* was shown to carry out dephosphorylation of its own tyrosine residue (at 428th position; Tyr^{428}) by engaging a PP member CIPP1 (CERK1-interacting protein phosphatase 1) [454]. Generally, in the absence of MAMP (chitin)-induced trigger CERK1 undergoes autoposphorylation at Tyr^{428} residue and remains inactivated. However, as soon as chitin-mediated elicitations ensue, CERK1 is activated and PTI signalling is relayed downstream. Alongside, CIPP1 is deployed by CERK1 to dephosphorylate the latter's Tyr^{428} residue which consequently marks the attenuation of chitin-induced trigger perpetuating through CERK1 [454]. Besides, certain protein phosphatases can regulate plant immunity by modifying transcriptional activities of RNA polymerase II enzyme [455, 456]. Specific amino acid residues at the C-Terminal Domain (CTD) of RNA polymerase II (RNAP II) are generally reversibly phosphorylated by certain transcriptional co-activators. Meanwhile, the transcriptional co-activators (e.g., cyclin-dependent kinase C) can be induced by MAMP-mediated elicitations. CTD-phosphorylation in RNAP II was positively correlated with expression of defence-associated genes after MAMP induction. Interestingly, CPL3 (C-TERMINAL DOMAIN PHOSPHATASE-LIKE 3) phosphatase have been implicated in the regulation of

RNAP II transcriptional functions and thus down-regulation of MAMP-triggered immunity by dephosphorylating CTD of RNAPII [455]. Recently, CPL1 (a homologue of CPL3) phosphatase was reported to dampen flg22-triggered PTI responses in *Arabidopsis* [456]. CPL1 was distinctly shown to attenuate the appearance of typical PTI-related markers such as MAPK phosphorylation, ROS generation, and synthesis *PR* gene product etc and thereby compromise plant resistance to pathogenic *P. syringae* [456].

2.12.2 MicroRNA mediated regulation

Micro RNAs (miRNAs) are a class of small RNA which can repress expression of target genes by inducing mRNA decay or via halting of mRNA translation through recruitment of multi-protein-RNA complexes referred to as RISC (RNA-induced silencing complex) [457]. MiRNAs are prudent post-transcriptional regulators of numerous genes and play multifaceted role in plant life processes that include biotic stress management [457, 458]. Several studies have concluded essential roles of miRNAs in MAMP/pathogen-triggered plant immunity outcomes. For instance, upon induced by flg22 MAMP, *Arabidopsis* micro RNA candidate *miR393* dampens auxin hormone signalling through down-regulation of TIR1/AFB1 auxin receptors that in turn promote SA-hormone associated defence responses along with build-up of glucosinolate antimicrobials [459]. Earlier findings indicated auxin hormone signalling mediated repression of SA-related defence sector. *MiR393* was hence found to positively regulate flg22-elicited PTI signalling [459]. Likewise, two rice miRNA candidates, namely *miR160a* and *miR398b*, were positively correlated with PTI responses (accompanied by generation of H_2O_2, induction of defence-related genes etc) to MAMP (flg22) and *M. oryzae* elicitations [460]. The *Arabidopsis* miR172b was revealed to positively regulate transcription of *FLS2* (LRR RLK PRR gene) receptor by repressing accretion of TOE1 and TOE2 (TARGET OF EAT1/2) transcription factors during seedling development stage. Notably, TOE1 and TOE2 were found to directly associate with *FLS2* promoter for suppression of the latter's expression [461]. Micro RNAs can also subdue MAMP/pathogen-triggered immunity in plants. For example, *Oryza sativa miR1871* was shown to attenuate key PTI markers (such as H_2O_2 generation, callose accumulation, defence gene expressions etc) in response to flg22 and chitin triggers in rice [462]. *Arabidopsis miR773* also influences PTI responses negatively as it was shown to repress *METHYLTRANSFERASE 2* (*MET2*) gene expression that eventually obstructed ROS generation, reduced callose deposition and dampened host plant resistance against fungal phytopathogens such as *Plectosphaerrella cucumerina*, *F. oxysporum* and *Colletotrichum higginianum* [463].

2.12.3 PRR mediated regulation

Significance of phosphorylation/transphosphorylation events occurring amidst PRR, co-receptor RLK and RLCK members that participates in the pattern-triggered signalling processes have already been established [448, 464]. In fact, PRR mediated inhibition of co-receptor phosphorylation as a means for PTI signalling regulation has been realized recently [464]. It was revealed that by adding phosphoryl group at Serine^{286} residue of BAK1, FLS2 could prevent BIK1-mediated phosphorylation of BAK1 at six different amino acid positions. Likewise, through phosphorylation at Threonine^{455} residue of BAK1, EFR too inhibited BAK1 from getting phosphorylated by BIK1 [464]. Generally, BIK1 driven phosphorylation is considered to further stimulate BAK1 activities leading to immune triggering. On the other hand, inhibition of BIK1-mediated phosphorylation of BAK1 by FLS2/EFR receptors has been contemplated as an effective mechanism to subdue already initiated immune signalling triggered by MAMP elicitation at the surface [448].

2.12.4 RLK-mediated regulation

NIK1 (nuclear shuttle protein (NSP)-interacting kinase 1) is a member of LRR RLK sub-family II and also a close relative of SERKs (somatic embryogenesis receptor kinases) like SERK3/SERK4 [465]. NIK1 was shown to be essential for plant viral immunity but, it dampens PTI signalling induced by bacterial pathogens via impediment of FLS2/BAK1 complex formation. NIK1 shows affinity for FLS2/BAK1 complex, which is further enhanced upon flg22 elicitations. In fact, NIK1 was shown to be a negative regulator of FLS2/BAK1 complex in both the states, i.e. before and after flg22 elicitations [465]. Eventually, concentration of NIK1 was revealed to determine FLS2/BAK1 complex formations as well as downstream signalling after flg22-induced trigger [465]. Two other LRR RLK members BIR2 and BIR3 (BAK1-interacting RLK 2 and 3) were earlier shown to interact with BAK1 and prevent the latter's association with FLS2 PRR in absence of flg22 elicitor, leading to inhibition of PTI signalling. However, unlike NIK1, after flg22 binding, the interaction of BIR2 and BIR3 with BAK1 was eliminated and BAK1 was set free to unite with FLS2 for subsequent pattern-triggered signalling [465, 466].

The plasma membrane localized malectin-like RLKs have been further shown to regulate PRR receptor functions during PTI signalling. These RLKs can influence PRR activities both positively as well as negatively. For instance, the *Arabidopsis* malectin-like domain containing LRR RLK member IMPAIRED OOMYCETE SUSCEPTIBILITY1 (IOS1) was revealed to boost FLS2 signalling upon flg22 elicitation [467]. Likewise, membrane resident malectin-like RLK members such as FERONIA (FER), GmLMM1 (*Glycine max* LMM1) etc can also regulate PRR signalling. FER promotes PTI signalling triggered through FLS2 as well as EFR RLKs by aiding BAK1 association with both [245]. Contrary to this, GmLMM1 was shown to inhibit flg22-induced PTI triggering by regulating BAK1 and FLS2 interactions [468].

2.12.5 RLCK mediated regulation

BIK1, a crucial RLCK member engaged in several pattern-triggered signalling initiated via varied PRRs (like CERK1 [for chitin], RLP23 [for nlp20], FLS2 [for flg22], EFR [for elf18] etc) has been divulged to have contrasting roles in PTI signalling [310]. BIK1's discordant role may correspond to PRR-type (RLK or

RLP) and/or the latter's cognate MAMPs. For instance, loss-of-function mutation in *bik1* was correlated with elevated ROS burst upon nlp20 and PG3 (both patterns being detected by LRR RLP PRRs) mediated triggering. On the other hand, BIK1 has been shown to act as a positive regulator of flg22 and elf18 (both MAMPs recognized by LRR RLK PRRs) induced immune signalling pathways [310].

2.12.6 ETI mediated regulation of PTI

Gradually evidence is being added in regard to interdependences persisting amidst PTI and ETI signalling pathways in plants during immunity responses [62, 63, 274]. This implies, participatory components of one of the pathways might influence signalling mechanisms of the other. Accordingly, elements of ETI routes have been discovered to effect functioning of PTI pathway and thus outcomes of overall PTI signalling [63, 274]. Likewise, reports on PTI mediated regulation of ETI signalling processes have also emerged [276].

The plant ETI pathway relies on intracellular receptors (NLRs), one of the constituent members of which is sensor NLRs (and the other one being helper NLRs) [67]. Sensor NLRs (sNLRs) would generally comprise two distinct types of domains according to which they are classified as CNLs (NLRs containing a coiled-coil domain) and TNLs (NLRs bearing a TIR domain) [63]. One of the E3 (ubiquitin) ligases (enzymes involved in protein degradation through ubiquitination and proteasome complex), SNIPER1 is known to regulate maintenance of sNLRs within the cell [63]. Overexpression of SNIPER1 in *Arabidopsis* subjected to flg22 and nlp20 MAMP elicitations was found to have significantly reduced levels of *SARD1* and *FMO1*gene products as compared to wild-type plants. Importantly, these two gene products are considered crucial indicators of MAMP-triggered PTI responses in plants [63]. Notably, *SARD1* codes for a vital transcription factor involved in plant defence, whereas *FMO1* specifies an essential enzyme participating in synthesis of N-hydroxypipecolic acid, a signalling compound required in plant defence [63]. Besides conspicuous reduction in *SARD1* and *FMO1*gene products, the level of synthesized SA hormone was also greatly reduced upon flg22- and nlp20-mediated elicitation of SNIPER1 overexpressing plant [63]. Destabilization of sNLRs (in fact the TNLs) via SNIPER1 was thus accounting for inhibition of certain PTI signalling routes (i.e., SARD1, FMO1, SA hormonal pathways) in plants, even after flg22- and nlp20-mediated elicitations. These clearly indicate dependence of PTI pathway on ETI machinery for amplification of meaningful immune signalling.

A further support to the ETI-mediated regulation of PTI signalling came from the revelations that the EDS1-PAD4-ADR1 node of ETI pathway is critical for nlp20-triggered PTI responses via RLP23 receptor protein and also certain other RLP-ligand pairs [274]. Mutations in the *eds1* and *pad4* genes could be correlated with significantly reduced levels of PTI responses after nlp20 elicitations, manifested by substantial reduction in ROS generation, ET production, callose deposition etc, generally considered distinct PTI response markers [274].

2.12.7 Regulation by MAP kinases

MAP kinases acting downstream to PRR-elicitor immune complexes are generally said to amplify immune responses via activation of a myriad of signalling components [366, 379]. For instance, MAP4K4 (a member of the *Arabidopsis* MAP kinase cascade) positively modulates BIK1 function and thus its downstream PTI signalling after specific MAMP elicitations [469]. Moreover, findings have started to crop up gradually, revealing negative regulatory role of MAP kinases in the PTI responses as well. The *Arabidopsis* MAP kinase member, MKKK7 was revealed to subdue downstream immune signalling beyond FLS2 post flg22 triggering [470]. MKKK7 interacts directly with FLS2 and the former necessitates phosphorylation induced modifications for its defence inhibitory function. Inhibition of ROS upsurge, a notable outcome of MKKK7-mediated immunity suppression further indicated MKKK7's direct influence on FLS2 immune complex at the plasma membrane. In addition, MKK7 was also found to alleviate MPK6-mediated immunity functions downstream, thereby compromising host basal immunity [470]. This further suggests that upstream MAP kinase components not only aid in activating downstream MAP kinase elements, but are also capable enough of thwarting the latter's immunity-related functions. Very recently, *St*MKK1 (*S. tuberosum* MAP kinase kinase 1) was shown to attenuate SA-hormone-associated immune signalling after *P. infestans* effector-triggered activation via recruitment of *St*PTP1a (a potato protein phosphatase) [453].

2.12.8 Regulation by calcium-responsive proteins

MAP kinase cascades can be regulated by certain calcium-responsive kinases that in turn modulate PTI responses in their downstream. CIPK14 (CBL-interacting protein kinase 14), for instance, was shown to subdue immune responses downstream of *Arabidopsis* MPK3/MPK6 members after flg22 induction. CIPK14 was in fact, correlated with dephosphorylated status of MPK3/MPK6 that resulted in diminished expression of PTI marker genes such as *FRK1*, *CYP81F2*, *WAK2* and *FOX* [360].

In the absence of MAMP elicitations, i.e. (resting state), rice *Os*CPK4 associates with *Os*RLCK176 (an *At*BIK1 homologue in rice) and depletes *Os*RLCK176 build-up, thereby controlling downstream immune responses negatively [471]. In fact, *Os*CPK4 lacking kinase function was found to accelerate *Os*RLCK176 removal more pronouncedly than when *Os*CPK4 kinase function was intact. However, as soon as MAMP elicitation is sensed, mutual phosphorylation events between *Os*CPK4 and *Os*RLCK176 occur, which then render stability to *Os*RLCK176 and subsequently turn on enhanced immune responses downstream. In a sense, phosphorylation status of *Os*CPK4 modulates build-up or depletion of *Os*RLCK176 protein, and thus regulates downstream defence signalling in rice [471].

2.12.9 Pathogen induced regulation

In order to ensure successful invasion and colonization in host plants, phytopathogens can deploy numerous strategies to diminish host defences. In particular, host

PTI responses, emanating from a pathogen's initial interaction events with the plant, emerge as a formidable barrier to pathogens. Accordingly, subjugating a plant's frontline tier of defence becomes indispensable for the pathogen's success. To achieve this end, various tactics recruited by phytopathogens have been uncovered and related new aspects are also being updated at regular intervals [25, 472, 473]. Among the bountiful illustrations described, a few examples are portrayed here.

X. oryzae pv. *oryzae* (*Xoo*) suppresses rice PTI network by targeting rice RLCK member *Os*RLCK185 that functions downstream of *Os*CERK1 [474]. *Xoo* deploys one of its T3SS (type III secretion system) secreted effectors called *Xoo*1488 (a 282 amino acid residue protein) to disable functioning of *Os*RLCK185 [474]. *Xoo*1488-mediated PTI signalling inhibition was revealed to be via repression of *Os*CERK1-mediated phosphorylation of *Os*RLCK185 [474]. *Xoo* deployment of another T3SS-secreted effector XopR was found to dampen PTI responses in rice. XopR suppressed ROS generation, callose deposition as well as down-regulating expression of essential PTI-related genes like *PR2*, *PR5*, *RAC1*, *WRKY29*, *WRKY7* etc [475]. Similarly, several of the T3SS-secreted effectors from *X. campestris* pv. *campestris* (*Xcc*) were found to down-regulate MAMP-elicited PTI responses through suppression of key response markers like ROS generation, callose deposition, expression of immunity-related genes along with inhibition of MAP kinase cascades [476]. The list of these *Xcc* effectors is long and some of them include XopK, XopQ, HrpW, XopN, XopAC, XopD, XopL etc [476].

Phytopathogenic *R. solanacearum* secrets type III effector proteins such as RipV2 [472], RipBT [477] and RipN [684] for subverting PTI responses in respective plant hosts. RipV2 was revealed to possess a novel E3 ubiquitin ligase (NEL) function [472]. On the other hand, RipN is said to retain ADP ribose/NADH pyrophosphorylase activity and carry a Nudix hydrolase domain within [478]. All these three effectors were found to dampen pattern (i.e., flg22)-elicited defence responses in hosts by suppressing expression of key PTI marker genes. Further, RipV2 [472] and RipBT [477] were shown to attenuate flg22-triggered ROS generation. Besides, RipN [478] could inhibit accumulation of callose and also modulate NADH/NAD^+ ratio in the plant after being elicited by flg22. Recently, seven new type III effectors from *R. solanacearum* RS1000 strain were revealed to suppress flg22-induced PTI responses in host plants [479]. One of these seven effectors, RipE1 showed significant homology with HopX1 effector from *P. syringae* pv. *tabaci*. RipE1 was subsequently divulged to repress SA-hormone-induced defence signalling, but derepress JA-hormone signalling sector of the host defence [479].

The type III secretion system apparatus-associated protein HrpP from *P. syringae* pv. *tomato* DC3000 was reported to pacify PTI responses via modulation of MAP kinase kinase 2 (MKK2) function in *N. benthamiana* and *Arabidopsis* [480]. Pertinently, MKK2 is a constituent member of the MPK4 pathway that is implicated in conferring basal resistance to plants after MAMP/pathogen elicitation [299]. MKK2 was revealed to be stimulated by HrpP through direct interaction that leads to augmentation of SA-hormone synthesis and subsequent defence signalling, although JA hormonal sector is put to silence. HrpP was further shown to dampen flg22-triggered ROS generation (in *N. benthamiana*, *Arabidopsis* and tomato), callose build-up (in *N. benthamiana*). Down-regulation of PTI marker gene (such as *FRK1*, *WRKY22*) expression further confirmed HrpP-mediated suppression of pattern-induced immunity remarkably [480]. A model for *P. syringae* HrpP-mediated inhibition of PTI signalling is presented in figure 2.6.

Cytospora chrysosperma, the fungal agent for canker disease in numerous woody plants, deploys a virulent effector-like protein CcCAP1 to subdue PTI responses in *N. benthamiana* [481]. CcCAP1 activity represses ROS generation, callose build-up as well as expression of immunity-related genes such as *NbPR1* and *NbPR4*.

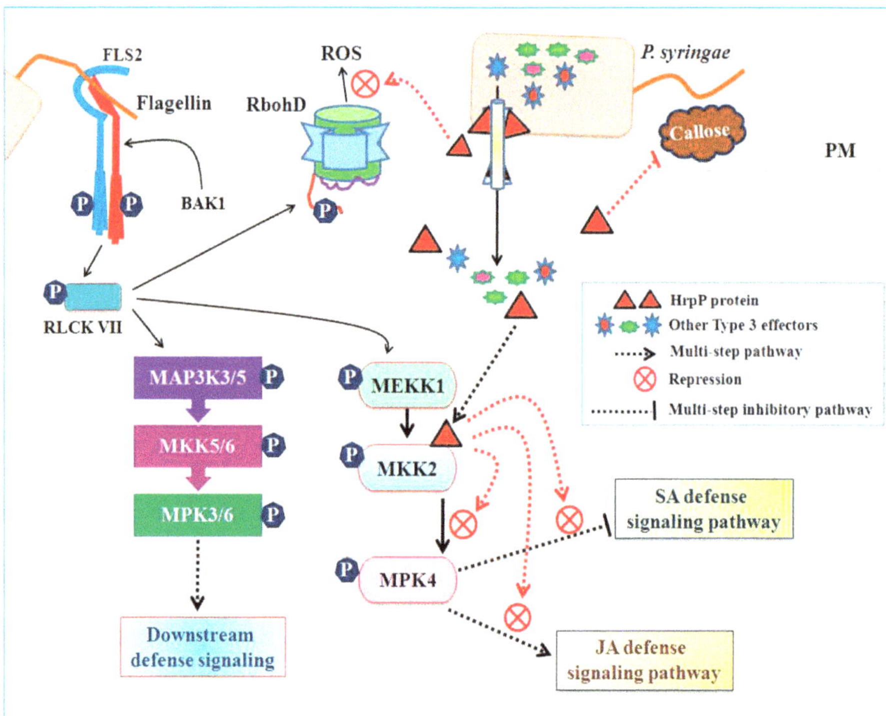

Figure 2.6. An illustration of pathogen-induced regulation of PTI signalling. Phytopathogenic *P. syringae* bacteria recruit several type III effectors to subdue host defences and for this purpose they assemble a sophisticated protein secretory apparatus that permeates through the plant cell membrane. Their type III secretory apparatus comprises various constituent proteins including HrpP protein [480]. HrpP protein was shown to suppress flagellin-triggered PTI responses by hindering ROS generation and callose deposition in tested plant cells [480]. Besides persistence in the apoplastic region, HrpP was further revealed to drift towards the plant cell cytoplasm. HrpP interacts with one of the MAP kinase members called MKK2 that belong to the MPK4 cascade and HrpP-MKK2 contact is predicted to occur either in the plasma membrane or in the cytoplasm. Subsequently, HrpP inhibits MKK2 function downstream that in turn dampens MPK4-mediated signalling. HrpP-mediated inhibition of MPK4 activity leads to repression of the JA-hormone defence signalling route, but, in contrast it facilitates the SA-hormone defence sector to persist [480]. Pertinently, MPK4 pathway is known to repress SA-mediated immune signalling in response to biotrophic pathogens (such as *P. syringae*) and maintain basal defence [270]. *P. syringae* may thus deploy HrpP to subdue MPK4-aided plant basal resistance and can further lead to derepression of autoimmune responses such as incidences of cell deaths [480]. The image was drawn following the findings of Jin *et al* (2023) [480].

CcCAP1 was further shown to dampen INF1 (oomycete MAMP)-triggered cell death indicating significant PTI suppressive potential of this fungal effector-like protein [481]. Tomato root and vascular fungal pathogen *F. oxysporum* subdue flg22-induced PTI responses by inhibiting BIK1 (RLCK essential for several PRRs including FLS2-triggered signal relay) functions through its effector protein Avr2 [473]. Avr2 hinders the monoubiquitination process of BIK1 which subsequently prevents BIK1 movement to nucleocytoplasmic locations and thus hampers the latter's downstream immune signalling steps [473].

Oomycete phytopathogen *P. infestans* manipulates an RxLR-type effector called PiAvr3b to dampen PTI responses in *N. benthamiana* and potato [482]. Two isoforms of PiAvr3b effector were shown to repress immunity signalling elicited by three MAMPs, namely PiNpp, PiINF1, and PsXeg1 along with an effector. However, both PiAvr3b isoforms are recognized by potato resistance protein *St*R3b that in turn triggers hypersensitive response (HR) reactions as part of ETI signalling [482]. Two other RxLR-type effectors from *P. infestans*, namely PITG20300 and PITG20303 were revealed to pacify PTI signalling by enhancing stability of *St*MKK1 (*S. tuberosum* MAP Kinase Kinase 1) which is already stated to be a negative regulator of pattern-elicited immunity [483]. *St*MKK1 suppresses SA-hormone-mediated defence signalling through activation of another protein *St*PTP1a (a potato protein phosphatase) [453]. Importantly, the phosphatase functions of *St*PTP1a are essential for attenuating downstream signalling activities of two other MAP kinase members—*St*MPK4 and *St*MPK7, which are generally positive regulators of SA-related defence signalling [453]. Likewise, a number of other RxLR category effectors from *Phytophthora* spp. such as Pi04089 [484], SFI5 [485], RxLR207 [486], have been recently described to negatively modulate immune signalling via targeting of essential components in PTI machinery. In the same way, *P. capsici* RxLR effector candidate RxLR48 attenuates PTI responses in *N. benthamiana* [487]. RxLR48 dampens ROS generation, callose deposition, and also down-regulates expression of the *FRK1* gene, suggesting suppression of typical PTI signalling markers by the former. Besides, RxLR48 is further capable of arresting SA-hormone signalling route through manipulation of a key master regulator termed as NPR1 in respective hosts [487].

Plant pathogens may further possess some effectors that mimic the host plant's endogenous DAMP molecules. These effectors can be used by the pathogens to subdue host PTI signalling by competing for DAMP receptors like PEPRs [488]. The smut fungal species *Sporisorium scitamineum* and *Ustilago maydis* for instance, exploit their specific effectors (SsPele1 and UmPele1, respectively) to dampen host PTI triggering. Both SsPele1 (*S. scitamineum* plant elicitor peptide-like effector 1) and UmPele1 (*U. maydis* plant elicitor peptide-like effector 1) carries Pep1 like domain at their C-termini. Importantly, SsPele1 and UmPele1 could compete with host-derived Pep1 molecules (e.g., ScPep1 and ZmPep1) for binding with Pep receptors ScPEPR1 and ZmPEPR1, respectively. *S. scitamineum* and *U. maydis* effectors could thereby attenuate host defence signalling and render the host susceptible to infections [488].

The ability of certain plant pathogens to regulate PTI responses via targeting of ETI pathway components appear to be another intriguing tactic. The oomycete phytopathogen *Phytophthora capsici* secreted RxLR homologue effector PcAvh103 was resolved to target one of the important components in the ETI signalling pathway while establishing its virulence in *N. benthamiana*. The ETI pathway component targeted by PcAvh103 was revealed to be EDS1 [489]. EDS1 (Enhancer of Disease Susceptibility 1), which carries an N-terminal lipase domain, forms an integral part of two prominent immune signalling modules (i.e. EDS1-PAD4-ADR1 and EDS1-SAG101-NRG1) through which TNL (TIR-NLR) sector of the ETI pathway generally executes immune signalling [489]. Importantly, EDS1-PAD4-ADR1 module was recently established to be a crucial determinant of PTI signalling and subsequent defence responses [274]. PcAvh103 is thought to inhibit PTI responses by disrupting the EDS1-PAD4 route of immune signalling [489].

2.12.10 Nematode- and insect herbivore-mediated regulation

Down-regulation of host immune responses is also a key strategy employed by several parasitic nematodes and herbivore insect pests for their successful infestations in the host plants. Like plant pathogenic bacteria/fungi, parasitic nematodes and insect pests too recruit different effectors (generally present in oral secretions of these organisms) to disable important components of plant PTI signalling pathways that could, for example, inhibit ROS generation, paralyse defence hormone synthesis and their signalling, diminish callose depositions and hinder function of immunity regulators like ubiquitin ligases etc [490–496]. Given below are a few illustrations from different experimental studies.

Plant parasitic nematodes include root-knot nematode (RKN) member *M. incognita* that maintains an obligate biotrophic interaction in the plant root region for a few to several weeks [490]. During this long interaction process, *M. incognita* requires to subdue imminent threats posed by nematode-induced plant defence responses quite smartly. In order to achieve this, *M. incognita* secrets an effector protein called Mi-CRT (*M. incognita* Calrecticulin) into apoplasm of the infected root region that subsequently represses PTI responses in *Arabidopsis* substantially. Analyses of the elf18 (MAMP)-triggered PTI marker gene expressions along with callose deposition phenotype under the effect of over-expressed Mi-CRT protein levels in *Arabidopsis* revealed down-regulation of most of the SA, JA/ET hormone responsive genes as well as absence of callose accumulation [490]. Notably, even though exact Mi-CRT effector target in the *Arabidopsis* PTI signalling machinery is unknown, yet this effector could down-regulate PTI responses quite remarkably. Recently, a new effector protein MiPDCD6 from *M. incognita* was shown to dampen host PTI responses via inhibition of SA-regulated defence signalling and thus ensure the former's parasitic lifestyle in tomato root [496]. In efforts to sustain parasitism, *M. javanica* (another species of *Meloidogyne*) on the other hand, tries to pacify host PTI responses by diminishing ROS generation through activation of a crucial player in the host anti-oxidant machinery with the aid of its own effector MjTTL5 [497]. Likewise, a member of the parasitic cereal cyst nematode,

Heterodera avenae recruits its effector Ha18764 to suppress both PTI and ETI sectors of host defence to favour its parasitic lifestyle [491]. *Bursaphelenchus xylophilus*, the parasitic nematode causing pine wilt disease (PWD) deploys one of its effectors BxSCD1 to attenuate PTI responses induced by BxCDP1 elicitor in *Pinus thunbergii* and *N. benthamiana* hosts [498]. Notably, BxCDP1 is a conserved NAMP carried by *B. xylophilus*. BxSCD1 was shown to interact with 1-amino-cyclopropane-1-carboxylate oxidase 1 (ACO1) proteins of *P. thunbergii* and modulated immune responses in the host [498]. In addition, BxSCD3 is a newly identified effector from *B. xylophilus* that attenuates pattern-elicited defence responses triggered by MAMPs (such as PsXEG1and INF1) in *N. benthamiana*. BxSCD3 further regulated expression of *P. thunbergii* defence-related genes-PtPR-3 and PtPR-6 [499].

Polyphagous herbivore insect pests like *Apolygus lucorum* (mirid bug) has been posing a severe threat to agricultural practices and especially toward Bt-cotton cultivated fields [492]. One of the successful mechanisms adopted by this insect is to introduce an effector protein reported as Al6 in its oral secretions into its host. Al6 is a glutathione peroxidase that can dampen ROS generation in the plant host upon MAMP (such as flg22 and INF1)-triggering [492]. Thus, *A. lucorum* can negatively regulate host PTI responses through ROS suppression activity of the Al6 effector. Another effector Al106 from *A. lucorum* OS has been subsequently revealed to attenuate PTI responses in *N. benthamiana* and *Arabidopsis* plants by halting ubiquitination of plant U-Box protein 33 (PUB33) [495]. PUB33 was shown to be a positive regulator of pattern (i.e. flg22)-induced defence in *Arabidopsis*, as it could be correlated with emergence of PTI markers such as ROS production, expression of *FRK1* and *NHL10* defence genes. Meanwhile, in *N. benthamiana* leaves, silencing of *NbPUB33* gene enhanced susceptibility of the leaves toward infection by oomycete pathogen *P. capsici* and also favoured cotton boll worm-mediated herbivory [495], indicative of the crucial affirmative role played by PUB33 in *Arabidopsis, N. benthamiana* immunity. Notably, Al106 was identified as a cyclophilin class of protein that could perform an important function as a molecular chaperone aiding in protein folding. Al106 was further shown to bear an intrinsic peptidyl-prolyl cis-trans isomerase (PPIase) domain which was shown to be essential for Al106-mediated inhibition of PUB33 ubiquitination [495]. Likewise, salivary gland effector proteins from two commonly observed herbivorous spider mite species (*Tetranychidae urticae* and *T. evansi*) infesting upon various crops including tomato, could repress SA-induced defence signalling in their host *N. benthamiana* [500]. Greenbug aphid (*Schizaphis graminum*), a serious pest of cereals, secrets a salivary effector termed Sg2204 into its hosts such as wheat during their feeding processes [494]. Greenbug salivary effector Sg2204 was shown to dampen PTI responses in wheat via inhibition of callose accumulation, accompanied by down-regulation of SA/JA hormone synthesis genes and repression of SA/JA responsive immunity-related gene expressions. Sg2204 mediated suppression of wheat PTI signalling positively facilitated greenbug feeding behaviour and also improved reproduction potential in this pest [494]. A prominent cereal pest *Laodelphax striatellus* (small brown plant hopper or SBPH) was revealed to introduce one of its salivary effectors

VgC (polypeptide fragment at the C-terminal end of Vitellogenin) into rice plant which then arrested H_2O_2 generation via direct regulation of *Os*WRKY71 transcription factor in the same host. VgC-driven abolition of H_2O_2 production (a well-defined hallmark of rice defence) assisted in improved herbivory of SBPH [493].

2.13 Hormonal influence on PTI

Phytohormones play an essential role in plant defence. Among others, SA, JA and ET hormones contribute significantly towards pathogen/pest, MAMP, DAMP as well as wound-induced immune responses in plants [221]. These hormones are synthesized within plant cells through specific biochemical pathways and maintained at their basal level concentrations. Essentially, pathogen/elicitor as well as physical wound-mediated stimulations can further evoke their synthesis considerably [235, 401, 501]. These hormones can act in the downstream of pathogen/pattern-triggered PRR-immune complexes and induce appropriate transcriptional regulators to control defence gene expression, antimicrobial metabolite production etc that in turn confer resistance phenotype in plants [239]. Nevertheless, phytohormone-mediated downstream defence signalling routes are not straightforward and may frequently involve synergistic or antagonistic relationships amidst participating hormonal pathways. For instance, cooperative participation of ET and JA pathways (along with MAP kinase network) was linked to induction of defence-related camalexin synthesis in *Arabidopsis* while responding to *B. cinerea* and chitin elicitors [305]. Likewise, in response to flg22 elicitation, the SA-mediated defence sector was revealed to promote ROS production and callose accumulation in *Arabidopsis*, which was markedly repressed by the JA-signalling sector [502]. Similarly, accumulation of callose and lignin in *Arabidopsis* roots, which was augmented by Pep1 (an iDAMP) triggered JA-signalling pathway, however, was opposed by the ET hormonal pathway [401]. Again, the same endogenous DAMP family, i.e. *At*Peps (*At*Pep1, *At*Pep2) further induced expression of defensin protein PDF1.2 in *Arabidopsis* leaves through collaboration of both JA and ET hormone signalling routes [237]. On occasion, hormones may further assist an elicitor to promote generation of another pattern molecule from plant and reinforce immune responses. For example, in respect of elicitor-evoked stomatal immunity and stomatal closure in *Arabidopsis*, positive influence of SA-signalling pathway on the PIP1 (PAMP-Induced Peptide 1) DAMP-mediated immune triggering mechanism was reported [225]. In fact, SA hormonal cascade was shown to facilitate flg22-induced expression of prePIP1 which forms the precursor of mature PIP1 DAMP [225]. Apart from these illustrations, there are numerous occasions when hormones influence key outcomes of PTI signalling in intriguing ways.

Chitosan oligosaccharide (COs; an established MAMP), for instance, evokes both SA- and JA-mediated immune networks to facilitate *Arabidopsis* resistance against the bacterial phytopathogen *P. syringae* pv. *tomato* DC300 (*Pst*) [503]. Meanwhile, variations in the requirements for hormone responsive networks to induce immunity outcomes have also been noticed. For example, in *Arabidopsis*, oligogalacturonide (OG; a classical DAMP) was shown to elicit resistance responses against the fungal pathogen *B. cinerea* (a necrotroph) via a route involving *PAD3* (*PHYTOALEXIN DEFICIENT3*) gene expression product and was independent of ET, JA and SA hormone pathways [301]. In contrast, in a recent study involving the *Arabidopsis*-OG system, OG was found to trigger defence responses against a hemibiotrophic bacterial pathogen *Pst* via induction of SA and JA hormonal pathways which was also reported to act redundantly [504]. OGs triggering activation of two distinct defence pathways against two dissimilar phytopathogens indicate presence of discrete defence mechanisms in *Arabidopsis* for restricting phytopathogens.

Pertinently, hormonal impact on PTI responses is just not limited to MAMP- or DAMP-mediated triggers and extends to elicitations by other immunogenic biological patterns such as NAMPs and HAMPs as well [135–137, 330, 335, 505]. Exhaustive details on hormonal regulation of pattern immunity signalling can be found in numerous published works and readers may refer to these [221, 287, 292, 506].

2.14 Improvement of PTI signalling via PRR manipulation

The plant cell surface PRR recognition system (that detects varied danger patterns) is a crucial link to the intracellular immune signal relay network that in turn stimulates a myriad of signalling intermediates to generate effective and multifaceted defence responses termed as PTI. Therefore, PRR-triggered PTI responses can essentially confer necessary defences to plants against numerous external invaders. However, genome-wide studies conducted so far have revealed that all plant species may not possess certain crucial PRR types and in fact particular kind(s) of PRRs might be restricted to specific plant species/family or few plant families only. For instance, EFR receptor is confined to the *Brassicaceae* family [507], whereas, CORE receptor may be restricted to only a few solanaceous plant species [129]. With the advent of advanced genomic tools, inter-family transfer of intact as well as modified PRR systems by heterologous expression in target plant species has been feasible [508–513]. These strategies undoubtedly carry immense prospects for disease-resistant plant breeding programs. In this context, a couple of experimentally feasible strategies are outlined below.

2.14.1 Heterologous expression of PRR receptors

Heterologous expression/co-expression of certain intact PRR receptors (derived from another plant species) in target plant species that naturally lack those PRRs can impart resistance phenotype against important plant pathogens. Some of the examples that are being cited below can substantiate significance of this strategy.

Sweet orange (*Citrus sinensis*) varieties are in general susceptible to bacterial phytopathogens such as *X. citri* subsp. *citri* (*Xcc*) and *X. fastidiosa* subsp. *pauca* (*Xfp*). *Xcc* and *Xfp* are responsible for causing two serious diseases in sweet orange, i.e. citrus canker (CC) and citrus variegated chlorosis (CVC), respectively, leading to considerable economic losses [512]. Importantly, sweet orange cultivars lack natural resistance means to both these pathogens [512]. The *Brassicaceae* family specific LRR RLK EFR recognizes elf18 peptide derived from bacterial Ef-Tu MAMP and confers resistance responses to varied phytopathogenic bacterial species [507].

Both *Xcc* and *Xfp* are reported to release Ef-Tu factors in their biofilms and outer membrane vesicles, respectively [514]. However, due to lack of corresponding EFR receptor, no EFR-elf peptide (from bacterial pathogen's Ef-Tu)- mediated defence signalling occurs in sweet orange plants. In this context, contemplation of engineering EFR-mediated resistance in sweet orange against CC and CVC seemed quite rational. Accordingly, transgenic Valencia sweet orange lines were developed by heterologous expression of EFR RLK gene from *Arabidopsis* [512]. Upon tested through specific pathogen inoculation methodologies for *Xcc* and *Xfp* separately, the transgenic sweet orange plants expressing EFR RLK receptors demonstrated marked reduction in colonization and spread of the corresponding pathogens. Further, conspicuous symptoms for CC and CVC were also minimally visible as compared to wild-type sweet orange plants, indicating stimulation of effective EFR-mediated resistance responses in the transgenic lines [512]. Another interesting finding gained from this work was that the immune signalling cascades downstream of heterologously expressed EFR receptor in sweet orange displayed responsiveness to respective elf peptide/intact pathogen-mediated elicitations [512]. The appearances of PTI response markers such as ROS generation, MAP kinase network activation, expression of defence genes (such as *SGT1*, *EDS1*, *WRKY23* and *NPR2*) etc signified induced resistance responses to the pathogens [512].

Similarly, transgenic apple root-stocks developed by heterologous expression of EFR receptor (from *Arabidopsis*) was found to confer resistance toward *Erwinia amylovora* pathogen [513]. *E. amylovora* is the causal agent of fire-blight disease in apple that leads to severe losses in apple production globally [513]. Apple root-stocks expressing EFR responded to *E. amylovora* supernatant and emergence of PTI responses including ROS generation, defence gene expressions indicated resistance phenotype in the transgenic lines towards fire-blight disease. Moreover, transgenic apple root-stock had minimal tissue necrosis when compared to wild-type apple plants [513]. Inter-family EFR receptor transfer can therefore stand out to be an efficient measure against *E. amylovora* while infusing a fire-blight resistance trait in susceptible apple varieties.

Likewise, *At*EFR expression in several other plant species including valuable crops such as rice, wheat, potato, tomato etc was found to facilitate significant protection against many host-specific phytopathogens [515]. In potato, *At*EFR expression could provide striking resistance to bacterial wilt pathogen *R. solanacearum* [516]. Similarly, transgenic rice and wheat gained substantial protection against bacterial phytopathogens *X. oryzae* pv. *oryzae* (*Xoo*) and *P. syringae* pv. *oryzae*, respectively, in the presence of expressed *At*EFR in both the crops [517, 518].

Rice PRR Xa21 recognizes a sulphated protein RaxX from *Xoo* to initiate PTI signalling downstream that subsequently confers strong protection against *Xoo* [82, 120]. Rice Xa21 was transformed into Anliucheng sweet orange (*C. sinensis* Osbeck) plants for developing transgenic lines. Importantly, these transgenic Anliucheng sweet orange plants were observed to manifest enhanced protection against a devastating citrus pathogen *Xanthomonas axonopodis* pv. *citri* (*Xac*) [510]. Notably, *Xac* is responsible for causing 'citrus canker' which is considered a severely distressing disease of citrus globally [510].

R. solanacearum is a devastating phytopathogen causing bacterial wilt disease in numerous plant species including several ones belonging to the *Solanaceae* family [440]. Unlike many bacterial phytopathogens, flagellin component of *R. solanacearum* cannot be detected by FLS2 PRR of numerous infected plant hosts. *Arabidopsis* FLS2 too cannot recognize flg22 peptide from *R. solanacearum*. Intriguingly, *R. solanacearum* flg22 peptide manifests a polymorphic trait because of which the pathogen can remarkably avoid detection by host FLS2 receptors and shuns subsequent host immune responses against it [342, 511]. Interestingly, *G. max* (soybean) appears to be resistant to lethal *R. solanacearum* pathogenicity and the basis for resistance might correspond to an effective immunity mechanism persisting in the crop plant. One of the possible advantages could be offered by presence of two homologues of *At*FLS2 receptor kinase proteins in soybean, namely *Gm*FLS2a and *Gm*FLS2b [519] which can surprisingly recognize polymorphic *R. solanacearum* flg22 peptides along with flg22 epitopes from other bacteria as well [511]. Recently, through adequate mutational studies, *Gm*FLS2a/2b was shown to be effective in conferring resistance to *R. solanacearum* when the pathogen was *in vitro* inoculated (e.g. via stem inoculation) into soybean plants [520]. Nevertheless, additional defence mechanisms persisting in soybean are also expected to contribute towards bacterial wilt resistance in this legume [520]. Interestingly, discovery of *Gm*FLS2 conferring significant resistance to *R. solanacearum* can have crucial implications, since compatible *in planta* expression of these PRRs might be able to bestow bacterial wilt resistance/tolerance in susceptible plant species of agronomic importance [511, 520]. In this context, efforts for achieving heterologous co-expression of *Gm*FLS2 along with *Gm*BAK1 proteins in tomato and *N. benthamiana* led to some exciting findings [511]. Leaf-inoculation of *R. solanacearum* into *Gm*FLS2/*Gm*BAK1 co-expressed *N. benthamiana* plants indicated reduction in pathogenic population and even an elevated reduction in *R. solanacearum* population could be recorded when these plants were pre-treated with flg22 peptides from the former. Likewise, co-expression of *Gm*FLS2/*Gm*BAK1 components in tomato roots conferred significant resistance in these tomato plants against *R. solanacearum* when inoculated via soil-drenching methodology [511]. Pertinently, *R. solanacearum* is considered to infect most of its plant hosts via root wounds that can be pathogen driven or caused by physical/mechanical distresses. It is noteworthy that *Gm*FLS2/*Gm*BAK1 was also effective in detection of flg22 epitopes from other pathogenic bacteria and could induce downstream immune responses in the transgenic tomato lines [511]. Henceforth, *Gm*FLS2 can certainly be an important candidate for PRR engineering in other bacterial wilt susceptible crop species as well.

In another instance, inter-family transfer of an L-type lectin receptor kinase LecRK-I.9 (from *Arabidopsis*) into potato and *N. benthamiana* plants conferred substantial protection against late-blight disease caused by oomycete pathogen *P. infestans* [509]. Notably, *Arabidopsis* LecRK-I.9 is a PRR that recognizes RXLR effector protein IPI-O from *P. infestans* as MAMP [521]. In a sense, *Phytophthora* late-blight resistance rendered by *Arabidopsis* LecRK-I.9 in these two solanaceaous plant species clearly indicates feasibility and advantages of inter-family PRR transfer strategy remarkably. Likewise, transgenic tobacco and cotton plants expressing LRR RLP gene *Ve1* from tomato gained significant resistance toward wilt disease caused by *Verticillium* genus [522]. Notably, tomato Ve1 RLP PRR recognizes Ave1 effectors secreted by members of *Verticillium* genus to trigger downstream defence responses [522]. Adding to the list, *Arabidopsis* LRR RLP RLP23 expression in transgenic potato plants enabled the latter to sense nlp20 MAMP and subsequently initiate strong defence responses against phytopathogenic *P. infestans* and *Sclerotinia sclerotiorum* [147].

Potato (*S. tuberosum*) cultivars belonging to Atlantic lines are known to be susceptible to oomycete late-blight pathogen *P. infestans* [523]. Further, Atlantic lines generally do not respond to 'Pep13' MAMP of phytopathogenic oomycete *Phytophthora* species. Constitutive heterologous expression of a newly discovered LRR RLK PRR called PERU (that perceives Pep13 elicitor) in potato Atlantic lines conferred significant resistance towards pathogenic *P. infestans* strain [163] indicating the possibility of generating late-blight resistant potato cultivars in breeding programs.

2.14.2 Expression of chimeric PRR proteins

Combination of essential domains from different/similar PRR types and generation of chimeric PRR receptors can be another effective strategy to confer disease resistance phenotypes in plants [515]. For instance, CRXA is a chimeric PRR which was generated by combining extracellular LysM domains from rice Chitin elicitor-binding protein (CEBiP; a LysM RLP) and the internal serine/threonine kinase domain from Xa21 (a rice LRR RLK), respectively [508]. Subsequent expression of CRXA in transgenic rice lines imparted strong resistance phenotype in the latter against rice blast pathogen *M. oryzae*, thereby conferring substantial protection to rice blast disease [508]. Another chimeric PRR constituting CEBiP and intracellular kinase domain of Pi-d2 (a PRR found in indica rice) was generated and designated as a CRPi receptor [524]. Transgenic rice with expressed CRPi hybrid receptor conferred immunity against *M. oryzae* significantly. CRPi was shown to be actively elicited by chitin oligosaccharide fraction N-acetylchitoheptaose [524].

Likewise, when a fusion RLP protein generated by connecting vital domains from two RLP PRRs, namely ReMAX (detects eMax MAMP from *Xanthomonas*) and Eix2 (recognizes Xylanase MAMP from pathogens) was introduced into *N. benthamiana* plants, the transgenic tobacco lines manifested induction of immune responses to eMax mediated elicitations effectively [525, 526]. Pertinently, *N. benthamiana* does not possess surface receptor systems for eMax and xylanase MAMPs naturally [525, 526].

Recently, a hybrid PRR receptor EFR-Cf-9 was created by combining *Arabidopsis* EFR RLK extracellular domain with transmembrane and cytoplasmic domains of tomato Cf-9 RLP protein [527]. When expressed in transgenic tobacco plants, EFR-Cf-9 chimeric receptor was found to induce strong resistance responses upon elf18 peptide mediated elicitation. Transgenic tobacco harbouring EFR-Cf-9 further gained significant protection against phytopathogenic *Pseudomonas amygdali* pv. *tabaci* (*Pta*) 11 528 and *P. syringae* pv. *tomato* DC3000 (*Pst*) strains [527].

Notably, immune responses in tobacco comprised generation of effective HR (hypersensitive responses) [527].

2.14.3 Approaches to engineering DAMP-detecting PRRs

Plant surface-localized DAMP-recognizing PRR receptors also play a crucial role in a plant's local and systemic defence [223, 528]. Experiments involving inter-family transfer of DAMP-perceiving PRRs or creation of chimeric PRR proteins containing essential domains from MAMP- and DAMP-recognizing PRRs have been demonstrated [126, 528]. It is anticipated that the early experimental outcomes and other ongoing trials would push the research goals of developing disease-resistant plant/crop varieties sufficiently forward.

About eight years ago, heterologous expression of *At*PEPR1 (PEPR1 receptor of *Arabidopsis*) and *Zm*PEPR1a (PEPR1a receptor of *Z. mays*) receptors was carried out to generate transgenic *N. benthamiana* lines [529]. *N. benthamiana* lines harbouring *At*PEPR1 and *Zm*PEPR1a were able to detect plant elicitor peptides (Peps) derived from *Brassicaceae* or *Poaceae* plant families, respectively. Recognition of Pep elicitors from either of the families also triggered active downstream defence signalling in the transgenic *N. benthamiana* lines. However, Pep elicitors derived from other plant families couldn't be detected by *At*PEPR1 and *Zm*PEPR1a expressed *N. benthamiana* lines nor there was induction of immunity signalling in them. These experiments revealed challenges to inter-family transfer of PEPR receptors due to incompatibility of PEPR-Pep systems across different plant families (and that's because of dissimilar Pep sequences). However, an interesting aspect is downstream signalling components of PEPRs are largely analogous [529]. Nevertheless, it is possible that within families, plant species that lack or have defective Pep-perceiving PEPRs may be incorporated with efficient receptors from compatible plant varieties and thus would be reinforced with resistant phenotype.

Creation of a chimeric version of PRRs to facilitate improved defence responses in target plant species has been evident [515]. In a proof-of-concept study, two chimeric variants of PRRs were developed by exchanging vital domains of EFR and WAK1 receptors [126]. WAK1 (Wall-Associated Kinase 1 PRR of *Arabidopsis*) detects endogenous oligogalacturonide (OG) fragments and hence is an established DAMP receptor. However, EFR (also a PRR of *Arabidopsis*) detects a proteinaceous MAMP, i.e. elf18 (derived from bacterial Ef-Tu). One of the chimeric PRRs was created by combining the ectodomain region from WAK1 and the kinase domain of EFR receptors. This hybrid PRR was observed to sense OG (DAMP) and thereby trigger kinase domain of EFR receptor. The second chimera was generated by joining the EFR ectodomain with the kinase domain of WAK1. The second chimeric PRR was able to recognize elf18 MAMP and subsequently activate the kinase domain of WAK1. In the case of both the hybrid PRRs, respective elicitors (i.e. DAMP or MAMP) could trigger downstream immunity signalling confirming feasibility of domain exchanges amidst different PRR members without hindering their normal functioning [126].

2.15 Conclusions

PTI responses triggered by exogenous (MAMP/NAMP/HAMP) and endogenous (DAMPs) patterns both sufficiently converge at crucial junctures. Essentially, both pathways share common signalling components of PTI machinery and have been found to augment defence responses quite significantly [223, 255]. Besides known resemblances, a number of distinctions between both the pathways has also been illuminated. For instance, MAMP-triggered immunity seems to be weaker in the *Arabidopsis* root region as compared to DAMP-elicited defence [412, 413]. This further highlights conspicuous attributes of MAMP- and DAMP-triggered signalling in plants. In days to come, more details in this regard are expected to crop up.

Apart from continuing discoveries relating newer MAMP/DAMP-PRR pairs, recent revelations of specific plant PRRs against molecular patterns from parasitic nematodes (NAMPs) [135, 136] and insect herbivores (HAMPs) [134], are anticipated to enrich understanding of PTI mechanisms elicited by diverse biological organisms. In future, it might be possible to comprehend how in a natural milieu, multiple biotic agents could influence plant defence mechanisms at given time points. This could be instrumental in designing effective countermeasures to infuse resistance phenotypes in important crop plant species. The persistent evolution of pathogen effectors targeted at plant PTI machinery further necessitates innovations in plant defence mechanisms. However, plant-mediated innovations are hard to come by at the same pace [530]. It is possible that due to sharing of similar PTI machinery components by MAMP and DAMP elicitors, pathogen effectors might be targeting the shared components selectively, so that PTI responses emanating from MAMP or DAMP triggers might be pacified simultaneously. Notably, pathogen effectors have also aided remarkably in the elucidation of components of PTI apparatus and their specific attributes [473, 474, 482]. Moreover, investigations are on to elucidate the missing links in the PTI signalling machinery and their corresponding pathways.

Lastly, proper understandings of PTI signalling mechanism with appropriate receptor (PRR)-ligand (pattern) coupling can significantly assist in engineering disease resistance phenotypes across important crop plant species. The experimental success of heterologous PRR expressions [511–513], design of effective chimeric PRRs [515, 527] etc for instance, are expected to provision alternative genomic approaches for breeding future disease-resistant crops which is also a need of the hour.

Acknowledgements

The author (A Barman) would like to humbly acknowledge the financial support received from Science and Engineering Research Board (SERB) in the form of SERB-TARE (Teachers Associateship for Research Excellence) grant (File no: TAR/2022/000 349). The author further wishes to thank Dr Anirudh Kumar, Department of Botany, Central Tribal University of A.P., Vizianagaram, India for his kind support and considerations that helped in completion of the manuscript. The author would further like to acknowledge Professor Suvendra Kumar Ray, Department of Molecular Biology and Biotechnology, Tezpur University, Assam, India, for his kind and inspiring words.

References

[1] Rodríguez H and Fraga R 1999 Phosphate solubilizing bacteria and their role in plant growth promotion *Biotechnol. Adv.* **17** 319–39

[2] Ollerton J, Winfree R and Tarrant S 2011 How many flowering plants are pollinated by animals? *Oikos* **120** 321–6

[3] Mittler R 2006 Abiotic stress, the field environment and stress combination *Trends Plant Sci.* **11** 15–9

[4] Moura J C M S, Bonine C A V, de Oliveira Fernandes Viana J, Dornelas M C and Mazzafera P 2010 Abiotic and biotic stresses and changes in the lignin content and composition in plants *J. Integr. Plant Biol.* **52** 360–76

[5] Bai Y, Kissoudis C, Yan Z, Visser R G and van der Linden G 2018 Plant behaviour under combined stress: tomato responses to combined salinity and pathogen stress *Plant* J. **93** 781–93

[6] Franche C, Lindström K and Elmerich C 2009 Nitrogen-fixing bacteria associated with leguminous and non-leguminous plants *Plant Soil* **321** 35–59

[7] Lucas J A 2009 *Plant Pathology and Plant Pathogens* (New York: Wiley)

[8] Scholthof K B G, Adkins S, Czosnek H, Palukaitis P, Jacquot E, Hohn T *et al* 2011 Top 10 plant viruses in molecular plant pathology *Mol. Plant Pathol.* **12** 938–54

[9] Mansfield J, Genin S, Magori S, Citovsky V, Sriariyanum M, Ronald P *et al* 2012 Top 10 plant pathogenic bacteria in molecular plant pathology *Mol. Plant Pathol.* **13** 614–29

[10] Dean R, Van Kan J A, Pretorius Z A, Hammond-Kosack K E, Di Pietro A, Spanu P D *et al* 2012 The top 10 fungal pathogens in molecular plant pathology *Mol. Plant Pathol.* **13** 414–30

[11] Jones J T, Haegeman A, Danchin E G, Gaur H S, Helder J, Jones M G *et al* 2013 Top 10 plant-parasitic nematodes in molecular plant pathology *Mol. Plant Pathol.* **14** 946–61

[12] Oksanen L and Olofsson J 2009 Vertebrate herbivory and its ecosystem consequences *eLS* 1–11

[13] Barman A 2014 A brief perspective on Gmelina tree insect pest *Craspedonta leayana J. Entomol. Zool. Stud.* **2** 276–8

[14] Ng J C and Perry K L 2004 Transmission of plant viruses by aphid vectors *Mol. Plant Pathol.* **5** 505–11

[15] Perilla-Henao L M and Casteel C L 2016 Vector-borne bacterial plant pathogens: interactions with hemipteran insects and plants *Front. Plant Sci.* **7** 1163

[16] Delaux P M and Schornack S 2021 Plant evolution driven by interactions with symbiotic and pathogenic microbes *Science* **371** eaba6605

[17] Teixeira P J P L, Colaianni N R, Fitzpatrick C R and Dangl J L 2019 Beyond pathogens: microbiota interactions with the plant immune system *Curr. Opin. Microbiol.* **49** 7–17

[18] Zhang M and Kong X 2022 How plants discern friends from foes *Trends Plant Sci.* **27** 107–9

[19] Abdelfattah , Tack A, Wasserman A J, Liu B, Berg J, Norelli G *et al* 2022 Evidence for host–microbiome co-evolution in apple *New Phytol.* **234** 2088–100

[20] Barrett L G, Kniskern J M, Bodenhausen N, Zhang W and Bergelson J 2009 Continua of specificity and virulence in plant host–pathogen interactions: causes and consequences *New Phytol.* **183** 513–29

[21] Bennett R N and Wallsgrove R M 1994 Secondary metabolites in plant defence mechanisms *New Phytol.* **127** 617–33

[22] Dodds P N and Rathjen J P 2010 Plant immunity: towards an integrated view of plant–pathogen interactions *Nat. Rev. Genet.* **11** 539–48
[23] Voigt C A 2014 Callose-mediated resistance to pathogenic intruders in plant defense-related papillae *Front. Plant Sci.* **5** 168
[24] Barman A and Ray S K 2020 Protein phosphatase mediated responses in plant host–pathogen interactions *Protein Phosphatases and Stress Management in Plants* ed G K Pandey (Cham: Springer) pp 289–330
[25] Buscaill P and van der Hoorn R A 2021 Defeated by the nines: nine extracellular strategies to avoid microbe-associated molecular patterns recognition in plants *Plant Cell* **33** 2116–30
[26] Boller T and He S Y 2009 Innate immunity in plants: an arms race between pattern recognition receptors in plants and effectors in microbial pathogens *Science* **324** 742–4
[27] Koch K, Bhushan B and Barthlott W 2008 Diversity of structure, morphology and wetting of plant surfaces *Soft Matter* **4** 1943–63
[28] Javelle M, Vernoud V, Rogowsky P M and Ingram G C 2011 Epidermis: the formation and functions of a fundamental plant tissue *New Phytol.* **189** 17–39
[29] Cluzet S, Mérillon J M and Ramawat K G 2020 Specialized metabolites and plant defence ed J M Mérillon and K G Ramawat *Plant Defence: Biological Control* (Cham: Springer) pp 45–80
[30] Castilleux R, Plancot B, Ropitaux M, Carreras A, Leprince J, Boulogne I *et al* 2018 Cell wall extensins in root–microbe interactions and root secretions *J. Exp. Bot.* **69** 4235–47
[31] Goyal S, Lambert C, Cluzet S, Merillon J M and Ramawat K G 2012 Secondary metabolites and plant defence ed J M Merillon and K G Ramawat *Plant Defence: Biological Control* (Berlin: Springer) pp 109–38
[32] Holbein J, Franke R B, Marhavý P, Fujita S, Górecka M *et al* 2019 Root endodermal barrier system contributes to defence against plant-parasitic cyst and root-knot nematodes *Plant* J. **100** 221–36
[33] Driouich A, Follet-Gueye M L, Vicré-Gibouin M and Hawes M 2013 Root border cells and secretions as critical elements in plant host defense *Curr. Opin. Plant Biol.* **16** 489–95
[34] Driouich A, Gaudry A, Pawlak B and Moore J P 2021 Root cap–derived cells and mucilage: a protective network at the root tip *Protoplasma* **258** 1179–85
[35] Driouich A, Smith C, Ropitaux M, Chambard M, Boulogne I, Bernard S *et al* 2019 Root extracellular traps versus neutrophil extracellular traps in host defence, a case of functional convergence? *Biol. Rev.* **94** 1685–700
[36] Chambard , Plasson M, Derambure C, Coutant C, Tournier S, Lefranc I *et al* 2021 New insights into plant extracellular DNA. A study in soybean root extracellular trap *Cells* **10** 69
[37] Malinovsky F G, Fangel J U and Willats W G 2014 The role of the cell wall in plant immunity *Front. Plant Sci.* **5** 178
[38] Scheller H V and Ulvskov P 2010 Hemicelluloses *Annu. Rev. Plant Biol.* **61** 263–89
[39] Pauly M, Gille S, Liu L F, Mansoori N, De Souza A, Schultink A *et al* 2013 Hemicellulose biosynthesis *Planta* **238** 627–42
[40] Hernandez-Blanco C, Feng D X, Hu J, Sanchez-Vallet A, Deslandes L, Llorente F *et al* 2007 Impairment of cellulose synthases required for Arabidopsis secondary cell wall formation enhances disease resistance *Plant Cell* **19** 890–903
[41] Hamann T 2012 Plant cell wall integrity maintenance as an essential component of biotic stress response mechanisms *Front. Plant Sci.* **3** 77

[42] Endler A and Persson S 2011 Cellulose synthases and synthesis in *Arabidopsis Mol. Plant* **4** 199–211

[43] Lee M H, Jeon H S, Kim S H, Chung J H, Roppolo D, Lee H J *et al* 2019 Lignin based barrier restricts pathogens to the infection site and confers resistance in plants *EMBO J.* **38** e101948

[44] Bhuiyan N H, Selvaraj G, Wei Y and King J 2009 Role of lignification in plant defense *Plant Signal. Behav.* **4** 158–9

[45] Phillips M A and Rodney B 1999 Croteau resin-based defenses in conifers *Trends Plant Sci.* **4** 184–90

[46] Hagel J M, Yeung E C and Facchini P J 2008 Got milk? The secret life of laticifers *Trends Plant Sci.* **13** 631–9

[47] Castelblanque L, García-Andrade J, Martínez-Arias C, Rodríguez J J, Escaray F J, Aguilar-Fenollosa E *et al* 2021 Opposing roles of plant laticifer cells in the resistance to insect herbivores and fungal pathogens *Plant Commun.* **2** 100112

[48] Konno K 2011 Plant latex and other exudates as plant defense systems: roles of various defense chemicals and proteins contained therein *Phytochem.* **72** 1510–30

[49] Huckelhoven R 2007 Cell wall-associated mechanisms of disease resistance and susceptibility *Annu. Rev. Phytopathol.* **45** 101–27

[50] Valueva T A and Mosolov V V 2004 Role of inhibitors of proteolytic enzymes in plant defense against phytopathogenic microorganisms *Biochemistry* **69** 1305–9

[51] Ryan C A 1990 Proteinase inhibitors in plants: genes for improving defenses against insects and pathogens *Annu. Rev. Phytopathol.* **28** 425–49

[52] Chandrasekaran M, Thangavelu B, Chun S C and Sathiyabama M 2016 Proteases from phytopathogenic fungi and their importance in phytopathogenicity *J. Gen. Plant Pathol.* **82** 233–9

[53] Figaj D, Ambroziak P, Przepiora T and Skorko-Glonek J 2019 The role of proteases in the virulence of plant pathogenic bacteria *Int. J. Mol. Sci.* **20** 672

[54] Huma H and Khalid M F 2007 Plant protease inhibitors: a defense strategy in plants *Biotechnol. Mol. Biol. Rev.* **2** 68–85

[55] Mosolov V V, Loginova M D, Fedurkina N V and Benken I I 1976 The biological significance of proteinase inhibitors in plants *Plant Sci. Lett.* **7** 77–80

[56] Newman M A, Sundelin T, Nielsen J T and Erbs G 2013 MAMP (microbe-associated molecular pattern) triggered immunity in plants *Front. Plant Sci.* **4** 139

[57] Choi H W and Klessig D F 2016 DAMPs, MAMPs, and NAMPs in plant innate immunity *BMC Plant Biol.* **16** 1–10

[58] DeFalco T A and Zipfel C 2021 Molecular mechanisms of early plant pattern-triggered immune signaling *Mol. Cell* **81** 3449–67

[59] Wang J and Chai J 2020 Structural insights into the plant immune receptors PRRs and NLRs *Plant Physiol.* **182** 1566–81

[60] Bentham A R, De la Concepcion J C, Mukhi N, Zdrzałek R, Draeger M, Gorenkin D *et al* 2020 A molecular roadmap to the plant immune system *J. Biol. Chem.* **295** 14916–35

[61] Duxbury Z, Wu C H and Ding P 2021 A comparative overview of the intracellular guardians of plants and animals: NLRs in innate immunity and beyond *Annu. Rev. Plant Biol.* **72** 155–84

[62] Ngou B P M, Ahn H K, Ding P and Jones J D 2021 Mutual potentiation of plant immunity by cell-surface and intracellular receptors *Nature* **592** 110–5

[63] Tian H, Wu Z, Chen S, Ao K, Huang W, Yaghmaiean H *et al* 2021 Activation of TIR signalling boosts pattern-triggered immunity *Nature* **598** 500–3
[64] Yuan M, Ngou B P M, Ding P and Xin X F 2021 PTI-ETI crosstalk: an integrative view of plant immunity *Curr. Opin. Plant Biol.* **62** 102030
[65] Yuan M, Jiang Z, Bi G, Nomura K, Liu M, Wang Y *et al* 2021 Pattern-recognition receptors are required for NLR-mediated plant immunity *Nature* **592** 105–9
[66] Cui H, Tsuda K and Parker J E 2015 Effector-triggered immunity: from pathogen perception to robust defense *Annu. Rev. Plant Bio.* **66** 10–1146
[67] Jubic L M, Saile S, Furzer O J, El Kasmi F and Dangl J L 2019 Help wanted: helper NLRs and plant immune responses *Curr. Opin. Plant Biol.* **50** 82–94
[68] Sun Y, Zhu Y X, Balint-Kurti P J and Wang G F 2020 Fine-tuning immunity: players and regulators for plant NLRs *Trends Plant Sci.* **25** 695–713
[69] Martin G B, Bogdanove A J and Sessa G 2003 Understanding the functions of plant disease resistance proteins *Annu. Rev. Plant Biol.* **54** 23–61
[70] Zipfel C 2008 Pattern-recognition receptors in plant innate immunity *Curr. Opin. Immunol.* **20** 10–16
[71] Mermigka G, Amprazi M, Mentzelopoulou A, Amartolou A and Sarris P F 2020 Plant and animal innate immunity complexes: fighting different enemies with similar weapons *Trends Plant Sci.* **25** 80–91
[72] Ausubel F M 2005 Are innate immune signaling pathways in plants and animals conserved? *Nat. Immunol.* **6** 973–9
[73] Boller T and Felix G 2009 A renaissance of elicitors: perception of microbe-associated molecular patterns and danger signals by pattern-recognition receptors *Annu. Rev. Plant Biol.* **60** 379–406
[74] Gómez-Gómez L and Boller T 2002 Flagellin perception: a paradigm for innate immunity *Trends Plant Sci.* **7** 251–6
[75] Spoel S H and Dong X 2012 How do plants achieve immunity? Defence without specialized immune cells *Nat. Rev. Immunol.* **12** 89–100
[76] Saijo Y and Loo E P I 2020 Plant immunity in signal integration between biotic and abiotic stress responses *New Phytol.* **225** 87–104
[77] Zipfel C and Oldroyd G E 2017 Plant signalling in symbiosis and immunity *Nature* **543** 328–36
[78] Uroz S, Courty P E and Oger P 2019 Plant symbionts are engineers of the plant-associated microbiome *Trends Plant Sci.* **24** 905–16
[79] Morelli M, Bahar O, Papadopoulou K K, Hopkins D L and Obradović A 2020 Role of endophytes in plant health and defense against pathogens *Front. Plant Sci.* **11** 1312
[80] Dangl J L and Jones J D 2001 Plant pathogens and integrated defence responses to infection *Nature* **411** 826–33
[81] Jones J D, Vance R E and Dangl J L 2016 Intracellular innate immune surveillance devices in plants and animals *Science* **354** aaf6395
[82] Song W Y, Wang G L, Chen L L, Kim H S, Pi L Y, Holsten T *et al* 1995 A receptor kinase-like protein encoded by the rice disease resistance gene, *Xa21 Science* **270** 1804–6
[83] Gómez-Gómez L and Boller T 2000 FLS2: an LRR receptor–like kinase involved in the perception of the bacterial elicitor flagellin in *Arabidopsis Mol. Cell* **5** 1003–11
[84] Stone J M and Walker J C 1995 Plant protein kinase families and signal transduction *Plant Physiol.* **108** 451–7

[85] Lehti-Shiu M D and Shiu S H 2012 Diversity, classification and function of the plant protein kinase superfamily *Philos. Trans. R. Soc. Lond., B, Biol. Sci.* **367** 2619–39
[86] Gish L A and Clark S E 2011 The RLK/Pelle family of kinases *Plant* J. **66** 117–27
[87] Becraft P W 2002 Receptor kinase signaling in plant development *Annu. Rev. Cell Dev. Biol.* **18** 163–92
[88] Ye Y, Ding Y, Jiang Q, Wang F, Sun J and Zhu C 2017 The role of receptor-like protein kinases (RLKs) in abiotic stress response in plants *Plant Cell Rep.* **36** 235–42
[89] Dievart A, Gottin C, Périn C, Ranwez V and Chantret N 2020 Origin and diversity of plant receptor-like kinases *Annu. Rev. Plant Biol.* **71** 26.1–26
[90] Wu Y and Zhou J M 2013 Receptor like kinases in plant innate immunity *J. Integr. Plant Biol.* **55** 1271–86
[91] Escocard de Azevedo Manhães A M, Ortiz Morea F A, He P and Shan L 2021 Plant plasma membrane resident receptors: surveillance for infections and coordination for growth and development *J. Integr. Plant Biol.* **63** 79–101
[92] Walker J C and Zhang R 1990 Relationship of a putative receptor protein kinase from maize to the S-locus glycoproteins of *Brassica Nature* **345** 743–46
[93] Wang G, Ellendorff U, Kemp B, Mansfield J W, Forsyth A, Mitchell K *et al* 2008 A genome-wide functional investigation into the roles of receptor-like proteins in *Arabidopsis Plant Physiol.* **147** 503–17
[94] Jamieson P A, Shan L and He P 2018 Plant cell surface molecular cypher: receptor-like proteins and their roles in immunity and development *Plant Sci.* **274** 242–51
[95] Zipfel C 2014 Plant pattern-recognition receptors *Trends Immunol.* **35** 345–51
[96] Maruta N, Burdett H, Lim B Y, Hu X, Desa S, Manik M K *et al* 2022 Structural basis of NLR activation and innate immune signalling in plants *Immunogenetics* **74** 5–26
[97] Kapos P, Devendrakumar K T and Li X 2019 Plant NLRs: from discovery to application *Plant Sci.* **279** 3–18
[98] Ma X, Xu G, He P and Shan L 2016 SERKing coreceptors for receptors *Trends Plant Sci.* **21** 1017–33
[99] Yasuda S, Okada K and Saijo Y 2017 A look at plant immunity through the window of the multitasking coreceptor BAK1 *Curr. Opin. Plant Biol.* **38** 10–8
[100] Heese A, Hann D R, Gimenez-Ibanez S, Jones A M, He K, Li J *et al* 2007 The receptor-like kinase SERK3/BAK1 is a central regulator of innate immunity in plants *Proc. Natl Acad. Sci. USA* **104** 12217–22
[101] Liebrand T W, van den Burg H A and Joosten M H 2014 Two for all: receptor-associated kinases SOBIR1 and BAK1 *Trends Plant Sci.* **19** 123–32
[102] Liang X and Zhou J M 2018 Receptor-like cytoplasmic kinases: central players in plant receptor kinase–mediated signaling *Annu. Rev. Plant Biol.* **69** 267–99
[103] Sun L and Zhang J 2020 Regulatory role of receptor-like cytoplasmic kinases in early immune signaling events in plants *FEMS Microbiol. Rev.* **44** 845–56
[104] Gao X, Cox K L Jr and He P 2014 Functions of calcium-dependent protein kinases in plant innate immunity *Plants* **3** 160–76
[105] Ngou B P M, Ding P and Jones J D 2022 Thirty years of resistance: zig-zag through the plant immune system *Plant Cell* **34** 1447–78
[106] Shiu S H and Bleecker A B 2003 Expansion of the receptor-like kinase/Pelle gene family and receptor-like proteins in *Arabidopsis Plant Physiol.* **132** 530–43

[107] Lee D H, Lee H S and Belkhadir Y 2021 Coding of plant immune signals by surface receptors *Curr. Opin. Plant Biol.* **62** 102044

[108] Shiu S H and Bleecker A B 2001 Receptor-like kinases from *Arabidopsis* form a monophyletic gene family related to animal receptor kinases *Proc. Natl Acad. Sci. USA* **98** 10763–8

[109] Macho A P, Lozano-Durán R and Zipfel C 2015 Importance of tyrosine phosphorylation in receptor kinase complexes *Trends Plant Sci.* **20** 269–72

[110] Johnson L N, Noble M E M and Owen D J 1996 Active and inactive protein kinases: structural basis for regulation *Cell* **85** 149–58

[111] Dardick C and Ronald P 2006 Plant and animal pathogen recognition receptors signal through non-RD kinases *PLoS Pathog.* **2** e2

[112] Couto D and Zipfel C 2016 Regulation of pattern recognition receptor signalling in plants *Nat. Rev. Immunol.* **16** 537–52

[113] Tör M, Lotze M T and Holton N 2009 Receptor-mediated signalling in plants: molecular patterns and programmes *J. Exp. Bot.* **60** 3645–54

[114] Song W, Forderer A, Yu D and Chai J 2021 Structural biology of plant defence *New Phytol.* **229** 692–711

[115] He Z H, Cheeseman I, He D and Kohorn B D 1999 A cluster of five cell wall-associated receptor kinase genes, Wak1–5, are expressed in specific organs of *Arabidopsis Plant Mol. Biol.* **39** 1189–96

[116] Catanzariti A M, Lim G T and Jones D A 2015 The tomato *I3* gene: a novel gene for resistance to *Fusarium* wilt disease *New Phytol.* **207** 106–18

[117] Takayama S and Isogai A 2005 Self-incompatibiltiy in plants *Annu. Rev. Plant Biol.* **56** 467–89

[118] Saijo Y, Loo E P I and Yasuda S 2018 Pattern recognition receptors and signaling in plant–microbe interactions *Plant* J. **93** 592–613

[119] Yamaguchi Y, Huffaker A, Bryan A C, Tax F E and Ryan C A 2010 PEPR2 is a second receptor for the Pep1 and Pep2 peptides and contributes to defense responses in *Arabidopsis Plant Cell* **22** 508–22

[120] Pruitt R N, Schwessinger B, Joe A, Thomas N, Liu F, Albert M *et al* 2015 The rice immune receptor XA21 recognizes a tyrosine-sulfated protein from a Gram-negative bacterium *Sci. Adv.* **1** e1500245

[121] Ron M and Avni A 2004 The receptor for the fungal elicitor ethylene-inducing xylanase is a member of a resistance-like gene family in tomato *Plant Cell* **16** 1604–15

[122] Kutschera A and Ranf S 2019 The multifaceted functions of lipopolysaccharide in plant–bacteria interactions *Biochimie* **159** 93–8

[123] Lu Y and Tsuda K 2021 Intimate association of PRR-and NLR-mediated signaling in plant immunity *Mol. Plant-Microbe Interact.* **34** 3–14

[124] Siddique S, Coomer A, Baum T and Williamson V M 2022 Recognition and response in plant–nematode interactions *Annu. Rev. Phytopathol.* **60** 143–62

[125] Ferrari S, Savatin D V, Sicilia F, Gramegna G, Cervone F and Lorenzo G D 2013 Oligogalacturonides: plant damage-associated molecular patterns and regulators of growth and development *Front. Plant Sci.* **4** 49

[126] Brutus A, Sicillia F, Macone A, Cervone F and De Lorenzo G 2010 A domain swap approach reveals a role of the plant wall-associated kinase 1 (WAK1) as a receptor for oligogalacturonides *Proc. Natl Acad. Sci. USA* **107** 9452–7

[127] Yamaguchi Y and Huffaker A 2011 Endogenous peptide elicitors in higher plants *Curr. Opin. Plant Biol.* **14** 351–7

[128] Bartels S and Boller T 2015 Quo vadis, Pep? Plant elicitor peptides at the crossroads of immunity, stress, and development *J. Exp. Bot.* **66** 5183–93

[129] Wang L, Albert M, Einig E, Fürst U, Krust D and Felix G 2016 The pattern-recognition receptor CORE of *Solanaceae* detects bacterial cold-shock protein *Nat. Plants* **2** 1–9

[130] Desaki , Kouzai Y, Ninomiya Y, Iwase Y, Shimizu R, Seko Y *et al* 2018 OsCERK 1 plays a crucial role in the lipopolysaccharide induced immune response of rice *New Phytol.* **217** 1042–9

[131] Teixeira R M, Ferreira M A, Raimundo G A, Loriato V A, Reis P A and Fontes E P 2019 Virus perception at the cell surface: revisiting the roles of receptor-like kinases as viral pattern recognition receptors *Mol. Plant Pathol.* **20** 1196–202

[132] Zhang H, Chen C, Li L, Tan X, Wei Z, Li Y *et al* 2021 A rice LRR receptor like protein associates with its adaptor kinase *Os*SOBIR1 to mediate plant immunity against viral infection *Plant Biotechnol. J.* **19** 2319–32

[133] Reymond P 2021 Receptor kinases in plant responses to herbivory *Curr. Opin. Biotechnol.* **70** 143–50

[134] Steinbrenner A D, Muñoz-Amatriaín M, Chaparro A F, Aguilar-Venegas J M, Lo S, Okuda S *et al* 2020 A receptor-like protein mediates plant immune responses to herbivore-associated molecular patterns *Proc. Natl Acad. Sci. USA* **117** 31510–8

[135] Mendy B, Wang'ombe M W, Radakovic Z S, Holbein J, Ilyas M, Chopra D *et al* 2017 *Arabidopsis* leucine-rich repeat receptor–like kinase NILR1 is required for induction of innate immunity to parasitic nematodes *PLoS Pathog.* **13** e1006284

[136] Huang L, Yuan Y, Lewis C, Kud J, Kuhl J C, Caplan A *et al* 2023 NILR1 perceives a nematode ascaroside triggering immune signaling and resistance *Curr. Biol.* **33** 1–6

[137] Manosalva P, Manohar M, Von Reuss S H, Chen S, Koch A, Kaplan F *et al* 2015 Conscrvcd ncmatodc signalling molcculcs clicit plant dcfcnscs and pathogcn rcsistancc *Nat. Commun.* **6** 7795

[138] Klessig D F, Manohar M, Baby S, Koch A, Danquah W B, Luna E *et al* 2019 Nematode ascaroside enhances resistance in a broad spectrum of plant–pathogen systems *J. Phytopathol.* **167** 265–72

[139] Fishman M R and Shirasu K 2021 How to resist parasitic plants: pre-and post-attachment strategies *Curr. Opin. Plant Biol.* **62** 102004

[140] Schellenberger R, Touchard M, Clément C, Baillieul F, Cordelier S, Crouzet J *et al* 2019 Apoplastic invasion patterns triggering plant immunity: plasma membrane sensing at the frontline *Mol. Plant Pathol.* **20** 1602–16

[141] Luderer R, Takken F L, de Wit P J and Joosten M H 2002 *Cladosporium fulvum* overcomes Cf-2-mediated resistance by producing truncated AVR2 elicitor proteins *Mol. Microbiol.* **45** 875–84

[142] Rooney H C, Van't Klooster J W, van der Hoorn R A, Joosten M H, Jones J D and de Wit P J 2005 *Cladosporium* Avr2 inhibits tomato Rcr3 protease required for Cf-2-dependent disease resistance *Science* **308** 1783–6

[143] Rivas S and Thomas C M 2005 Molecular interactions between tomato and the leaf mold pathogen *Cladosporium fulvum Annu. Rev. Phytopathol.* **43** 395

[144] Gong B Q, Xue J, Zhang N, Xu L, Yao X, Yang Q J *et al* 2017 Rice chitin receptor OsCEBiP is not a transmembrane protein but targets the plasma membrane via a GPI anchor *Mol. Plant* **10** 767–70

[145] Cao Y, Liang Y, Tanaka K, Nguyen C T, Jedrzejczak R P, Joachimiak A *et al* 2014 The kinase LYK5 is a major chitin receptor in and forms a chitin-induced complex with related kinase CERK1 *eLife* **3** e03766

[146] Hayafune M, Berisio R, Marchetti R, Silipo A, Kayama M, Desaki Y *et al* 2014 Chitin-induced activation of immune signaling by the rice receptor CEBiP relies on a unique sandwich-type dimerization *Proc. Natl Acad. Sci. USA* **111** 404–13

[147] Albert I, Böhm H, Albert M, Feiler C E, Imkampe J, Wallmeroth N *et al* 2015 An RLP23–SOBIR1–BAK1 complex mediates NLP-triggered immunity *Nat. Plants* **1** 1–9

[148] Cheung A Y, Qu L J, Russinova E, Zhao Y and Zipfel C 2020 Update on receptors and signaling *Plant Physiol.* **182** 1527–30

[149] Dardick C, Schwessinger B and Ronald P 2012 Non-arginine-aspartate (non-RD) kinases are associated with innate immune receptors that recognize conserved microbial signatures *Curr. Opin. Plant Biol.* **15** 358–66

[150] Domazakis E, Wouters D, Visser R G, Kamoun S, Joosten M H and Vleeshouwers V G 2018 The ELR-SOBIR1 complex functions as a two-component receptor-like kinase to mount defense against *Phytophthora infestans Mol Plant-Microbe Interact.* **31** 795–802

[151] Van Der Burgh A M, Postma J, Robatzek S and Joosten M H 2019 Kinase activity of SOBIR1 and BAK1 is required for immune signalling *Mol. Plant Pathol.* **20** 410–22

[152] Mott G A, Thakur S, Smakowska E, Wang P W, Belkhadir Y, Desveaux D *et al* 2016 Genomic screens identify a new phytobacterial microbe-associated molecular pattern and the cognate *Arabidopsis* receptor-like kinase that mediates its immune elicitation *Genome Biol.* **17** 98

[153] Li J, Wen J, Lease K A, Doke J T, Tax F E and Walker J C 2002 BAK1, an Arabidopsis LRR receptor-like protein kinase, interacts with BRI1 and modulates brassinosteroid signaling *Cell* **110** 213–22

[154] Fàbregas N, Li N, Boeren S, Nash T E, Goshe M B, Clouse S D *et al* 2013 The BRASSINOSTEROID INSENSITIVE1–LIKE3 signalosome complex regulates *Arabidopsis* root development *Plant Cell* **25** 3377–88

[155] Chinchilla D, Shan L, He P, de Vries S and Kemmerling B 2009 One for all: the receptor-associated kinase BAK1 *Trends Plant Sci.* **14** 535–41

[156] Liang X and Zhang J 2022 Regulation of plant responses to biotic and abiotic stress by receptor-like cytoplasmic kinases *Stress Biol.* **2** 1–12

[157] Rao S, Zhou Z, Miao P, Bi G, Hu M, Wu Y *et al* 2018 Roles of receptor-like cytoplasmic kinase VII members in pattern-triggered immune signaling *Plant Physiol.* **177** 1679–90

[158] Li P, Zhao L, Qi F, Htwe N M P S, Li Q, Zhang D *et al* 2021 The receptor-like cytoplasmic kinase RIPK regulates broad-spectrum ROS signaling in multiple layers of plant immune system *Mol. Plant* **14** 1652–67

[159] Jaiswal N, Liao C J, Mengesha B, Han H, Lee S, Sharon A *et al* 2022 Regulation of plant immunity and growth by tomato receptor like cytoplasmic kinase TRK1 *New Phytol.* **233** 458–78

[160] Ranf S 2017 Sensing of molecular patterns through cell surface immune receptors *Curr. Opin. Plant Biol.* **38** 68–77

[161] Chinchilla D, Bauer Z, Regenass M, Boller T and Felix G 2006 The *Arabidopsis* receptor kinase FLS2 binds flg22 and determines the specificity of flagellin perception *Plant Cell* **18** 465–76

[162] Liu Z, Wu Y, Yang F, Zhang Y, Chen S, Xie Q *et al* 2013 BIK1 interacts with PEPRs to mediate ethylene-induced immunity *Proc. Natl Acad. Sci. USA* **110** 6205–10
[163] Torres Ascurra Y C, Zhang L, Toghani A, Hua C, Rangegowda N J, Posbeyikian A *et al* 2023 Functional diversification of a wild potato immune receptor at its center of origin *Science* **381** 891–7
[164] Gust A A 2015 Peptidoglycan perception in plants *PLoS Pathog.* **11** e1005275
[165] Liu B, Li J F, Ao Y, Qu J, Li Z, Su J *et al* 2012 Lysin motif-containing proteins LYP4 and LYP6 play dual roles in peptidoglycan and chitin perception in rice innate immunity *Plant Cell* **24** 3406–19
[166] Choi J, Tanaka K, Cao Y, Qi Y, Qiu J, Liang Y *et al* 2014 Identification of a plant receptor for extracellular ATP *Science* **343** 290–4
[167] Wang C, Zhou M, Zhang X, Yao J, Zhang Y and Mou Z 2017 A lectin receptor kinase as a potential sensor for extracellular nicotinamide adenine dinucleotide in *Arabidopsis thaliana Elife* **6** e25474
[168] Kutschera A, Dawid C, Gisch N, Schmid C, Raasch L, Gerster T *et al* 2019 Bacterial medium-chain 3-hydroxy fatty acid metabolites trigger immunity in *Arabidopsis* plants *Science* **364** 178–81
[169] Ranf S, Gisch N, Schäffer M, Illig T, Westphal L, Knirel Y A *et al* 2015 A lectin S-domain receptor kinase mediates lipopolysaccharide sensing in *Arabidopsis thaliana Nat. Immunol.* **16** 426–33
[170] Yang C, Wang E and Liu J 2022 CERK1, more than a co-receptor in plant–microbe interactions *New Phytol.* **234** 1606–13
[171] Schellenberger R, Crouzet J, Nickzad A, Shu L J, Kutschera A, Gerster T *et al* 2021 Bacterial rhamnolipids and their 3-hydroxyalkanoate precursors activate *Arabidopsis* innate immunity through two independent mechanisms *Proc. Natl Acad. Sci. USA* **118** e2101366118
[172] Gust A A, Pruitt R and Nürnberger T 2017 Sensing danger: key to activating plant immunity *Trends Plant Sci.* **22** 779–91
[173] Nie J, Yin Z, Li Z, Wu Y and Huang L 2019 A small cysteine-rich protein from two kingdoms of microbes is recognized as a novel pathogen-associated molecular pattern *New Phytol.* **222** 995–1011
[174] Conrath U 2011 Molecular aspects of defence priming *Trends Plant Sci.* **16** 524–31
[175] Felix G, Duran J D, Volko S and Boller T 1999 Plants have a sensitive perception system for the most conserved domain of bacterial flagellin *Plant* J. **18** 265–76
[176] Cai R, Lewis J, Yan S, Liu H, Clarke C R, Campanile F *et al* 2011 The plant pathogen *Pseudomonas syringae* pv. *tomato* is genetically monomorphic and under strong selection to evade tomato immunity *PLoS Pathog.* **7** e1002130
[177] Harvey K L, Jarocki V M, Charles I G and Djordjevic S P 2019 The diverse functional roles of elongation factor Tu (EF-Tu) in microbial pathogenesis *Front. Microbiol.* **10** 2351
[178] Kunze G, Zipfel C, Robatzek S, Niehaus K, Boller T and Felix G 2004 The N terminus of bacterial elongation factor Tu elicits innate immunity in *Arabidopsis* plants *Plant Cell* **16** 3496–507
[179] Gust A A, Biswas R, Lenz H D, Rauhut T, Ranf S, Kemmerling B *et al* 2007 Bacteria-derived peptidoglycans constitute pathogen-associated molecular patterns triggering innate immunity in *Arabidopsis J. Biol. Chem.* **282** 32338–48

[180] Erbs G, Silipo A, Aslam S, De Castro C, Liparoti V, Flagiello A *et al* 2008 Peptidoglycan and muropeptides from pathogens *Agrobacterium* and *Xanthomonas* elicit plant innate immunity: structure and activity *Chem. Biol.* **15** 438–48

[181] Zhang G E, Meredith T C and Kahne D 2013 On the essentiality of lipopolysaccharide to Gram-negative bacteria *Curr. Opin. Microbiol.* **16** 779–85

[182] Bertani B and Ruiz N 2018 Function and biogenesis of lipopolysaccharides *Ecosal Plus* **8** 10–1128

[183] Ranf S 2016 Immune sensing of lipopolysaccharide in plants and animals: same but different *PLoS Pathog.* **12** e1005596

[184] Böhm H, Albert I, Fan L, Reinhard A and Nürnberger T 2014 Immune receptor complexes at the plant cell surface *Curr. Opin. Plant Biol.* **20** 47–54

[185] Oome S, Raaymakers T M, Cabral A, Samwel S, Böhm H, Albert I *et al* 2014 Nep1-like proteins from three kingdoms of life act as a microbe-associated molecular pattern in *Arabidopsis Proc. Natl Acad. Sci. USA* **111** 16955–60

[186] Felix G and Boller T 2003 Molecular sensing of bacteria in plants: the highly conserved RNA-binding motif RNP-1 of bacterial cold shock proteins is recognized as an elicitor signal in tobacco *J. Biol. Chem.* **278** 6201–8

[187] Meng F, Altier C and Martin G B 2013 *Salmonella* colonization activates the plant immune system and benefits from association with plant pathogenic bacteria *Environ. Microbiol.* **15** 2418–30

[188] Brunner F, Rosahl S, Lee J, Rudd J J, Geiler C, Kauppinen S *et al* 2002 Pep13, a plant defense inducing pathogen associated pattern from *Phytophthora transglutaminases EMBO J.* **21** 6681–8

[189] Lin X, Torres Ascurra Y C, Fillianti H, Dethier L, De Rond L, Domazakis E *et al* 2023 Recognition of Pep-13/25 MAMPs of *Phytophthora* localizes to an RLK locus in *Solanum microdontum Front. Plant Sci.* **13** 1037030

[190] Kamoun S, Young M, Glascock C B and Tyler B M 1993 Extracellular protein elicitors from *Phytophthora*: host-specificity and induction of resistance to bacterial and fungal phytopathogens *Mol. Plant Microbe Interact.* **6** 15–5

[191] DeFalco T A 2023 Convergent evolution of elicitin perception by divergent pattern-recognition receptors *Plant Cell* **35** 1165–6

[192] Du J, Verzaux E, Chaparro-Garcia A, Bijsterbosch G, Keizer L C, Zhou J I *et al* 2015 Elicitin recognition confers enhanced resistance to *Phytophthora infestans* in potato *Nat. Plants* **1** 1–5

[193] Chen Z, Liu F, Zeng M, Wang L, Liu H, Sun Y *et al* 2023 Convergent evolution of immune receptors underpins distinct elicitin recognition in closely related Solanaceous plants *Plant Cell* **35** 1186–201

[194] Gaulin E, Drame N, Lafitte C, Torto-Alalibo T, Martinez Y, Ameline-Torregrosa C *et al* 2006 Cellulose binding domains of a *Phytophthora* cell wall protein are novel pathogen-associated molecular patterns *Plant Cell* **18** 1766–77

[195] Mateos F V, Rickauer M and Esquerré-Tugayé M T 1997 Cloning and characterization of a cDNA encoding an elicitor of *Phytophthora parasitica* var. *nicotianae* that shows cellulose-binding and lectin-like activities *Mol. Plant Microbe Interact.* **10** 1045–53

[196] Khatib M, Lafitte C, Esquerré-Tugayé M T, Bottin A and Rickauer M 2004 The CBEL elicitor of *Phytophthora parasitica* var. *nicotianae* activates defence in *Arabidopsis thaliana* via three different signalling pathways *New Phytol.* **162** 501–10

[197] Durkin C A, Mock T and Armbrust E V 2009 Chitin in diatoms and its association with the cell wall *Eukaryot. Cell* **8** 1038–50
[198] Kuchitsu K, Kikuyama M and Shibuya N 1993 N-acetylchitooligosaccharides, biotic elicitor for phytoalexin production, induce transient membrane depolarization in suspension-cultured rice cells *Protoplasma* **174** 79–81
[199] Gubaeva E, Gubaev A, Melcher R L, Cord-Landwehr S, Singh R, El Gueddari N E *et al* 2018 Slipped sandwich'model for chitin and chitosan perception in *Arabidopsis Mol. Plant Microbe Interact.* **31** 1145–53
[200] Lombard V, Golaconda Ramulu H, Drula E, Coutinho P M and Henrissat B 2014 The carbohydrate-active enzymes database (CAZy) in 2013 *Nucleic Acids Res.* **42** D490–5
[201] Valls M, Genin S and Boucher C 2006 Integrated regulation of the type III secretion system and other virulence determinants in *Ralstonia solanacearum PLoS Pathog.* **2** e82
[202] Guo X, Liu N, Zhang Y and Chen J 2022 Pathogen-associated molecular pattern active sites of GH45 endoglucanohydrolase from *Rhizoctonia solani Phytopathol.* **112** 355–63
[203] Nühse T S 2012 Cell wall integrity signaling and innate immunity in plants *Front. Plant Sci.* **3** 280
[204] Ma Z, Song T, Zhu L, Ye W, Wang Y, Shao Y *et al* 2015 A *Phytophthora sojae* glycoside hydrolase 12 protein is a major virulence factor during soybean infection and is recognized as a PAMP *Plant Cell* **27** 2057–72
[205] Zhang L, Yan J, Fu Z, Shi W, Ninkuu V, Li G *et al* 2021 FoEG1, a secreted glycoside hydrolase family 12 protein from *Fusarium oxysporum*, triggers cell death and modulates plant immunity *Mol. Plant Pathol.* **22** 522–38
[206] Enkerli J, Felix G and Boller T 1999 The enzymatic activity of fungal xylanase is not necessary for its elicitor activity *Plant Physiol.* **121** 391–8
[207] Fuchs Y, Saxena A, Gamble H R and Anderson J D 1989 Ethylene biosynthesis-inducing protein from cellulysin is an endoxylanase *Plant Physiol.* **89** 138–43
[208] Yang Y, Yang X, Dong Y and Qiu D 2018 Thc *Botrytis cinerea* xylanase BcXyl1 modulates plant immunity *Front. Microbiol.* **9** 417320
[209] Fesel P H and Zuccaro A 2016 β-Glucan: crucial component of the fungal cell wall and elusive MAMP in plants *Fungal Genet. Biol.* **90** 53–60
[210] Melida H, Sopeña-Torres S, Bacete L, Garrido-Arandia M, Jordá L, Lopez G *et al* 2018 Non-branched β-1,3-glucan oligosaccharides trigger immune responses in *Arabidopsis Plant* J. **93** 34–49
[211] de Pinto M C, Lavermicocca P, Evidente A, Corsaro M M, Lazzaroni S and De Gara L 2003 Exopolysaccharides produced by plant pathogenic bacteria affect ascorbate metabolism in *Nicotiana tabacum Plant Cell Physiol.* **44** 803–10
[212] ElOirdi M, El Rahman T A, Rigano L, El Hadrami A, Rodriguez M C, Daayf F *et al* 2011 *Botrytis cinerea* manipulates the antagonistic effects between immune pathways to promote disease development in tomato *Plant Cell* **23** 2405–21
[213] Bostock R M, Kuc J A and Laine R A 1981 Eicosapentaenoic and arachidonic acids from *Phytophthora infestans* elicit fungitoxic sesquiterpenes in the potato *Science* **212** 67–9
[214] Klemptner R L, Sherwood J S, Tugizimana F, Dubery I A and Piater L A 2014 Ergosterol, an orphan fungal microbe associated molecular pattern (MAMP) *Mol. Plant Pathol.* **15** 747–61
[215] Kaloshian I and Teixeira M 2019 Advances in plant–nematode interactions with emphasis on the notorious nematode genus *Meloidogyne Phytopathology* **109** 1988–96
[216] Jones A C, Felton G W and Tumlinson J H 2022 The dual function of elicitors and effectors from insects: reviewing the 'arms race' against plant defenses *Plant Mol. Biol.* **109** 427–45

[217] Snoeck S, Guayazán-Palacios N and Steinbrenner A D 2022 Molecular tug-of-war: plant immune recognition of herbivory *Plant Cell* **34** 1497–513
[218] Basu S, Varsani S and Louis J 2018 Altering plant defenses: herbivore-associated molecular patterns and effector arsenal of chewing herbivores *Mol. Plant-Microbe Interact.* **31** 13–21
[219] Gao H, Zou J, Lin X, Zhang H, Yu N and Liu Z 2022 *Nilaparvata lugens* salivary protein NlG14 triggers defense response in plants *J. Exp. Bot.* **73** 7477–87
[220] Gao H, Lin X, Yuan X, Zou J, Zhang H, Zhang Y *et al* 2023 The salivary chaperone protein NlDNAJB9 of *Nilaparvata lugens* activates plant immune responses *J. Exp. Bot.* **74** 6874–88
[221] Pieterse C M, Van der Does D, Zamioudis C, Leon-Reyes A and Van Wees S C 2012 Hormonal modulation of plant immunity *Annu. Rev. Cell Dev. Biol.* **28** 489–521
[222] Lazebnik J, Frago E, Dicke M and Van Loon J J 2014 Phytohormone mediation of interactions between herbivores and plant pathogens *J. Chem. Ecol.* **40** 730–41
[223] Tanaka K and Heil M 2021 Damage-associated molecular patterns (DAMPs) in plant innate immunity: applying the danger model and evolutionary perspectives *Annu. Rev. Phytopathol.* **59** 53–75
[224] Hou S, Liu D and He P 2021 Phytocytokines function as immunological modulators of plant immunity *Stress Biol.* **1** 8
[225] Hou S, Liu Z, Shen H and Wu D 2019 Damage-associated molecular pattern-triggered immunity in plants *Front. Plant Sci.* **10** 646
[226] McGurl B, Pearce G, Orozco-Cardenas M and Ryan C A 1992 Structure, expression, and antisense inhibition of the systemin precursor gene *Science* **255** 1570–3
[227] Côté F and Hahn M G 1994 Oligosaccharins: structures and signal transduction *Signals and Signal Transduction Pathways in Plants* (Springer) pp 143–75
[228] Pearce G, Moura D S, Stratmann J and Ryan C A 2001 Production of multiple plant hormones from a single polyprotein precursor *Nature* **411** 817–20
[229] Bartels S, Lori M, Mbengue M, Van Verk M, Klauser D, Hander T *et al* 2013 The family of Peps and their precursors in *Arabidopsis*: differential expression and localization but similar induction of pattern-triggered immune responses *J. Exp. Bot.* **64** 5309–21
[230] Hou S, Wang X, Chen D, Yang X, Wang M, Turrà D *et al* 2014 The secreted peptide PIP1 amplifies immunity through receptor-like kinase 7 *PLoS Pathog.* **10** e1004331
[231] Duran-Flores D and Heil M 2018 Extracellular self-DNA as a damage-associated molecular pattern (DAMP) that triggers self-specific immunity induction in plants *Brain Behav. Immun.* **72** 78–88
[232] De Lorenzo G and Cervone F 2022 Plant immunity by damage-associated molecular patterns (DAMPs) *Essays Biochem.* **66** 459–69
[233] O'Neill M A, Albersheim P and Darvill A 1990 The pectic polysaccharides of primary cell walls *Methods in Plant Biochemistry* **2** (New York: Academic) pp 415–41
[234] Ferrusquía-Jiménez N I, Chandrakasan G, Torres-Pacheco I, Rico-Garcia E, Feregrino-Perez A A and Guevara-González R G 2021 Extracellular DNA: a relevant plant damage-associated molecular pattern (DAMP) for crop protection against pests—a review *J. Plant Growth Regul.* **40** 451–63
[235] Zhou X, Gao H, Zhang X, Khashi u Rahman M, Mazzoleni , Du S, M *et al* 2023 Plant extracellular self-DNA inhibits growth and induces immunity via the jasmonate signaling pathway *Plant Physiol.* **192** 2475–91

[236] Rassizadeh L, Cervero R, Flors V and Gamir J 2021 Extracellular DNA as an elicitor of broad-spectrum resistance in *Arabidopsis thaliana Plant Sci.* **312** 111036

[237] Huffaker A and Ryan C A 2007 Endogenous peptide defense signals in *Arabidopsis* differentially amplify signaling for the innate immune response *Proc. Natl. Acad. Sci. USA* **104** 10732–6

[238] Klauser D, Desurmont G A, Glauser G, Vallat A, Flury P, Boller T *et al* 2015 The *Arabidopsis* Pep-PEPR system is induced by herbivore feeding and contributes to JA-mediated plant defence against herbivory *J. Exp. Bot.* **66** 5327–36

[239] Huffaker A, Pearce G and Ryan C A 2006 An endogenous peptide signal in Arabidopsis activates components of the innate immune response *Proc. Natl. Acad. Sci. USA* **103** 10098–103

[240] Huffaker A, Dafoe N J and Schmelz E A 2011 *Zm*Pep1, an ortholog of *Arabidopsis* elicitor peptide 1, regulates maize innate immunity and enhances disease resistance *Plant Physiol.* **155** 1325–38

[241] Huffaker A, Pearce G, Veyrat N, Erb M, Turlings T C, Sartor R *et al* 2013 Plant elicitor peptides are conserved signals regulating direct and indirect antiherbivore defense *Proc. Natl. Acad. Sci. USA* **110** 5707–12

[242] Zhang H, Zhang H and Lin J 2020 Systemin mediated long distance systemic defense responses *New Phytol.* **226** 1573–82

[243] Constabel C P, Yip L and Ryan C A 1998 Prosystemin from potato, black nightshade, and bell pepper: primary structure and biological activity of predicted systemin polypeptides *Plant Mol. Biol.* **36** 55–62

[244] He Y H, Zhang Z R, Xu Y P, Chen S Y and Cai X Z 2022 Genome-wide identification of rapid alkalinization factor family in *Brassica napus* and functional analysis of BnRALF10 in immunity to *Sclerotinia sclerotiorum Front. Plant Sci.* **13** 877404

[245] Stegmann M, Monaghan J, Smakowska-Luzan E, Rovenich H, Lehner A, Holton N *et al* 2017 The receptor kinase FER is a RALF-regulated scaffold controlling plant immune signaling *Science* **355** 287–9

[246] Vie A K, Najafi J, Liu B, Winge P, Butenko M A, Hornslien K S *et al* 2015 The IDA/IDA-LIKE and PIP/PIP-LIKE gene families in *Arabidopsis*: phylogenetic relationship, expression patterns, and transcriptional effect of the PIPL3 peptide *J. Exp. Bot.* **66** 5351–65

[247] Combest M M, Moroz N, Tanaka K, Rogan C J, Anderson J C, Thura L *et al* 2021 *St*PIP1, a PAMP-induced peptide in potato, elicits plant defenses and is associated with disease symptom severity in a compatible interaction with Potato virus *Y J. Exp. Bot.* **72** 4472–88

[248] Duong H N, Cho S H, Wang L, Pham A Q, Davies J M and Stacey G 2022 Cyclic nucleotide gated ion channel 6 is involved in extracellular ATP signaling and plant immunity *Plant* J. **109** 1386–96

[249] Chen Y L, Lee C Y, Cheng K T, Chang W H, Huang R N, Nam H G *et al* 2014 Quantitative peptidomics study reveals that a wound-induced peptide from PR-1 regulates immune signaling in tomato *Plant Cell* **26** 4135–48

[250] Wang X, Hou S, Wu Q, Lin M, Acharya B R, Wu D *et al* 2017 IDL6-HAE/HSL2 impacts pectin degradation and resistance to *Pseudomonas syringae*pv*tomato* DC 3000 in *Arabidopsis* leaves *Plant* J. **89** 250–63

[251] Schmelz E A, LeClere S, Carroll M J, Alborn H T and Teal P E 2007 Cowpea chloroplastic ATP synthase is the source of multiple plant defense elicitors during insect herbivory *Plant Physiol.* **144** 793–805

[252] Zhang H, Hu Z, Lei C, Zheng C, Wang J, Shao S *et al* 2018 A plant phytosulfokine peptide initiates auxin-dependent immunity through cytosolic Ca^{2+} signaling in tomato *Plant Cell* **30** 652–67
[253] Bethke G, Grundman R E, Sreekanta S, Truman W, Katagiri F and Glazebrook J 2014 *Arabidopsis* PECTIN METHYLESTERASEs contribute to immunity against *Pseudomonas syringae Plant Physiol.* **164** 1093–107
[254] Rhodes J, Yang H, Moussu S, Boutrot F, Santiago J and Zipfel C 2021 Perception of a divergent family of phytocytokines by the *Arabidopsis* receptor kinase MIK2 *Nat. Commun.* **12** 705
[255] Pastor V, Cervero R and Gamir J 2022 The simultaneous perception of self-and non-self-danger signals potentiates plant innate immunity responses *Planta* **256** 10
[256] Yu X, Feng B, He P and Shan L 2017 From chaos to harmony: responses and signaling upon microbial pattern recognition *Annu. Rev. Phytopathol.* **55** 109–37
[257] Bi G, Zhou Z, Wang W, Li L, Rao S, Wu Y *et al* 2018 Receptor-like cytoplasmic kinases directly link diverse pattern recognition receptors to the activation of mitogen-activated protein kinase cascades in *Arabidopsis Plant Cell* **30** 1543–61
[258] Kadota Y, Sklenar J, Derbyshire P, Stransfeld L, Asai S, Ntoukakis V *et al* 2014 Direct regulation of the NADPH oxidase RBOHD by the PRR-associated kinase BIK1 during plant immunity *Mol. Cell* **54** 43–55
[259] Lee D, Lal N K, Lin Z J D, Ma S, Liu J, Castro B *et al* 2020 Regulation of reactive oxygen species during plant immunity through phosphorylation and ubiquitination of RBOHD *Nat. Commun.* **11** 1838
[260] Ranf S, Eschen-Lippold L, Fröhlich K, Westphal L, Scheel D and Lee J 2014 Microbe-associated molecular pattern-induced calcium signaling requires the receptor-like cytoplasmic kinases, PBL1 and BIK1 *BMC Plant Biol.* **14** 1–15
[261] Bredow M and Monaghan J 2019 Regulation of plant immune signaling by calcium-dependent protein kinases *Mol. Plant-Microbe Interact.* **32** 6–19
[262] Cui F, Sun W and Kong X 2018 RLCKs bridge plant immune receptors and MAPK cascades *Trends Plant Sci.* **23** 1039–41
[263] Wu S, Shan L and He P 2014 Microbial signature-triggered plant defense responses and early signaling mechanisms *Plant Sci.* **228** 118–26
[264] Yamada K, Yamaguchi K, Shirakawa T, Nakagami H, Mine A, Ishikawa K *et al* 2016 The *Arabidopsis* CERK 1-associated kinase PBL 27 connects chitin perception to MAPK activation *EMBO J.* **35** 2468–83
[265] Wang C, Wang G, Zhang C, Zhu P, Dai H, Yu N *et al* 2017 OsCERK1-mediated chitin perception and immune signaling requires receptor-like cytoplasmic kinase 185 to activate an MAPK cascade in rice *Mol. Plant* **10** 619–33
[266] Liu J, Ding P, Sun T, Nitta Y, Dong O, Huang X *et al* 2013 Heterotrimeric G proteins serve as a converging point in plant defense signaling activated by multiple receptor-like kinases *Plant Physiol.* **161** 2146–58
[267] Meng X and Zhang S 2013 MAPK cascades in plant disease resistance signaling *Annu. Rev. Phytopathol.* **51** 245–66
[268] Petersen M, Brodersen P, Naested H, Andreasson E, Lindhart U, Johansen B *et al* 2000 *Arabidopsis* MAP kinase 4 negatively regulates systemic acquired resistance *Cell* **103** 1111–20
[269] Andreasson E, Jenkins T, Brodersen P, Thorgrimsen S, Petersen N H, Zhu S *et al* 2005 The MAP kinase substrate MKS1 is a regulator of plant defense responses *EMBO J.* **24** 2579–89

[270] Qiu J L, Fiil B K, Petersen K, Nielsen H B, Botanga C J, Thorgrimsen S *et al* 2008a *Arabidopsis* MAP kinase 4 regulates gene expression through transcription factor release in the nucleus *EMBO J.* **27** 2214–21

[271] Li B, Meng X, Shan L and He P 2016 Transcriptional regulation of pattern-triggered immunity in plants *Cell Host Microbe* **19** 641–50

[272] Jing Y and Lin R 2015 The VQ motif-containing protein family of plant-specific transcriptional regulators *Plant Physiol.* **169** 371–8

[273] Pecher P, Eschen-Lippold L, Herklotz S, Kuhle K, Naumann K, Bethke G *et al* 2014 The *Arabidopsis thaliana* mitogen activated protein kinases MPK3 and MPK6 target a subclass of 'VQ-motif' containing proteins to regulate immune responses *New Phytol.* **203** 592–606

[274] Pruitt R N, Locci F, Wanke F, Zhang L, Saile S C, Joe A *et al* 2021 The EDS1–PAD4–ADR1 node mediates *Arabidopsis* pattern-triggered immunity *Nature* **598** 495–9

[275] Aerts N, Chhillar H, Ding P and Van Wees S C 2022 Transcriptional regulation of plant innate immunity *Essays Biochem.* **66** 607–20

[276] Wang D, Wei L, Liu T, Ma J, Huang K, Guo H *et al* 2023 Suppression of ETI by PTI priming to balance plant growth and defense through an MPK3/MPK6-WRKYs-PP2Cs module *Mol. Plant* **16** 903–18

[277] Kishi-Kaboshi M, Seo S, Takahashi A and Hirochika H 2018 The MAMP-responsive MYB transcription factors MYB30, MYB55 and MYB110 activate the HCAA synthesis pathway and enhance immunity in rice *Plant Cell Physiol.* **59** 903–15

[278] Huang W, Miao M, Kud J, Niu X, Ouyang B, Zhang J *et al* 2013 SlNAC 1, a stress related transcription factor, is fine tuned on both the transcriptional and the post translational level *New Phytol.* **197** 1214–24

[279] Kim S H, Oikawa T, Kyozuka J, Wong H L, Umemura K, Kishi-Kaboshi M *et al* 2012 The bHLH Rac Immunity1 (RAI1) is activated by OsRac1 via OsMAPK3 and OsMAPK6 in rice immunity *Plant Cell Physiol.* **53** 740–54

[280] Spears B J, Howton T C, Gao F, Garner C M, Mukhtar M S and Gassmann W 2019 Direct regulation of the EFR-dependent immune response by Arabidopsis TCP transcription factors *Mol. Plant-Microbe Interact.* **32** 540–9

[281] Lu W, Deng F, Jia J, Chen X, Li J, Wen Q *et al* 2020 The *Arabidopsis thaliana* gene AtERF019 negatively regulates plant resistance to *Phytophthora parasitica* by suppressing PAMP triggered immunity *Mol. Plant Pathol.* **21** 1179–93

[282] Rahman H, Yang J, Xu Y P, Munyampundu J P and Cai X Z 2016 Phylogeny of plant CAMTAs and role of AtCAMTAs in nonhost resistance to *Xanthomonas oryzae* pv. *oryzae* *Front. Plant Sci.* **7** 166981

[283] Tsuda K and Somssich I E 2015 Transcriptional networks in plant immunity *New Phytol.* **206** 932–47

[284] Leba L J, Cheval C, Ortiz Martín I, Ranty B, Beuzón C R, Galaud J P *et al* 2012 CML9, an *Arabidopsis* calmodulin-like protein, contributes to plant innate immunity through a flagellin dependent signalling pathway *Plant* J. **71** 976–89

[285] Wang C and Luan S 2024 Calcium homeostasis and signaling in plant immunity *Curr. Opin. Plant Biol.* **77** 102485

[286] Zhang Y, Xu S, Ding P, Wang D, Cheng Y T, He J *et al* 2010 Control of salicylic acid synthesis and systemic acquired resistance by two members of a plant-specific family of transcription factors *Proc. Natl. Acad. Sci. USA* **107** 18220–5

[287] Zhang Y and Li X 2019 Salicylic acid: biosynthesis, perception, and contributions to plant immunity *Curr. Opin. Plant Biol.* **50** 29–36

[288] Seo J S, Sun H X, Park B S, Huang C H, Yeh S D, Jung C *et al* 2017 ELF18-induced long-noncoding RNA associates with mediator to enhance expression of innate immune response genes in *Arabidopsis Plant Cell* **29** 1024–38

[289] Ding Y, Sun T, Ao K, Peng Y, Zhang Y, Li X *et al* 2018 Opposite roles of salicylic acid receptors NPR1 and NPR3/NPR4 in transcriptional regulation of plant immunity *Cell* **173** 1454–67

[290] Saini R and Nandi A K 2022 TOPLESS in the regulation of plant immunity *Plant Mol. Biol.* **109** 1–12

[291] Chen H, Li M, Qi G, Zhao M, Liu L, Zhang J *et al* 2021 Two interacting transcriptional coactivators cooperatively control plant immune responses *Sci. Adv.* **7** eabl7173

[292] Campos M L, Kang J H and Howe G A 2014 Jasmonate-triggered plant immunity *J. Chem. Ecol.* **40** 657–75

[293] Malik S and Roeder R G 2010 The metazoan mediator co-activator complex as an integrative hub for transcriptional regulation *Nat. Rev. Genet.* **11** 761–72

[294] Samanta S and Thakur J K 2015 Importance of mediator complex in the regulation and integration of diverse signaling pathways in plants *Front. Plant Sci.* **6** 757

[295] Ding B, Bellizzi M D R, Ning Y, Meyers B C and Wang G L 2012 HDT701, a histone H4 deacetylase, negatively regulates plant innate immunity by modulating histone H4 acetylation of defense-related genes in rice *Plant Cell* **24** 3783–94

[296] Halter T, Wang J, Amesefe D, Lastrucci E, Charvin M, Singla Rastogi M *et al* 2021 The *Arabidopsis* active demethylase ROS1 cis-regulates defence genes by erasing DNA methylation at promoter-regulatory regions *eLife* **10** e62994

[297] Huang C Y and Jin H 2022 Coordinated epigenetic regulation in plants: a potent managerial tool to conquer biotic stress *Front. Plant Sci.* **12** 795274

[298] Lee S, Fu F, Xu S, Lee S Y, Yun D J and Mengiste T 2016 Global regulation of plant immunity by histone lysine methyl transferases *Plant Cell* **28** 1640–61

[299] Chen J, Clinton M, Qi G, Wang D, Liu F and Fu Z Q 2020 Reprogramming and remodeling: transcriptional and epigenetic regulation of salicylic acid-mediated plant defense *J. Exp. Bot.* **71** 5256–68

[300] Xie S S and Duan C G 2023 Epigenetic regulation of plant immunity: from chromatin codes to plant disease resistance *aBIOTECH* **4** 124–39

[301] Ferrari S, Galletti R, Denoux C, De Lorenzo G, Ausubel F M and Dewdney J 2007 Resistance to *Botrytis cinerea* induced in *Arabidopsis* by elicitors is independent of salicylic acid, ethylene, or jasmonate signaling but requires Phytoalexin Deficient3 *Plant Physiol.* **144** 367–79

[302] Clay N K, Adio A M, Denoux C, Jander G and Ausubel F M 2009 Glucosinolate metabolites required for an *Arabidopsis* innate immune response *Science* **323** 95–101

[303] Ahmad S, Veyrat N, Gordon-Weeks R, Zhang Y, Martin J, Smart L *et al* 2011 Benzoxazinoid metabolites regulate innate immunity against aphids and fungi in maize *Plant Physiol.* **157** 317–27

[304] Yamaguchi Y, Barona G, Ryan C A and Pearce G 2011 GmPep914, an eight-amino acid peptide isolated from soybean leaves, activates defense-related genes *Plant Physiol.* **156** 932–42

[305] Zhou J, Mu Q, Wang X, Zhang J, Yu H, Huang T *et al* 2022 Multilayered synergistic regulation of phytoalexin biosynthesis by ethylene, jasmonate, and MAPK signaling pathways in *Arabidopsis Plant Cell* **34** 3066–87

[306] Ekanayake G, Leslie M E, Smith J M and Heese A 2023 *Arabidopsis* dynamin-related protein AtDRP2A contributes to late flg22-signaling and effective immunity against *Pseudomonas syringae* bacteria *Mol. Plant. Microbe Interact.* **36** 201–7

[307] Kliebenstein D J, Rowe H C and Denby K J 2005 Secondary metabolites influence *Arabidopsis/Botrytis* interactions: variation in host production and pathogen sensitivity *Plant J.* **44** 25–36

[308] Sanchez-Vallet A, Ramos B, Bednarek P, López G, Piślewska Bednarek M, Schulze Lefert P *et al* 2010 Tryptophan derived secondary metabolites in *Arabidopsis thaliana* confer non-host resistance to necrotrophic *Plectosphaerella cucumerina* fungi *Plant* J. **63** 115–27

[309] Schlaeppi K, Abou-Mansour E, Buchala A and Mauch F 2010 Disease resistance of *Arabidopsis* to *Phytophthora brassicae* is established by the sequential action of indole glucosinolates and camalexin *Plant* J. **62** 840–51

[310] Wan W L, Zhang L, Pruitt R, Zaidem M, Brugman R, Ma X *et al* 2019 Comparing Arabidopsis receptor kinase and receptor protein mediated immune signaling reveals BIK1 dependent differences *New Phytol.* **221** 2080–95

[311] Luna E, Pastor V, Robert J, Flors V, Mauch-Mani B and Ton J 2011 Callose deposition: a multifaceted plant defense response *Mol. Plant-Microbe Interact.* **24** 183–93

[312] Holbein J, Grundler F M and Siddique S 2016 Plant basal resistance to nematodes: an update *J. Exp. Bot.* **67** 2049–61

[313] Ning S, Zhang L, Ma J, Chen L, Zeng G, Yang C *et al* 2020 Modular and scalable synthesis of nematode pheromone ascarosides: implications in eliciting plant defense response *Org. Biomol. Chem.* **18** 4956–61

[314] Zeng J, Ye W, Hu W, Jin X, Kuai P, Xiao W *et al* 2023 The N-terminal subunit of vitellogenin in planthopper eggs and saliva acts as a reliable elicitor that induces defenses in rice *New Phytol.* **238** 1230–44

[315] Huang L F, Lin K H, He S L, Chen J L, Jiang J Z, Chen B H *et al* 2016 Multiple patterns of regulation and overexpression of a ribonuclease-like pathogenesis-related protein gene, *OsPR10a*, conferring disease resistance in rice and *Arabidopsis PLoS One* **11** e0156414

[316] McCormick A C, Unsicker S B and Gershenzon J 2012 The specificity of herbivore-induced plant volatiles in attracting herbivore enemies *Trends Plant Sci.* **17** 303–10

[317] Turlings T C, Tumlinson J H and Lewis W J 1990 Exploitation of herbivore-induced plant odors by host-seeking parasitic wasps *Science* **250** 1251–3

[318] Moraes C M, Lewis W J, Pare P W, Alborn H T and Tumlinson J H 1998 Herbivore-infested plants selectively attract parasitoids *Nature* **393** 570–3

[319] Kessler A and T. Baldwin I 2004 Herbivore-induced plant vaccination. Part I. The orchestration of plant defenses in nature and their fitness consequences in the wild tobacco *Nicotiana attenuata Plant* J. **38** 639–49

[320] Xiu C, Dai W, Pan H, Zhang W, Luo S, Wyckhuys K A *et al* 2019 Herbivore-induced plant volatiles enhance field-level parasitism of the mirid bug *Apolygus lucorum Biol. Control* **135** 41–7

[321] Pan Y, Wang Z, Zhao S W, Wang X, Li Y S, Liu J N *et al* 2022 The herbivore induced plant volatile tetradecane enhances plant resistance to *Holotrichia parallela* larvae in maize roots *Pest Manag. Sci.* **78** 550–60

[322] Takabayashi J 2022 Herbivory-induced plant volatiles mediate multitrophic relationships in ecosystems *Plant Cell Physiol.* **63** 1344–55

[323] Peng H C and Kaloshian I 2014 The tomato leucine-rich repeat receptor-like kinases SlSERK3A and SlSERK3B have overlapping functions in bacterial and nematode innate immunity *PLoS One* **9** e93302

[324] Teixeira M A, Wei L and Kaloshian I 2016 Root knot nematodes induce pattern triggered immunity in *Arabidopsis thaliana* roots *New Phytol.* **211** 276–87

[325] Wu J, Hettenhausen C, Meldau S and Baldwin I T 2007 Herbivory rapidly activates MAPK signaling in attacked and unattacked leaf regions but not between leaves of *Nicotiana attenuata Plant Cell* **19** 1096–122

[326] Meldau S, Wu J and Baldwin I T 2009 Silencing two herbivory activated MAP kinases, SIPK and WIPK, does not increase *Nicotiana attenuata*'s susceptibility to herbivores in the glasshouse and in nature *New Phytol.* **181** 161–73

[327] Dong Y, Huang X, Yang Y, Li J, Zhang M, Shen H *et al* 2022 Characterization of salivary secreted proteins that induce cell death from *Riptortuspedestris* (Fabricius) and their roles in insect–plant interactions *Front. Plant Sci.* **13** 912603

[328] de Moura S M, Babilonia K, de Macedo L L P, Grossi-de-Sá M F, Shan L, He P *et al* 2022 The oral secretion from cotton boll weevil (*Anthonomus grandis*) induces defense responses in cotton (*Gossypium* spp.) and *Arabidopsis thaliana. Curr Plant Biol.* **31** 100250

[329] Xu J, Wang X, Zu H, Zeng X, Baldwin I T, Lou Y *et al* 2021 Molecular dissection of rice phytohormone signaling involved in resistance to a piercing-sucking herbivore *New Phytol.* **230** 1639–52

[330] Erb M, Meldau S and Howe G A 2012 Role of phytohormones in insect-specific plant reactions *Trends Plant Sci.* **17** 250–9

[331] Yang D H, Hettenhausen C, Baldwin I T and Wu J 2011 BAK1 regulates the accumulation of jasmonic acid and the levels of trypsin proteinase inhibitors in *Nicotiana attenuata*'s responses to herbivory *J. Exp. Bot.* **62** 641–52

[332] Brodersen P, Petersen M, Bjørn Nielsen H, Zhu S, Newman M A, Shokat K M *et al* 2006 *Arabidopsis* MAP kinase 4 regulates salicylic acid and jasmonic acid/ethylene dependent responses via EDS1 and PAD4 *Plant* J. **47** 532–46

[333] Hettenhausen C, Baldwin I T and Wu J 2013 *Nicotiana attenuata* MPK 4 suppresses a novel jasmonic acid (JA) signaling-independent defense pathway against the specialist insect *Manduca sexta*, but is not required for the resistance to the generalist *Spodoptera littoralis New Phytol.* **199** 787–99

[334] Schmelz E A 2015 Impacts of insect oral secretions on defoliation-induced plant defense *Curr. Opin. Insect. Sci.* **9** 7–15

[335] Erb M and Reymond P 2019 Molecular interactions between plants and insect herbivores *Annu. Rev. Plant Biol.* **70** 527–57

[336] Shen W, Zhang X, Liu J, Tao K, Li C, Xiao S *et al* 2022 Plant elicitor peptide signalling confers rice resistance to piercing-sucking insect herbivores and pathogens *Plant Biotechnol. J.* **20** 991–1005

[337] Huang J, Zhang N, Shan J, Peng Y, Guo J, Zhou C *et al* 2020 Salivary protein 1 of brown planthopper is required for survival and induces immunity response in plants *Front. Plant Sci.* **11** 571280

[338] Shah S J, Anjam M S, Mendy B, Anwer M A, Habash S S, Lozano-Torres J L *et al* 2017 Damage-associated responses of the host contribute to defence against cyst nematodes but not root-knot nematodes *J. Exp. Bot.* **68** 5949–60

[339] Cheng Q, Xiao H and Xiong Q 2020 Conserved exitrons of FLAGELLIN-SENSING 2 (FLS2) across dicot plants and their functions *Plant Sci.* **296** 110507

[340] Roberts R, Liu A E, Wan L, Geiger A M, Hind S R, Rosli H G *et al* 2020 Molecular characterization of differences between the tomato immune receptors flagellin sensing 3 and flagellin sensing 2 *Plant Physiol.* **183** 1825–37

[341] Chinchilla D, Zipfel C, Robatzek S, Kemmerling B, Nürnberger T, Jones J D *et al* 2007 A flagellin-induced complex of the receptor FLS2 and BAK1 initiates plant defence *Nature* **448** 497–500

[342] Sun Y, Li L, Macho A P, Han Z, Hu Z, Zipfel C *et al* 2013 Structural basis for flg22-induced activation of the *Arabidopsis* FLS2-BAK1 immune complex *Science* **342** 624–8

[343] Lu D, Wu S, Gao X, Zhang Y, Shan L and He P 2010 A receptor-like cytoplasmic kinase, BIK1, associates with a flagellin receptor complex to initiate plant innate immunity *Proc. Natl Acad. Sci. USA* **107** 496–501

[344] Zhang J, Li W, Xiang T, Liu Z, Laluk K, Ding X *et al* 2010 Receptor-like cytoplasmic kinases integrate signaling from multiple plant immune receptors and are targeted by a *Pseudomonas syringae* effector *Cell Host Microbe* **7** 290–301

[345] Xu J, Wei X, Yan L, Liu D, Ma Y, Guo Y *et al* 2013 Identification and functional analysis of phosphorylation residues of the *Arabidopsis* Botrytis-Induced Kinase1 *Protein Cell* **4** 771–81

[346] Shi H, Shen Q, Qi Y, Yan H, Nie H, Chen Y *et al* 2013 BR-Signaling Kinase1 physically associates with Flagellin Sensing2 and regulates plant innate immunity in *Arabidopsis Plant Cell* **25** 1143–57

[347] Yan H, Zhao Y, Shi H, Li J, Wang Y and Tang D 2018 Brassinosteroid-Signaling Kinase 1 phosphorylates MAPKKK5 to regulate immunity in Arabidopsis *Plant Physiol.* **176** 2991–3002

[348] Majhi B B, Sobol G, Gachie S, Sreeramulu S and Sessa G 2021 BRASSINOSTEROID SIGNALLING KINASES 7 and 8 associate with the FLS2 immune receptor and are required for flg22-induced PTI responses *Mol. Plant Pathol.* **22** 786–99

[349] Sobol G, Majhi B B, Pasmanik-Chor M, Zhang N, Roberts H M, Martin G B *et al* 2023 Tomato receptor-like cytoplasmic kinase Fir1 is involved in flagellin signaling and preinvasion immunity *Plant Physiol.* **192** 565–81

[350] Ranf S, Eschen Lippold L, Pecher P, Lee J and Scheel D 2011 Interplay between calcium signalling and early signalling elements during defence responses to microbe-or damage associated molecular patterns *Plant* J. **68** 100–13

[351] Kwaaitaal M, Huisman R, Maintz J, Reinstädler A and Panstruga R 2011 Ionotropic glutamate receptor (iGluR)-like channels mediate MAMP-induced calcium influx in *Arabidopsis thaliana Biochem.* J. **440** 355–73

[352] Thor K, Jiang S, Michard E, George J, Scherzer S, Huang S *et al* 2020 The calcium-permeable channel OSCA1.3 regulates plant stomatal immunity *Nature* **585** 569–73

[353] Ma Y, Walker R K, Zhao Y and Berkowitz G A 2012 Linking ligand perception by PEPR pattern recognition receptors to cytosolic Ca^{2+} elevation and downstream immune signaling in plants *Proc. Natl Acad. Sci. USA* **109** 19852–7

[354] Jeworutzki E, Roelfsema M R G, Anschütz U, Krol E, Elzenga J T M, Felix G *et al* 2010 Early signaling through the *Arabidopsis* pattern recognition receptors FLS2 and EFR involves Ca^{2+} associated opening of plasma membrane anion channels *Plant* J. **62** 367–78

[355] Li L, Li M, Yu L, Zhou Z, Liang X, Liu Z *et al* 2014 The FLS2-associated kinase BIK1 directly phosphorylates the NADPH oxidase RbohD to control plant immunity *Cell Host Microbe* **15** 329–38
[356] Torres M A, Dangl J L and Jones J D 2002 *Arabidopsis* gp91^{phox} homologues *AtrbohD* and *AtrbohF* are required for accumulation of reactive oxygen intermediates in the plant defense response *Proc. Natl Acad. Sci. USA* **99** 517–22
[357] Dubiella U, Seybold H, Durian G, Komander E, Lassig R, Witte C P *et al* 2013 Calcium-dependent protein kinase/NADPH oxidase activation circuit is required for rapid defense signal propagation *Proc. Natl Acad. Sci. USA* **110** 8744–9
[358] Cheval C, Aldon D, Galaud J P and Ranty B 2013 Calcium/calmodulin-mediated regulation of plant immunity *Biochim. Biophys. Acta* **1833** 1766–71
[359] Sardar A, Nandi A K and Chattopadhyay D 2017 CBL-interacting protein kinase 6 negatively regulates immune response to *Pseudomonas syringae* in Arabidopsis *J. Exp. Bot.* **68** 3573–84
[360] Ma Y, Chen Q, He J, Cao J, Liu Z, Wang J *et al* 2021 The kinase CIPK14 functions as a negative regulator of plant immune responses to *Pseudomonas syringae* in *Arabidopsis Plant Sci.* **312** 111017
[361] Monaghan J, Matschi S, Shorinola O, Rovenich H, Matei A, Segonzac C *et al* 2014 The calcium-dependent protein kinase CPK28 buffers plant immunity and regulates BIK1 turnover *Cell Host Microbe* **16** 605–15
[362] Boudsocq M, Willmann M R, McCormack M, Lee H, Shan L, He P *et al* 2010 Differential innate immune signalling via Ca^{2+} sensor protein kinases *Nature* **464** 418–22
[363] Thilmony R, Underwood W and He S Y 2006 Genome wide transcriptional analysis of the *Arabidopsis thaliana* interaction with the plant pathogen *Pseudomonas syringae* pv. *tomato* DC3000 and the human pathogen *Escherichia coli* O157: H7 *Plant J.* **46** 34–53
[364] Petersen L N, Ingle R A, Knight M R and Denby K J 2009 OXI1 protein kinase is required for plant immunity against *Pseudomonas syringae* in *Arabidopsis J. Exp. Bot.* **60** 3727–35
[365] Jacobs S, Zechmann B, Molitor A, Trujillo M, Petutschnig E, Lipka V *et al* 2011 Broad-spectrum suppression of innate immunity is required for colonization of *Arabidopsis* roots by the fungus *Piriformospora indica Plant Physiol.* **156** 726–40
[366] Asai T, Tena G, Plotnikova J, Willmann M R, Chiu W L, Gomez-Gomez L *et al* 2002 MAP kinase signalling cascade in *Arabidopsis* innate immunity *Nature* **415** 977–83
[367] Ichimura K, Casais C, Peck S C, Shinozaki K and Shirasu K 2006 MEKK1 is required for MPK4 activation and regulates tissue-specific and temperature-dependent cell death in Arabidopsis *J. Biol. Chem.* **281** 36969–76
[368] Sun T, Nitta Y, Zhang Q, Wu D, Tian H, Lee J S *et al* 2018 Antagonistic interactions between two MAP kinase cascades in plant development and immune signaling *EMBO Rep.* **19** e45324
[369] Lian K, Gao F, Sun T, van Wersch R, Ao K, Kong Q *et al* 2018 MKK6 functions in two parallel MAP kinase cascades in immune signaling *Plant Physiol.* **178** 1284–95
[370] Zhang M and Zhang S 2022 Mitogen activated protein kinase cascades in plant signaling *J. Integr. Plant Biol.* **64** 301–41
[371] Su J, Zhang M, Zhang L, Sun T, Liu Y, Lukowitz W *et al* 2017 Regulation of stomatal immunity by interdependent functions of a pathogen-responsive MPK3/MPK6 cascade and abscisic acid *Plant Cell* **29** 526–42

[372] Bharath P, Gahir S and Raghavendra A S 2021 Abscisic acid-induced stomatal closure: an important component of plant defense against abiotic and biotic stress *Front. Plant Sci.* **12** 615114

[373] Wang Z, Li X, Yao X, Ma J, Lu K, An Y *et al* 2023 MYB44 regulates PTI by promoting the expression of EIN2 and MPK3/6 in *Arabidopsis Plant Commun.* **4** 100628

[374] Xu G, Greene G H, Yoo H, Liu L, Marqués J, Motley J *et al* 2017 Global translational reprogramming is a fundamental layer of immune regulation in plants *Nature* **545** 487–90

[375] Freidit Frey N, Garcia A V, Bigeard J, Zaag R, Bueso E, Garmier M *et al* 2014 Functional analysis of *Arabidopsis* immune-related MAPKs uncovers a role for MPK3 as negative regulator of inducible defences *Genome Biol.* **15** 1–22

[376] Bethke G, Unthan T, Uhrig J F, Pöschl Y, Gust A A, Scheel D *et al* 2009 Flg22 regulates the release of an ethylene response factor substrate from MAP kinase 6 in *Arabidopsis thaliana* via ethylene signaling *Proc. Natl Acad. Sci. USA* **106** 8067–72

[377] Kang S, Yang F, Li L, Chen H, Chen S and Zhang J 2015 The *Arabidopsis* transcription factor Brassinosteroid Insensitive1-Ethyl Methanesulfonate-Suppressor1 is a direct substrate of Mitogen-Activated Protein Kinase6 and regulates immunity *Plant Physiol.* **167** 1076–86

[378] Galletti R, Ferrari S and De Lorenzo G 2011 *Arabidopsis* MPK3 and MPK6 play different roles in basal and oligogalacturonide-or flagellin-induced resistance against *Botrytis cinerea Plant Physiol.* **157** 804–14

[379] Suarez-Rodriguez M C, Adams-Phillips L, Liu Y, Wang H, Su S H, Jester P J *et al* 2007 MEKK1 is required for flg22-induced MPK4 activation in *Arabidopsis* plants *Plant Physiol.* **143** 661–9

[380] Gao M, Liu J, Bi D, Zhang Z, Cheng F, Chen S *et al* 2008 MEKK1, MKK1/MKK2 and MPK4 function together in a mitogen-activated protein kinase cascade to regulate innate immunity in plants *Cell Res.* **18** 1190–8

[381] Zhang Z, Wu Y, Gao M, Zhang J, Kong Q, Liu Y *et al* 2012 Disruption of PAMP-induced MAP kinase cascade by a *Pseudomonas syringae* effector activates plant immunity mediated by the NB-LRR protein SUMM2 *Cell Host Microbe* **11** 253–63

[382] Kong Q, Qu N, Gao M, Zhang Z, Ding X, Yang F *et al* 2012 The MEKK1-MKK1/MKK2-MPK4 kinase cascade negatively regulates immunity mediated by a mitogen-activated protein kinase kinase kinase in *Arabidopsis Plant Cell* **24** 2225–36

[383] Nitta Y, Qiu Y, Yaghmaiean H, Zhang Q, Huang J, Adams K *et al* 2020 MEKK2 inhibits activation of MAP kinases in *Arabidopsis Plant* J. **103** 705–14

[384] Zhang Z, Liu Y, Huang H, Gao M, Wu D, Kong Q *et al* 2017 The NLR protein SUMM2 senses the disruption of an immune signaling MAP kinase cascade via CRCK 3 *EMBO Rep.* **18** 292–302

[385] Yang Y, Liu J, Yin C, de Souza Vespoli L, Ge D, Huang Y *et al* 2020 RNA interference-based screen reveals concerted functions of MEKK2 and CRCK3 in plant cell death regulation *Plant Physiol.* **183** 331–44

[386] Saga H, Ogawa T, Kai K, Suzuki H, Ogata Y, Sakurai N *et al* 2012 Identification and characterization of ANAC042, a transcription factor family gene involved in the regulation of camalexin biosynthesis in *Arabidopsis Mol. Plant-Microbe Interact.* **25** 684–96

[387] Chezem W R, Memon A, Li F S, Weng J K and Clay N K 2017 SG2-type R2R3-MYB transcription factor MYB15 controls defense-induced lignification and basal immunity in *Arabidopsis Plant Cell* **29** 1907–26

[388] Hu L, Kvitko B H, Severns P M and Yang L 2023 Shoot maturation strengthens FLS2-mediated resistance to *Pseudomonas syringae Mol. Plant-Microbe Interact. (ja)*
[389] Matern A, Böttcher C, Eschen-Lippold L, Westermann B, Smolka U, Döll S *et al* 2019 A substrate of the ABC transporter PEN3 stimulates bacterial flagellin (flg22)-induced callose deposition in *Arabidopsis thaliana J. Biol. Chem.* **294** 6857–70
[390] Shi Q, Febres V J, Jones J B and Moore G A 2015 Responsiveness of different citrus genotypes to the *Xanthomonas citri* ssp. *citri* derived pathogen associated molecular pattern (PAMP) flg22 correlates with resistance to citrus canker *Mol. Plant Pathol.* **16** 507–20
[391] Tang J, Han Z, Sun Y, Zhang H, Gong X and Chai J 2015 Structural basis for recognition of an endogenous peptide by the plant receptor kinase PEPR1 *Cell Res.* **25** 110–20
[392] Shen W, Liu J and Li J F 2019 Type-II metacaspases mediate the processing of plant elicitor peptides in *Arabidopsis Mol. Plant* **12** 1524–33
[393] Yamaguchi Y, Pearce G and Ryan C A 2006 The cell surface leucine-rich repeat receptor for AtPep1, an endogenous peptide elicitor in *Arabidopsis*, is functional in transgenic tobacco cells *Proc. Natl Acad. Sci. USA* **103** 10104–9
[394] Postel S, Küfner I, Beuter C, Mazzotta S, Schwedt A, Borlotti A *et al* 2010 The multifunctional leucine-rich repeat receptor kinase BAK1 is implicated in *Arabidopsis* development and immunity *Eur. J. Cell Biol.* **89** 169–74
[395] Qi Z, Verma R, Gehring C, Yamaguchi Y, Zhao Y, Ryan C A *et al* 2010 Ca^{2+} signaling by plant *Arabidopsis thaliana* Pep peptides depends on AtPepR1, a receptor with guanylyl cyclase activity, and cGMP-activated Ca^{2+} channels *Proc. Natl Acad. Sci. USA* **107** 21193–8
[396] Flury P, Klauser D, Schulze B, Boller T and Bartels S 2013 The anticipation of danger: microbe-associated molecular pattern perception enhances AtPep-triggered oxidative burst *Plant Physiol.* **161** 2023–35
[397] Ma Y, Zhao Y, Walker R K and Berkowitz G A 2013 Molecular steps in the immune signaling pathway evoked by plant elicitor peptides: Ca^{2+} -dependent protein kinases, nitric oxide, and reactive oxygen species are downstream from the early Ca^{2+} signal *Plant Physiol.* **163** 1459–71
[398] Dressano K, Weckwerth P R, Poretsky E, Takahashi Y, Villarreal C, Shen Z *et al* 2020 Dynamic regulation of Pep-induced immunity through post-translational control of defence transcript splicing *Nat. Plants* **6** 1008–19
[399] Rich-Griffin C, Eichmann R, Reitz M U, Hermann S, Woolley-Allen K, Brown P E *et al* 2020 Regulation of cell type-specific immunity networks in *Arabidopsis* roots *Plant Cell* **32** 2742–62
[400] Sreekanta S, Bethke G, Hatsugai N, Tsuda K, Thao A, Wang L *et al* 2015 The receptor like cytoplasmic kinase PCRK 1 contributes to pattern triggered immunity against *Pseudomonas syringae* in *Arabidopsis thaliana New Phytol.* **207** 78–90
[401] Zhang J, Li Y, Bao Q, Wang H and Hou S 2022 Plant elicitor peptide 1 fortifies root cell walls and triggers a systemic root-to-shoot immune signaling in *Arabidopsis Plant Signal. Behav.* **17** 2034270
[402] Zelman A K and Berkowitz G A 2023 Plant elicitor peptide (Pep) signaling and pathogen defense in tomato *Plants* **12** 2856
[403] Bazin J, Mariappan K, Jiang Y, Blein T, Voelz R, Crespi M *et al* 2020 Role of MPK4 in pathogen-associated molecular pattern-triggered alternative splicing in *Arabidopsis PLoS Pathog.* **16** e1008401

[404] Liu X, Zhou Y, Du M, Liang X, Fan F, Huang G *et al* 2022 The calcium-dependent protein kinase CPK28 is targeted by the ubiquitin ligases ATL31 and ATL6 for proteasome-mediated degradation to fine-tune immune signaling in *Arabidopsis Plant Cell* **34** 679–97

[405] Liu X, Zhou Y, Chen K, Xiao Z, Liang X and Lu D 2023 Phosphorylation status of CPK28 affects its ubiquitination and protein stability *New Phytol.* **237** 1270–84

[406] Dindas J, DeFalco T A, Yu G, Zhang L, David P, Bjornson M *et al* 2022 Direct inhibition of phosphate transport by immune signaling in Arabidopsis *Curr. Biol.* **32** 488–95

[407] Safaeizadeh M and Boller T 2019 Differential and tissue-specific activation pattern of the *AtPROPEP* and *AtPEPR* genes in response to biotic and abiotic stress in *Arabidopsis thaliana Plant Signal. Behav.* **14** e1590094

[408] Ortiz-Morea F A and Reyes-Bermudez A A 2019 Endogenous peptides: key modulators of plant immunity *Bioactive Molecules in Plant Defense* ed S Jogaiah and M Abdelrahman (Cham: Springer) pp 159–77

[409] Kohorn B D and Kohorn S L 2012 The cell wall-associated kinases, WAKs, as pectin receptors *Front. Plant Sci.* **3** 88

[410] de Oliveira L F V, Christoff A P, de Lima J C, de Ross B C F, Sachetto-Martins G, Margis-Pinheiro M *et al* 2014 The Wall-associated Kinase gene family in rice genomes *Plant Sci.* **229** 181–92

[411] Yu T Y, Sun M K and Liang L K 2021 Receptors in the induction of the plant innate immunity *Mol. Plant-Microbe Interact.* **34** 587–601

[412] Poncini L, Wyrsch I, Dénervaud Tendon V, Vorley T, Boller T, Geldner N *et al* 2017 In roots of *Arabidopsis thaliana*, the damage-associated molecular pattern *At*Pep1 is a stronger elicitor of immune signalling than flg22 or the chitin heptamer *PLoS One* **12** e0185808

[413] Huang Y, Cui J, Li M, Yang R, Hu Y, Yu X *et al* 2023 Conservation and divergence of flg22, pep1 and nlp20 in activation of immune response and inhibition of root development *Plant Sci.* **331** 111686

[414] Wyrsch I, Domínguez-Ferreras A, Geldner N and Boller T 2015 Tissue specific Flagellin Sensing 2 (FLS 2) expression in roots restores immune responses in *Arabidopsis* fls2 mutants *New Phytol.* **206** 774–84

[415] Pontiggia D, Benedetti M, Costantini S, De Lorenzo G and Cervone F 2020 Dampening the DAMPs: how plants maintain the homeostasis of cell wall molecular patterns and avoid hyper-immunity *Front. Plant Sci.* **11** 613259

[416] Locci F, Benedetti M, Pontiggia D, Citterico M, Caprari C, Mattei B *et al* 2019 An *Arabidopsis* berberine bridge enzyme like protein specifically oxidizes cellulose oligomers and plays a role in immunity *Plant* J. **98** 540–54

[417] Yang F, Kimberlin A N, Elowsky C G, Liu Y, Gonzalez-Solis A, Cahoon E B *et al* 2019 A plant immune receptor degraded by selective autophagy *Mol. Plant* **12** 113–23

[418] Robatzek S, Chinchilla D and Boller T 2006 Ligand-induced endocytosis of the pattern recognition receptor FLS2 in *Arabidopsis. Genes Dev.* **20** 537–42

[419] Lu D, Lin W, Gao X, Wu S, Cheng C, Avila J *et al* 2011 Direct ubiquitination of pattern recognition receptor FLS2 attenuates plant innate immunity *Science* **332** 1439–42

[420] Beck M, Zhou J, Faulkner C, MacLean D and Robatzek S 2012 Spatio-temporal cellular dynamics of the *Arabidopsis* flagellin receptor reveal activation status-dependent endosomal sorting *Plant Cell* **24** 4205–19

[421] Mbengue M, Bourdais G, Gervasi F, Beck M, Zhou J, Spallek T *et al* 2016 Clathrin-dependent endocytosis is required for immunity mediated by pattern recognition receptor kinases *Proc. Natl Acad. Sci. USA* **113** 11034–9

[422] Ortiz-Morea F A, Savatin D V, Dejonghe W, Kumar R, Luo Y, Adamowski M *et al* 2016 Danger-associated peptide signaling in *Arabidopsis* requires clathrin *Proc. Natl Acad. Sci. USA* **113** 11028–33

[423] Erwig J, Ghareeb H, Kopischke M, Hacke R, Matei A, Petutschnig E *et al* 2017 Chitin induced and Chitin Elicitor Receptor Kinase1 (CERK1) phosphorylation dependent endocytosis of *Arabidopsis thaliana* Lysin Motif Containing Receptor Like Kinase5 (LYK5) *New Phytol.* **215** 382–96

[424] Gu Y, Zavaliev R and Dong X 2017 Membrane trafficking in plant immunity *Mol. Plant* **10** 1026–34

[425] Choi S W, Tamaki T, Ebine K, Uemura T, Ueda T and Nakano A 2013 RABA members act in distinct steps of subcellular trafficking of the Flagellin Sensing2 receptor *Plant Cell* **25** 1174–87

[426] Spallek T, Beck M, Ben Khaled S, Salomon S, Bourdais G, Schellmann S *et al* 2013 ESCRT-I mediates FLS2 endosomal sorting and plant immunity *PLoS Genet.* **9** e1004035

[427] Claus L A N, Savatin D V and Russinova E 2018 The crossroads of receptor mediated signaling and endocytosis in plants *J. Integr. Plant Biol.* **60** 827–40

[428] Trujillo M 2021 Ubiquitin signalling: controlling the message of surface immune receptors *New Phytol.* **231** 47–53

[429] Stegmann M, Anderson R G, Ichimura K, Pecenkova T, Reuter P, Žárský V *et al* 2012 The ubiquitin ligase PUB22 targets a subunit of the exocyst complex required for PAMP-triggered responses in *Arabidopsis Plant Cell* **24** 4703–16

[430] Wang W, Liu N, Gao C, Cai H, Romeis T and Tang D 2020 The *Arabidopsis* exocyst subunits EXO70B1 and EXO70B2 regulate FLS2 homeostasis at the plasma membrane *New Phytol.* **227** 529–44

[431] Harman G, Khadka R, Doni F and Uphoff N 2021 Benefits to plant health and productivity from enhancing plant microbial symbionts *Front. Plant Sci.* **11** 610065

[432] Chandan R K, Kumar R, Kabyashree K, Yadav S K, Roy M, Swain D M *et al* 2023 A prophage tail like protein facilitates the endophytic growth of *Burkholderia gladioli* and mounting immunity in tomato *New Phytol.* **240** 1202–18

[433] Cao Y, Halane M K, Gassmann W and Stacey G 2017 The role of plant innate immunity in the legume-rhizobium symbiosis *Annu. Rev. Plant Biol.* **68** 535–61

[434] Calla B 2022 Friend or foe: how plants discriminate between pathogenic and mutualistic bacteria *Plant Physiol.* **189** 1893–5

[435] Colaianni N R, Parys K, Lee H S, Conway J M, Kim N H, Edelbacher N *et al* 2021 A complex immune response to flagellin epitope variation in commensal communities *Cell Host Microbe* **29** 635–49

[436] Cheng J H, Bredow M, Monaghan J and DiCenzo G C 2021 Proteobacteria contain diverse flg22 epitopes that elicit varying immune responses in *Arabidopsis thaliana Mol. Plant-Microbe Interact.* **34** 504–10

[437] Lopez-Gomez M, Sandal N, Stougaard J and Boller T 2012 Interplay of flg22-induced defence responses and nodulation in *Lotus japonicus J. Exp. Bot.* **63** 393–401

[438] Teulet A, Busset N, Fardoux J, Gully D, Chaintreuil C, Cartieaux F *et al* 2019 The rhizobial type III effector ErnA confers the ability to form nodules in legumes *Proc. Natl Acad. Sci. USA* **116** 21758–68

[439] Piromyou P, Nguyen H P, Songwattana P, Boonchuen P, Teamtisong K, Tittabutr P *et al* 2021 The *Bradyrhizobium diazoefficiens* type III effector NopE modulates the regulation of plant hormones towards nodulation in *Vigna radiata Sci Rep.* **11** 16604

[440] Kumar R, Barman A, Phukan T, Kabyashree K, Singh N, Jha G *et al* 2017 *Ralstonia solanacearum* virulence in tomato seedlings inoculated by leaf clipping *Plant Pathol.* **66** 835–41

[441] Yu K, Pieterse C M, Bakker P A and Berendsen R L 2019 Beneficial microbes going underground of root immunity *Plant Cell Environ.* **42** 2860–70

[442] Thoms D, Liang Y and Haney C H 2021 Maintaining symbiotic homeostasis: how do plants engage with beneficial microorganisms while at the same time restricting pathogens? *Mol. Plant-Microbe Interact.* **34** 462–9

[443] Kongala S I and Kondreddy A 2023 A review on plant and pathogen derived carbohydrates, oligosaccharides and their role in plant's immunity *Carbohydr. Polym. Technol. Appl.* 100330

[444] He J, Zhang C, Dai H, Liu H, Zhang X, Yang J *et al* 2019 A LysM receptor heteromer mediates perception of arbuscular mycorrhizal symbiotic signal in rice *Mol. Plant* **12** 1561–76

[445] Bozsoki Z, Cheng J, Feng F, Gysel K, Vinther M, Andersen K R *et al* 2017 Receptor-mediated chitin perception in legume roots is functionally separable from Nod factor perception *Proc. Natl Acad. Sci. USA* **114** E8118–27

[446] Leppyanen I V, Shakhnazarova V Y, Shtark O Y, Vishnevskaya N A, Tikhonovich I A and Dolgikh E A 2017 Receptor-like kinase LYK9 in *Pisum sativum* L. is the CERK1-like receptor that controls both plant immunity and AM symbiosis development *Int. J. Mol. Sci.* **19** 8

[447] Gibelin Viala C, Amblard E, Puech Pages V, Bonhomme M, Garcia M, Bascaules Bedin A *et al* 2019 The *Medicago truncatula* LysM receptor like kinase LYK9 plays a dual role in immunity and the arbuscular mycorrhizal symbiosis *New Phytol.* **223** 1516–29

[448] Withers J and Dong X 2017 Post-translational regulation of plant immunity *Curr. Opin. Plant Biol.* **38** 124–32

[449] Segonzac C, Macho A P, Sanmartín M, Ntoukakis V, Sánchez-Serrano J J and Zipfel C 2014 Negative control of BAK 1 by protein phosphatase 2A during plant innate immunity *EMBO J.* **33** 2069–79

[450] Couto D, Niebergall R, Liang X, Bücherl C A, Sklenar J, Macho A P *et al* 2016 The *Arabidopsis* protein phosphatase PP2C38 negatively regulates the central immune kinase BIK1 *PLoS Pathog.* **12** e1005811

[451] Shubchynskyy V, Boniecka J, Schweighofer A, Simulis J, Kvederaviciute K, Stumpe M *et al* 2017 Protein phosphatase AP2C1 negatively regulates basal resistance and defense responses to *Pseudomonas syringae J. Exp. Bot.* **68** 1169–83

[452] Wang B, Huang M, He W, Wang Y, Yu L, Zhou D *et al* 2023 Protein phosphatase StTOPP6 negatively regulates potato bacterial wilt resistance by modulating MAPK signaling *J. Exp. Bot.* **74** 4208–24

[453] Li F, Chen X, Yang R, Zhang K, Shan W, Joosten M H *et al* 2023 Potato protein tyrosine phosphatase StPTP1a is activated by StMKK1 to negatively regulate plant immunity *Plant Biotechnol. J.* **21** 646–61

[454] Liu J, Liu B, Chen S, Gong B Q, Chen L, Zhou Q *et al* 2018 A tyrosine phosphorylation cycle regulates fungal activation of a plant receptor Ser/Thr kinase *Cell Host Microbe* **23** 241–53

[455] Li F, Cheng C, Cui F, de Oliveira M V, Yu X, Meng X *et al* 2014 Modulation of RNA polymerase II phosphorylation downstream of pathogen perception orchestrates plant immunity *Cell Host Microbe* **16** 748–58

[456] Wei J, Sun W, Zheng X, Qiu S, Jiao S, Babilonia K *et al* 2023 *Arabidopsis* RNA polymerase II C-terminal domain phosphatase like 1 (CPL1) targets MAPK cascades to suppress plant immunity *J. Integr. Plant Biol.* **65** 2380–94

[457] Sunkar R, Li Y F and Jagadeeswaran G 2012 Functions of microRNAs in plant stress responses *Trends Plant Sci.* **17** 196–203

[458] Barman A, Phukan T and Ray S K 2021 Harnessing perks of miRNA principles for betterment of agriculture and food security *Omics Technologies for Sustainable Agriculture and Global Food Security* **vol II** ed A Kumar, R Kumar, P Shukla and H K Patel (Singapore: Springer) pp 123–91

[459] RobertSeilaniantz A, MacLean D, Jikumaru Y, Hill L, Yamaguchi S, Kamiya Y *et al* 2011 The microRNA miR393 redirects secondary metabolite biosynthesis away from camalexin and towards glucosinolates *Plant* J. **67** 218–31

[460] Li Y, Lu Y G, Shi Y, Wu L, Xu Y J, Huang F *et al* 2014 Multiple rice microRNAs are involved in immunity against the blast fungus *Magnaporthe oryzae Plant Physiol.* **164** 1077–92

[461] Zou Y, Wang S, Zhou Y, Bai J, Huang G, Liu X *et al* 2018 Transcriptional regulation of the immune receptor FLS2 controls the ontogeny of plant innate immunity *Plant Cell* **30** 2779–94

[462] Li Y, Li T T, He X R, Zhu Y, Feng Q, Yang X M *et al* 2022 Blocking Osa-miR1871 enhances rice resistance against *Magnaporthe oryzae* and yield *Plant Biotechnol. J.* **20** 646–59

[463] Salvador-Guirao R, Baldrich P, Weigel D, Rubio-Somoza I and San Segundo B 2018 The microRNA miR773 is involved in the *Arabidopsis* immune response to fungal pathogens *Mol. Plant-Microbe Interact.* **31** 249–59

[464] Wang Y, Li Z, Liu D, Xu J, Wei X, Yan L *et al* 2014 Assessment of BAK1 activity in different plant receptor-like kinase complexes by quantitative profiling of phosphorylation patterns *J. Proteomics* **108** 484–93

[465] Li B, Ferreira M A, Huang M, Camargos L F, Yu X, Teixeira R M *et al* 2019 The receptor-like kinase NIK1 targets FLS2/BAK1 immune complex and inversely modulates antiviral and antibacterial immunity *Nat. Commun.* **10** 4996

[466] Imkampe J, Halter T, Huang S, Schulze S, Mazzotta S, Schmidt N *et al* 2017 The *Arabidopsis* leucine-rich repeat receptor kinase BIR3 negatively regulates BAK1 receptor complex formation and stabilizes BAK1 *Plant Cell* **29** 2285–303

[467] Yeh Y H, Panzeri D, Kadota Y, Huang Y C, Huang P Y, Tao C N *et al* 2016 The *Arabidopsis*malectin-like/LRR-RLK IOS1 is critical for BAK1-dependent and BAK1-independent pattern-triggered immunity *Plant Cell* **28** 1701–21

[468] Wang D, Liang X, Bao Y, Yang S, Zhang X, Yu H *et al* 2020 A malectin like receptor kinase regulates cell death and pattern triggered immunity in soybean *EMBO Rep.* **21** e50442

[469] Jiang Y, Han B, Zhang H, Mariappan K G, Bigeard J, Colcombet J *et al* 2019 MAP4K4 associates with BIK1 to regulate plant innate immunity *EMBO Rep.* **20** e47965

[470] Mithoe S C, Ludwig C, Pel M J, Cucinotta M, Casartelli A, Mbengue M *et al* 2016 Attenuation of pattern recognition receptor signaling is mediated by a MAP kinase kinasekinase *EMBO Rep.* **17** 441–54

[471] Wang J, Wang S, Hu K, Yang J, Xin X, Zhou W *et al* 2018 The kinase OsCPK4 regulates a buffering mechanism that fine-tunes innate immunity *Plant Physiol.* **176** 1835–49

[472] Cheng D, Zhou D, Wang Y, Wang B, He Q, Song B *et al* 2021 *Ralstonia solanacearum* type III effector RipV2 encoding a novel E3 ubiquitin ligase (NEL) is required for full virulence by suppressing plant PAMP-triggered immunity *Biochem. Biophys. Res. Commun.* **550** 120–6

[473] Blekemolen M C, Liu Z, Stegman M, Zipfel C, Shan L and Takken F L 2023 The PTI suppressing Avr2 effector from *Fusarium oxysporum* suppresses mono ubiquitination and plasma membrane dissociation of BIK1 *Mol. Plant Pathol.* **24** 1273–86

[474] Yamaguchi K, Yamada K, Ishikawa K, Yoshimura S, Hayashi N, Uchihashi K *et al* 2013 A receptor-like cytoplasmic kinase targeted by a plant pathogen effector is directly phosphorylated by the chitin receptor and mediates rice immunity *Cell Host Microbe* **13** 347–57

[475] Verma G, Sharma M and Mondal K K 2018 XopR TTSS-effector regulates in planta growth, virulence of Indian strain of *Xanthomonas oryzae* pv. *oryzae* via suppressing reactive oxygen species production and cell wall-associated rice immune responses during blight induction *Funct. Plant Biol.* **45** 561–74

[476] Huang Y, Li T, Xu T, Tang Z, Guo J and Cai Y 2020 Multiple *Xanthomonas campestris* pv. *campestris* 8004 type III effectors inhibit immunity induced by flg22 *Planta* **252** 1–9

[477] Qiu H, Wang B, Huang M, Sun X, Yu L, Cheng D *et al* 2023 A novel effector RipBT contributes to *Ralstonia solanacearum* virulence on potato *Mol. Plant Pathol.* **24** 947–60

[478] Sun Y, Li P, Shen D, Wei Q, He J and Lu Y 2019 The *Ralstonia solanacearum* effector RipN suppresses plant PAMP triggered immunity, localizes to the endoplasmic reticulum and nucleus, and alters the NADH/NAD $^{+}$ ratio in *Arabidopsis Mol. Plant Pathol.* **20** 533–46

[479] Nakano M and Mukaihara T 2019 Comprehensive identification of PTI suppressors in type III effector repertoire reveals that *Ralstonia solanacearum* activates jasmonate signaling at two different steps *Int. J. Mol. Sci.* **20** 5992

[480] Jin Y, Zhang W, Cong S, Zhuang Q G, Gu Y L, Ma Y N *et al* 2023 *Pseudomonas syringae* type III secretion protein HrpP manipulates plant immunity to promote infection *Microbiol. Spectr* **11** e05148–22

[481] Han Z, Xiong D, Xu Z, Liu T and Tian C 2021 The *Cytospora chrysosperma* virulence effector CcCAP1 mainly localizes to the plant nucleus to suppress plant immune responses *mSphere* **6** 10–1128

[482] Gu B, Gao W, Liu Z, Shao G, Peng Q, Mu Y *et al* 2023 A single region of the *Phytophthora infestans* avirulence effector Avr3b functions in both cell death induction and plant immunity suppression *Mol. Plant Pathol.* **24** 317–30

[483] Du Y, Chen X, Guo Y, Zhang X, Zhang H, Li F *et al* 2021 *Phytophthora infestans* RXLR effector PITG20303 targets a potato MKK1 protein to suppress plant immunity *New Phytol.* **229** 501–15

[484] Luo M, Sun X, Qi Y, Zhou J, Wu X and Tian Z 2021 *Phytophthora infestans* RXLR effector Pi04089 perturbs diverse defense-related genes to suppress host immunity *BMC Plant Biol.* **21** 1–15

[485] Zheng X, Wagener N, McLellan H, Boevink P C, Hua C, Birch P R *et al* 2018 *Phytophthora infestans* RXLR effector SFI 5 requires association with calmodulin for PTI/MTI suppressing activity *New Phytol.* **219** 1433–46

[486] Li Q, Chen Y, Wang J, Zou F, Jia Y, Shen D *et al* 2019 A *Phytophthora capsici* virulence effector associates with NPR1 and suppresses plant immune responses *Phytopathol. Res.* **1** 1–11

[487] Li Q, Ai G, Shen D, Zou F, Wang J, Bai T *et al* 2019 A *Phytophthora capsici* effector targets ACD11 binding partners that regulate ROS-mediated defense response in *Arabidopsis Mol. Plant* **12** 565–81

[488] Ling H, Fu X, Huang N, Zhong Z, Su W, Lin W *et al* 2022 A sugarcane smut fungus effector simulates the host endogenous elicitor peptide to suppress plant immunity *New Phytol.* **233** 919–33

[489] Li Q, Wang J, Bai T, Zhang M, Jia Y, Shen D *et al* 2020 A *Phytophthora capsici* effector suppresses plant immunity via interaction with EDS1 *Mol. Plant Pathol.* **21** 502–11

[490] Jaouannet M, Magliano M, Arguel M J, Gourgues M, Evangelisti E, Abad P *et al* 2013 The root-knot nematode calreticulin Mi-CRT is a key effector in plant defense suppression *Mol. Plant-Microbe Interact.* **26** 97–105

[491] Yang S, Dai Y, Chen Y, Yang J, Yang D, Liu Q *et al* 2019 A novel G16B09-like effector from *Heteroderaavenae* suppresses plant defenses and promotes parasitism. *Front Plant Sci.* **10** 66

[492] Dong Y, Jing M, Shen D, Wang C, Zhang M, Liang D *et al* 2020 The mirid bug *Apolyguslucorum* deploys a glutathione peroxidase as a candidate effector to enhance plant susceptibility *J. Exp. Bot.* **71** 2701–12

[493] Ji R, Fu J, Shi Y, Li J, Jing M, Wang L *et al* 2021 Vitellogenin from planthopper oral secretion acts as a novel effector to impair plant defenses *New Phytol.* **232** 802–17

[494] Zhang Y, Liu X, Francis F, Xie H, Fan J, Wang Q *et al* 2022 The salivary effector protein Sg2204 in the greenbug *Schizaphisgraminum* suppresses wheat defence and is essential for enabling aphid feeding on host plants *Plant Biotechnol. J.* **20** 2187–201

[495] Dong Y, Zhou J, Yang Y, Lu W, Jin Y, Huang X *et al* 2023 Cyclophilin effector Al106 of mirid bug *Apolygus lucorum* inhibits plant immunity and promotes insect feeding by targeting PUB33 *New Phytol.* **237** 2388–403

[496] Kamaruzzaman M, Zhao L F, Zhang J A, Zhu L T, Li Y, Deng X D *et al* 2023 MiPDCD6 effector suppresses host PAMP-triggered immunity to facilitate *Meloidogyne incognita* parasitism in tomato *Plant Pathol.* **72** 195–206

[497] Lin B, Zhuo K, Chen S, Hu L, Sun L, Wang X *et al* 2016 A novel nematode effector suppresses plant immunity by activating host reactive oxygen species scavenging system *New Phytol.* **209** 1159–73

[498] Wen T Y, Wu X Q, Hu L J, Qiu Y J, Rui L, Zhang Y *et al* 2021 A novel pine wood nematode effector, BxSCD1, suppresses plant immunity and interacts with an ethylene forming enzyme in pine *Mol. Plant Pathol.* **22** 1399–412

[499] Hu L J, Wu X Q, Wen T Y, Qiu Y J, Rui L, Zhang Y *et al* 2022 A *Bursaphelenchus xylophilus* effector, BxSCD3, suppresses plant defense and contributes to virulence *Int. J. Mol. Sci.* **23** 6417

[500] Villarroel C A, Jonckheere W, Alba J M, Glas J J, Dermauw W, Haring M A *et al* 2016 Salivary proteins of spider mites suppress defenses in *Nicotiana benthamiana* and promote mite reproduction *Plant* J. **86** 119–31

[501] Wang L, Tsuda K, Truman W, Sato M, Nguyen L V, Katagiri F *et al* 2011 CBP60g and SARD1 play partially redundant critical roles in salicylic acid signaling *Plant* J. **67** 1029–41

[502] Yi S Y, Shirasu K, Moon J S, Lee S G and Kwon S Y 2014 The activated SA and JA signaling pathways have an influence on flg22-triggered oxidative burst and callose deposition *PLoS One* **9** e88951
[503] Jia X, Zeng H, Wang W, Zhang F and Yin H 2018 Chitosan oligosaccharide induces resistance to *Pseudomonas syringae* pv. *tomato* DC3000 in *Arabidopsis thaliana* by activating both salicylic acid–and jasmonic acid–mediated pathways *Mol. Plant-Microbe Interact* **31** 1271–9
[504] Howlader P, Bose S K, Jia X, Zhang C, Wang W and Yin H 2020 Oligogalacturonides induce resistance in *Arabidopsis thaliana* by triggering salicylic acid and jasmonic acid pathways against *Pst* DC3000 *Int. J. Biol. Macromol.* **164** 4054–64
[505] Schuman M C and Baldwin I T 2016 The layers of plant responses to insect herbivores *Annu. Rev. Entomol.* **61** 373–94
[506] Aerts N, Pereira Mendes M and Van Wees S C 2021 Multiple levels of crosstalk in hormone networks regulating plant defense *Plant* J. **105** 489–504
[507] Boutrot F and Zipfel C 2017 Function, discovery, and exploitation of plant pattern recognition receptors for broad-spectrum disease resistance *Annu. Rev. Phytopathol.* **55** 257–86
[508] Kishimoto K, Kouzai Y, Kaku H, Shibuya N, Minami E and Nishizawa Y 2010 Perception of the chitin oligosaccharides contributes to disease resistance to blast fungus *Magnaporthe oryzae* in rice *Plant* J. **64** 343–54
[509] Bouwmeester K, Han M, Blanco Portales R, Song W, Weide R, Guo L Y *et al* 2014 The *Arabidopsis* lectin receptor kinase Lec RKI. 9 enhances resistance to *Phytophthora infestans* in Solanaceous plants *Plant Biotechnol. J.* **12** 10–6
[510] Li D L, Xiao X and Guo W W 2014 Production of transgenic anliucheng sweet orange (*Citrus sinensis* Osbeck) with *Xa21* gene for potential canker resistance *J. Integr. Agric.* **13** 2370–7
[511] Wei Y, Balaceanu A, Rufian J S, Segonzac C, Zhao A, Morcillo R J *et al* 2020 An immune receptor complex evolved in soybean to perceive a polymorphic bacterial flagellin *Nat. Commun.* **11** 3763
[512] Mitre L K, Teixeira-Silva N S, Rybak K, Magalhães D M, de Souza Neto R R, Robatzek S *et al* 2021 The *Arabidopsis* immune receptor EFR increases resistance to the bacterial pathogens *Xanthomonas* and *Xylella* in transgenic sweet orange *Plant Biotechnol. J.* **19** 1294
[513] Piazza S, Campa M, Pompili V, Costa L D, Salvagnin U, Nekrasov V *et al* 2021 The *Arabidopsis* pattern recognition receptor EFR enhances fire blight resistance in apple *Hortic. Res.* **8** 204
[514] Zimaro T, Thomas L, Marondedze C, Garavaglia B S, Gehring C, Ottado J *et al* 2013 Insights into *Xanthomonas axonopodis* pv. *citri* biofilm through proteomics *BMC Microbiol.* **13** 1–14
[515] Vuong U T, Iswanto A B B, Nguyen Q M, Kang H, Lee J, Moon J *et al* 2023 Engineering plant immune circuit: walking to the bright future with a novel toolbox *Plant Biotechnol. J.* **21** 17–45
[516] Boschi F, Schvartzman C, Murchio S, Ferreira V, Siri M I, Galván G A *et al* 2017 Enhanced bacterial wilt resistance in potato through expression of *Arabidopsis* EFR and introgression of quantitative resistance from *Solanum commersonii Front. Plant Sci.* **8** 1642
[517] Lu F, Wang H, Wang S, Jiang W, Shan C, Li B *et al* 2015 Enhancement of innate immune system in monocot rice by transferring the dicotyledonous elongation factor Tu receptor EFR *J. Integr. Plant Biol.* **57** 641–52

[518] Schoonbeek H J, Wang H H, Stefanato F L, Craze M, Bowden S, Wallington E *et al* 2015 *Arabidopsis* EF-Tu receptor enhances bacterial disease resistance in transgenic wheat *New Phytol.* **206** 606–13

[519] Tian S N, Liu D D, Zhong C L, Xu H Y, Yang S, Fang Y *et al* 2020 Silencing GmFLS2 enhances the susceptibility of soybean to bacterial pathogen through attenuating the activation of GmMAPK signaling pathway *Plant Sci.* **292** 110386

[520] Chen Y, Zhao A, Wei Y, Mao Y, Zhu J K and Macho A P 2023 GmFLS2 contributes to soybean resistance to *Ralstonia solanacearum New Phytol.* **240** 17–22

[521] Gouget A, Senchou V, Govers F, Sanson A, Barre A, Rougé P *et al* 2006 Lectin receptor kinases participate in protein–protein interactions to mediate plasma membrane-cell wall adhesions in *Arabidopsis Plant Physiol.* **140** 81–90

[522] Song Y, Liu L, Wang Y, Valkenburg D J, Zhang X, Zhu L *et al* 2018 Transfer of tomato immune receptor Ve1 confers Ave1 dependent *Verticillium* resistance in tobacco and cotton *Plant Biotechnol. J.* **16** 638–48

[523] Felcher K J, Douches D S, Kirk W W, Hammerschmidt R and Li W 2003 Expression of a fungal glucose oxidase gene in three potato cultivars with different susceptibility to late blight (*Phytophthora infestans* Mont. deBary) *J. Am. Soc. Hortic. Sci.* **128** 238–45

[524] Kouzai Y, Kaku H, Shibuya N, Minami E and Nishizawa Y 2013 Expression of the chimeric receptor between the chitin elicitor receptor CEBiP and the receptor-like protein kinase Pi-d2 leads to enhanced responses to the chitin elicitor and disease resistance against *Magnaporthe oryzae* in rice *Plant Mol. Biol.* **81** 287–95

[525] Jehle A K, Fürst U, Lipschis M, Albert M and Felix G 2013 Perception of the novel MAMP eMax from different *Xanthomonas* species requires the *Arabidopsis* receptor-like protein ReMAX and the receptor kinase SOBIR *Plant Signal. Behav.* **8** e27408

[526] Jehle A K, Lipschis M, Albert M, Fallahzadeh-Mamaghani V, Fürst U, Mueller K *et al* 2013 The receptor-like protein ReMAX of *Arabidopsis* detects the microbe-associated molecular pattern eMax from *Xanthomonas Plant Cell* **25** 2330–40

[527] Wu J, Reca I B, Spinelli F, Lironi D, De Lorenzo G, Poltronieri P *et al* 2019 An EFR-Cf-9 chimera confers enhanced resistance to bacterial pathogens by SOBIR1 and BAK1 dependent recognition of elf18 *Mol. Plant Pathol.* **20** 751–64

[528] Ross A, Yamada K, Hiruma K, Yamashita Yamada M, Lu X, Takano Y *et al* 2014 The *Arabidopsis* PEPR pathway couples local and systemic plant immunity *EMBO J.* **33** 62–75

[529] Lori M, Van Verk M C, Hander T, Schatowitz H, Klauser D, Flury P *et al* 2015 Evolutionary divergence of the plant elicitor peptides (Peps) and their receptors: interfamily incompatibility of perception but compatibility of downstream signalling *J. Exp. Bot.* **66** 5315–25

[530] Kourelis J, Marchal C, Posbeyikian A, Harant A and Kamoun S 2023 NLR immune receptor–nanobody fusions confer plant disease resistance *Science* **379** 934–9

[531] DeFalco T A, Anne P, James S R, Willoughby A C, Schwanke F, Johanndrees O *et al* 2022 A conserved module regulates receptor kinase signalling in immunity and development *Nat. Plants* **8** 356–65

[532] Wang W, Feng B, Zhou J M and Tang D 2020 Plant immune signaling: advancing on two frontiers *J. Integr. Plant Biol.* **62** 2–24

[533] Desaki Y, Kohari M, Shibuya N and Kaku H 2019 MAMP-triggered plant immunity mediated by the LysM-receptor kinase CERK1 *J. Gen. Plant Pathol.* **85** 1–11

[534] Yuan P, Jauregui E, Du L, Tanaka K and Poovaiah B W 2017 Calcium signatures and signaling events orchestrate plant–microbe interactions *Curr. Opin. Plant Biol.* **38** 173–83
[535] Singh N and Pandey G K 2020 Calcium signatures and signal transduction schemes during microbe interactions in *Arabidopsis thaliana J. Plant Biochem. Biotechnol.* **29** 675–86
[536] Yuan P, Jewell J B, Behera S, Tanaka K and Poovaiah B W 2020 Distinct molecular pattern-induced calcium signatures lead to different downstream transcriptional regulations via AtSR1/CAMTA3 *Int. J. Mol. Sci.* **21** 8163
[537] Wang Y, Xu Y, Sun Y, Wang H, Qi J, Wan B *et al* 2018 Leucine-rich repeat receptor-like gene screen reveals that Nicotiana RXEG1 regulates glycoside hydrolase 12 MAMP detection *Nat. Commun.* **9** 594
[538] Sun Y, Wang Y, Zhang X, Chen Z, Xia Y, Wang L *et al* 2022 Plant receptor-like protein activation by a microbial glycoside hydrolase *Nature* **610** 335–42
[539] Zhang L, Hua C, Pruitt R N, Qin S, Wang L, Albert I *et al* 2021 Distinct immune sensor systems for fungal endopolygalacturonases in closely related *Brassicaceae Nat. Plants* **7** 1254–63
[540] Fan L, Fröhlich K, Melzer E, Pruitt R N, Albert I, Zhang L *et al* 2022 Genotyping-by-sequencing-based identification of *Arabidopsis* pattern recognition receptor RLP32 recognizing proteobacterial translation initiation factor IF1 *Nat. Commun.* **13** 1294
[541] Yang Y, Steidele C E, Rössner C, Löffelhardt B, Kolb D, Leisen T *et al* 2023 Convergent evolution of plant pattern recognition receptors sensing cysteine-rich patterns from three microbial kingdoms *Nat. Commun.* **14** 3621
[542] Pei Y, Ji P, Si J, Zhao H, Zhang S, Xu R *et al* 2023 A *Phytophthora* receptor-like kinase regulates oospore development and can activate pattern-triggered plant immunity *Nat. Commun.* **14** 4593
[543] Jiang H, Xia Y, Zhang S, Zhang Z, Feng H, Zhang Q *et al* 2023 The CAP superfamily protein PsCAP1 secreted by *Phytophthora* triggers immune responses in *Nicotiana benthamiana* through a leucine-rich repeat receptor-like protein *New Phytol.* **242** 784–801
[544] Duriez P, Vautrin S, Auriac M C, Bazerque J, Boniface M C, Callot C *et al* 2019 A receptor-like kinase enhances sunflower resistance to *Orobanche cumana Nat. Plants* **5** 1211–5

IOP Publishing

Advances in Biochemical and Molecular Mechanisms of Plant–Pathogen Interaction

Hitendra K Patel and Anirudh Kumar

Chapter 3

Fine-tuning the responses: role of phytohormones and their crosstalk in plant defence

Shakuntala E Pillai and Ravinayak Patlavath

Plant defence hormones and their derivatives play an indispensable role in the rapid activation and transmission of the stress signal in infected and uninfected tissues. Jasmonic acid (JA), salicylic acid (SA), ethylene (ET) and abscisic acid (ABA) are the major plant defence hormones. In response to a changing environment a plant synthesizes, degrades, or modifies its hormones as per the need of the time. Each hormone carries out a particular set of functions. The modulation in the levels of these hormones helps the plant in switching from growth state to defence state and vice versa. The signalling cascades induced by the defence hormones may interact at multiple levels with each other. The plant hormone signalling is highly complex where multiple hormones have intense crosstalk and regulate a single or multiple responses. In this chapter, we discuss signalling of the major defence hormones and their crosstalk with other hormones.

3.1 Introduction

Plants have evolved an immune system different from that of humans. Plants lack dedicated immune cells like lymphocytes, macrophages, natural killer cells, etc and lack a circulatory system to raise a systemic response. In contrast, like animal cells, each plant cell can raise a defence response; hence the plant immune system is often referred to as the 'innate immune system' [1]. Plants perceive pathogen presence by recognition of the 'self' and 'non-self' molecules released during the invasion. The 'non-self' molecules are carbohydrates, lipids, and peptides of pathogen origin, also known as pathogen-associated molecular patterns (PAMPs), are sensed by the plant cells (figure 3.1). Among PAMPs, bacterial flagellin protein [Flg22], elongation

doi:10.1088/978-0-7503-5673-2ch3

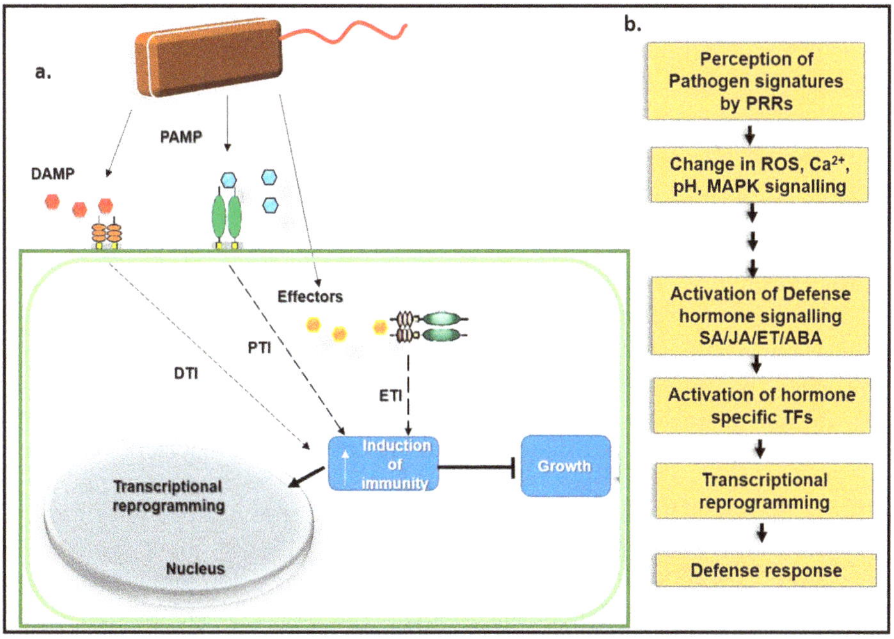

Figure 3.1. Plant perception of pathogen and induction of immunity. (a) Plant perceives molecular signatures of pathogen (PAMPs and effector) or self-damaged products (DAMPs) which leads to induction of plant immunity (PTI, ETI, DTI) and suppression of growth. (b)The perception of pathogen by PRRs induces a cascade of signalling events that leads to activation of specific defence hormone signalling and transcriptional reprograming. PAMP—pathogen-associated molecular patterns; DAMP—damage-associated molecular patterns; PTI—PAMP-triggered immunity; DTI-DAMP-triggered immunity; ETI—effector-triggered immunity, PRR—pattern recognition receptors.

factor-Tu (EF-Tu), the sulphated peptide of *Xanthomonas oryzae* pv. *oryzae* (Ax21), peptidoglycan, chitin, and glucan are a few well-studied PAMPs [2–4]. During invasion, the pathogens secrete a different class of cell-wall-degrading enzymes (CWDEs) either to break the plant's physical barrier or to use the cell wall as a source of nutrients. This results in release of the plant's self-molecules which is perceived by plants as damage-associated molecular patterns (DAMPs). Oligosaccharides, oligogalacturonides (OGs), cellobiose, systemin, hydroxyproline rich peptide, plant elicitor peptide-1 (Pep-1), extracellular adenosine triphosphate, and High Mobility Group Box-1 are a few well-studied DAMP molecules [5–8]. Recognition of PAMP and DAMP molecules in the extracellular space or apoplast activates an immune response also known as 'PAMP-triggered immunity'(PTI) and 'DAMP-triggered immunity'(DTI), respectively [9, 10]. However, plant pathogens have developed new strategies to overcome the host defence by secreting 'effector molecules' which suppress the host defence barrier [11]. Plants have evolved further to recognize these effector molecules which induces an effector-triggered immunity (ETI). PAMPs and DAMPs are recognised by plasma membrane receptors termed as the pattern recognition receptors (PRRs) [10]. Effectors are recognized by a group of plant proteins named as 'R proteins'. In either case perception of the threat triggers changes in intracellular signalling which leads to change in gene expression profile. The transcriptional reprogramming is highly variable and it differs with the pathogen, type of elicitor, exposure period, etc. For activation of a strong plant defence, the transcription of genes involved the synthesis of particular defence hormones and the defence genes are enhanced. In Nature, a plant cell rarely encounters a single PAMP, DAMP or an effector alone, it is usually in a combination of two or all together. Hence, it should be clear that in Nature PTI, DTI and ETI may or may not be an independent event. The strength of the stimuli decides the outcome of the plant defence response.

The plant raises varied defence responses depending on the type of threat. However, increased reactive oxygen species (ROS), secretion of antimicrobial compounds, closing of stomata, callose deposition, secretion of lytic enzymes etc are a few common defence strategies used by the plant to prevent pathogen entry and its growth. The next immediate precaution taken by the plant is to communicate the signal regarding pathogen invasion to the neighbouring cells and distant uninfected tissues. This leads to activation of a systemic defence response in the entire plant body which is also known as priming [1, 12, 13]. Plant hormones and their derivatives play an important role in the rapid activation and transmission of stress signal in infected and uninfected tissues. Earlier, only SA, JA and ET were considered as plant defence hormones. However, recent studies show that other plant hormones like ABA, auxins, and gibberellins can also affect plant defence signalling. In this chapter, we discuss signalling regulated by plant defence hormones and the crosstalk between them.

3.2 Plant defence hormones

All hormone levels are kept in homeostasis by maintaining a balance between their biosynthesis and metabolism. In the absence of threat, defence hormones are maintained at a basal level. This is achieved either by hydrolysis of the hormone or by converting them to an inactive form. Plant defence hormones are often conjugated with other molecules which affect their native property. The conjugated forms of hormone serve as inactive stock that can be immediately hydrolysed and used when required. Also, some of the conjugated forms of hormone help them mobilize from one tissue to another and a few volatile forms of hormones are used for long-distance communication [14]. Elevated defence hormone levels trigger specific transcriptional reprogramming, which leads to the activation of hormone-specific defence response that helps the plant cell to overcome the threat [15]. Each hormone has their own intracellular receptor and hormone-responsive transcription factors, chromatin regulators which carry out the specific transcriptional reprogramming. As induction of an immune response is an energy-consuming process, the activity of the receptors and their transcription factors are controlled by negative regulators. The receptors of many hormones are located inside the nucleus where binding of a repressor triggers receptor degradation in the absence of the hormone [16]. In the section below, for each plant defence hormone, we discuss the stimuli that trigger the hormone synthesis, the key biosynthetic enzymes, hormones conjugate forms, hormone receptors, hormone-specific marker genes, and hormone-specific transcription factors.

3.2.1 Salicylic acid

SA plays a vital role in regulating the plant defence responses against biotrophic and hemibiotrophic pathogens. Perception of biotrophic pathogens or their molecular signatures (PAMP) leads to induction of SA-mediated signalling which prevents pathogen growth. Also, the recognition of pathogen-secreted effectors induces SA-mediated ETI [11]. SA also promotes the closing of stomata which prevents pathogen entry. *Arabidopsis* plants deficient in SA, *sid2* and NahG plants (a transgenic that expresses a bacterial enzyme that degrades SA) both fail to close stomata upon treatment with SA or on pathogen infection [17]. SA also prevents spreading of viruses, by promoting the closure of plasmodesmata [18]. Plants synthesize SA from shikimate or chorismate in plastids by the enzyme ISOCHORISMATE SYNTHASE (ICS) [19]. Plants produce different derivatives of SA, each with a different function. The enzyme SALICYLIC ACID METHYL TRANSFERASE-1 (BSMT-1) converts SA into methyl SA (MeSA) which is a volatile inactive form of SA used to mobilize the molecule to other plant parts where it is converted back to SA. Other derivatives of SA are SA glucose ester (SGE) and SA-O-β-glucoside (SAG), which are products of SA GLUCOSYL TRANSFERASE -1(SAGT-1) enzyme [20]. SAG is stored in vacuoles and is converted to SA upon recognition of the threat. SA can also be catabolized to 2,5-Dihydroxybenzoic acid (2,5-DHBA) by the enzyme SA-5-HYDROXYLASE (S5H) [21] BSMT-1 and SAGT-1 enzymes help in maintaining homeostasis of the cytoplasmic levels of SA.

Activation of SA-mediated defence response triggers ROS production, and expression of SA biosynthetic and defence genes [22]. Levels of SA are perceived by its intracellular receptors like *Non-expresser of PR*1 (*NPR-1*) gene (figure 3.2(a)). Ligand binding studies indicate that NPR1 has high affinity for SA. In the absence of SA, NPR1 is found in cytosol and forms oligomers. In the presence of SA, SA binds to NPR1 which prevents oligomerization and leads to translocation of NPR1 into the nucleus. NPR1 lacks a DNA-binding motif so it executes the SA-mediated transcriptional reprogramming by directly binding to the TGA family of transcription factors, predominantly by TGA2/TGA5/TGA6 proteins [23–25]. In *Arabidopsis*, TGA1, TGA4, TGA2, TGA5, TGA6, TGA3, and TGA7 are key regulators of SA-mediated response. In *Arabidopsis* NPR1 paralogs, NPR1 lIKE PROTEIN-3 and -4 (NPR3/4) are also identified as the intracellular receptors for SA but these two are negative regulators of SA-mediated signalling [1, 26]. The NPR1 acts as an activator and its binding to SA induces the expression of SA-responsive genes while NPR3/4 acts as a repressor and these two in the absence of SA binds to TGA and negatively regulates expression of SA-responsive genes [16, 26, 27]. The binding of SA to NPR3/4 prevents their repressor activity and allows NPR1 binding. Thus, the balance of gene expression in the presence and absence of any threat is regulated by binding of SA to the NPR1 (activator) and NPR3/4 (suppressor). The *Arabidopsis* mutants that either fail to synthesize SA (*isochorismate mutant; sid2*) or

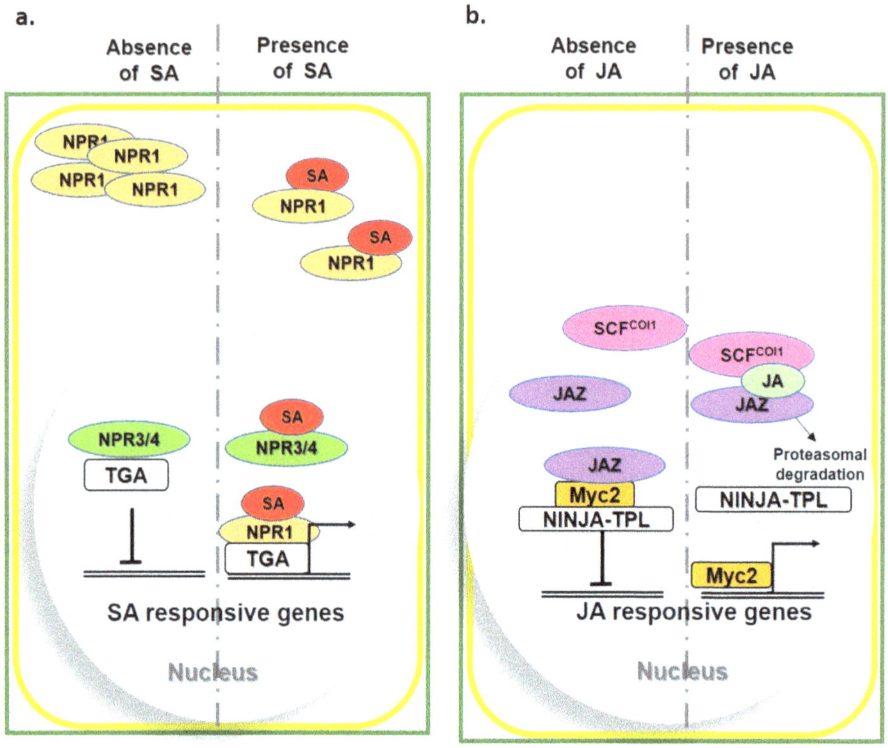

Figure 3.2. Activation of SA and JA receptors. (a) In the absence of SA, NPR1 oligomer is found in cytosol. In the presence of SA, SA binds to NPR1 and leads to translocation of NPR1 into the nucleus. NPR1 binds to the TGA transcription factors which executes SA-mediated transcriptional reprograming. NPR3/4) are negative regulators that prevent their repressor activity. (b) In the absence of JA-Ile, binding of MYC2 is prevented by JAZ which triggers binding of other co-repressors TPL, TRP and NINJA. In the presence of JA-Ile, COI1 binds to JA, the SCF^{COI1} complex polyubiquitinates JAZ followed by its proteasomal degradation; as a result MYC2 is released and it initiates the expression of JA-responsive genes.

sense SA levels (*npr1*/*npr2*/3 mutants) display enhanced susceptibility to biotrophic infection. SA-mediated cellular signalling involves the expression of SA biosynthesis genes like *ICS, BSMT*-1, *SAGT-1, Phenylalanine Ammonia Lyase (PAL)* genes, etc. Other defence genes like *Pathogenesis Related genes (PR1,* -3, -5*), Phytoalexin Deficient*-4 *(PAD4), PBS*3 *(avrPphB Susceptible-3), enhanced disease susceptibility (EDS)*-1, and -5 are a few well-studied SA-responsive genes [28]. These defence genes also serve as markers for studying the induction of SA-mediated defence responses in many plants.

The cell that perceives pathogen attack sends an alarm signal to the neighbouring tissue, which leads to activation of plant immunity even in the uninfected tissues and the entire plant becomes alert for subsequent attack. This is termed as systemic acquired resistance (SAR). SA is essential and indispensable for activation of SAR. The *Arabidopsis* mutants *sid2* and *npr1*/*npr2*/3 mutants fail to show SAR upon infection. During SAR, activated cells use small mobile molecules to communicate the threat signal to distant tissue. N-hydroxypipecolic acid (NHP), azelaic acid, MeSA, lipid transfer protein (DIR1), Glycerol 3-phosphate, and Abietane diterpenoid dehydroabietinal are a few mobile molecules tested for their direct or indirect involvement in the induction of SAR [28]. Among these molecules NHP is essential for SAR and SA is required for its biosynthesis.

3.2.2 Jasmonic acid

JA is required for development, and tolerance to biotic as well as abiotic stress. Activation of a JA-mediated response is observed during infection with hemibiotropic as well as necrotrophic pathogen, cell-wall damage, wound, herbivore attack and also upon touch stimuli [29–33]. Root colonising rhizobacteria activates a systemic immune response which is different from SA-induced SAR known as induced systemic resistance (ISR). This ISR involves activation of JA-mediated defence response [34]. JA is biosynthesized by the enzymes PHOSPHOLIPASE-α (PLA) or DEFECTIVE ANTHER DEHISCENCE-1(DAD1) using the fatty acids (α-linolenic acid) present in the chloroplast membrane. The JA biosynthetic enzymes that are involved downstream of PLA are LIPOXYGENASE (LOX2), ALLENE OXIDE SYNTHASE (AOS), ALLENE OXIDASE CYCLASE (AOC), and 12-OXO-PHYTODIENOIC ACID (OPDA) REDUCTASE (OPR3). The bioactive form of JA is (+)7-iso-Jasmonyl-L-isoleucine (JA-Ile) formed by the enzyme JASMONIC ACID RESISTANCE-1 (JAR1). The initial steps of JA synthesis occur in the chloroplast followed by β-oxidation in the peroxisome while the action of JAR is in the cytoplasm. There are two more pathways reported for JA biosynthesis; the hexadecane pathway and OPR3 independent pathway, both follow the same sequence from chloroplast to peroxisome to cytoplasm [35].

JA-mediated signalling is activated after recognition of JA-Ile in the nucleus by its receptor CORONATINE INSENSITIVE-1 (COI1) (figure 3.2(b)). COI1 is one of the many proteins present in SKP1-CULLIN1-F-box-type (SCF^{COI1}), and is an E3 ubiquitin ligase [36]. Most of the JA-regulated genes are transcribed by basic helix-loop-helix (bHLH) type of DNA-binding protein, the MYC2 transcription factor [37]. When there is no JA-Ile, the function of MYC2 is prevented by direct binding of a JA transcriptional repressor, JASMONATE ZIM DOMAIN (JAZ) proteins. Binding of JAZ to MYC2 triggers binding of many other co-repressors like TOPLESS (TPL), TOPL-related protein (TRP), NOVEL INTERACTOR OF JAZ (NINJA) [32]. All these repressors together recruit histone deacetylase that silence the expression of the JA-responsive gene by chromatin condensation. However, in the presence of JA-Ile, the SCF^{COI1} complex polyubiquitinates JAZ protein followed by its proteasomal degradation [38, 39]. In the absence of JAZ, MYC2 is released and it initiates the expression of JA-responsive genes. The JA-induced defence proteins are PR3, PR4, PR12 and PLANT DEFENSIN-1.2 (PDF1.2). The transcription of JA biosynthetic genes, *COI1, MYC2*, and the repressor *JAZ* proteins are positively regulated by JA levels. The JA biosynthetic genes (*LOX2, AOS, AOC, OPDA, OPR3, JAR1*), receptor (*COI1*), and repressor (*JAZ*s) are all JA-responsive genes and are extensively used as makers for studying activation of JA signalling.

Other than JA-Ile, intermediates of JA pathways are also reported to have a role in plant defence responses. Accumulation of OPDA in *jar1* tomato plant provides enhanced tolerance to *Botrytis infection* [40]. Cis-jasmone, another derivative of JA, repels herbivorous insects and attracts predators of the pests. Like MeSA, Methyl jasmonic acid (MeJA) is a volatile form of JA which induces expression of *PDF1.2* and *VEGETATIVE STATE PROTEIN (VSP)*, and it also provides resistance to *Botrytis cinera* infection in *Arabidopsis* [14].

3.2.3 Ethylene

ET was the first gaseous hormone to be discovered, which due to its volatile nature is easily transported through membranes. Like JA, ET also regulates plant development, biotic and abiotic stress responses [41, 42]. Commercially, ET is well known for its use in artificial fruit ripening. In plants, ET biosynthesis involves only two steps. S-adenosyl methionine, the precursor molecule, is converted into 1-aminocylopropane-1-carboxylic acid (ACC) by the action of ACC SYNTHASE (ACS) enzyme. The action of ACC OXIDASE enzyme releases the ET molecule from ACC. Unlike other plant hormone receptors where the receptor is present inside the nucleus, ET receptor ETHYLENE RESPONSE-1 (ETR1) is embedded in the membrane of endoplasmic reticulum (figure 3.3(a)). The receptor is associated with a kinase CONSTITUTIVE TRIPLE RESPONSE 1 (CTR1) which keeps another ER membrane protein ETHYLENE INSENSITIVE (EIN2) in phosphorylated state in the absence of ET. In the presence of ET, the dephosphorylated form of EIN2 cleaves and releases the C-terminal domain also known as CEND which translocates into the nucleus and activates EIN3, which initiates expression of ETHYLENE RESPONSIVE FACTOR-1 (ERF1) which is a DNA-binding protein, a transcription factor. This ERF1 carries out the expression of the ET-responsive genes. In *Arabidopsis*, overexpression of *AtERF1* provides resistance to *Botrytis cinerea* infection. In the absence of ET, EIN3 is negatively regulated by F-box proteins, namely EIN3 binding F-box-1 and -2 proteins (EBF1/2), which promotes proteasomal degradation of EIN3.

3.2.4 Abscisic acid

ABA is a sesquiterpene which regulates biotic as well as abiotic stress responses. ABA response is activated upon infection with a few biotrophic pathogens like wheat rust fungi as well as upon infection with necrotrophic pathogens like *B. cinerea* [43, 44]. ABA induces stomata closure which prevents water loss under abiotic stress and prevents pathogen entry during infection [45–47]. Additionally, activation of receptors of Pep1 peptide (a DAMP) and flg22 (a PAMP) triggers ABA-mediated stomata closure. Upon the perception of threat, plants quickly accumulate ABA through the carotenoid pathway using beta-carotene as the precursor molecule [48]. ABA is conjugated by glycosylation with the help of an enzyme, UDP-glucosyltransferase, to form the ABA-glucose ester which is an inactive form stored in plant cells and under stress, a betaglucosidase enzyme converts it to active ABA [49, 50]. The intracellular ABA receptor is PYRABACTIN RESISTANCE-1-LIKE (PYL) proteins (figure 3.3(b)). There are

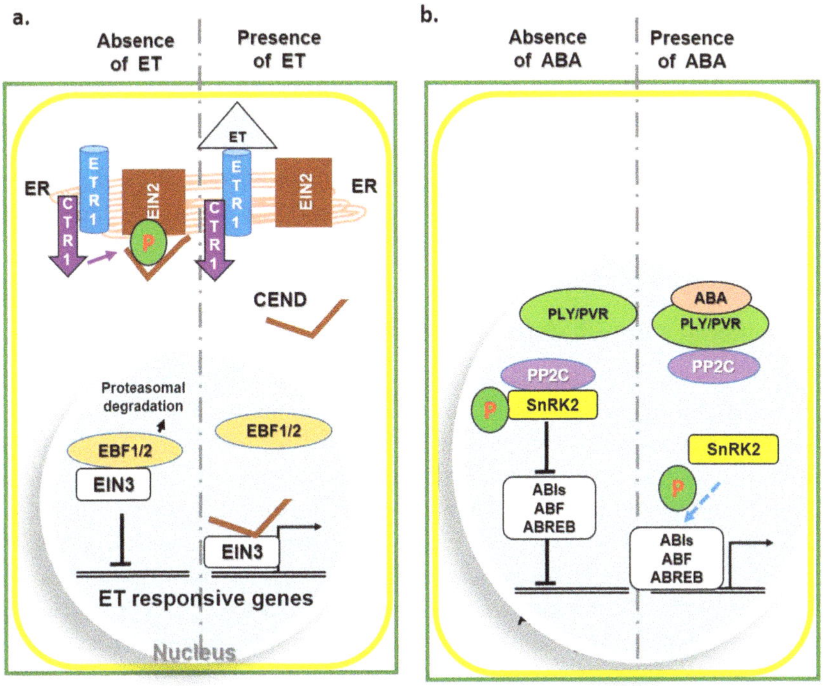

Figure 3.3. Activation of ET and ABA receptors. (a) ETR1 is embedded in the membrane of endoplasmic reticulum and it is associated with a kinase CTR1) which keeps EIN2 in phosphorylated state (inactive) in the absence of ET. In the presence of ET, the dephosphorylated from releases its C-terminal domain, CEND which activates EIN3 in nucleus. In absence of ET, EIN3 is negatively regulated by EBF1/2 which promotes its proteasomal degradation. (b) The ABA bound receptor PYL interacts with PP2C which in absence of ABA remains bound to a protein kinase, SnRK2. The release of SnRK2 activates signalling events that result in the phosphorylation of transcription factors like ABIs, ABF, ABREBs etc.

thirteen PLYs identified in *Arabidopsis*. The binding of ABA to PYL promotes interaction with another signalling component PP2C, a phosphatase, which in the absence of ABA remains bound to a protein kinase, SnRK2. The release of SnRK2 activates signalling events that result in the phosphorylation of ABSISIC ACID INSENSITIVE (ABI) transcription factors [51]. ABI3 and ABI5 are key transcription factors that positively regulate the ABA-mediated plant defence response. ABA induces the expression of genes like AIL1, RD29B, RAB18, EM1, and EM6 which are well-studied ABA-responsive marker genes [51].

3.3 The plant hormone crosstalk

The signalling cascades induced by plant hormones may interact in a synergistic or antagonistic manner. The crosstalk usually occurs at multiple levels in the signalling cascade. In the antagonistic type, one hormone may interfere with the signalling of the other hormone by either degradation of a positive regulator or enhancing the stability of the repressor, or competing for a common regulator (examples are discussed in the following section). The synergistic crosstalk is quite simple where the two hormones in interaction induce or activate a common regulator which controls the final response. It is observed that the crosstalk between any two plant hormones is not fixed; it may change from synergistic to antagonistic depending upon the state of the plant. Also, Aerts *et al* [52] in their review have classified the hormone crosstalk based on the type of molecular interaction viz, crosstalk involving protein-level interaction, gene expression level and hormone homeostasis level [52]. Here we discuss the crosstalk reported extensively during the plant defence responses excluding those interactions observed during growth and development.

3.3.1 Crosstalk in a growth–defence trade-off

In a state of threat, an organism's complete energy is channelled to overcome the stress and it makes its whole effort to survive. Analogous to animals, induction of the defence response causes suppression of growth and vice versa. Upon activation of plant immunity, the transcription reprogramming attack involves upregulation of defence gene expression and simultaneous downregulation of regular cellular activity like photosynthesis and other growth-related processes. The suppression of growth is achieved by the crosstalk between the defence and growth hormones. Auxin, gibberellic acids (GAs) and brassinosteroids are the major growth hormones that are targeted by plant defence hormones. Auxin is reported as a negative regulator of plant defence [53]. Hence, upon induction of plant defence, the auxin-mediated responses are downregulated. Auxin is shown to interact with JA in an antagonistic manner during necrotrophic infection. In *Arabidopsis*, auxin induces the expression of transcription factor, *AtWRKY57* which enhances the expression of the negative regulators of JA, *AtJAZ1* and *AtJAZ5* [54]. Similarly, activation of the SA-mediated defence response inhibits plant growth by suppressing auxin-mediated function. Treatment with BTH, an SA analogue, suppresses the expression of almost 21 auxin-responsive genes [55]. However, it should be made clear that SA and JA are also required for growth and development in the absence of biotic stress [28]. Similarly, ABA receptors AtPLY8 and AtPLY9 interact with auxin response receptor [56].

Another plant growth hormone that is targeted by defence hormones to suppress growth is GA. GA is required for cell elongation and growth. Similar to JAZ repressors, GA-mediated response is regulated by the DELLA repressor proteins. In the absence of GA, DELLA prevents the expression of GA-regulated genes by directly binding to the positive regulators like GIBBERELLIC ACID INSENSITIVE (GAI), RGA-like protein (RGL1, -2, -3), and REPRESSOR OF GA (RGA) transcription factors. JAZ1 directly binds to DELLA in the absence of JA promoting GA-responsive growth. During infection when the JA levels are elevated, JAZ1 is degraded, thus releasing DELLA free to bind to its targets and suppress the GA-mediated growth [57]. GA also interacts with ABA in an antagonistic manner during seed germination and abiotic stress [51]. ABA interferes with GA response through positive interaction between ABI3-ABI5 with DELLA proteins. Brassinosteroid (BR) is another plant growth hormone required for aerial growth of plants. High levels of BR inhibit JA biosynthesis and a high level of JA suppresses BR biosynthesis [58].

3.3.2 Crosstalk between defence hormones

The SA-mediated defence response acts against a wide range of biotrophic/hemi-biotrophic infection and the JA-mediated defence response is triggered against most necrotrophic infection. Also, SA and JA interact in an antagonistic manner, which means activation of SA-mediated defence response would suppress the expression of JA-regulated genes and vice versa. This statement is supported by extensive mutant studies. In *Arabidopsis*, JA-insensitive plants, *coi1*, and *jin1* mutants, accumulate a high level of SA [59, 60]. Likewise, plants that accumulate low SA, like *sid1–2* mutants and NahG plants, display high levels of JA. Gene expression studies reveal that the SA-responsive WRKY genes (*WRKY70* and *WRKY62*) negatively regulate the expression of *PDF1.2* [61, 62]. The SA-responsive transcription factor, TGA2, TGA5, and TGA6 also negatively regulates the activation of the JA-mediated pathway [63]. There are multiple examples where this antagonistic theory stands true, but there are also examples where, SA and JA function synergistically. A high concentration of SA and JA leads to antagonistic regulation while at their low concentration (10–100micromole) synergistic effect is observed [64].

Another well-explored hormonal crosstalk is the JA and ET signalling interaction which displays a noticeable synergistic response. This is achieved by activation of a common transcription factor, OCTADECANOID-RESPONSIVE AP2/ERF-59 (ORA). *ORA59* gene is expressed in response to elevated levels of JA as well as ET. Overexpression of *ORA59* renders resistance against *B. cinerea* infection and induces expression of the defence gene, *PDF1*.2 which is a JA-responsive gene [42]. JAZ proteins, the repressors of JA signalling, physically interact with ET-responsive genes, EIN3 and EIL1, and suppress ET-regulated response [65]. Increased JA level promotes JAZ degradation and releases suppression on ET-responsive transcription factors thus acting synergistically.

In contrast, ET and SA interact antagonistically in plant defence and growth. The mutant *ethylene overproducer*-1 which has high levels of ET accumulates a significantly low level of SA [66]. Likewise, ET-deficient double-mutant, *ein3*-1 *eil*1-1 exhibits a high expression of levels of *ICS-1*, a key enzyme involved in synthesis of SA [67]. SA is also shown to suppress expression of the genes involved in jasmonate and ET signalling through WRKY70 transcription factor [61, 62]. However, under certain circumstances, SA, JA, and ET, are all activated together. For example, the perception of Flg22, a bacterial PAMP, leads to activation of all three, JA-, ET- and SA-regulated genes in *Arabidopsis* [68]. Similarly, the over-expression of a DAMP-induced rice transcription factor, *OsAP2/ERF152* induces the expression of SA and JA–ET-responsive genes, and leads to enhanced tolerance against both biotrophic (*Pst*) and necrotrophic (*Rhizoctonia solani*) pathogens in *Arabidopsis* [69]. Likewise, the NAC family of transcription factors is expressed in response to ABA, ET, SA, or JA hormones [63]. Such transcription factors like *OsAP2/ERF152* and *NAC* family members act as a common responsive gene that participates in plant immunity in response to any of the defence hormones. Identifying such common players will help in the development of plant varieties displaying tolerance to a wide range of pathogens.

ABA crosstalk is comparatively less explored in biotic stress responses. ABA and JA act synergistically in herbivory-mediated stress response. The activity of the positive regulators of ABA, ABI5, and ABI9 transcription factors, are suppressed by the binding of the JA repressor, JAZ. The accumulation of JA causes the degradation of the repressor JAZ and the release of these two ABA-responsive transcription factors [32]. However, ABA–JA is also reported to act antagonistically through the interaction between ABA receptor PYL6 and JA-responsive transcription factor, MYC2.

3.3.3 Pathogens exploit phytohormone crosstalk

The plant homeostasis and activation of successful defence relies on perception of the pathogen and on the fine-tuning of the crosstalk between the plant hormones. When the plant fails to perceive a pathogen correctly disease is developed. During evolution in plant–pathogen interaction, many pathogens have evolved strategies to take advantage of phytohormonal crosstalk in the favour of the pathogen virulence. A famous example is of a bacterial toxin, coronatine, produced by *P. syringae* which is a molecular mimic of JA-isoleucine. Coronatine has a high affinity for the JA-Ile receptor, SCF^{COI1}, and it directly binds to the receptor. This, binding triggers degradation of the repressor JAZ and causes the release of MCY2 the positive regulator of JA response. In the absence of coronatine, the *P. syringe* induces an SA-mediated response which prevents the progression of infection by the closing of stomata. Thus, by secreting coronatine, the pathogen diverts the SA–JA crosstalk in its favour which causes stomatal reopening and enhances the bacterial growth in tomato leaves. *P. syringe* also secretes AvrB and HopZa which also interfere in JA signalling by disturbing functions of COI-1 and JAZ proteins [70, 71]. Another such example of pathogen interference is where *Pst* produces indole acetic acid (IAA), a form of auxin, which promotes pathogen growth by suppressing the SA-regulated response in *Arabidopsis* [72]. IAA also regulates the expression of many bacterial virulence genes which are required by the pathogen for their growth and survival within the host.

Pathogens can modulate plant defence by degrading plant defence hormones. For example, *Ralstonia solanacearum* transforms SA into gentisic acid, which is further broken down to pyruvate and fumarate [73]. Some plant pathogens are capable of modifying plant hormones. Many pathogens carry NahG gene in their genome which codes for enzyme SA hydroxylases that degrade SA [74]. Similarly, the pathogen causing rice blast disease, *Magnaporthe oryzae*, secretes antibiotic biosynthetic monooxygenase (Abm) which inhibits JA-mediated signalling by converting JA to 12-OH-JA which is an inactive form [75].

3.4 Conclusions

With multiple plant hormones form multiple signalling networks, it becomes very difficult to compartmentalize which plant hormone regulates which type of pathogen. To be clear, during plant–pathogen interaction, 'one hormonal signal ON and all others OFF' is never the case. As stated by Jones and Dangl [12], if the above statement were true, then the defence response raised against one type of pathogen (biotrophic) would leave the plant more susceptible to infection with other types of pathogens (say necrotrophic). However, we do not observe such examples in the natural environment. Recent studies indicate that a fine fine-tuning between all the types of hormones leads to the final outcome. Thus, the present picture of plant hormone signalling is highly complex where multiple hormones have intense crosstalk and regulate a single response [22]. Thus, a time-lapse quantitative approach using sensors for multiple hormones would help us to better understand the status of each hormone change upon challenge [76]. Another research, shows that the SA and JA crosstalk that occurs during plant defence is not systemic, it is localized to the infected region [77]. Interestingly, when an *Arabidopsis* plant with JA and SA marker genes (VSP and PR1, respectively) tagged with fluorescent markers was infected with *P. syringe*, a clear spatial separation of JA- and SA-mediated signalling was noted. The infected region displayed programmed cell death, surrounding cells displayed expression of SA marker genes and, next to this region, the outermost layer exhibited expression of JA marker genes [78]. This concentric spatial separation of hormonal regulation within the same infected organ adds a completely new area to explore. Thus, time-lapse or spatiotemporal studies of plant–pathogen interaction is essential to improve our understanding of hormonal crosstalk (figure 3.4) [79].

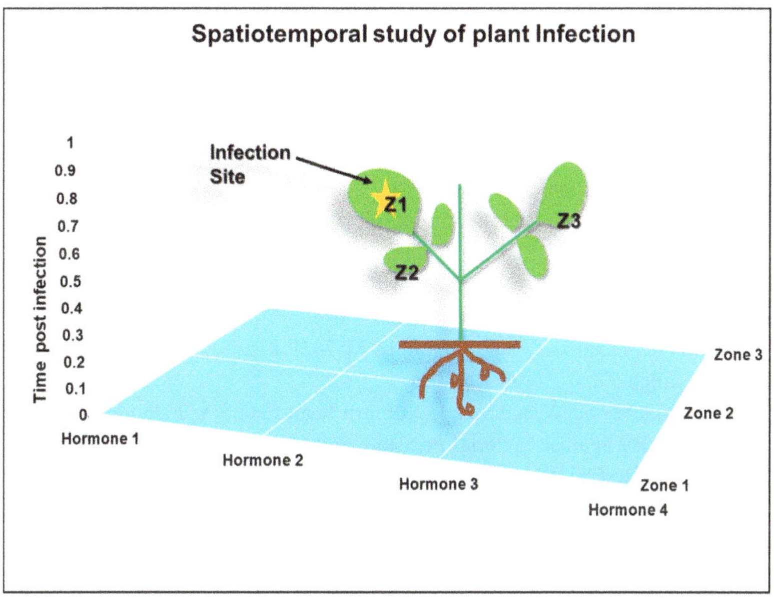

Figure 3.4. Spatiotemporal study of plant during infection. Using present knowledge and technology available, a spatiotemporal study of plant defence signalling will help us to better understand the crosstalk between the phytohormones. In the image Z1, Z2 and Z3 refers to different zone on the plant based on the distance from the site of infection. The hormones 1, 2, 3 and 4 represent the expression a marker gene representing a particular hormone. A spatiotemporal study may involve studying the expression of multiple defence hormone genes at different time points post-infection in the different parts of the plant body.

References

[1] Spoel S H and Dong X 2012 How do plants achieve immunity? Defence without specialized immune cells *Nat. Rev. Immunol.* **12** 89–100

[2] Liu W, Liu J, Ning Y, Ding B, Wang X, Wang Z and Wang G L 2013 Recent progress in understanding PAMP- and effector-triggered immunity against the rice blast fungus *Magnaporthe oryzae Mol. Plant* **6** 605–20

[3] Choi H W and Klessig D F 2016 DAMPs, MAMPs, and NAMPs in plant innate immunity *BMC Plant Biol.* **16** 1–10

[4] Yakushiji S, Ishiga Y, Inagaki Y, Toyoda K, Shiraishi T and Ichinose Y 2009 Bacterial DNA activates immunity in *Arabidopsis thaliana J. Gen. Plant Pathol.* **75** 227–34

[5] Souza C D A, Li S, Lin A Z, Boutrot F, Grossmann G, Zipfel C and Somerville S C 2017 Cellulose-derived oligomers act as damage-associated molecular patterns and trigger defense-like responses *Plant Physiol.* **173** 2383–98

[6] Duran-Flores D and Heil M 2016 Sources of specificity in plant damaged-self recognition *Curr. Opin. Plant Biol.* **32** 77–87

[7] Heil M and Land W G 2014 Danger signals–damaged-self recognition across the tree of life *Front. Plant Sci.* **5** 578

[8] Tanaka K, Choi J, Cao Y and Stacey G 2014 Extracellular ATP acts as a damage-associated molecular pattern (DAMP) signal in plants *Front. Plant Sci.* **5** 446

[9] Göhre V and Robatzek S 2008 Breaking the barriers: microbial effector molecules subvert plant immunity *Annu. Rev. Phytopathol.* **46** 189–215

[10] Monaghan J and Zipfel C 2012 Plant pattern recognition receptor complexes at the plasma membrane *Curr. Opin. Plant Biol.* **15** 349–57

[11] van Wersch S, Tian L, Hoy R and Li X 2020 Plant NLRs: the whistleblowers of plant immunity *Plant Commun.* **1** 100016

[12] Jones J D G and Dangl J L 2006 The plant immune system *Nature* **444** 323–9

[13] Bigeard J, Colcombet J and Hirt H 2015 Signaling mechanisms in pattern-triggered immunity (PTI) *Mol. Plant* **8** 521–39

[14] Dani K G S and Loreto F 2022 Plant volatiles as regulators of hormone homeostasis *New Phytol.* **234** 804–12

[15] Li B, Meng X, Shan L and He P 2016 Transcriptional regulation of pattern-triggered immunity in plants *Cell HostMicrobe* **19** 641–50

[16] Fukui J, Seto K, Takaoka Y and Okamoto M 2021 Ligand–receptor interactions in plant hormone signaling *Plant J.* **105** 290–306

[17] Melotto M, Underwood W, Koczan J, Nomura K and He S Y 2006 Plant stomata function in innate immunity against bacterial invasion *Cell* **126** 969–80

[18] Wang X, Sager R, Cui W, Zhang C, Lu H and Lee J Y 2013 Salicylic acid regulates plasmodesmata closure during innate immune responses in Arabidopsis *Plant Cell* **25** 2315–29

[19] Huang W, Wang Y, Li X and Zhang Y 2020 Biosynthesis and regulation of salicylic acid and N-hydroxypipecolic acid in plant immunity *Mol. Plant* **13** 31–41

[20] Dean J V and Delaney S P 2008 Metabolism of salicylic acid in wild-type, ugt74f1 and ugt74f2 glucosyltransferase mutants of *Arabidopsis thaliana Physiol. Plant.* **132** 417–25

[21] Zhang Y, Zhao L, Zhao J, Li Y, Wang J, Guo R and Zhang K 2017 S5H/DMR6 encodes a salicylic acid 5-hydroxylase that fine-tunes salicylic acid homeostasis *Plant Physiol.* **175** 1082–93

[22] Robert-Seilaniantz A, Grant M and Jones J D 2011 Hormone crosstalk in plant disease and defense: more than just jasmonate-salicylate antagonism *Annu. Rev. Phytopathol.* **49** 317–43
[23] Zhang Y, Fan W, Kinkema M, Li X and Dong X 1999 Interaction of NPR1 with basic leucine zipper protein transcription factors that bind sequences required for salicylic acid induction of the PR-1 gene *Proc. Natl Acad. Sci.* **96** 6523–8
[24] Zhang Y, Cheng Y T, Qu N, Zhao Q, Bi D and Li X 2006 Negative regulation of defense responses in Arabidopsis by two NPR1 paralogs *Plant J.* **48** 647–56
[25] Zhou J M, Trifa Y, Silva H, Pontier D, Lam E, Shah J and Klessig D F 2000 NPR1 differentially interacts with members of the TGA/OBF family of transcription factors that bind an element of the PR-1 gene required for induction by salicylic acid *Mol. Plant-Microbe Interact.* **13** 191–202
[26] Liu Y, Sun T, Sun Y, Zhang Y, Radojičić A, Ding Y and Zhang Y 2020 Diverse roles of the salicylic acid receptors NPR1 and NPR3/NPR4 in plant immunity *Plant Cell* **32** 4002–16
[27] Ding Y, Sun T, Ao K, Peng Y, Zhang Y, Li X and Zhang Y 2018 Opposite roles of salicylic acid receptors NPR1 and NPR3/NPR4 in transcriptional regulation of plant immunity *Cell* **173** 1454–67
[28] Peng Y, Yang J, Li X and Zhang Y 2021 Salicylic acid: biosynthesis and signaling *Annu. Rev. Plant Biol.* **72** 761–91
[29] Farmer E E and Ryan C A 1992 Octadecanoid precursors of jasmonic acid activate the synthesis of wound-inducible proteinase inhibitors *Plant Cell* **4** 129–34
[30] Braam J 2005 In touch: plant responses to mechanical stimuli *New Phytol.* **165** 373–89
[31] Jha G, Patel H K, Dasgupta M, Palaparthi R and Sonti R V 2010 Transcriptional profiling of rice leaves undergoing a hypersensitive response like reaction induced by *Xanthomonas oryzae* pv. oryzae cellulase *Rice* **3** 1–21
[32] Liu H and Timko M P 2914 2021 Jasmonic acid signaling and molecular crosstalk with other phytohormones *Int. J. Mol. Sci.* **22**
[33] Patlavath R, Pillai S E, Gandhi D and Albert S 2022 Cajanus cajan shows multiple novel adaptations in response to regular mechanical stress *J. Plant Res.* **135** 809–21
[34] Martínez-Medina A, Van Wees S C and Pieterse C M 2017 Airborne signals from Trichoderma fungi stimulate iron uptake responses in roots resulting in priming of jasmonic acid-dependent defences in shoots of *Arabidopsis thaliana* and *Solanum lycopersicum Plant Cell Environ.* **40** 2691–705
[35] Chini A, Monte I, Zamarreno A M, Hamberg M, Lassueur S, Reymond P and Solano R 2018 An OPR3-independent pathway uses 4, 5-didehydrojasmonate for jasmonate synthesis *Nat. Chem. Biol.* **14** 171–8
[36] Sheard L B, Tan X, Mao H, Withers J, Ben-Nissan G, Hinds T R and Zheng N 2010 Jasmonate perception by inositol-phosphate-potentiated COI1–JAZ co-receptor *Nature* **468** 400–5
[37] Kazan K and Manners J M 2013 MYC2: the master in action *Mol. Plant* **6** 686–703
[38] Thines B, Katsir L, Melotto M, Niu Y, Mandaokar A, Liu G and Browse J 2007 JAZ repressor proteins are targets of the SCFCOI1 complex during jasmonate signalling *Nature* **448** 661–5
[39] Pauwels L and Goossens A 2011 The JAZ proteins: a crucial interface in the jasmonate signaling cascade *Plant Cell* **23** 3089–100
[40] Larrieu A and Vernoux T 2016 QandA: how does jasmonate signaling enable plants to adapt and survive? *BMC Biol.* **14** 1–8

[41] Bleecker A B and Kende H 2000 Ethylene: a gaseous signal molecule in plants *Annu. Rev. Cell Dev. Biol.* **16** 1–18

[42] Huang P Y, Catinot J and Zimmerli L 2016 Ethylene response factors in Arabidopsis immunity *J. Exp. Bot.* **67** 1231–41

[43] Huai B, Yang Q, Qian Y, Qian W, Kang Z and Liu J 2019 ABA-induced sugar transporter TaSTP6 promotes wheat susceptibility to stripe rust *Plant Physiol.* **181** 1328–43

[44] Bharath P, Gahir S and Raghavendra A S 2021 Abscisic acid-induced stomatal closure: an important component of plant defense against abiotic and biotic stress *Front. Plant Sci.* **12** 615114

[45] Gudesblat G E, Torres P S and Vojno A A 2009 Stomata and pathogens: warfare at the gates *Plant Signal. Behav.* **4** 1114–6

[46] Sussmilch F C and McAdam S A 2017 Surviving a dry future: abscisic acid (ABA)-mediated plant mechanisms for conserving water under low humidity *Plants* **6** 54

[47] Agurla S and Raghavendra A S 2016 Convergence and divergence of signaling events in guard cells during stomatal closure by plant hormones or microbial elicitors *Front. Plant Sci.* **7** 1332

[48] Nambara E and Marion-Poll A 2005 Abscisic acid biosynthesis and catabolism *Annu. Rev. Plant Biol.* **56** 165

[49] Xu Z Y, Lee K H, Dong T, Jeong J C, Jin J B, Kanno Y and Hwang I 2012 A vacuolar β-glucosidase homolog that possesses glucose-conjugated abscisic acid hydrolyzing activity plays an important role in osmotic stress responses in Arabidopsis *Plant Cell* **24** 2184–99

[50] Liu S, Kracher B, Ziegler J, Birkenbihl R P and Somssich I E 2015 Negative regulation of ABA signaling by WRKY33 is critical for Arabidopsis immunity towards *Botrytis cinerea* 2100 *Elife* **4** e07295

[51] Chen K, Li G J, Bressan R A, Song C P, Zhu J K and Zhao Y 2020 Abscisic acid dynamics, signaling, and functions in plants *J. Integr. Plant Biol.* **62** 25–54

[52] Aerts N, Pereira Mendes M and Van Wees S C 2021 Multiple levels of crosstalk in hormone networks regulating plant defense *Plant J.* **105** 489–504

[53] Seo P J and Park C M 2009 Auxin homeostasis during lateral root development under drought condition *Plant Signal. Behav.* **4** 1002–4

[54] Jiang Y and Yu D 2016 The WRKY57 transcription factor affects the expression of jasmonate ZIM-domain genes transcriptionally to compromise *Botrytis cinerea* resistance *Plant Physiol.* **171** 2771–82

[55] Wang D, Pajerowska-Mukhtar K, Culler A H and Dong X 2007 Salicylic acid inhibits pathogen growth in plants through repression of the auxin signaling pathway *Curr. Biol.* **17** 1784–90

[56] Xing L, Zhao Y, Gao J, Xiang C and Zhu J K 2016 The ABA receptor PYL9 together with PYL8 plays an important role in regulating lateral root growth *Sci. Rep.* **6** 1–13

[57] Yang D L, Yao J, Mei C S, Tong X H, Zeng L J, Li Q and He S Y 2012 Plant hormone jasmonate prioritizes defense over growth by interfering with gibberellin signaling cascade *Proc. Natl. Acad. Sci.* **109** E1192–200

[58] Choudhary S P, Yu J Q, Yamaguchi-Shinozaki K, Shinozaki K and Tran L S P 2012 Benefits of brassinosteroid crosstalk *Trends Plant Sci.* **17** 594–605

[59] Kloek A P, Verbsky M L, Sharma S B, Schoelz J E, Vogel J, Klessig D F and Kunkel B N 2001 Resistance to *Pseudomonas syringae* conferred by an *Arabidopsis thaliana* coronatine-insensitive (coi1) mutation occurs through two distinct mechanisms *Plant* J. **26** 509–22

[60] Nickstadt A, Thomma B P, Feussner I V O, Kangasjärvi J, Zeier J, Loeffler C and Berger S 2004 The jasmonate-insensitive mutant jin1 shows increased resistance to biotrophic as well as necrotrophic pathogens *Mol. Plant Pathol.* **5** 425–34
[61] Li J, Brader G and Palva E T 2004 The WRKY70 transcription factor: a node of convergence for jasmonate-mediated and salicylate-mediated signals in plant defense *Plant Cell* **16** 319–31
[62] Li J, Brader G, Kariola T and Tapio Palva E 2006 WRKY70 modulates the selection of signaling pathways in plant defense *Plant J.* **46** 477–91
[63] Seo E and Choi D 2015 Functional studies of transcription factors involved in plant defenses in the genomics era *Brief. Funct. Genomics* **14** 260–7
[64] Mur L A, Kenton P, Atzorn R, Miersch O and Wasternack C 2006 The outcomes of concentration-specific interactions between salicylate and jasmonate signaling include synergy, antagonism, and oxidative stress leading to cell death *Plant Physiol.* **140** 249–62
[65] Zhu Z and Lee B 2015 Friends or foes: new insights in jasmonate and ethylene co-actions *Plant Cell Physiol.* **56** 414–20
[66] Li W, Nishiyama R, Watanabe Y, Van Ha C, Kojima M, An P and Tran L S P 2018 Effects of overproduced ethylene on the contents of other phytohormones and expression of their key biosynthetic genes *Plant Physiol. Biochem.* **128** 170–7
[67] Chen H, Xue L, Chintamanani S, Germain H, Lin H, Cui H and Zhou J M 2009 Ethylene Insensitive3 and Ethylene Insensitive3-Like1 repress salicylic ACID Induction Deficient2 expression to negatively regulate plant innate immunity in Arabidopsis *Plant Cell* **21** 2527–40
[68] Denoux C, Galletti R, Mammarella N, Gopalan S, Werck D, De Lorenzo G and Dewdney J 2008 Activation of defense response pathways by OGs and Flg22 elicitors in Arabidopsis seedlings *Mol. Plant* **1** 423–45
[69] Pillai S E, Kumar C, Dasgupta M, Kumar B K, Vungarala S, Patel H K and Sonti R V 2020 Ectopic expression of a cell-wall-degrading enzyme-induced OsAP2/ERF152 leads to resistance against bacterial and fungal infection in *Arabidopsis Phytopathology* **110** 726–33
[70] Zhou Z, Wu Y, Yang Y, Du M, Zhang X, Guo Y and Zhou J M 2015 An Arabidopsis plasma membrane proton ATPase modulates JA signaling and is exploited by the *Pseudomonas syringae* effector protein AvrB for stomatal invasion *Plant Cell* **27** 2032–41
[71] Jiang S, Yao J, Ma K W, Zhou H, Song J, He S Y and Ma W 2013 Bacterial effector activates jasmonate signaling by directly targeting JAZ transcriptional repressors *PLoS Pathog.* **9** e1003715
[72] Djami-Tchatchou A T, Harrison G A, Harper C P, Wang R, Prigge M J, Estelle M and Kunkel B N 2020 Dual role of auxin in regulating plant defense and bacterial virulence gene expression during *Pseudomonas syringae* PtoDC3000 pathogenesis *Mol. Plant-Microbe Interact.* **33** 1059–71
[73] Lowe-Power T M, Jacobs J M, Ailloud F, Fochs B, Prior P and Allen C 2016 Degradation of the plant defense signal salicylic acid protects *Ralstonia solanacearum* from toxicity and enhances virulence on tobacco *MBio* **7** e00656–16
[74] Qi G, Chen J, Chang M, Chen H, Hall K, Korin J and Fu Z Q 2018 Pandemonium breaks out: disruption of salicylic acid-mediated defense by plant pathogens *Mol. Plant* **11** 1427–39
[75] Patkar R N, Benke P I, Qu Z, Constance Chen Y Y, Yang F, Swarup S and Naqvi N I 2015 A fungal monooxygenase-derived jasmonate attenuates host innate immunity *Nat. Chem. Biol.* **11** 733–40

[76] Isoda R, Yoshinari A, Ishikawa Y, Sadoine M, Simon R, Frommer W B and Nakamura M 2021 Sensors for the quantification, localization and analysis of the dynamics of plant hormones *Plant J.* **105** 542–57

[77] Spoel S H, Johnson J S and Dong X 2007 Regulation of tradeoffs between plant defenses against pathogens with different lifestyles *Proc. Natl. Acad. Sci.* **104** 18842–7

[78] Betsuyaku S, Katou S, Takebayashi Y, Sakakibara H, Nomura N and Fukuda H 2018 Salicylic acid and jasmonic acid pathways are activated in spatially different domains around the infection site during effector-triggered immunity in *Arabidopsis thaliana Plant Cell Physiol.* **59** 8–16

[79] Tsuda K 2018 Division of tasks: defense by the spatial separation of antagonistic hormone activities *Plant Cell Physiol.* **59** 3–4

IOP Publishing

Advances in Biochemical and Molecular Mechanisms of Plant–Pathogen Interaction

Hitendra K Patel and Anirudh Kumar

Chapter 4

Secret of success: effector-triggered host susceptibility by phytopathogens

Anjana Sharma, Praveen Kumar Nayak and Hitendra K Patel

Plants encounter various pathogenic microorganisms and represent a primary obstacle in the breeding of cash and food crops. Thus, plants possess an arsenal of immune receptors that can recognize these pathogens and trigger immune response. Pathogens such as oomycetes, fungi, and bacteria, deploy effectors as weaponry for invasion, enabling successful colonization and proliferation within host plants. Pathogens deploy a repertoire of effectors to surpass physical barriers, evade or suppress immune detection, and obtain nutrients from host tissues. Here we discuss the effectors' capacity to aid pathogen entry into the host interior, dampen plant immune responses, and modify host physiology to benefit the pathogen, thus leading to susceptibility.

4.1 Introduction

Plants are attacked by numerous pathogenic microorganisms, which has the potential to cause an extensive yield loss in agriculture. When a pathogen invades a plant, it will lead either to immune activation which renders the plant resistant to pathogens, or it may lead to delayed activation or no activation of immunity thus making a plant susceptible [1]. The mechanism of immune response in plants has been extensively studied and it is well explained by the zigzag model of plant immunity given by Jones and Dangl. According to the model, plants respond to pathogen invasion using two layers of innate immune system, namely pattern triggered immunity (PTI) and effector-triggered immunity (ETI). The pathogen-specific patterns, which include nucleic acids (DNA, dsRNA, ssRNA, and 5′-triphosphate RNA), surface glycoproteins (GP), lipoproteins (LP), and cell-wall/membrane components (peptidoglycans [PG], lipoteichoic acid), are recognized by pattern recognition receptors (PRRs) [2]. These molecular patterns of pathogens

doi:10.1088/978-0-7503-5673-2ch4

upon recognition by PRRs, present on the plasma membrane of plants trigger the first layer of immunity, PTI—a non-specific type of immunity that can be activated against any kind of microbe. When PTI is activated, extensive downstream signalling occurs, which includes the generation of phytohormones, a significant transcriptional and metabolic reprogramming, a fast burst of Ca^{2+} and reactive oxygen species (ROS), and the activation of Ca^{2+}-dependent protein kinases (CDPKs) and mitogen-activated protein kinases (MAPKs) [3]. A successful pathogen would attempt to suppress this immunity by secreting effector molecules [4]. Effectors function by mimicking or inhibiting different plant cellular pathways or can also function by protecting the pathogen-secreted proteins [5]. These effectors are recognized by plant nucleotide-binding domain and leucine-rich repeat protein (NB-LRR) receptors [6]. Upon recognition of effector, the initiation of hypersensitive reaction will occur causing rapid local cell death which is the hallmark of ETI and, in addition, extensive downstream signalling will take place to limit the growth of the pathogen [7]. However, when a plant is susceptible to a particular pathogen, it fails to recognize these effectors or the pathogen will cleverly mask the effect or utilizes the effector to cause the disease. In this chapter we will discuss in detail the types of effectors and how theses effectors render the host plants susceptible.

Effectors can be divided based on their nature as proteinaceous and non-proteinaceous (pathogen-secreted secondary metabolites and small interfering RNA) or based on the location in the cell where they are functioning [8]. Proteinaceous effectors fall under two categories, namely apoplastic and cytoplasmic effectors that are secreted using the type II secretion system (T2SS) and the type III secretion system (T3SS), respectively, by many phytopathogenic bacteria. Many *Xanthomonas* species secrete two classes of type III effectors (T3Es), namely transcription activator-like (TAL) effectors (TALEs) and non-TAL effector proteins. On the other hand, many fungi and oomycete pathogens secrete effectors via a special structure called an appressorium. TALEs consist of a DNA binding domain, activation domain and a nuclear localization signal as a result of which they can localize to the nucleus, bind to DNA and induce the expression of certain genes that favour pathogen survival.On the other hand, non-TAL effectors target signalling pathways either by directly or indirectly binding with host immune components or manipulating them thus leading to dampening of the immune system [9]. Non-proteinaceous effectors are pathogen-secreted secondary metabolites which mostly include toxins that function by inhibiting translation, chromatin modifications, induction of programmed cell death etc and small interfering RNAs that lead to the suppression of target genes involved in immune responses [10]. During incompatible pant–pathogen interaction the pathogen not only supresses the host immune responses, but it also utilizes plant genes which provide the pathogen with an advantage. Such genes are referred to as susceptibility genes (S-genes) and the associated responses are called as susceptible responses. The induction of S-genes provides signals for pathogen infection as well as aiding in aggressiveness and accommodation in the host [1]. Plant S-genes facilitate pathogen establishment by serving three major functions: (i) aid host recognition and pathogen penetration, (ii) supression of immune regulators, (iii) facilitating pathogen proliferation through metabolic pathways [11]. The actions of effectors and S-genes cumulatively leads to susceptibility. Different processes utilized by pathogens to hijack host machinery using effectors and S-genes to establish infection *in planta* are discussed below.

4.2 Breaking down the physical barriers

4.2.1 Effectors manipulating stomatal defence and plant cell wall

In order to get entry inside the host, a pathogen must defeat the physical barriers which include cell wall and epidermis, or it can also make its entry through natural openings like stomata or wounds [12]. Pathogens have evolved diverse effectors to facilitate their entry; for example, effector Pep1 from *Ustilago maydis* is essential to make entry by penetrating the epidermis [13]. Similarly, some effectors have evolved to manipulate stomata for effective entry either by mimicking the hormone that helps stomatal opening (jasmonic acid (JA)) or by inhibiting the expression of hormone that leads to stomatal closure (salicylic acid (SA)) as these hormones are antagonistic in nature. Plant pathogens have the ability to manipulate stomatal opening using different mechanisms [14]. For instance, *Pseudomonas syringae* pv. *tomato* utilizes coronatine which can mimic JA-isoleucine and inhibit the function of SA thus leading to stomatal opening [14] (figure 4.1). The effector HopX1 of *P. syringae* pv. *tabaci* degrades JAZ transcription repressor which regulates JA signalling [15]. Another physical barrier that a pathogen has to overcome is the plant cell wall (PCW), which is composed of complex polysaccharide. PCWs serve as a physical barrier to the pathogen as well as being involved in immune signalling. For successful entry into a host plant, the pathogen has to alter the properties and structure of the PCW [16]. In order to damage host cells and encourage colonization, many plant infections, particularly those that lack particular penetrating features, release a variety of cell-wall-degrading enzymes (CWDEs), including glycoside hydrolases, glycosyltransferases, and pectin lyases, which hydrolyse cell wall components [17] (figure 4.1). This method is especially prevalent in necrotrophic infections, such as *Botrytis cinerea*, *Verticillium dahliae*, and *Mycosphaerella graminicola*, where specific CWDEs have a positive correlation with virulence [18]. Plants contain the repertoire of PCW loosening enzyme but pathogen does not contain enough enzyme to alter PCW as its genome size is small, hence it strategically utilizes enzymes from the plant [19]. PCW loosening and secretion allows easy movement and spread of pathogens and the component obtained from the degradation of PCW serves as a nutrient for pathogens [20]. *Pectobacterium atrosepticum* a vascular pathogen induces the release of plant rhamnogalacturonan I (PCW component) which is utilized by bacteria to form an extracellular matrix for bacterial assembly [20]. Plant methyl esterases and acetylesterases have been found to be exploited during *Pectobacterium carotovorum* and *B. cinerea* infection while esterase inhibition exhibited reduced susceptibility [21]. Thus, pathogens have evolved military tactics to modulate PCW composition for their benefits.

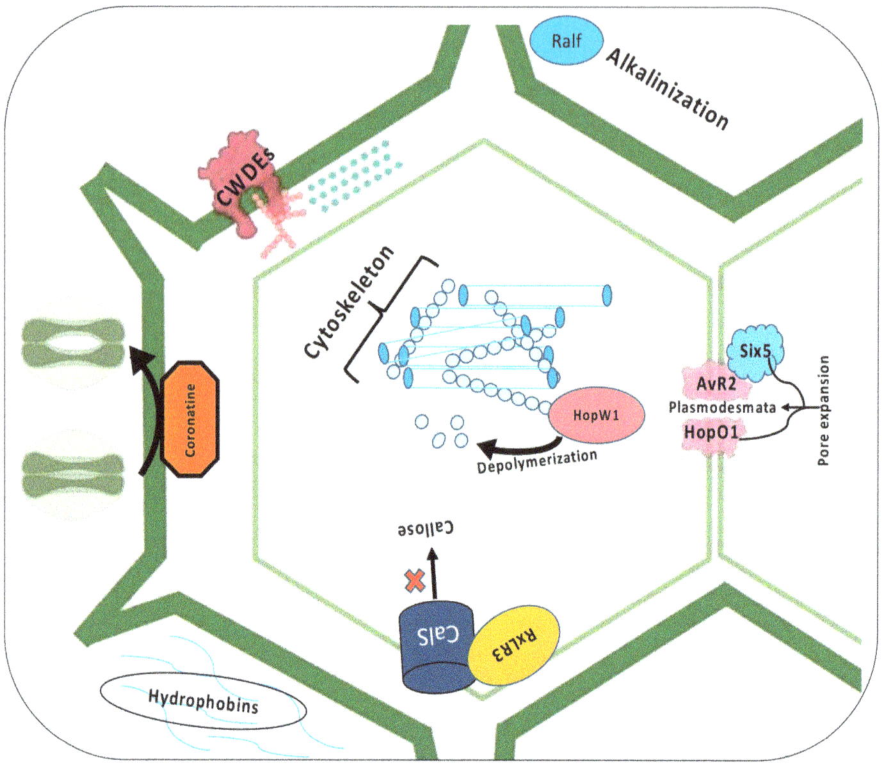

Figure 4.1. Entry and Niche establishment by pathogen using effectors. CWDEs—cell-wall-degrading enzymes, CalS—callose synthase. Coronatine, rapid alkalinization factor (RALF), HopW1, RxLR3, AvR2, Six5, HopO1 are effectors. Pathogens make their entry into the host by manipulating stomatal opening by inhibiting hormone SA which otherwise helps in stomatal closure. Once they are inside, the pathogen will secrete hydrophobins which will help in the attachment to cells and they also secrete effector-like RALF, which causes alkalinization, which allows the stability of secreted effectors in apoplast. Pathogens secrete CWDEs to cross the physical barrier and the components released are also utilized as nutrients by pathogens. RxLR3 interferes with callose synthase to avoid immune activation. HopW10 interferes with cytoskeleton to inhibit the movement of cargo-containing immune components. Effectors like AvR2, Six5 and Hop01 widen the plasmodesmata for easy movement of effector from one cell to another.

4.2.2 Effectors attacking plasmodesmata

Plasmodesmata (PD) are pores lined with membranes that establish connections between neighbouring cells, facilitating symplastic communication within plants. Upon immune activation, plants close PD as part of their defence mechanism, a strategy exploited by pathogens to aid infection. The transfer of effector through PD is essential for successful colonization of the plant tissues by phytopathogens. To neutralize closure of PD and the build-up of callose, the effectors target and enlarge the size of PD pore thus controlling cytoplasmic continuity. Avr2 and Six5 effectors from *Fusarium oxysporum* have been found to interact with PD in order to expand pore size [22] (figure 4.1). The effector AVR2 interacts with Six5, leading to the opening of PD, allowing AVR2 to traverse between cells via PD. *Phytophthora brassicae*'s RxLR3 effector interacts with CalS1, CalS2, and CalS3 which are callose synthases [23] thus, inhibiting callose deposition in PD (figure 4.1). The effector HopO1-1 of *P. syringae* resides in PD during infection and regulates PD permeability. HopO1-1 expression in *Arabidopsis* increases the distance of PD-dependent molecular flux of two neighbouring plant cells [24] (figure 4.1). Moreover, HopO1-1 interacts with PD-localized proteins PDLP7 and possibly PDLP5 and destabilizes them. Mutant plants lacking PDLP7 or PDLP5 exhibit significantly increased bacterial proliferation, suggesting the involvement of PDLP7 and PDLP5 in plant immunity against bacteria [25].

4.2.3 Effectors aiding in destruction of the host plant cytoskeleton

The cytoskeleton is involved in response to various stimuli, including both biotic and abiotic factors. For plant cells encountering pathogens, there are rapid localized changes of the typical cytoskeletal structure thus facilitating the transport of cargo-containing defence-related compounds [26]. It has been found that certain effectors interfere with plant metabolic and physiological processes by affecting the host cytoskeleton formation. For instance, HopW1, a type III effector of *P. syringae* inhibits endocytosis and the transport of specific proteins to vesicles by interacting with actin and destroying the actin cytoskeleton (figure 4.1). The degradation of F-actin *in vitro* by shortening of actin filaments is done by the C-terminal region of HopW1 [27]. An effector ROPIP1 of powdery mildew fungus can interact with microtubule-associated ROP GTPase HvRACB of barley and destabilize microtubules [28]. Notably, T3E XopR of *Xanthomonas oleifera* go through liquid–liquid phase separation (LLPS) by exploiting multivalent intrinsically disordered region (IDR)-mediated interactions of *Arabidopsis* actin cytoskeleton. During infection, XopR slowly moves into the host cell to form macromolecular complexes with actin-binding proteins in the cell cortex, thereby disrupting several stages of actin assembly and the host actin cytoskeleton [29].

4.3 Creating conditions favourable for infection

4.3.1 Effectors involved in construction of hydrophobic space

Mycelium or spores often face the challenge of transitioning from an aqueous environment to growing in air while infecting plants. Certain effectors facilitate infection by formation of hydrophobic spaces between host plants and pathogens. Hydrophobins (HP) are tiny fungal proteins secreted by fungi, characterized by eight conserved cysteine residues. These proteins are known to self-assemble as amphiphilic monomolecular membranes at the interface between hydrophobic and hydrophilic regions. They function by forming a hydrophobic coating in mycelium or spores, helping in attachment to surfaces which are hydrophobic, interplay with the environment, defence against the host, and other processes required in aerial growth of mycelium and spore dispersal for their transition from aqueous environments [30] (figure 4.1). In the early stages of infection in *Magnaporthe oryzae*, the hydrophobin MPG is significantly induced and functions as a sensor on the hydrophobic surface of plants, initiating the attachment development process [31]. Mpgl deletion mutants have decreased pathogenicity because of problems with attachment formation and attachment deficits on hydrophobic surfaces. Cutinase 2 is effectively recruited and retained when MPG1 is coated on hyphae, which aids in appressorium penetration and differentiation. *Fusarium graminearum* has one class II gene (FgHyd5) and four class I hydrophobin genes (FgHyd1–FgHyd4) in its genome. ΔFgHyd2, ΔFgHyd3, and ΔFgHyd4 are examples of single gene deletion mutants that exhibit decreased pathogenicity, which is explained by decreased hyphal penetration of the water–air interface and decreased adhesion to hydrophobic plant surfaces [32]. Furthermore, the rice blast fungus's extracellular matrix protein EMP1 functions similar to hydrophobin. EMP1 plays a critical role in *M. oryzae* appressorium production, as evidenced by the dramatically decreased pathogenicity and appressorium formation of knockout mutants, although mycelium growth and sporulation are intact [33].

4.3.2 Effectors inducing extracellular alkalinisation

Fungal pathogenicity is largely controlled by environmental pH, which also has a major impact on the growth and development of pathogenic fungus. Extracellular alkalinization, which is a pH increase in the surrounding host tissues during fungal-induced plant infection, is frequently observed and is thought to be connected to fungal pathogenesis [34]. Widespread in fungi, rapid alkalinization factor (RALF) homologues stimulate the activation of conserved mitogen-activated protein kinases (MAPKs) crucial for pathogenicity, raising extracellular pH levels and encouraging invasive fungal growth (figure 4.1). Plant cell elongation is impeded by the inhibition of plasma membrane H^+-ATPase AHA2 caused by endogenous RALF–Feronia signalling [35]. The fungus that infects roots *F. oxysporum* utilizes a functional homologue of the plant regulatory peptide RALF to cause alkalinization and induce plant disease. This peptide hormone raises the pH of the apoplastic environment around fruit tissue or the rhizosphere by more than two units, which promotes fungal colonization [36]. Mutants of *F. oxysporum* that lack a functional RALF peptide show markedly decreased virulence in tomato plants and are unable to cause host alkalinization, instead inducing a strong host immunological response. The plant receptor-like kinase Feronia, which also modulates the reaction to endogenous plant RALF peptides, appears to be the target of F-Ralf. Increased tolerance to *F. oxysporum* is seen in *Arabidopsis* plants deficient in Feronia, the receptor-like kinase that is responsible for RALF-triggered alkalinization response [37].

4.4 Protecting or masking themselves

4.4.1 Effectors targeting pattern recognition receptor

Plants possess PRR at the plasm membrane, which act as an innate immune barrier by recognizing patterns associated with pathogens. The recognition of patterns and its subsequent activation results in stomatal closure and inhibition of initial pathogen proliferation [38]. Pathogens consist of effectors that can interfere with the PRR complex. Effector AvrPtoB from *P.syringae* is an E3 ubiquitin ligase which mediates the degradation of several PRRs [39] (figure 4.2). FLS2 (Flagellin-sensing 2)

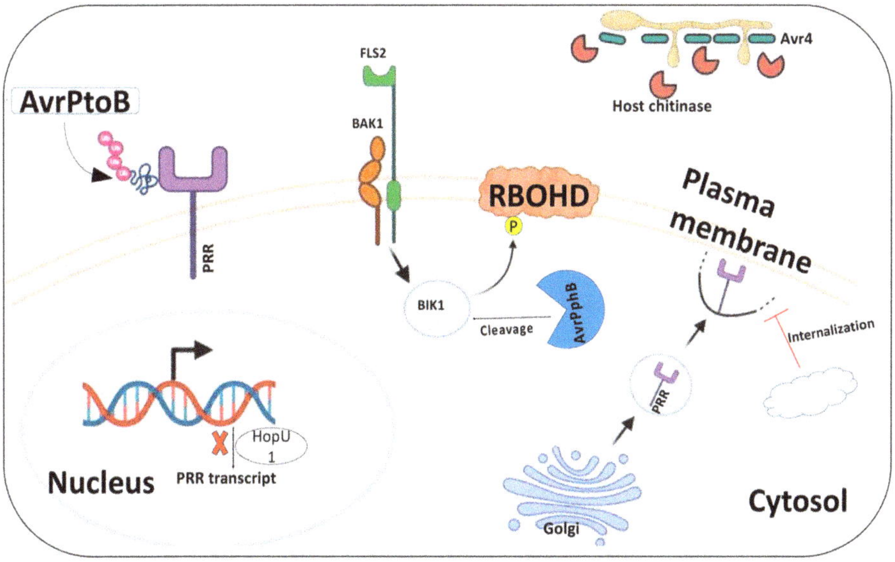

Figure 4.2. Masking of effectors from recognition by host. PRR—pattern recognition receptor; RBOHD—respiratory burst oxidase homologue D; FLS—flagellin sensing 2; BAK1—BRI1 associated receptor kinase 1; BIK1—brassinosteroid insensitive 1.AvrPtoB, Avr4, AvrPphB, AVR3a, HopU1 are effectors. Once inside the host, pathogens will deploy different strategies to mask themselves using effectors in order to inhibit immune initiation. Fungal pathogen utilizes Avr4 effector, which binds to chitin thus protecting it from host chitinases. AvrPtoB does the ubiquitination of PRR receptor to degrade it thus masking its PAMP recognition. HopU1 causes the degradation of protein GRP7 which is an RNA-binding protein subsequently lowering the PPR transcript. Avr3a inhibits the internalization of vesicle to plasma membrane containing PPR thus reducing its number in the plasma membrane. AvrPphB is a protease which does the cleavage of BIK1, which is the receptor-like cytoplasmic kinase and it plays a central role in signalling during PAMP-triggered immunity. BIK1 phosphorylates RBOHD as a result of which there will be a reactive oxygen burst initiating immune signalling which will reinforce cell wall strengthening and production of antimicrobial compound.

and the EFR (elongation factor Tu receptor) are two well characterized PRRs whose kinase activity has been found to be inhibited by effector AvrPto, thus enhancing bacterial virulence [40]. Certain effectors, instead of targeting PRR degradation, interfere with translation of PRRs. A mono-ADP ribosyltransferase called HopU1 *of P. syringae* targets GRP7 which is an RNA-binding protein. HopU1 functions by inhibiting the binding of GRP7 to PRR mRNAs, thus resulting in reduced PRRs number at the plasma membrane and subsequently reduced immune responses [41] (figure 4.2). Some effectors can hinder the PRR complex components and downstream signalling. Members of the plant receptor-like cytoplasmic kinase (RLCK) which belongs to subgroup VII has been shown to play a positive role in immune responses [42]. A member of the PRR complex BIK1 RLCK transphosphorylates primary immune receptors and its coreceptor BAK1 and also transphosphorylates the RBOHD NADPH oxidase, leading to an oxidative burst thus acting as an antimicrobial compound and strengthening the plant cell wall [43] (figure 4.2). Many effectors of bacteria have been recognized to target RLCK. AvrPphB effector from *P. syringae* is a protease which has the ability to cleave BIK1 and other RLCKs (figure 4.2) [44].

4.4.2 Effectors masking PAMP perception

PAMPs are the molecules that are recognized by PRR but some pathogens have evolved the way to actively mask PAMPs (peptidoglycan, β-glucan, chitin, chitosan, cellulose) which serve as a potent immune elicitor, thus restricting its recognition and ultimately the suppression of immune signalling which otherwise upon recognition by PRR induces the production of several protease, pathogenesis-related (PR) proteins, phytoalexin [45]. Pathogen *Cladosporium fulvum* consists of a small cysteine rich apoplastic effector Avr2 that binds and inhibits the papain-like cytoplasmic protease Rcr3 and PiP1 secreted by the host during infection [46]. The oomycete pathogen *Phytophthora infestans* secreted effectors EPIC1 and EPIC2B also targets same protease in tomato and potato [47]. Apart from defending themselves from proteolytic actions, filamentous pathogens are able to release effectors that give passive defence. The *C. fulvum* consists of apoplastic effector Avr4 which is a carbohydrate-binding module family 14 (CBM14). CBM14 shows specific binding to chitin and protects fungal chitin from host chitinases [48] (figure 4.2). The expression of AvR4 is extensively seen during the colonization phase and localizes on the surface of hyphae thus protecting it from host chitinases. In this way the pathogen masks its PAMP from being perceived and mutation inAvR4 is associated with reduced virulence [48].

4.5 Interfering with host plant cell physiological activities and manipulating plant downstream immune responses

4.5.1 Effectors that mimic and interfere with plant hormones

Different plant hormones form a complex network to regulate growth, development and defence. During biotrophic and hemibiotrophic pathogen attack SA signalling is activated, whereas for necrotrophs and herbivore ET and JA, induction is required [49] (figure 4.3). SA, JA and ET signalling play an important role for PTI and ETI induction. Hence, pathogens have developed different strategies to tackle with the plant hormone-mediated defence signalling. Many biotrophic pathogen effectors have been found to target SA biosynthesis and perception. For instance, *U. maydis* secretes the effector Cmu1, which is a chorismate mutase. In plants, chorismate mutase functions in the shikimic acid pathway which produces SA. Chorismate mutase effectors exploit the shikimic acid pathway to promote virulence [50] (figure 4.3). When compared to wild-type strains, maize plants infected with ΔcmuI of *U. maydis* strains exhibit fewer disease symptoms and enhanced SA levels. Additionally, Cmu effectors have been found in a number of nematodes, including the soybean cyst nematode *Heterodera glycines*, the sugar beet cyst nematode, and the potato cyst nematodes *Globodera rostochiensis* and *Globodera pallida*. Some effectors like Pslscl (*Phytophthora sojae*) and Vdlscl (*Verticillium dahlia*) function by hydrolysing SA precursor isochorismate into 2,3-dihydro-2,3-

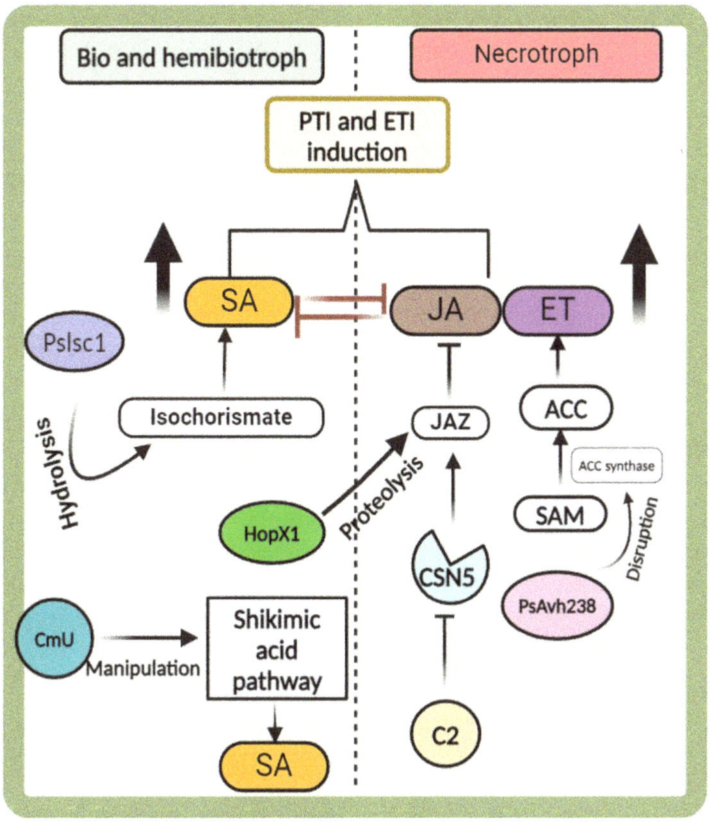

Figure 4.3. Schematic representation of hormone signalling during pathogen infection. PTI—pattern triggered immunity; ETI—effector-triggered immunity; SA—salicylic acid; JA—jasmonic acid; ET—ethylene; ACC—1-Aminocyclopropane 1-carboxylic acid; SAM—S′adenosylmethionine; JAZ—Jasmonate ZIM (zinc-finger inflorescence meristem); CSN5.COP9 signalosome complex. Pslscl, HopX1, CmU, PsAvh238, C2 are effectors. Pslscl causes degradation of isochorismate which is the component of shikimic acid pathway which results in SA production. Similarly, cmU is a chorismate mutase, which causes manipulation of shikimic acid pathway. HopX1 induces the degradation of JAZ, which is the repressor of JA-responsive gene thereby increasing JA signalling in cell. Once JA signalling is activated, it will inhibit the SA-mediated response, which is favourable for biotrophic pathogen. C2 protein inhibits CSN5, which degrades JAZ leading JA signalling. PsAvh238 disrupts ACC synthase, which converts SAM to ACC, which finally produces ET, hence supressing the ET-mediated immune response.

dihydroxybenzoate (DDHB) and pyruvate, making isochorismate unavailable for SA biosynthesis [51] (figure 4.3). As already discussed, necrotrophic infection induces JA in the host; it has been shown that SA and JA are antagonistic in nature. Effectors like HopX1 from *P. syringae* induce the JA pathway by degrading the transcription inhibitor JAZ thus inhibiting SA signalling [52] (figure 4.3). The pathogen of *Arabidopsis Hyaloperonospora arabidopsidis* secretes an effector HaRxL44 which causes the degradation of mediator subunit 19a which is a conserved multiprotein complex found in eukaryote, and it bridges the RNA polymerase II with diverse transcription factors. Mediator degradation results in

JA elevation and reduced SA responses [53]. JA mediated responses are also suppressed by targeting COP9 signalosome CSN5 through C2 protein released from *Tomato yellow leaf curl virus* which has been found to be conserved across Gemini viruses [54] (figure 4.3). Pathogens can mimic plant hormones in addition to interfering with the sensing and signalling of plant hormones. As an illustration, certain *P. syringae* strains generate coronatine, a JA-isoleucine mimic [55]. Effectors that imitate plant CLAVATA3/endosperm surrounding region-related (CLE) peptide hormones are secreted by cyst nematodes. RLKs sense, transport, and secrete plant CLE peptides. In plants, CLE perception regulates a variety of developmental processes and is essential for cell-to-cell communication [56]. After post-translational processing, CLE-like effectors show a mature peptide ligand consisting of 12–13 amino acids that resembles plant CLE peptides. In *Arabidopsis*, overexpression of CLE-like effectors causes early meristem termination and phenocopies plants that overexpress CLE peptides. A number of plants' RLKs are able to identify CLE-like effectors, which are necessary for the growth of nematode-induced syncytiums (giant multinucleated cells) and for the effective infection of nematodes. It is proposed that in order to rewire plant tissues as feeding areas for their own advantage, cyst nematodes release CLE-like effectors [57]. Additionally, ET is a traditional plant defence hormone that, when activated, confers resistance on plants against harmful bacterial invasion [58]. The synthesis of ET is dependent on the activity of ACC synthase (ACS), which catalyses the formation of S-adenosylmethionine (S-ADOMet) [59] (figure 4.3). While PsAvh238, a soybean blast RxLR polymorphic effector, promotes soybean blast infection by disrupting GmACSs and blocking ET biosynthesis, *P. syringae* targets the methionine cycle (precursor of ET) through HopAF1 [60]. On the other hand, rice brown spot disease is caused by the necrotrophic fungus *Cochliobolus miyabeanus*, which requires the activation of ET signalling [61]. The colonization of rice leaves by *C. miyabeanus* was significantly reduced upon blocking the mechanism that hijacks the ET signal, indicating that *C. miyabeanus* likely uses ET as a metabolite effector to promote virulence [62]. Other plant hormones, such as auxin and gibberellins, can also be changed by pathogen effectors to increase pathogenicity. What is still unclear, though, are the timing and amplitude of pathogen hormone modification [49]. Further studies in this field will improve our comprehension of the dynamics of plant hormones during pathogen infection.

4.5.2 Effectors inducing host cell death

Effectors from hemibiotrophs and other necrotrophic fungi purposefully cause necrosis to increase virulence, in contrast to strict biotrophs. These necrotrophic effectors can contribute to the host range of the infection by acting as host-selective toxins or by being non-specific [63]. The protein superfamily known as ET-inducing peptide 1 (NEP1)-like proteins (NLPs) is made by oomycetes, fungi, and bacteria. In numerous dicots, NLPs cause robust immunity-related reactions in addition to inducing cell death. More than 70% of NLPs originate from pathogens that show hemibiotrophic or necrotrophic lifestyle. These NLPs may stimulate immunological responses that lead to cell death and uptake of nutrients [64]. NLPs are viewed as PAMPs in *Arabidopsis*, and the RLP23 immune receptor recognizes a conserved 20 amino acid fragment that triggers canonical PTI reactions, which induces extracellular ROS burst, generation of ethylene, activation of MAPK, and the expression of defence-related genes [65]. In addition to inducing necrosis, Crinkler (CRN) effectors also produce the distinctive leaf crinkling phenotype. CRNs are widely dispersed throughout oomycetes, and some of them localize to the plant nuclei. They also have a distinctive N-terminal LxFLAK translocation domain. Because it localizes to plant nuclei and exhibits kinase activity, the *P. infestans* CRN8 effector disrupts host signalling pathways during infection [66]. Some necrotrophic effectors cause severe necrosis on specific plant genotypes with dominant sensitivity genes, despite the fact that the majority of them lack host selectivity. Upon cloning many sensitivity genes, it was discovered that they shared architecture with LRR domain and NB sites with intracellular plant immune receptors [67]. As an alternative to ETI, necrotrophic effectors can cause effector-triggered sensitivity by taking advantage of the activation of plant NB-LRRs. *Stagonospora nodorum* (syn. *Parastagonospora nodorum*) and *Pyrenophora tritici-repentis* release many effectors that cause heavy necrosis in wheat genotypes with appropriate sensitivity genes [68]. In wheat genotypes carrying Tsn1, the effector PtrToxA generated by *P. tritici-repentis* causes chlorosis or necrosis. Tsn1 has a serine/threonine protein kinase domain as well as an NB-LRR domain architecture. Only wheat genotypes with Tsn1 exhibit internalization of the PtrToxA effector upon fungal secretion. Later it localizes to the chloroplast and interacts with ToxAB1, which may be important in the production of thylakoids. PtrToxA orthologs have now been found in the maize pathogen *Cochliobolus heterostrophus* as well as the sister species *Stagonospora avenaria tritici* [69].

4.5.3 Effectors inhibiting cell-to-cell movement

Recent research has demonstrated that the effectors of rice blast fungus *M. oryzae* can migrate from one cell to another cell to promote further colonization in addition to their ability to target other cellular processes [70]. Upon entering the plant, *M. oryzae* uses intracellular invasive hyphae to colonize rice in a biotrophic manner [71]. Fungi that have been genetically modified to produce fluorescently labelled effectors have facilitated the exploration of how these effectors are expressed, released, and distributed during the natural infection process in rice. Identified effectors and secreted proteins linked to the biotrophic phase tend to gather in a specialized structure known as the biotrophic interfacial complex, and certain effectors have been observed within the cells of the host plant. In contrast, the invasive hyphae's apoplastic effectors were expressed consistently throughout. It is interesting to note that certain cytosolic effectors, such BAS1 and PWL2, may use plasmodesmata to migrate up to four cells away from invasive hyphae. Effector molecules capable of intercellular movement likely contribute to preparing host cells for colonization by pathogens. In addition to their capacity to migrate between cells, effectors can have distinct intracellular locations [70]. Hence, effectors serve as exceptional tools for probing cellular processes, with specific ones found to localize within nearly every subcellular region of plants. Among filamentous pathogen effectors, some exhibit concentrated accumulation around the biotrophic interface. For instance, *Colletotrichum orbiculare*, a pathogen affecting cucumbers, releases multiple effectors that gather around the neck of biotrophic hyphae. Notably, a component of *C. orbiculare*'s exocyst, SEC4, also localizes to this compartment. Deletion of SEC4 within the pathogen resulted in compromised delivery of effectors to the plant–pathogen interface. These findings underscore the critical role of targeted effector secretion in facilitating specific suppression of plant immune responses. [72]. These findings demonstrate how crucial localized effector secretion is for the precise inhibition of immune responses.

4.5.4 Effector manipulation of vesicular trafficking

Vesicular trafficking is the biological process which involves the movement of cargo from one cell compartment to another or to the cellular environment. During pathogen infection it leads to the delivery of a cargo-containing plasma membrane immune receptor, antimicrobial compounds and phytoalexin [73]. It has been found that if these cellular pathways are disturbed plants show compromised immunity. *P. syringae* effector HopM1 has been found to target vesicular trafficking in *Arabidopsis* by inducing AtMIN7 proteasomal degradation, which is an adenine diphosphate ribosylation factor–guanine nucleotide exchange factor (ARF-GEF) that functions in vesicular trafficking. AtMIN7 regulates endocytosis of plasma membrane proteins and HopM1 has been shown to colocalize with AtMIN7 at the *trans*-Golgi network/early endosome compartment (figure 4.4). Furthermore, a mutant of AtMIN7 exhibited compromised resistance [74]. Some effectors like AVR3a inhibit the internalization of receptor FLS, a PRR thus reducing its number on PM [75]. Another well-investigated effector AVRblb2 from *P. infestans* hinders the secretion of immune cargo. AVRblb2 localizes around haustoria and inhibits the secretion of papain-like CPC14 into apoplast and upon silencing and overexpression of this protein in *Nicotiana benthamiana* it resulted in enhanced susceptibility and resistance, respectively, to *P. infestans* [75]. Thus, this cellular process is exploited by pathogens to decrease the receptor abundance at the plasma membrane.

4.6 Reprogramming the host: effectors manipulating host gene expression

Upon pathogen invasion a plant shifts its transcriptional landscape towards defence. Transcription analyses in *Arabidopsis* indicates 44% genes are differentially expressed upon *P. syringae* infection [76]. In the same manner in maize 21% of genes are differentially expressed upon *U. maydis* infection [77]. In order to proliferate inside the host, pathogens have developed various strategies to reprogramme the host gene expression such as susceptibility genes expression, targeting host transcription factors and interference with RNAi machinery.

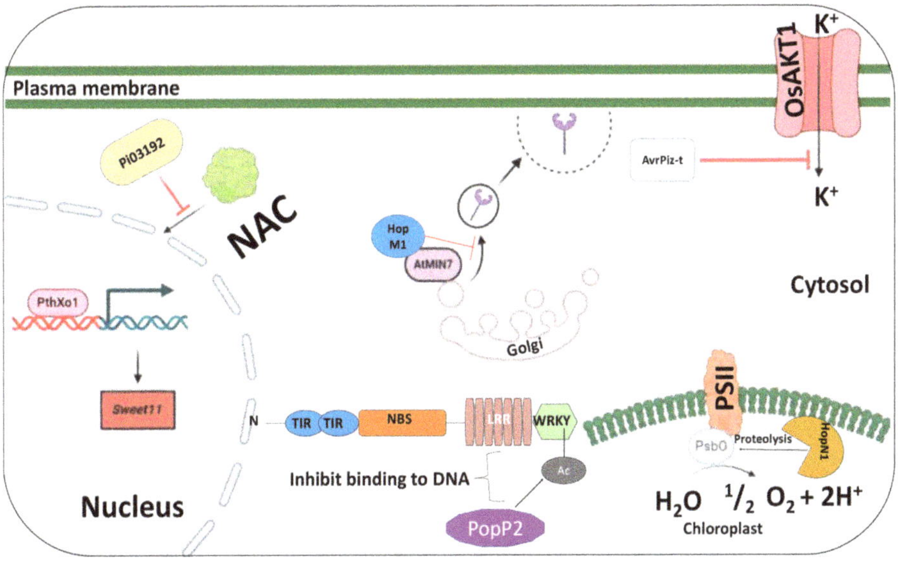

Figure 4.4. Pictorial representation of effectors manipulating different cellular processes. Pi03192, AvrPiz-t, HopM1, PthXo1, HopN1, PopP2 are effectors. Diverse effectors are utilized by pathogens to disturb the normal cellular environment to cause susceptibility. Pi03192 inhibits the translocation of transcription factor NAC which his involved in biotic stress response. HopM1 causes the degradation of AtMIN7, which is an adenine diphosphate ribosylation factor–guanine nucleotide exchange factor (ARF-GEF), and functions in vesicular trafficking and by doing so the pathogen is ensuring less immune receptor at the plasma membrane. Pthxo1 is a TAL (transcription activator-like) effector that induces the expression of SWEET11 gene which a transporter and transports sugar to the apoplast for pathogen nutrient. AvrPiz-t inhibits potassium channel thus impairing downstream immune signalling. HopN1 causes proteolysis of photosystem II associated protein PsbO, impairing water photolysis chloroplast. PopP2 targets intracellular immune receptor NB-LRR containing WRKY motif by acetyl group to lysine residue thereby inhibiting its binding to DNA and subsequently inhibiting expression of immune-related gene (tables 4.1 and 4.2).

4.6.1 Effector modulation of plant transcription factors

Several pathogen effectors seem to target host transcription factors differently, acting as hubs. Effectors have the ability to specifically target transcription factors or transcription repressors, changing their subcellular position and stability as well as preventing them from activating. For instance, the coat protein of the turnip wrinkling virus and the *P. infestans* RxLR effector Pi03192 interact with plant NAC transcription factors, limiting its localization to the nucleus and thus impairing its function [78] (figure 4.4). Furthermore, transcription factors TCP seem to be important for susceptibility in plant immunological signalling and are attacked by a number of pathogens, such as *Phytophthora* and *Phytoplasmas* [79]. JAZ proteins are targeted by many pathogen effectors and transcriptionally suppress JA signals, making them another weak point in plant immune signalling [80]. Recognized for its DNA binding domain containing the WRKYGQK sequence motif, WRKY transcription factors are necessary proteins that control the defence response to many infections [81]. The primary intracellular immune receptor, RRS1-R, a NB-LLR is

Table 4.1. Bacterial effectors with their virulence action in respective hosts.

Bacterial effectors	Organism	Proposed virulence action (references)
HopAB2	*Pseudomonas syringae* (tomato)	Changes responses of ethylene and suppression of cell death. Inhibition of HopPsyA-dependent Hypersensitive response (HR). Suppression of basal defence [104].
AvrB1	*Pseudomonas syringae* (*glycinea*)	Suppression of basal defences, induction of JA-responsive genes [105].
AvrBs2	*Xanthomonas campestris* (many pathovars)	Required for full virulence [106].
AvrBs3	*Xanthomonas spp.*	Suppression of non-host HR [107].
AvrRpt2	*Pseudomonas syringae* (maculicola)	Suppression of basal defences and induction of JA-responsive genes [108].
AvrXv4	*Xanthomonas campestris* (*vesicatoria*)	DeSUMOylate proteins [109].
GALA1-7	*Ralstonia solanacearum*	Targeting of proteins for ubiquitination [110].
HopAI1	*Pseudomonas syringae* (*tomato*)	Inhibition of MAPK-signalling, inhibit flagellin-induced NHO1 expression [111].
HopAO1	*Pseudomonas syringae* (*tomato*)	Suppression of cell death and induction of JA-responsive genes [112].
HopAR1	*Pseudomonas syringae* (*phaseolicola*)	Induction of JA-responsive genes [113].
HopC1	*Pseudomonas syringae* (*tomato*)	Inhibition of flagellin-induced *NHO1* expression [114].
HopN1	*Pseudomonas syringae* (*tomato*)	Suppression of cell death [115].
HopX1	*Pseudomonas syringae* (*many pathovars*)	Induction of JA-responsive genes and suppression of HopPsyA-dependent HR [116].
HopZ2	*Pseudomonas syringae* (*pisi*)	Target host SUMOylated protein [117].
HsvG/HsvB	*Pantoea agglomerans*	Formation of gall [118].
XopD	*Xanthomonas campestris* (*vesicatoria*)	DeSUMOylate proteins [119].
Tal8	*Xanthomonas translucens pv. undulosa*	Increases the expression of the host gene 9-cis-epoxycarotenoid dioxygenase (TaNCED-5BS) which is involved in the ABA biosynthesis and decreases expression of defence gene TaNPR1 [120]).
PthXo3JXOV	*Xanthomonas oryzae* pv. *oryzae*	Involved in the upregulation of the susceptibility gene OsSWEET14 which triggers sugar release and the effector also inhibits HR and callose deposition [121].
RipAL	*Ralstonia solanacearum*	Putative lipase which catalyses linoleic acid release from chloroplast lipids; induction of JA production and suppression of SA signalling [122].

Table 4.2. Fungal effectors with their virulence action in respective hosts.

Fungal effectors	Organism	Proposed virulence action (references)
Umrip1	*Ustilago maydis*	Induces the susceptibility factor ZmLox3, ZmLox3 is involved in repression of ROS burst [123].
Pi02860	*Phytophthora infestans*	Induces the susceptibility factor NRL1. NLR1 promotes degradation of StSWAP70 which is a positive regulator of immunity [124].
Pi04314/RD24	*Phytophthora infestans*	Relocalizes PP1 catalytic subunits from the nucleolus to the nucleoplasm by targeting them; Pi04314–0PP1c holoenzymes inhibit the routes for SA and JA [125].
PsAvh52	*Phytophthora sojae*	Targets GmTAP1, the susceptibility factor, which results in a transfer from the cytoplasm to the nucleus. H3K9 acetylation is encouraged by GmTAP1 to increase disease vulnerability [126].
PiAvr2	*Phytophthora infestans*	Interacts with BSL1, BSL2, and BSL3 (BRI1-SUPPRESSOR1-like). BSL1, BSL3, and BSL1 and BSL3 inhibit INF1-triggered cell death [127].
ToxA (Host-selective toxin)	*Pyrenophora tritici-repentis*	Targets Tsn1, a susceptibility factor that promotes necrotrophy by ToxA-mediated cell death [128].
SnTox1 (Host-selective toxin)	*Parastagonospora nodorum*	Targets Snn1, a susceptibility factor that promotes necrotrophy and is implicated in SnTox1-triggered cell death; it also shields the fungus from host chitinases [129].
PtrToxB (Host-selective)	*Pyrenophora tritici-repentis*	Targets Tsc2, a susceptibility factor involved in PtrToxB-mediated cell death that favours necrotrophy [130].
PSR2	*Phytophthora sp.*	Increases the sensitivity of *Arabidopsis* to pathogens by inhibiting the generation of secondary siRNAs (PPR-siRNAs) [131].
HaRxL21	*Hyaloperonospora arabidopsidis*	Interacting with TPL/TPR1 *Arabidopsis* proteins to inhibit transcription [132].
SsITL	*Sclerotinia sclerotiorum*	Prevents SA build-up by interacting with the chloroplast's CAS receptor [91].
Pst_12806	*Puccinia striiformis f. sp. tritici*	Decreases ROS build-up and photosynthesis; interacts with TaISP, which is a subunit of Cyt b6/f in the chloroplast [90].
PstGSRE1	*Puccinia striiformis f. sp. tritici*	Suppresses ROS-mediated cell death by altering the nuclear localization of the transcription factor TaLOL2, which is related with ROS [96].
PstGSRE4	*Puccinia striiformis f. sp. tritici*	Reduces H_2O_2 build-up and HR by inhibiting the activity of the wheat copper zinc superoxide dismutase enzyme TaCZSOD2 [133].
Pst18363	*Puccinia striiformis f. sp. tritici*	TaNUDX23 is stabilized by Pst18363, which inhibits ROS accumulation that increases susceptibility [134].
SCRE6	*Ustilaginoidea virens*	Suppresses plant immunity by interacting with and dephosphorylating the target OsMPK6 to stabilize it [135].

present in *Arabidopsis*, and it possesses the WRKY domain that targets pathogen effectors AvrRps4 or PopP2 and initiates for immune activation. But these effectors have evolved an ability to acetylate the lysine residue and inhibit its binding to DNA [82] (figure 4.4). It was recently discovered that the type III effector RipAB, which is released by the *Ralstonia solanacearum*, targets the transcription factor TGA and directly promotes the production of RBOHD and RBOHF and binds specifically to the TGACG sequence in *Arabidopsis*. RipAB suppresses TGA expression by interfering with RNA polymerase II recruitment and suppressing SA signalling, successfully infecting the host [83]. Host transcription is also impacted by nematode infestations. Plant kinase IPK phosphorylates the effector 10A07 of *H. schachtii*—the sugar beet nematode, enabling it to localize to the nucleus and interact with the auxin-responsive transcription factor INDOLE-3-ACETIC AID INDUCIBLE16 (IAA16) [84]. Therefore, 10A07 intervenes with auxin signalling via binding to the transcription factor IAA16.

4.6.2 Glorious TALEs: effectors acting as plant transcription factors

The expression of genes linked to host vulnerability can be induced by pathogen effectors through direct transcription factor action. The type-3 secreted effectors known as transcriptional activator-like effectors (TALEs) are present in multiple *Xanthomonas* and *Ralstonia* species. Determining the precise mechanism by which TALE attaches itself to the target gene promoters made it easier to pinpoint effector targets [85]. The rice bacterial blight agent, *Xanthomonas oryzae* pv. *oryzae* (*Xoo*), delivers the TALE PthXo1, which shows the ability to bind to the promoter of *OsSWEET11*—a sucrose transporter in order to enhance its expression and increase pathogenicity [86]. TALEs AvrXa7 and PthXo3 in *Xoo* promote *OsSWEET14* expression (figure 4.4). Therefore, the *SWEET* gene family's sugar transporters are the virulence targets of a number of TALEs, and they probably help with sugar export for bacterial consumption [87]. The prediction of such target genes that lead to bacterial susceptibility and targets in resistant plants that regulate the production of defence genes was made possible by the unravelling of the TALE-DNA binding specificity. This information also made it possible to create artificial TALEs coupled to nucleases, which were extensively employed to modify the genomes of human, animal, and plant cells [88]. The necessity of examining the molecular underpinnings of effector action is demonstrated by the extensive application of TALE for genome editing.

4.6.3 Effectors' interference with host plant protein function

Certain pathogen effectors specifically target functional proteins in the cytosol of the host plant, frequently by changing the proteins' location, which results in pathogenic consequences and functional disturbances. The *P. syringae* effector HopI1, which transfers the host protein Hsp70 to the chloroplast, is a prime example. This process modifies the structure of thylakoids and prevents the build-up of SA, an essential hormone for plant defence [89]. Another *P. syringae* effector, HopNI, acts as a cysteine protease to cleave PsbQ, a component of the photosystem II in tomato, impairing water photolysis (figure 4.4). The haustorium-specific effector Pst_12806 from *Puccinia striiformis f. sp. tritici* targets wheat chloroplasts, interacting with the wheat TaISP protein's C-terminal Rieske domain. This interaction disrupts electron transport, reducing ROS accumulation and dampening defence gene expression [90]. Additionally, the SsITL effector targets the chloroplast calcium-sensitive receptor (CAS), affecting SA-mediated immunity [91]. The *P. sojae* effector PsAvh262 stabilizes BIPS in the endoplasmic reticulum (ER), a negative regulator of plant resistance, thus inhibiting cell death induced by ER stress and facilitating infection. This strategy of targeting ER stress regulators might be a general tactic for microbial manipulation of hosts [92]. *M. oryzae* effector MoCDIP4 disrupts the OsDjA9-OsDRP1E protein complex associated with mitochondrial division, causing abnormal mitochondrial development and compromised immunity in rice [93]. AvrPiz-t, another *M. oryzae* effector, disrupts the interaction between the rice plasma membrane K^+ channel OsAKT1 and its regulator OsCIPK23, affecting K^+ signalling and host immunity [94] (figure 4.4). *V. dahliae*'s PevD1 effector manipulates cryptochrome signalling by antagonizing the interaction between CRY2 and the asparagine-rich protein NRP, affecting early flowering in Cotton and *Arabidopsis* [95]. *P. infestans*' effector Pi04314 interacts with host PP1c isoforms, relocating them within the nucleus to promote pathogen colonization and suppress JA and SA-responsive gene induction [78]. The effector PstGSRE1 from wheat stripe rust binds to the ROS-associated zinc-finger transcription factor TaLOL2, preventing its nuclear localization and thereby undermining plant defences by inhibiting ROS-mediated responses [96]. These examples illustrate the sophisticated mechanisms by which pathogens manipulate host cellular processes to facilitate infection and overcome plant defences.

4.6.4 Effectors targeting RNA silencing machinery

In eukaryotes, RNA silencing, also known as post-transcriptional gene silencing, happens to be a conserved regulatory mechanism. Additionally, hosts employ the RNA silencing apparatus to protect themselves against viral infections. Numerous plant viruses replicate by means of a double-stranded RNA intermediate and have an RNA genome. Double-stranded RNA enables the host to selectively target viral RNA for destruction. Suppressors of RNA silencing have been developed by a number of plant viruses, which are necessary for replication of the virus, in order to evade this defence mechanism [97]. Proteinase P1/HC-Pro, an auxiliary component of the *Tobacco etch potyvirus*, is one of the earliest known viral suppressors of RNA silencing [98]. P1/HC-Pro performs a number of tasks, including digesting polyproteins, aphid dispersal, cell-to-cell transfer, and viral genome amplification [99]. All of these actions, meanwhile, are connected to its suppression activity of RNA silencing. Numerous viral suppressors of RNA silencing, such as P19 from *Tomato bushy stunt virus* and 2b from *Cucumber mosaic virus*, target distinct elements of the RNA silencing pathway. The development of viral induced susceptibility in plants depends on these suppressors as vital effectors [97]. Other groups of pathogens possess effectors that impede post-transcriptional processes, apart from viruses. The RxLR effectors PSR1 and PSR2 (*Phytophthora* suppressor of RNA silencing 1 and

2), which function as RNA silencing suppressors, are sourced from *P. sojae*. The pathogenicity of *Potato viruses X* and *P. infestans* is increased by PSR1 expression in *N. benthamiana*, highlighting the significance of this effector in pathogen reproduction. PSR1-expressing *Arabidopsis* plants accumulate fewer small RNAs, such as small interfering RNAs (siRNAs) and precursor microRNAs (miRNAs) [100]. PSR1 binds to PSR1-interacting protein 1, a nuclear protein that controls the accumulation of siRNA and miRNA and has an RNA helicase domain. Thus, PSR1 likely has a Dicer-dependent effect on the synthesis of short RNAs [101]. PSR2 decreases the accumulation of particular trans-acting siRNAs and suppresses different aspects of RNA silencing [100]. It is unknown if *P. syringae* directly affects the miRNA pathway, but it can supply effectors that do so [102]. Fungal pathogens are also capable of manipulating components of RNA silencing. The fungus *B. cinerea* produces short RNAs that suppress the expression of immune-related genes by targeting the host's silencing apparatus [103]. These findings suggest that one common method employed by pathogen effectors to stifle plant defence responses and promote pathogen reproduction is modification of host transcriptional mechanisms.

4.7 Conclusions

Effectors play a crucial role in mediating interactions between plants and pathogens. The effective delivery of effectors empowers pathogens to successfully colonize their host organisms. Various pathogens possess the capability to manipulate shared components of plant immunity. For instance, stealthy pathogens release effectors that suppress cell death and immune responses, while necrotrophic pathogens exploit these components to induce cell death and enhance susceptibility. The tactics employed by the pathogen to induce disease in the host signify its ongoing coevolutionary dynamic with plant defence mechanisms, a process essential for its survival and proliferation. Understanding how the pathogen utilizes susceptibility factor of host in order to proliferate would give an insight into the development of new resistant varieties.

References

[1] Lapin D and Van den Ackerveken G 2013 Susceptibility to plant disease: more than a failure of host immunity *Trends Plant Sci.* **18** 546–54

[2] Zipfel C and Felix G 2005 Plants and animals: a different taste for microbes? *Curr. Opin. Plant Biol.* **8** 353–60

[3] Chinchilla D, Zipfel C, Robatzek S, Kemmerling B, Nürnberger T, Jones J D G *et al* 2007 A flagellin-induced complex of the receptor FLS2 and BAK1 initiates plant defence *Nature* **448** 497–500

[4] Grant S R, Fisher E J, Chang J H, Mole B M and Dangl J L 2006 Subterfuge and manipulation: type III effector proteins of phytopathogenic bacteria *Annu. Rev. Microbiol.* **60** 425–49

[5] Abramovitch R B, Anderson J C and Martin G B 2006 Bacterial elicitation and evasion of plant innate immunity *Nat. Rev. Mol. Cell Biol.* **7** 601

[6] Bigeard J, Colcombet J and Hirt H 2015 Signaling mechanisms in pattern-triggered immunity (PTI) *Mol. Plant* **8** 521–39
[7] Balint-Kurti P 2019 The plant hypersensitive response: concepts, control and consequences *Mol. Plant Pathol.* **20** 1163–78
[8] Jaswal R, Kiran K, Rajarammohan S, Dubey H, Singh P K, Sharma Y *et al* 2020 Effector biology of biotrophic plant fungal pathogens: current advances and future prospects *Microbiol. Res.* **241** 126567
[9] Deb S, Madhavan V N, Gokulan C G, Patel H K and Sonti R V 2021 Arms and ammunitions: effectors at the interface of rice and it's pathogens and pests *Rice* **14** 94
[10] Mapuranga J, Chang J, Zhang L, Zhang N and Yang W 2022 Fungal secondary metabolites and small RNAs enhance pathogenicity during plant–fungal pathogen interactions *J. Fungi (Basel)* **9** 4
[11] van Schie C C N and Takken F L W 2014 Susceptibility genes 101: how to be a good host *Annu. Rev. Phytopathol.* **52** 551–81
[12] Serrano M, Coluccia F, Torres M, L'H F and Métraux J P 2014 The cuticle and plant defense to pathogens *Front. Plant Sci.* **5** 274
[13] Doehlemann G, van der Linde K, Aßmann D, Schwammbach D, Hof A, Mohanty A *et al* 2009 Pep1, a secreted effector protein of *Ustilago maydis*, is required for successful invasion of plant cells *PLoS Pathog.* **5** e1000290
[14] Melotto M, Underwood W, Koczan J, Nomura K and He S Y 2006 Plant stomata function in innate immunity against bacterial invasion *Cell* **126** 969–80
[15] Chini A, Fonseca S, Fernández G, Adie B, Chico J M, Lorenzo O *et al* 2007 The JAZ family of repressors is the missing link in jasmonate signalling *Nature* **448** 666–71
[16] Albersheim P, Darvill A, Roberts K, Sederoff R and Staehelin A 2010 *Plant cell Walls* (Garland Science)
[17] Kubicek C P, Starr T L and Glass N L 2014 Plant cell wall-degrading enzymes and their secretion in plant-pathogenic fungi *Annu. Rev. Phytopathol* **52** 427–51
[18] Kema G H J, van der Lee T A J, Mendes O, Verstappen E C P, Lankhorst R K, Sandbrink H *et al* 2008 Large-scale gene discovery in the septoria tritici blotch fungus *Mycosphaerella graminicola* with a focus on in planta expression *Mol. Plant Microbe. Interact.* **21** 1249–60
[19] Fry S C 1995 Polysaccharide-modifying enzymes in the plant cell wall *Annu. Rev. Plant Physiol. Plant Mol. Biol.* **46** 497–520
[20] Gorshkov V, Tsers I, Islamov B, Ageeva M, Gogoleva N, Mikshina P *et al* 2021 The modification of plant cell wall polysaccharides in potato plants during *Pectobacterium atrosepticum*-caused infection *Plants* **10** 1407
[21] An S H, Sohn K H, Choi H W, Hwang I S, Lee S C and Hwang B K 2008 Pepper pectin methylesterase inhibitor protein CaPMEI1 is required for antifungal activity, basal disease resistance and abiotic stress tolerance *Planta* **228** 61–78
[22] Cao L, Blekemolen M C, Tintor N, Cornelissen B J C and Takken F L W 2018 The *Fusarium oxysporum* Avr2-Six5 effector pair alters plasmodesmatal exclusion selectivity to facilitate cell-to-cell movement of Avr2 *Mol. Plant* **11** 691–705
[23] Tomczynska I, Stumpe M, Doan T G and Mauch F 2020 A Phytophthora effector protein promotes symplastic cell-to-cell trafficking by physical interaction with plasmodesmata-localised callose synthases *New Phytol.* **227** 1467–78
[24] Aung K, Kim P, Li Z, Joe A, Kvitko B, Alfano J R *et al* 2020 Pathogenic bacteria target plant plasmodesmata to colonize and invade surrounding tissues *Plant Cell* **32** 595–611

[25] Li Z, Variz H, Chen Y, Liu S L and Aung K 2021 Plasmodesmata-dependent intercellular movement of bacterial effectors *Front. Plant Sci.* **12** 640277

[26] Schmidt S M and Panstruga R 2007 Cytoskeleton functions in plant–microbe interactions *Physiol. Mol. Plant Pathol.* **71** 135–48

[27] Kang Y, Jelenska J, Cecchini N M, Li Y, Lee M W, Kovar D R *et al* 2014 HopW1 from *Pseudomonas syringae* disrupts the actin cytoskeleton to promote virulence in Arabidopsis *PLoS Pathog.* **10** e1004232

[28] Nottensteiner M, Zechmann B, McCollum C and Hückelhoven R 2018 A barley powdery mildew fungus non-autonomous retrotransposon encodes a peptide that supports penetration success on barley *J. Exp. Botany* **69** 3745–58

[29] Sun H, Zhu X, Li C, Ma Z, Han X, Luo Y *et al* 2021 *Xanthomonas* effector XopR hijacks host actin cytoskeleton via complex coacervation *Nat. Commun.* **12** 4064

[30] Bayry J, Aimanianda V, Guijarro J I, Sunde M and Latgé J P 2012 Hydrophobins—unique fungal proteins *PLoS Pathog.* **8** e1002700

[31] Talbot N J, Ebbole D J and Hamer J E 1993 Identification and characterization of MPG1, a gene involved in pathogenicity from the rice blast fungus *Magnaporthe grisea Plant Cell* **5** 1575–90

[32] Quarantin A, Hadeler B, Kröger C, Schäfer W, Favaron F, Sella L *et al* 2019 Different hydrophobins of *Fusarium graminearum* are involved in hyphal growth, attachment, water–air interface penetration and plant infection *Front. Microbiol.* **10** 751

[33] Ahn N, Kim S, Choi W, Im K H and Lee Y H 2004 Extracellular matrix protein gene, EMP1, is required for appressorium formation and pathogenicity of the rice blast fungus, *Magnaporthe grisea Mol. Cells* **17** 166–73

[34] Fernandes T R, Segorbe D, Prusky D and Di Pietro A 2017 How alkalinization drives fungal pathogenicity *PLoS Pathog.* **13** e1006621

[35] Haruta M, Sabat G, Stecker K, Minkoff B B and Sussman M R 2014 A peptide hormone and its receptor protein kinase regulate plant cell expansion *Science* **24343** 408–11

[36] Masachis S, Segorbe D, Turrà D, Leon-Ruiz M, Fürst U, El Ghalid M *et al* 2016 A fungal pathogen secretes plant alkalinizing peptides to increase infection *Nat. Microbiol.* **1** 16043

[37] Stegmann M, Monaghan J, Smakowska-Luzan E, Rovenich H, Lehner A, Holton N *et al* 2017 The receptor kinase FER is a RALF-regulated scaffold controlling plant immune signaling *Science* **20355** 287–9

[38] Zeng W and He S Y 2010 A prominent role of the flagellin receptor FLAGELLIN-SENSING2 in mediating stomatal response to *Pseudomonas syringae* pv tomato DC3000 in Arabidopsis *Plant Physiol.* **153** 1188–98

[39] Gimenez-Ibanez S, Hann D R, Ntoukakis V, Petutschnig E, Lipka V and Rathjen J P 2009 AvrPtoB targets the LysM receptor kinase CERK1 to promote bacterial virulence on plants *Curr. Biol.* **19** 423–9

[40] Göhre V, Spallek T, Häweker H, Mersmann S, Mentzel T, Boller T *et al* 2008 Plant pattern-recognition receptor FLS2 is directed for degradation by the bacterial ubiquitin ligase AvrPtoB *Curr. Biol.* **18** 1824–32

[41] Nicaise V, Joe A, ryool Jeong B, Korneli C, Boutrot F, Westedt I *et al* 2013 Pseudomonas HopU1 modulates plant immune receptor levels by blocking the interaction of their mRNAs with GRP7 *EMBO J.* **32** 701–12

[42] Guy E, Lautier M, Chabannes M, Roux B, Lauber E, Arlat M *et al* 2013 xopAC-triggered immunity against *Xanthomonas* depends on Arabidopsis receptor-like cytoplasmic kinase genes PBL2 and RIPK *PLoS One* **8** e73469

[43] Kadota Y, Sklenar J, Derbyshire P, Stransfeld L, Asai S, Ntoukakis V *et al* 2014 Direct regulation of the NADPH oxidase RBOHD by the PRR-associated kinase BIK1 during plant immunity *Mol. Cell* **54** 43–55

[44] Shao F, Golstein C, Ade J, Stoutemyer M, Dixon J E and Innes R W 2003 Cleavage of Arabidopsis PBS1 by a bacterial type III effector *Science* **29** 1230–3

[45] Sels J, Mathys J, De Coninck B M A, Cammue B P A and De Bolle M F C 2008 Plant pathogenesis-related (PR) proteins: a focus on PR peptides *Plant Physiol. Biochem.* **46** 941–50

[46] Shabab M, Shindo T, Gu C, Kaschani F, Pansuriya T, Chintha R *et al* 2008 Fungal effector protein AVR2 targets diversifying defense-related cys proteases of tomato *Plant Cell* **20** 1169–83

[47] Kaschani F, Shabab M, Bozkurt T, Shindo T, Schornack S, Gu C *et al* 2010 An effector-targeted protease contributes to defense against *Phytophthora infestans* and is under diversifying selection in natural hosts *Plant Physiol.* **154** 1794–804

[48] van den Burg H A, Harrison S J, Joosten M H A J, Vervoort J and de Wit P J G M 2006 *Cladosporium fulvum* Avr4 protects fungal cell walls against hydrolysis by plant chitinases accumulating during infection *Mol. Plant-Microbe Interact.* **19** 1420–30

[49] Pieterse C M J, Van der Does D, Zamioudis C, Leon-Reyes A and Van Wees S C M 2012 Hormonal modulation of plant immunity *Annu. Rev. Cell Dev. Biol.* **28** 489–521

[50] Djamei A, Schipper K, Rabe F, Ghosh A, Vincon V, Kahnt J *et al* 2011 Metabolic priming by a secreted fungal effector *Nature* **478** 395–8

[51] Liu T, Song T, Zhang X, Yuan H, Su L, Li W *et al* 2014 Unconventionally secreted effectors of two filamentous pathogens target plant salicylate biosynthesis *Nat. Commun.* **5** 4686

[52] Jiang S, Yao J, Ma K W, Zhou H, Song J, He S Y *et al* 2013 Bacterial effector activates jasmonate signaling by directly targeting JAZ transcriptional repressors *PLoS Pathog.* **9** e1003715

[53] Caillaud M C, Asai S, Rallapalli G, Piquerez S, Fabro G and Jones J D G 2013 A downy mildew effector attenuates salicylic acid-triggered immunity in Arabidopsis by interacting with the host mediator complex *PLoS Biol.* **11** e1001732

[54] Lozano-Durán R, Rosas-Díaz T, Gusmaroli G, Luna A P, Taconnat L, Deng X W *et al* 2011 Geminiviruses subvert ubiquitination by altering CSN-mediated derubylation of SCF E3 ligase complexes and inhibit jasmonate signaling in *Arabidopsis thaliana Plant Cell* **23** 1014–32

[55] Bender C L, Alarcón-Chaidez F and Gross D C 1999 *Pseudomonas syringae* phytotoxins: mode of action, regulation, and biosynthesis by peptide and polyketide synthetases *Microbiol. Mol. Biol. Rev.* **63** 266–92

[56] Leasure C D and He Z H 2012 CLE and RGF family peptide hormone signaling in plant development *Mol. Plant* **5** 1173–5

[57] Replogle A, Wang J, Paolillo V, Smeda J, Kinoshita A, Durbak A *et al* 2013 Synergistic interaction of CLAVATA1, CLAVATA2, and receptor-like protein kinase 2 in cyst nematode parasitism of Arabidopsis *Mol. Plant-Microbe Interact.* **26** 87–96

[58] Hoffman T, Schmidt J S, Zheng X and Bent A F 1999 Isolation of ethylene-insensitive soybean mutants that are altered in pathogen susceptibility and gene-for-gene disease resistance *Plant Physiol.* **119** 935–50

[59] Christians M J, Gingerich D J, Hansen M, Binder B M, Kieber J J and Vierstra R D 2009 The BTB ubiquitin ligases ETO1, EOL1 and EOL2 act collectively to regulate ethylene biosynthesis in Arabidopsis by controlling type-2 ACC synthase levels *Plant* J. **57** 332–45
[60] Washington E J, Mukhtar M S, Finkel O M, Wan L, Banfield M J, Kieber J J *et al* 2016 *Pseudomonas syringae* type III effector HopAF1 suppresses plant immunity by targeting methionine recycling to block ethylene induction *Proc. Natl Acad. Sci. USA* **113** E3577–3586
[61] Yang B, Wang Y, Guo B, Jing M, Zhou H, Li Y *et al* 2019 The *Phytophthora sojae* RXLR effector Avh238 destabilizes soybean Type2 GmACSs to suppress ethylene biosynthesis and promote infection *New Phytol.* **222** 425–37
[62] Van Bockhaven J, Spíchal L, Novák O, Strnad M, Asano T, Kikuchi S *et al* 2015 Silicon induces resistance to the brown spot fungus *Cochliobolus miyabeanus* by preventing the pathogen from hijacking the rice ethylene pathway *New Phytol.* **206** 761–73
[63] Wang X, Jiang N, Liu J, Liu W and Wang G L 2014 The role of effectors and host immunity in plant–necrotrophic fungal interactions *Virulence* **5** 722–32
[64] Pemberton C L and Salmond G P C 2004 The Nep1-like proteins-a growing family of microbial elicitors of plant necrosis *Mol. Plant Pathol.* **5** 353–9
[65] Oome S, Raaymakers T M, Cabral A, Samwel S, Böhm H, Albert I *et al* 2014 Nep1-like proteins from three kingdoms of life act as a microbe-associated molecular pattern in Arabidopsis *Proc. Natl Acad. Sci. USA* **111** 16955–60
[66] Schornack S, van Damme M, Bozkurt T O, Cano L M, Smoker M, Thines M *et al* 2010 Ancient class of translocated oomycete effectors targets the host nucleus *Proc. Natl Acad. Sci. USA* **107** 17421–6
[67] Tan K C, Oliver R P, Solomon P S and Moffat C S 2010 Proteinaceous necrotrophic effectors in fungal virulence *Funct. Plant Biol.* **37** 907–12
[68] Oliver R P, Friesen T L, Faris J D and Solomon P S 2012 *Stagonospora nodorum*: from pathology to genomics and host resistance *Annu. Rev. Phytopathol.* **50** 23–43
[69] Lu S, Gillian Turgeon B and Edwards M C 2015 A ToxA-like protein from *Cochliobolus heterostrophus* induces light-dependent leaf necrosis and acts as a virulence factor with host selectivity on maize *Fungal Genet. Biol.* **81** 12–24
[70] Khang C H, Berruyer R, Giraldo M C, Kankanala P, Park S Y, Czymmek K *et al* 2010 Translocation of magnaporthe oryzae effectors into rice cells and their subsequent cell-to-cell movement *Plant Cell* **22** 1388–403
[71] Howard R J and Valent B 1996 Breaking and entering: host penetration by the fungal rice blast pathogen *Magnaporthe grisea Annu. Rev, Microbiol.* **50** 491–512
[72] Irieda H, Maeda H, Akiyama K, Hagiwara A, Saitoh H, Uemura A *et al* 2014 *Colletotrichum orbiculare* secretes virulence effectors to a biotrophic interface at the primary hyphal neck via exocytosis coupled with SEC22-mediated traffic *Plant Cell* **26** 2265–81
[73] Teh O K and Hofius D 2014 Membrane trafficking and autophagy in pathogen-triggered cell death and immunity *J. Exp. Bot.* **65** 1297–312
[74] Nomura K, Mecey C, Lee Y N, Imboden L A, Chang J H and He S Y 2011 Effector-triggered immunity blocks pathogen degradation of an immunity-associated vesicle traffic regulator in Arabidopsis *Proc. Natl Acad. Sci. USA* **108** 10774–9
[75] Chaparro-Garcia A, Schwizer S, Sklenar J, Yoshida K, Petre B, Bos J I B *et al* 2015 *Phytophthora infestans* RXLR-WY effector AVR3a associates with dynaminrelated protein 2 required for endocytosis of the plant pattern recognition receptor FLS2 *PLoS One* **10** e0137071

[76] Lewis L A, Polanski K, de Torres-Zabala M, Jayaraman S, Bowden L, Moore J *et al* 2015 Transcriptional dynamics driving MAMP-triggered immunity and pathogen effector-mediated immunosuppression in arabidopsis leaves following infection with *Pseudomonas syringae* pv tomato DC3000 *Plant Cell* **27** 3038–64

[77] Doehlemann G, Wahl R, Horst R J, Voll L M, Usadel B, Poree F *et al* 2008 Reprogramming a maize plant: transcriptional and metabolic changes induced by the fungal biotroph *Ustilago maydis Plant* J. **56** 181–95

[78] McLellan H, Boevink P C, Armstrong M R, Pritchard L, Gomez S, Morales J *et al* 2013 An RxLR effector from *Phytophthora infestans* prevents re-localisation of two plant NAC transcription factors from the endoplasmic reticulum to the nucleus *PLoS Pathog.* **9** e1003670

[79] Stam R, Motion G B, Boevink P C and Huitema E 2021 A conserved oomycete CRN effector targets and modulates tomato TCP14-2 to enhance virulence *Mol. Plant-Microbe Interact.* **34** 309–18

[80] Gimenez-Ibanez S, Boter M, Fernández-Barbero G, Chini A, Rathjen J P and Solano R 2014 The bacterial effector HopX1 targets JAZ transcriptional repressors to activate jasmonate signaling and promote infection in Arabidopsis *PLoS Biol.* **12** e1001792

[81] Pandey S P and Somssich I E 2009 The role of WRKY transcription factors in plant immunity *Plant Physiol.* **150** 1648–55

[82] Le Roux C, Huet G, Jauneau A, Camborde L, Trémousaygue D, Kraut A *et al* 2015 A receptor pair with an integrated decoy converts pathogen disabling of transcription factors to immunity *Cell* **161** 1074–88

[83] Qi P, Huang M, Hu X, Zhang Y, Wang Y, Li P *et al* 2022 A *Ralstonia solanacearum* effector targets TGA transcription factors to subvert salicylic acid signaling *Plant Cell* **34** 1666–83

[84] Hewezi T, Juvale P S, Piya S, Maier T R, Rambani A, Rice J H *et al* 2015 The cyst nematode effector protein 10A07 targets and recruits host posttranslational machinery to mediate its nuclear trafficking and to promote parasitism in Arabidopsis *Plant Cell* **27** 891–907

[85] Boch J, Scholze H, Schornack S, Landgraf A, Hahn S, Kay S *et al* 2009 Breaking the code of DNA binding specificity of TAL-type III effectors *Science* **326** 1509–12

[86] Yang B, Sugio A and White F F 2006 Os8N3 is a host disease-susceptibility gene for bacterial blight of rice *Proc. Natl Acad. Sci. USA* **103** 10503–8

[87] Antony G, Zhou J, Huang S, Li T, Liu B, White F *et al* 2010 Rice xa13 recessive resistance to bacterial blight is defeated by induction of the disease susceptibility gene Os-11N3 *Plant Cell* **22** 3864–76

[88] Joung J K and Sander J D 2013 TALENs: a widely applicable technology for targeted genome editing *Nat. Rev. Mol. Cell Biol.* **14** 49–55

[89] Jelenska J, Yao N, Vinatzer B A, Wright C M, Brodsky J L and Greenberg J T 2007 A J domain virulence effector of *Pseudomonas syringae* remodels host chloroplasts and suppresses defenses *Curr. Biol.* **17** 499–508

[90] Xu Q, Tang C, Wang X, Sun S, Zhao J, Kang Z *et al* 2019 An effector protein of the wheat stripe rust fungus targets chloroplasts and suppresses chloroplast function *Nat. Commun.* **10** 5571

[91] Zhu W, Wei W, Fu Y, Cheng J, Xie J, Li G *et al* 2013 A secretory protein of necrotrophic fungus *Sclerotinia sclerotiorum* that suppresses host resistance *PLoS One* **8** e53901

[92] Jing M, Guo B, Li H, Yang B, Wang H, Kong G *et al* 2016 A *Phytophthora sojae* effector suppresses endoplasmic reticulum stress-mediated immunity by stabilizing plant binding immunoglobulin proteins *Nat. Commun.* **7** 11685
[93] Xu G, Zhong X, Shi Y, Liu Z, Jiang N, Liu J *et al* 2020 A fungal effector targets a heat shock-dynamin protein complex to modulate mitochondrial dynamics and reduce plant immunity *Sci. Adv.* **6** eabb7719
[94] Shi X, Long Y, He F, Zhang C, Wang R, Zhang T *et al* 2018 The fungal pathogen Magnaporthe oryzae suppresses innate immunity by modulating a host potassium channel *PLoS Pathog.* **14** e1006878
[95] Zhou R, Zhu T, Han L, Liu M, Xu M, Liu Y *et al* 2017 The asparagine-rich protein NRP interacts with the Verticillium effector PevD1 and regulates the subcellular localization of cryptochrome 2 *J. Exp. Bot.* **68** 3427–40
[96] Qi T, Guo J, Liu P, He F, Wan C, Islam M A *et al* 2019 Stripe rust effector PstGSRE1 disrupts nuclear localization of ROS-Promoting transcription factor TaLOL2 to defeat ROS-induced defense in wheat *Mol. Plant* **12** 1624–38
[97] Csorba T, Kontra L and Burgyán J 2015 viral silencing suppressors: tools forged to fine-tune host–pathogen coexistence *Virology* **479–480** 85–103
[98] Anandalakshmi R, Pruss G J, Ge X, Marathe R, Mallory A C, Smith T H *et al* 1998 A viral suppressor of gene silencing in plants *Proc. Natl Acad. Sci. USA* **95** 13079–84
[99] Urcuqui-Inchima S, Haenni A L and Bernardi F 2001 Potyvirus proteins: a wealth of functions *Virus Res.* **74** 157–75
[100] Qiao Y, Liu L, Xiong Q, Flores C, Wong J, Shi J *et al* 2013 Oomycete pathogens encode RNA silencing suppressors *Nat. Genet.* **45** 330–3
[101] Qiao Y, Shi J, Zhai Y, Hou Y and Ma W 2015 Phytophthora effector targets a novel component of small RNA pathway in plants to promote infection *Proc. Natl Acad. Sci. USA* **112** 5850–5
[102] Navarro L, Jay F, Nomura K, He S Y and Voinnet O 2008 Suppression of the microRNA pathway by bacterial effector proteins *Science* **321** 964–7
[103] Weiberg A, Wang M, Lin F M, Zhao H, Zhang Z, Kaloshian I *et al* 2013 Fungal small RNAs suppress plant immunity by hijacking host RNA interference pathways *Science* **342** 118–23
[104] Janjusevic R, Abramovitch R B, Martin G B and Stebbins C E 2006 A bacterial inhibitor of host programmed cell death defenses is an E3 ubiquitin ligase *Science* **311** 222–6
[105] Shang Y, Li X, Cui H, He P, Thilmony R, Chintamanani S *et al* 2006 RAR1, a central player in plant immunity, is targeted by *Pseudomonas syringae* effector AvrB *Proc. Natl Acad. Sci.* **103** 19200–5
[106] Kearney B and Staskawicz B J 1990 Widespread distribution and fitness contribution of *Xanthomonas campestris* avirulence gene avrBs2 *Nature* **346** 385–6
[107] Marois E, Van den Ackerveken G and Bonas U 2002 The *Xanthomonas* type III effector protein AvrBs3 modulates plant gene expression and induces cell hypertrophy in the susceptible host *Mol. Plant Microbe Interact.* **15** 637–46
[108] Axtell M J, Chisholm S T, Dahlbeck D and Staskawicz B J 2003 Genetic and molecular evidence that the *Pseudomonas syringae* type III effector protein AvrRpt2 is a cysteine protease *Mol. Microbiol.* **49** 1537–46
[109] Roden J, Eardley L, Hotson A, Cao Y and Mudgett M B 2004 Characterization of the *Xanthomonas* AvrXv4 effector, a SUMO protease translocated into plant cells *Mol. Plant Microbe Interact.* **17** 633–43

[110] Angot A, Peeters N, Lechner E, Vailleau F, Baud C, Gentzbittel L *et al* 2006 *Ralstonia solanacearum* requires F-box-like domain-containing type III effectors to promote disease on several host plants *Proc. Natl Acad. Sci.* **103** 14620–5
[111] Li H, Xu H, Zhou Y, Zhang J, Long C, Li S *et al* 2007 The *Phosphothreonine lyase* activity of a bacterial type III effector family *Science* **315** 1000–3
[112] Bretz J R, Mock N M, Charity J C, Zeyad S, Baker C J and Hutcheson S W 2003 A translocated protein tyrosine phosphatase of *Pseudomonas syringae* pv. tomato DC3000 modulates plant defence response to infection *Mol. Microbiol.* **49** 389–400
[113] Shao F, Merritt P M, Bao Z, Innes R W and Dixon J E 2002 A yersinia effector and a pseudomonas avirulence protein define a family of cysteine proteases functioning in bacterial pathogenesis *Cell* **109** 575–88
[114] Li X, Lin H, Zhang W, Zou Y, Zhang J, Tang X *et al* 2005 Flagellin induces innate immunity in nonhost interactions that is suppressed by *Pseudomonas syringae* effectors *Proc. Natl Acad. Sci.* **102** 12990–5
[115] López-Solanilla E, Bronstein P A, Schneider A R and Collmer A 2004 HopPtoN is a *Pseudomonas syringae* Hrp (type III secretion system) cysteine protease effector that suppresses pathogen-induced necrosis associated with both compatible and incompatible plant interactions *Mol. Microbiol.* **54** 353–65
[116] Nimchuk Z L, Fisher E J, Desveaux D, Chang J H and Dangl J L 2007 The hopX (AvrPphE) family of *Pseudomonas syringae* type III effectors require a catalytic triad and a novel N-terminal domain for function *Mol. Plant Microbe Interact.* **20** 346–57
[117] Arnold D L, Brown J, Jackson R W and Vivian A 1999 A dispensable region of the chromosome which is associated with an avirulence gene in *Pseudomonas syringae* pv. pisi *Microbiology* **145** 135–41
[118] Nissan G, Manulis-Sasson S, Weinthal D, Mor H, Sessa G and Barash I 2006 The type III effectors HsvG and HsvB of gall-forming Pantoea agglomerans determine host specificity and function as transcriptional activators *Mol. Microbiol.* **61** 1118–31
[119] Hotson A, Chosed R, Shu H, Orth K and Mudgett M B 2003 *Xanthomonas* type III effector XopD targets SUMO-conjugated proteins in planta *Mol. Microbiol.* **50** 377–89
[120] Peng Z, Hu Y, Zhang J, Huguet-Tapia J C, Block A K, Park S *et al* 2019 *Xanthomonas* translucens commandeers the host rate-limiting step in ABA biosynthesis for disease susceptibility *Proc. Natl Acad. Sci. USA* **116** 20938–46
[121] Li R, Wang S, Sun R, He X, Liu Y and Song C 2018 *Xanthomonas oryzae* pv. oryzae type III effector PthXo3JXOV suppresses innate immunity, induces susceptibility and binds to multiple targets in rice *FEMS Microbiol. Lett.* **365** fny037
[122] Nakano M and Mukaihara T 2018 *Ralstonia solanacearum* type III effector RipAL targets chloroplasts and induces jasmonic acid production to suppress salicylic acid-mediated defense responses in plants *Plant Cell Physiol.* **59** 2576–89
[123] Saado I, Chia K S, Betz R, Alcântara A, Pettkó-Szandtner A, Navarrete F *et al* 2022 Effector-mediated relocalization of a maize lipoxygenase protein triggers susceptibility to *Ustilago maydis Plant Cell* **34** 2785–805
[124] He Q, Naqvi S, McLellan H, Boevink P C, Champouret N, Hein I *et al* 2018 Plant pathogen effector utilizes host susceptibility factor NRL1 to degrade the immune regulator SWAP70 *Proc. Natl Acad. Sci. USA* **115** E7834–43
[125] Boevink P C, Wang X, McLellan H, He Q, Naqvi S, Armstrong M R *et al* 2016 A *Phytophthora infestans* RXLR effector targets plant PP1c isoforms that promote late blight disease *Nat. Commun.* **7** 10311

[126] Li H, Wang H, Jing M, Zhu J, Guo B, Wang Y *et al* A Phytophthora effector recruits a host cytoplasmic transacetylase into nuclear speckles to enhance plant susceptibility *eLife* **7** e40039

[127] Gilroy E M, Breen S, Whisson S C, Squires J, Hein I, Kaczmarek M *et al* 2011 Presence/absence, differential expression and sequence polymorphisms between PiAVR2 and PiAVR2-like in *Phytophthora infestans* determine virulence on R2 plants *New Phytol.* **191** 763–76

[128] Faris J D, Zhang Z, Lu H, Lu S, Reddy L, Cloutier S *et al* 2010 A unique wheat disease resistance-like gene governs effector-triggered susceptibility to necrotrophic pathogens *Proc. Natl Acad. Sci. USA* **107** 13544–9

[129] Liu Z, Zhang Z, Faris J D, Oliver R P, Syme R, McDonald M C *et al* 2012 The cysteine rich necrotrophic effector SnTox1 produced by *Stagonospora nodorum* triggers susceptibility of wheat lines harboring Snn1 *PLoS Pathog.* **8** e1002467

[130] Friesen T L and Faris J D 2004 Molecular mapping of resistance to *Pyrenophora tritici-repentis* race 5 and sensitivity to Ptr ToxB in wheat *Theor. Appl. Genet* **109** 464–71

[131] Hou Y, Zhai Y, Feng L, Karimi H Z, Rutter B D, Zeng L *et al* 2019 A *Phytophthora* effector suppresses trans-kingdom RNAi to promote disease susceptibility *Cell Host Microbe* **25** 153–165.e5

[132] Harvey S, Kumari P, Lapin D, Griebel T, Hickman R, Guo W *et al* 2020 Downy mildew effector HaRxL21 interacts with the transcriptional repressor TOPLESS to promote pathogen susceptibility *PLoS Pathog.* **16** e1008835

[133] Liu C, Wang Y, Wang Y, Du Y, Song C, Song P *et al* 2022 Glycine-serine-rich effector PstGSRE4 in *Puccinia striiformis f. sp.* tritici inhibits the activity of copper zinc superoxide dismutase to modulate immunity in wheat *PLoS Pathog.* **18** e1010702

[134] Yang Q, Huai B, Lu Y, Cai K, Guo J, Zhu X *et al* 2020 A stripe rust effector Pst18363 targets and stabilises TaNUDX23 that promotes stripe rust disease *New Phytol.* **225** 880–95

[135] Zheng X, Fang A, Qiu S, Zhao G, Wang J, Wang S *et al* 2022 Ustilaginoidea virens secretes a family of phosphatases that stabilize the negative immune regulator OsMPK6 and suppress plant immunity *Plant Cell* **34** 3088–109

IOP Publishing

Advances in Biochemical and Molecular Mechanisms of Plant–Pathogen Interaction

Hitendra K Patel and Anirudh Kumar

Chapter 5

Eliminating the enemy: effector-triggered host immunity to combat infection

Praveen Kumar Nayak, Anjana Sharma and Hitendra K Patel

Each plant cell is capable of sensing and responding to pathogen attacks. They have developed intracellular receptors called nucleotide-binding leucine-rich repeat receptors (NB-LRRs/NLRs) to perceive the effectors and effector-mediated degradation by direct or indirect interactions. NLRs are very diverse and sense effectors from a variety of phytopathogens. After effector recognition, NLRs initiate downstream signalling and mount a defence response. This effector-mediated defence response is very robust and requires a lot of energy. Autoactivation of NLRs leads to reduced plant growth and morphological malfunctioning. Therefore, to balance the NLR activity in plant cells, NLRs are regulated at various levels, such as transcriptional and translational levels. Effector recognition leads to oligomerization of NLRs, such as resistosomes, and these oligomerized NLR complexes cause effector-triggered immunity (ETI).

5.1 Introduction

Plants are immobile in nature and cannot escape from the different environmental stresses. They are surrounded by diverse groups of microbes like bacteria, viruses, fungi, etc. Plant-associated microbes can be beneficial, harmful, or neutral for plants [1]. These pathogenic microbes not only utilize nutrients but also alter developmental pathways. Plants have to recognize the harmful/pathogenic microbes and restrict their entry and growth [2]. Plants have evolved multilayered, sophisticated defence systems to tackle the diverse group of microbes. The first layer of defence is provided by physical barriers such as thick waxy cuticles, thorns, trichomes, and preformed chemicals to restrict pathogen entry and growth [3]. Most of the pathogen's entry is restricted by the physical barriers. Pathogens have evolved to break the physical barriers and have found many ways to enter inside the plant. The diverse group of

doi:10.1088/978-0-7503-5673-2ch5

pathogens are recognized by cell surface receptors present on the plasma membrane of host cells, known as pattern recognition receptors (PRRs). PRRs recognize the pathogen/microbe-associated molecular patterns (PAMPs/MAMPs) [4]. PAMPs are conserved motifs from the pathogens, such as flg22: a flagellar protein subunit, and EF-Tu: elongation factor. Pattern recognition by PRRs triggers an innate immune response known as pattern-triggered immunity (PTI). PTI leads to resistance against a variety of pathogens and is a very important defence layer for plant survival. This is an evolutionary process between pathogens and plants to cause and suppress the disease, respectively [5, 6]. To inhibit PTI, pathogens have evolved a group of effectors that inhibit the cellular pathways and make the host susceptible. This process is effector-triggered susceptibility (ETS). In turn, plants have also developed a sophisticated immune system to recognize either effectors or effector-mediated degraded/accumulated products leading to the plant immunity known as ETI [7]. ETI is initiated by intracellular receptors, NB-LRRs/NLRs. These NLRs are evolved in hosts to detect pathogen effectors and are encoded by the resistance (R) genes. This recognition initiates a cascade of signalling events that lead to the activation of defence responses. ETI is more enhanced than PTI and has a robust defence system. PTI is a more general defence mechanism initiated by the recognition of PAMPs by PRRs. ETI, on the other hand, provides a more specific and robust defence response by directly targeting the pathogen effectors [8]. ETI is often linked with localized plant cell death, a phenomenon referred to as the hypersensitive response (HR) and it also induces systemic acquired resistance (SAR). SAR primes uninfected tissues against subsequent pathogen attacks, providing the plant with enhanced resistance throughout its system [9]. For successful pathogenesis, pathogens need to evolve more effectors or lose the existing detectable effectors. This loss or gain of effectors in turn leads to the evolution of new R genes. This process is a continuous arms race between plant and pathogen that leads to the coevolution of different R-avirulence (Avr) pairs [3]. This interaction between PTI, ETS, and ETI is widely regarded as a 'zig-zag' intellectual framework [7]. In this chapter, we will focus on how NLRs recognize the pathogens/effectors, their regulation, and ultimately plant response mediated by NLRs, termed ETI.

5.2 Overview of NLR proteins

NLRs are the major R proteins, involved in plant immune responses. These proteins play a crucial role in recognizing specific pathogen-derived effectors and initiating defence responses in plants [10]. NLR genes are present in all land plants [11] and represent one of the largest gene families in plant genomes, often numbering several hundred in a single plant species (table 5.1). NLRs are even reported in basal angiosperms like *Amborella* and *Nymphaea* [12]. These genes are fast-evolving both in terms of sequence diversification and numbers. This rapid evolution is because of the constant arms race between plants and pathogens, where pathogens exert selection pressure on plants to detect and protect against diverse and rapidly changing pathogen effectors [13, 14].

Table 5.1. NLR genes reported in different plant species (adopted and modified from Jacob *et al* [8]).

Plant species	NLRs	References
Arabidopsis thaliana	151	[15]
Brachypodium distachyon	212	[16]
Brassica rapa	80	[17]
Glycine max	319	[18]
Oryza sativa	458	[16]
Sorghum bicolor	184	[16]
Solanum tuberosum	371	[19]
Zea mays	95	[16]

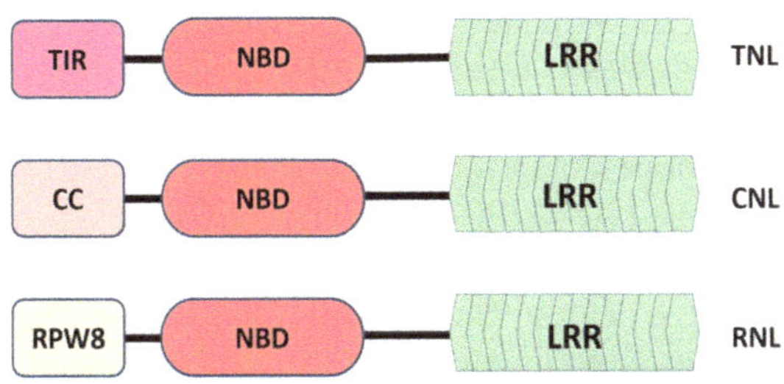

Figure 5.1. Subclasses of NB-LRRs based on the differences in their N-terminal domain. (TNL—toll/interleukin-1 receptor/resistance protein NLR, CNL—coiled-coil NLRs, RNL—RPW8-like CC domain NLR, TIR—toll/interleukin-1 receptor, CC—coiled-coil, RPW8—RESISTANCE TO POWDERY MILDEW 8, NBD—nucleotide-binding domain, LRR—leucine-rich repeats).

NLRs contain a nucleotide-binding (NB) domain at the N-termini and LRR domain at their C-termini. The NB domain consists of APAF-1 (apoptotic protease activating factor-1), different R proteins, and CED-4 (NB-ARC domain). Based on N-terminal domains, NLRs are categorized into three sub-classes [20, 21] as shown in figure 5.1.

5.2.1 Toll/interleukin-1 receptor/resistance protein NLRs (TNLs)

TNLs feature toll/interleukin-1 receptor (TIR) domains at their N-termini. TIR domains are common in proteins involved in signalling pathways, particularly those related to immune responses. TNLs are crucial for recognizing extracellular pathogens and initiating defence mechanisms.

5.2.2 Coiled-coil NLRs (CNLs)

These NLRs are characterized by the presence of coiled-coil (CC) domains at their N-termini. CC domains are known for their structural role in protein–protein interactions. CNLs are often involved in detecting intracellular pathogens and activating defence responses upon recognition.

5.2.3 RPW8-like CC domain NLRs (RNLs)

RNLs possess RPW8 (RESISTANCE TO POWDERY MILDEW 8)-like CC domains at their N-termini. These domains are structurally similar to CC domains and may play roles in protein–protein interactions and signal transduction. RNLs are implicated in plant defence against pathogens, though their specific mechanisms are still being elucidated.

Both CNLs and RNLs contain N-terminal CC domains, suggesting potential similarities in their modes of action or interaction partners despite their divergent classifications. CNLs are present in both dicots and monocots while TNLs are only present in dicots [8, 22]. These different domains have different functions in plants. NB-ARC domain has ADP-ATP binding site. It acts as an ADP-ATP exchanger and activates the NLRs. LRR domains are involved in the recognition of different effectors and this recognition leads to NLR oligomerization and the formation of resistosomes [23, 24]. After NLR activation, CC, TIR, and RPW8, the N-terminal domains initiate further downstream signalling [20, 25]. Few CC domains are known to interact directly with effectors like Avr-Pik and some function as guardee proteins like RIN4 [14, 26]. TIR domains show NADase activity and lead to the formation of variant-cyclic-ADP-ribose (v-cADPR) and they also perform 2′,3′-cAMP/cGMP synthetase activity [27, 28]. Overexpression of CC or TIR domain has been shown to be sufficient to induce a similar immune response like NLRs, indicating the importance of NLRs in providing immunity [29, 30]. So far, multiple R-genes have been identified from different plant–pathogen interaction studies and some of the important R-genes are enlisted in table 5.2.

5.3 Effector perception inside the plant cell

Effectors from diverse groups of organisms, such as viruses, bacteria, fungi, etc, are sensed by NLRs. These NLRs after detecting the effectors start the downstream signalling. This recognition by NLRs can be direct or indirect and involves binding to the effectors directly or with the help of some cofactors, as represented in figure 5.2.

5.3.1 Direct interaction of NLRs with pathogen effectors

The groundbreaking research done in the field of rust of flax disease resistance by H H Flor in 1942 put the foundation of the gene-for-gene hypothesis. This theory elucidates that pairs of genes, one in the host and the other cognate pair in the pathogen define the specificity of disease resistance [55]. This pivotal insight set the stage for further exploration into how receptors recognize and interact with effectors [56]. Regarding NLR-effector recognition in particular, one interpretation of the gene-for-gene paradigm proposes that the defensive mechanism of the plant is activated by the interaction between an effector and a receptor. It is shown by various molecular approaches like yeast two-hybrid assay and other *in vitro* assays, that the two complementary pairs (plant NLRs and pathogen effectors) interact and confer the disease resistance [57–59]. So far *in vitro* assays provide information about

Table 5.2. List of R genes (NB-LRR) identified in some plant–pathogen interactions.

Host	Pathogen	R gene	References
Arabidopsis	*Pseudomonas syringae DC3000*	RPM1	[31]
Arabidopsis	*Pseudomonas syringae pv. tomato*	RPS2	[32]
Arabidopsis	*Pseudomonas syringae pv. pisi*	RPs4	[33]
Arabidopsis	*Pseudomonas syringae pv tomato*	RPS5	[34]
Arabidopsis	*Peronospora parasitica*	RPP13	[35]
Arabidopsis	*Peronospora parasitica*	RPP8/HRT	[36]
Arabidopsis	*Peronospora parasitica*	RPP1	[37]
Arabidopsis	*Peronospora parasitica*	Rpp10	[38]
Arabidopsis	*Peronospora parasitica*	Rpp14	[38]
Arabidopsis	*Peronospora parasitica*	RPP4	[39]
Arabidopsis	*Peronospora parasitica*	RPP5	[40]
Maize	*Xanthomonas oryzae*	Rxo1	[41]
Pepper	*Xanthomonas campestris*	Bs2	[42]
Potato	*Phytophthora infestans*	R1	[43]
Potato	*Phytophthora infestans*	R3a	[44]
Potato	*Phytophthora infestans*	RB	[45]
Potato	*Phytophthora infestans*	Rpi-blb1	[46]
Rice	*Magnaporthe grisea*	Pi36	[47]
Rice	*Magnaporthe grisea*	Pib	[48]
Rice	*Magnaporthe grisea*	Pi-ta	[49]
Rice	*Magnaporthe oryzae*	Pi2	[50]
Rice	*Magnaporthe oryzae*	Pi-9	[50]
Rice	*Magnaporthe oryzae*	Piz-t	[50]
Rice	*Xanthomonas oryzae*	Xa1	[51]
Rice	*Xanthomonas oryzae*	Xa2	[52]
Rice	*Xanthomonas oryzae*	Xa3	[53]
Rice	*Xanthomonas oryzae*	Xa4	[54]
Rice	*Xanthomonas oryzae*	Xa21	[53]

cognate pairs but translating these findings within plant systems has proven challenging, likely due to the transient nature of protein interactions or the involvement of other intricate plant factors.

5.3.2 Indirect NLR-effector interactions

In the direct NLR-effector recognition model, NLR proteins directly recognize specific pathogen effectors and this is likely because of particular sequences present in both the pairs. This high level of NLR and effector diversity is observed in the direct NLR-effector model, indicating selection pressure on both the plant and the pathogen [14, 58]. The indirect model suggests that NLRs sense the diverse pathogen effectors by their actions on host proteins. NLRs detect the changes in host proteins caused by the effectors in the host and that leads to immune activation. This way of

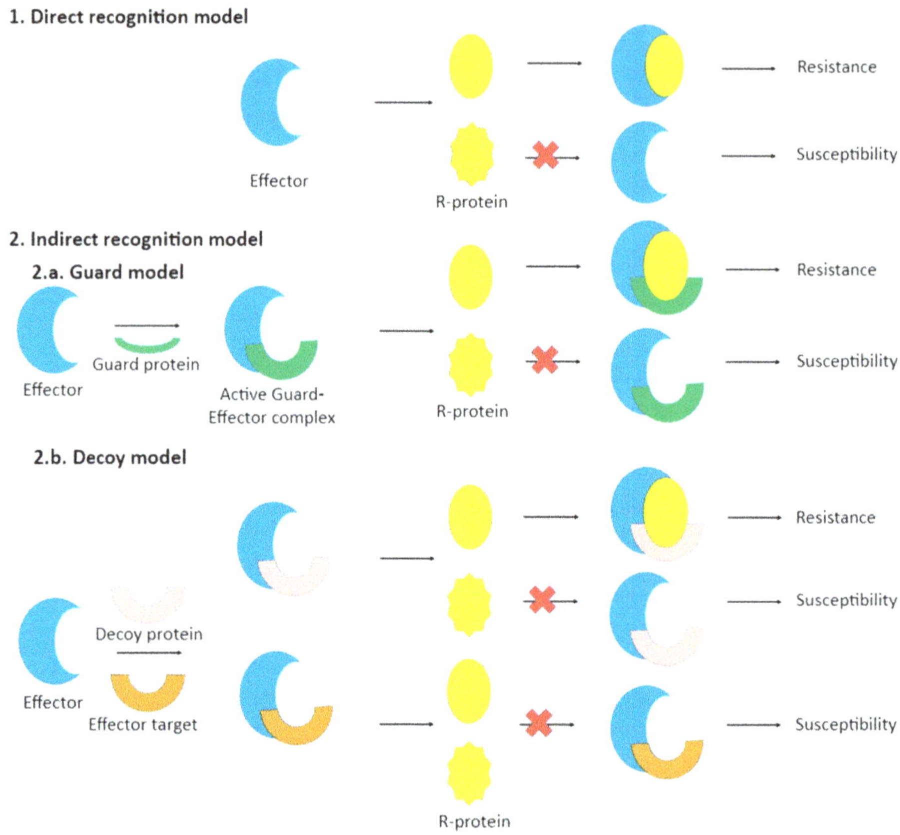

Figure 5.2. Effector recognition model. In the direct recognition model (1), the effector (Avr gene) is recognized by its cognate R-gene pair, leading to resistance, and the absence of that cognate pair will lead to susceptibility. In guard model (2. a), the effector will bind with the guard protein, and that modified guard protein (guard effector complex) is perceived by the NLRs and resistance. An example of this model is the RIN4 protein. A modification or advancement of the guard model is the decoy model. In the decoy model (2. b), the host plants have evolved decoy proteins. These decoy proteins are similar to effector target proteins and modification in these decoys leads to resistance.

pathogen recognition provides a wider range to the plant immune system to cope with rapidly evolving pathogens [60].

Such indirect NLR interactions are exemplified by the guard model. For instance, in *Pseudomonas–Arabidopsis* interaction, RIN4, a host protein is guarded by an NLR that detects the modification in the RIN4 protein by effector. This detection leads to the activation of defence response [61, 62].

Another example of indirect interaction is the decoy model. Here, the monitored host factor doesn't directly contribute to resistance but acts as bait to capture pathogen effectors targeting structurally similar basal defence components. This mechanism then initiates ETI [63]. The example of the decoy model is shown by tomato CNL Prf (for *Pseudomonas* resistance and fenthion sensitivity) to multiple bacterial effectors, including *Pseudomonas syringae*-AvrPto [64].

5.3.3 Bacterial recognition mediated by NLRs

Bacterial pathogens have evolved effectors to target key components of plant immune systems, suppressing the PTI. In response, plants have evolved an array of NLRs to sense the presence of these pathogens and trigger ETI. In tomato, the CNL protein Prf guards the decoy kinase Pto. When AvrPto interferes with Pto's kinase activity, Prf detects this alteration and initiates ETI [65]. Given the diversity of plant NLRs, some bacterial effectors are recognized by multiple NLRs across different plant species. For example, AvrB is detected by *At*TAO1, *At*RPM1, and *Glycine max Gm*RPG1b [66, 67]. Numerous bacterial effectors typically target the immune system's central hubs. As a result, NLRs, which are in charge of protecting these vital elements of immunological signalling, can identify several effectors [68]. For example, the receptor-like cytoplasmic kinase (RLCK) PBL2 is protected by the CNL *At*ZAR1, which works in tandem with the pseudokinase RKS1. By using this defensive mechanism, *At*ZAR1 can detect a number of effectors indirectly, such as AvrAC from *P. syringae* or *Xanthomonas campestris* and HopZ1a, HopF2, HopBA1, HopO1, HopX1, and maybe additional effectors that target RLCKs. *Nb*ZAR1 is also vital for recognizing XopJ4 from *Xanthomonas perforans* through the pseudokinase JIM2 [69–71].

Another way to detect the effector presence in plant cells is via TALEs (transcription activator-like effectors) study in *Xanthomonas* pathovars. TALEs have a DNA binding domain. TALEs localize inside the host nucleus and perform their function of regulating the host gene expression. Some of the known TALE targets that provide resistance against *Xanthomonas oryzae* pv. *oryzae* are *xa5, Xa7, xa13, Xa21, Xa27* etc [72]. Another example is AvrBs3 from *X. campestris* pv. *vesicatoria* that targets genes for cell enlargement [73]. In the host, the *Bs3* gene has evolved a similar effector binding sequence that is targeted by AvrBs3. In the presence of AvrBs3, an increase in *Bs3* gene expression leads to resistance [74, 75].

5.3.4 Fungal recognition mediated by NLRs

Plant NLRs are able to identify a variety of fungal pathogen effectors and compounds. RESISTANCE LOCUS A of *Hordeum vulgare* NLRs are capable of identifying a variety of *Blumeria graminis* effectors [76, 77], including races of wheat stripe rust [78]. LOCUS ORCHESTRATING VICTORIN EFFECTS 1 (LOV1) from *Arabidopsis* and *Phaseolus vulgaris* recognizes victorin, a secondary metabolite from *Cochliobolus victoriae* [79, 80]. The CNL *Cm*Fom-2 from *Cucumis melo* recognizes the AvrFom2 of *Fusarium oxysporum* [81]. Multiple TNLs from *Linum usitatissimum* identify effectors from *Melampsora lini* [82, 83]. Many CNLs from *Oryza sativa* recognize effectors from the rice blast fungus *Magnaporthe oryzae* [84, 85].

5.4 Regulation of resistance genes (NLRs)

R genes are very important for plant survival. In plants that are healthy and not under attack by pathogens, NLRs are expressed poorly and are usually not active. They play a basic role in monitoring the plant's surroundings. When plants lack the necessary R genes, they become vulnerable to specific pathogens, making them susceptible to diseases. However, if the regulation of NLRs is misbalanced, it can lead to a condition called autoimmunity in plants. Plants experiencing autoimmunity may survive, but they often exhibit stunted growth, appearing much smaller than healthy plants. Additionally, they may display abnormal physical characteristics such as twisted leaves and visible lesions on their surfaces. These changes in appearance are called morphological phenotypes and result from the plant's immune system malfunctioning [86]. One well-known example of an autoimmune mutant in plants is *snc1*. This mutant carries a specific point mutation that makes TNL more stable, leading to the development of autoimmunity. *bal1*, a variant due to genomic duplication incorporates an extra copy, and that results in increased transcription of SNC1 and increased plant immunity [87]. Several other mutations in NLRs have also been linked to autoimmunity, such as *ssi4* [88] and *chs1* [89]. Suppressor and enhancer screening with these mutant plants have helped us to better understand how NLRs are controlled and how they signal during immune responses. These studies have been crucial in uncovering the mechanisms that maintain the balance of NLRs and their downstream signalling pathways during ETI. The visible effects of autoimmunity, such as abnormal growth and morphological changes like twisted leaves and lesions, highlight the importance of tightly regulating NLRs in plants. Plants need to strike a balance between quickly detecting pathogens and maintaining normal growth and development. This delicate balance ensures that plants can defend themselves against invaders while still thriving and growing properly [86]. Because of their importance in plant growth and development, NLRs are regulated at multiple levels [90].

5.4.1 NLRs regulation at the transcriptional level

NLRs are important proteins in plant immune responses that, undergo strict regulation at the transcriptional level. After encountering pathogens, plants show a high-level change in the transcriptome. This change in the expression can be specific to tissues and organs [91]. Promoter enrichment analysis shows that the binding sites for some transcription factors (TFs) are enriched in NLRs. One of them is WRKYs. WRKYs, play a role in regulating NLR expression. It's not surprising because WRKY TFs are involved in various defence processes in plants [92]. However, the regulation of NLRs by WRKYs varies. Some NLRs, like Mla6 and Mla13, seem to have feedback loops specific to certain effectors, controlling their expression. On the other hand, some NLRs respond to changes triggered by the defence hormone like salicylic acid (SA) [88, 93]. This diversity in regulatory mechanisms highlights the complexity of how plants manage their immune responses. It shows how plants have developed complex defence mechanisms against infections that are adapted to the many circumstances and dangers they face in their surroundings.

5.4.2 Epigenetic regulation

DNA methylation plays a crucial role in regulating plant immunity. Generally, less methylation enhances defence mechanisms, while increased methylation can make plants more susceptible to pathogens. Mutations in proteins like DDM1 (Decrease in DNA Methylation 1) [94] and MOS1 (MODIFIER OF snc1) in *Arabidopsis* leads to decreased cytosine methylation, which affects the transcription of NLRs such as SNC1 (SUPPRESSOR OF npr1-1, CONSTITUTIVE1) [95]. DDM1 is particularly involved in chromatin remodelling, a process essential for regulating gene expression [96]. MUSE (mutant snc1-enhancing) and SPLAYED (SWI/SNF chromatin remodeller SYD) are identified as chromatin remodelling proteins [97]. In the case of CNL RPW8, pathogen infection can alter the methylation state of RPW8 locus, linking pathogen presence to changes in gene regulation directly [98]. Interestingly, genomic regions containing NLRs are often rich in transposons. Epigenetic modifications are prone to occur in these transposon-rich areas leading to reduced transcription [99]. This reduced transcription helps in overcoming higher expression levels of some NLRs and autoimmunity [100].

5.4.3 NLRs regulation at the post-transcriptional level

Post-transcriptional regulation plays a significant role in controlling the activity of NLR genes in plants. The transcribed mRNAs of NLRs undergo various regulatory processes after their transcription. One of the important aspects of mRNA regulation is alternative splicing. Mutations affecting splicing machinery often result in altered patterns of NLR gene splicing. Mutants with splicing defects, such as mos4 (MODIFIER OF snc1), cdc5 (cell division cycle 5), prl1(Pleiotropic Regulatory Locus 1), and mos12, often exhibit increased susceptibility to pathogens [101–103]. The different isoforms of some NLRs respond differently to pathogen infection. These transcripts are usually abnormal and are targeted for degradation, preventing the overaccumulation of NLR proteins in plant cells. Disruption of nonsense-mediated decay (NMD), a process that degrades abnormal transcripts, can lead to autoimmunity in plants [90].

5.4.4 Regulation of NLRs by small interfering RNAs

RNA silencing is a key mechanism that regulates gene expression at different levels: transcriptional gene silencing (TGS) and post-transcriptional gene silencing (PTGS) [104]. It was observed that certain R genes, like those found in the RPP5 (recognition of *Peronospora parasitica* 5) locus such as RPP4 and SNC1, are controlled by small interfering RNAs (siRNAs) through PTGS. When RNA silencing is disrupted in mutants lacking proteins like DCL4 (Dicer-like proteins) and AGO1(Argonaute family proteins), there is an increase in the transcription of SNC1, indicating the regulatory role of RNA silencing in these genes [105]. Moreover, a specific type of small RNA called microRNA miR482 has been identified to targeting numerous R genes across different plant species. This miRNA triggers the degradation of R gene mRNA, leading to the production of secondary siRNAs [106]. More recent research has shown that another RNA-related enzyme called RNA-dependent RNA polymerase 6 (RDR6) and miR472 are essential for both ETI and basal defence mechanisms by controlling PTGS of certain R genes [107]. Mutants lacking RDR6 and miR472 exhibit enhanced resistance against pathogens like RPS5-

mediated resistance, with increased levels of RPS5 transcripts. This suggests that RDR6 and miR472 act as negative regulators of ETI by controlling the expression of disease-resistance genes at the post-transcriptional level [105]. MicroRNAs, also play a crucial role in regulating NLR transcript levels. In spruce, for example, small RNAs contribute to the degradation of over 90% of TNLs [103]. MicroRNAs have been specifically associated with the regulation of numerous NLRs across different plant species, demonstrating their widespread involvement in fine-tuning plant immunity [104].

5.4.5 Regulation of NLRs at the post-translational level

Conformational changes play a crucial role in detecting effector activity and activating NLRs during plant immunity. Given this importance, it's not surprising that several chaperone proteins are required for NLR-triggered immunity. A key complex involved in this process is the RAR1–SGT1–HSP90 chaperone complex. This complex, composed of RAR1, SGT1, and HSP90 proteins, assists in the proper folding and functioning of NLRs during immune responses [108, 109]. However, it's also unsurprising that these chaperone proteins might become targets for pathogen effectors. Recent findings showed that a family of bacterial effectors called HopBF1 can specifically target and phosphorylate HSP90. This modification prevents HSP90 from properly activating NLRs, leading to disease symptoms in the plant [110]. It is notable that this targeting by HopBF1 is specific to HSP90 and can be observed even with HSP90 proteins from other eukaryotic organisms. This highlights the sophisticated strategies that pathogens employ to subvert plant immune responses and emphasizes the importance of chaperone proteins like HSP90 in plant defence mechanisms. Another mechanism that maintains the NLR protein levels is the ubiquitin–proteasome pathway and SUMO (small ubiquitin-like modification) pathway. In these pathways, proteins tagged with ubiquitin or SUMO are targeted for degradation by the proteasome or subjected to other regulatory processes [111, 112]. While components like E1s, E2s, and E4s in these pathways are generally nonspecific, certain E3 ligases play a crucial role in specifically targeting NLRs for degradation [113, 114]. For instance, the NLR protein SNC1 can be directly SUMOylated by the SIZ1 protein or influenced by the SUMOylation of an upstream positive regulator [115]. Specific E3 ligases have been identified to target particular NLRs for degradation. For example, the SCF^{CPR1} E3 complex targets SNC1 [116, 117], while simple RING-type E3 ligases like MUSE1/2 target SNC1's partners SIKIC1/2/3 for ubiquitination and degradation [118]. Recent advances in technology, such as Turbo-ID, have helped reveal new interactions within the ubiquitin–proteasome system. For example, the E3 ligase UBR7 has been found to negatively regulate the levels of the TNL-N [119]. Interestingly, some duplicated E3 ligases, like RIN2 and RIN3, serve as positive regulators of immunity by maintaining normal levels of defence response induced by some NLRs like RPM1 and RPS2 [120]. Upon immune activation, there is an observed increase in the ubiquitin–proteasome system, which leads to a decrease in many defence-related gene products after infection. The ubiquitination pathway can be attacked by effectors in a similar way to a chaperone, such as the SOC3 guardee SAUL1 acts as an E3 ligase, but there are many other examples of pathogen effectors targeting components of the ubiquitination pathway [121]. Overall, post-translational mechanisms, particularly ubiquitin–proteasome-mediated degradation and SUMOylation pathways, play critical roles in regulating NLR protein levels and modulating plant immunity. E3 ubiquitin ligases, while known to play roles in defence mechanisms, have diverse functions in both plants and other eukaryotic organisms [113]. These E3 ligases have undergone expansion in higher plants, resulting in genetic redundancy and making them challenging to study. However, an increasing number of E3 ligases are being recognized for their roles in ETI, particularly in regulating the levels of NLR proteins [122, 123]. Another protein called SRFR1 (SUPPRESSOR OF rps4-RLD), identified during a search for mutants with constant defence responses, partners with SGT1 to control the levels of SNC1, RPS2, and RPS4, preventing the overactivation of plant defence systems [116, 117]. Light also plays a role in regulating plant defence mediated by R proteins, with photoreceptors such as CRY2 and PHOT2 involved in this process [124]. Even though R proteins are essential for plants to mount a strong defence against pathogens, they remain inactive and tightly regulated under normal, non-pathogenic conditions to prevent unnecessary responses.

5.5 Association of intracellular receptors (NLRs) and signalling complex

5.5.1 NLRs association/oligomerization

NLRs are proteins that are involved in pathogen recognition and defence response, and how NLRs connect to basal resistance pathways in ETI remains unknown. In mammals, signalling adaptors typically initiate downstream pathways by interacting with receptors in a homotypic manner, but such proteins are not reported in plants [125]. In plants, there are truncated NLR genes present, that might function as adaptor proteins in plant resistance [8, 13]. The importance of truncated forms of NLRs, like the *Arabidopsis* proteins CHS1 (chilling sensitive 1) and CHL1 (CHS1-like 1), in regulating TNL immunity to maintain plant growth and fitness across various environmental conditions has been revealed through genetic and molecular studies [89, 126]. Additionally, alternative splicing of TNL mRNA produces truncated variants that enhance TNL resistance once ETI is initiated [127]. Studies on various plant NLR receptors, such as barley MLA10 (mildew A10), *Arabidopsis* RPS5, and tomato Prf CNL, suggest constitutive self-association, potentially as oligomers [64, 128]. However, tobacco TNL receptor N forms oligomers only in the presence of specific pathogens, indicating induced self-association [129]. Crystal structures of certain NLR domains have revealed the molecular details of their interactions, and disrupting these interactions abolishes their signalling activities, suggesting the importance of dimer formation for NLR receptor signalling [130]. Some NLRs can interact with each other. This interaction in NLRs and related signalling functions can be in the same complex. For instance, the *Arabidopsis* RRS1/RPS4 TNL receptors and the rice RGA5/RGA4 CNL pair form interacting pairs, emphasizing the significance of interactions between specific domains of NLRs in ETI [131, 132]. This concept is similar to the mechanism observed in mammalian innate immunity, where NLRs recognize PAMPs and assemble into complexes called inflammasomes to activate inflammatory responses [133].

5.5.2 Resistosome formation

When a plant senses a threat from a pathogen, a special protein called a monomeric NB-LRRs transforms into a resistosome complex (figure 5.3). This resistosome disrupts the plant cell's protective membrane, triggering a process called hypersensitive cell death. This drastic measure sacrifices infected cells to stop the pathogen from spreading further, protecting the plant [23, 134].

The formation of the resistosome complex is mediated by ZAR1 (HOPZ-ACTIVATED RESISTANCE 1). By connecting with the RLCK (receptor-like cytoplasmic kinase)—RKS1 (RESISTANCE RELATED KINASE 1), ZAR1, which is generally inactive in its ADP-bound monomeric state, remains inactive, primarily due to internal interactions within its leucine-rich repeat domain. On the other hand, the bacterial pathogen effector AvrAC attaches to RKS1 in the ZAR1–RKS1 complex after uridylating another RLCK, $PBL2^{UMP}$ (PBS1-LIKE PROTEIN 2). ZAR1's nucleotide-binding domain changes structurally as a result

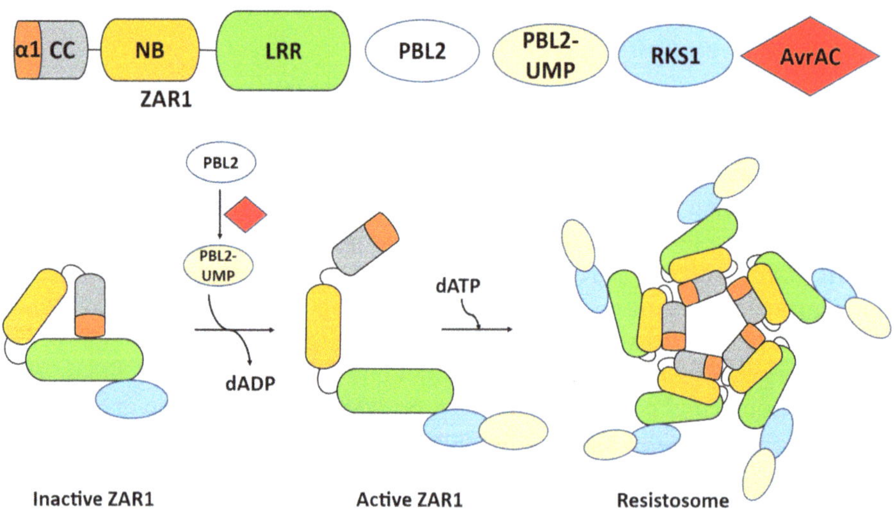

Figure 5.3. Resistosome formation (adopted and modified from [23, 134, 195]). All the individual components such as ZAR1 (HOPZ-ACTIVATED RESISTANCE1), PBL2 (PBS1-LIKE PROTEIN 2), $PBL2^{UMP}$ (uridylated PBL2), RKS1 (RESISTANCE RELATED KINASE 1), and AvrAC (*Xanthomonas campestris* effector) required for resistosome formation are represented in different colours. AvrAC effector uridylates the PBL2 to $PBL2^{UMP}$. NLR ZAR1 is present in inactive state bounded with RKS1. Uridylated $PBL2^{UMP}$ binds with the ZAR1–RKS1 complex resulting in conformational change in ZAR1 and dADP to dATP exchange. Active ZAR1–RKS1-$PBL2^{UMP}$ complex having exposed N-terminal α helix pentamerize and form a wheel-shaped structure. This pentameric complex is a resistosome that functions as a death switch.

of this interaction, releasing ADP and getting the complex ready for activation. Subsequently, the primed ZAR1–RKS1–PBL2^{UMP} complex binds to dATP/ATP, triggering the formation of a pentamer known as a resistosome (figure 5.3). The 'death switch,' which is caused by this pentamerization, is a major conformational shift that causes the α1 helices in ZAR1 to become active and protrude from the resistosome plane. For the resistosome to work, this visual fold switch is essential. Subsequent investigations reveal the pivotal functions of the α1 funnel, which include facilitating the AvrAC-induced accumulation of the ZAR1 complex in the plasma membrane and serving as a vital source of bacterial resistance. This complex process provides insights into the defence mechanisms plants use against bacterial infections by illuminating how molecular interactions and structural alterations influence plant immunity [23, 134]. The membrane pores and ion channels that arise during mammalian cell death during pyroptosis and necroptosis are similar to the ZAR1 resistosome fold switch. In fungus, a comparable process is also at work during NLR activation. When the fungus *Podospora anserina* induces structural remodelling in the NLR-type heterokaryon incompatibility protein S (HET-S), a buried N-terminal transmembrane region is released, embedding HET-S in the plasma membrane and transforming it into a pore-forming toxin [135, 136]. Regarding plant immune signalling, NLR-triggered immunity involves downstream signalling pathways like Ca^{2+} flux, mitogen-activated protein kinases (MAPK), and reactive oxygen species (ROS), suggesting that plasma membrane damage may activate these pathways, possibly aiding in cell-to-cell communication [137]. However, NLR resistosomes might also recruit downstream signalling components independently of membrane damage activities.

5.6 Plant defence response governed by ETI

5.6.1 Formation of reactive oxygen species and cell death

Elevated ROS levels are frequently linked to the start of the hypersensitive response (HR) in plants. Respiratory burst oxidase homologues (RBOHs), another name for NADPH oxidases in plants, are a major generator of ROS during plant–pathogen interactions. In *Arabidopsis*, it has been demonstrated that when avirulent bacterial pathogens infect mutants—such as *at*rbohD and *at*rbohD/F double mutants, these plants die after the wild-type plants. This suggests that RBOHs and ROS are essential for the cell death that these infections cause [138]. But as it turns out, there is more nuance to the link between ROS generation and cell death than first appears. Studies with avirulent oomycete pathogens have shown that even in the absence of ROS, *Arabidopsis* double mutants deficient in RBOHs exhibit increased cell death. This implies the presence of other channels that control cell death apart from ROS [138].

Furthermore, when exposed to the bacterial AvrRpm1, experiments including *Arabidopsis* mutants lacking catalase function, such as nca1 (no catalase activity 1) and cat2 (catalase 2) mutants, unexpectedly showed lower cell death. These mutants were expected to have higher ROS levels. This surprising discovery suggests that independent of ROS concentrations, additional mechanisms downstream of catalase are involved in controlling cell death [139]. These findings highlight how intricately

ROS and plant programmed cell death (PCD) are related biologically. They draw attention to the existence of ROS-independent pathways and stress the need for more investigation into the complex mechanisms controlling the regulation of cell death during plant–pathogen interactions.

5.6.2 Autophagy

One important finding in the field of ETI is the identification of autophagy as a regulator of PCD. It has been shown that when the tobacco mosaic virus (TMV) was introduced into *Nicotiana benthamiana*, having downregulated autophagy genes ATG6 and ATG7 showed a delayed cell death process. It shows the role of autophagy genes in limiting PCD to infection sites and suppressing cell death [140, 141]. Nevertheless, a different study found that ATG7 and ATG9 knockout mutants inhibited the start of PCD that was mediated by specific type R proteins, like the EDS1-dependent TNL type R proteins. On the other hand, NDR1-dependent CNL-mediated death is independent of autophagy [142]. Another research clarified the dual role of autophagy in PCD by emphasizing how it depends on the photoperiod circumstances and plant leaf age during trials. Particularly, older leaves of atg5 mutant plants showed extensive RPM1-mediated cell death, most likely as a result of SA build-up and high ROS generation [143]. Moreover, it was shown that ROS and catalase activity control autophagy-dependent cell death. Comprehending the complex molecular processes that underlie the regulating function of autophagy in PCD is of great significance for the domain of plant–microbe interactions.

5.6.3 Programmed cell death at infection site

The HR, which is defined by the quick start of PCD at the infection site is the distinguishing feature of ETI. This cellular reaction is essential for getting rid of biotrophic infections, which are organisms that feed on living host cells. PCD is classified into two primary types: autolytic and non-autolytic, each of which has unique properties [144]. Hydrolases are released from vacuoles in autolytic PCD, which causes fast cytoplasmic clearance and localized cell death. On the other hand, even in cases when tonoplast permeability (vacuole membranes) persistently increases, non-autolytic-PCD does not include the release of hydrolases [144]. The release of vacuolar antibacterial proteins into the apoplast is a consequence of the fusing of the plasma membrane and central vacuole membrane, which occurs when R protein signalling is activated. Since this mechanism does not entail the release of hydrolases to clear the cytoplasm, it is consistent with the features of non-autolytic PCD [145, 146].

5.6.4 Regulation of programmed cell death by salicylic acid

SA and its receptor paralogues like NPR, regulate PCD in *Arabidopsis*. Interestingly, the levels of SA affect whether or not cell death occurs: high levels of SA cause cell death, whilst low levels repress it. For instance, SA levels are noticeably higher on *Nicotiana tabacum* L. leaves infected with TMV at a radius of up to 3.5 mm from the centre of the cell death zone. When compared to areas 3.5–6.5 mm from the centre, this concentration is roughly eight times higher, and when compared to areas 6.5–10 mm from the centre, it is 27 times higher [147]. NPR3 and NPR4 proteins have BTB/POZ domain, with the help of that they interact with NPR1 and E3 ligase and serve as an adaptor protein. These proteins enable the proteasomal degradation of NPR1 protein, which is a PCD suppressor [148]. NPR3 has less affinity for SA than NPR4, which has a higher affinity. The interaction between NPR1 and NPR4 is interrupted at the core of the cell death zone where SA concentration is high, promoting the interaction between NPR1 and NPR3. As a result, NPR3 encourages NPR1 to be degraded to an extent that is insufficient to prevent cell death. On the other hand, in adjacent cells, the interaction between NPR1 and NPR4 is disrupted because of moderate SA level and also this moderate SA level is not sufficient to promote NPR1–NPR3 interaction leading to the prevention of NPR3/NPR4-mediated degradation of NPR1. This results in a high level of NPR1 in adjacent cells that inhibits the proliferation of cell death. NPR1 interacts with several TFs to initiate plant defence signalling responses [148].

5.7 Reprogramming of defence-related genes

A strong plant immune response known as ETI is elicited by a wide-ranging reprogramming of transcription, involving several transcriptional regulators [149, 150]. Upon the identification of Avr effectors during ETI, the dynamic recruitment of these regulators into R protein-mediated signalling pathways has become apparent as a critical event [151].

5.7.1 Regulation of WRKY transcription factors

It is known that a number of WRKY TFs are essential for controlling defence reactions [152]. For example, in the nucleus, the *R. solanacearum* type III effector PopP2 interacts with the TNL protein RRS1-R/WRKY52. RRS1-R/WRKY52 protein has a C-terminal WRKY domain [153, 154]. RRS1-R/WRKY52 changes from a transcriptional repressor to an inducer of resistance signalling upon detection of PopP2. In normal conditions, this protein is inactive and serves as a repressor of several resistance/defence genes [153]. More investigation into the target genes and RRS1-R/WRKY52/SLH1 interaction components will shed light on the regulatory function of this protein in plant defence signalling. The *Hv*WRKY1/2 TF, when activated by the avrA10 effector, in barley, it interacts with the R protein MLA10 [155]. *Hv*WRKY1/2 functions as transcriptional repressors of basal defence, as demonstrated by overexpression and silencing studies. This suggests that MLA-mediated defence responses may be triggered by the inhibition of these WRKY TFs [155]. *Os*WRKY45 in rice functions as a transcriptional activator, enhancing the plant's ability to withstand panicle blast. The CNL R protein Pb1 stabilizes *Os*WRKY45, increasing its transcriptional activity and blast resistance gene expression [156, 157]. Comparably, in tomato, resistance mediated by the R protein Motelle (Mi-1) and basal resistance, both are positively regulated by *Sl*WRKY72 TFs [158].

5.7.2 Regulation via mediator complex

In order to organize transcriptional reprogramming, TFs and RNA polymerase II (RNAPII) communicate through the mediator (MED) complex. It is a conserved transcription regulator in eukaryotes [159]. The MED complex, which is made up of more than 20 subunits arranged into head, middle, and tail modules, is essential for controlling the expression of certain genes. Numerous MED subunits in *Arabidopsis* have been functionally characterized, such as MED8, MED14, MED15, MED16, MED19a, MED21, and MED25 [160, 161]. Among these ETI is positively regulated by two tail module subunits, MED14 and MED16 [162, 163].

5.7.3 ETI mediated chromatin remodelling

Modulating chromatin configuration through several processes, such as methylation of cytosine residues in DNA, post-translational modifications of histones, and ATP-dependent chromatin remodelling, can have a significant impact on rapid transcriptional alterations in ETI. The *Arabidopsis* ELP2 (elongator complex subunit 2) is a genomic DNA demethylase and is required for the demethylation and induction of *NPR1* and *PAD4* in response to avirulent DC3000 (avrRpt2) infection [160]. Furthermore, ELP2 has histone acetyltransferase activity, which is necessary for the expression of a number of defence genes. It has been suggested that ELP2 is a significant positive regulator of ETI, possibly via histone modification, which also promotes RPS2- and RPS4-mediated resistance [164]. A Trithorax (TRX) family protein, *Arabidopsis* ATX1 functions as a histone methyltransferase that upregulates defence genes [165]. In *Arabidopsis*, TNL (RPS4-like) gene expression is epigenetically upregulated by another histone methyltransferase, SET DOMAIN GROUP 8 (SDG8), which contributes to RPM1-mediated resistance [166]. Pre-acetylation of nucleosomes may be regulated by the conserved TPR domain-containing protein SRFR1, which functions as a repressor of several NLR proteins [167]. In a similar vein, *At*TIP49a may operate as a negative regulator of two RPP (resistance to *Peronospora parasitica*)-dependent disease resistances through its interactions with the N-terminal half of RPM1 (CNL) and RPP5 (TNL). The possible function of chromatin-remodelling complexes in regulating plant defensive responses is demonstrated by these interactions [168].

5.7.4 Chromatin superstructure

DNA and histone methylation status remain unaltered when the ATPase genes of the Microrchidia (MORC) family, *At*MORC1 and *At*MORC6, are mutated. This results in the activation of transposable elements and DNA-methylated genes. The three-dimensional architecture of genomes was studied using Hi-C analysis, which showed that the pericentromeric DNA portions are decondensed in the *atmorc*1 and *atmorc*6 mutants [169]. One of the most important genes for disease resistance to the turnip crinkle virus (TCV) mediated by HRT has been found as *At*MORC1, the same as CRT1 [170]. Furthermore, CRT1 physically interacts with the R protein chaperone HSP90 as well as a number of R proteins, including RPS2, Rx, and

RPP8. When *N. benthamiana* CRT1 homologue was silenced, Pto and RPM1-induced cell death was lowered, while plants with *crt*1 mutations showed impaired R gene-mediated defence responses [171]. These results point to a major role for chromosomal superstructure regulation in the healthy operation of ETI.

5.8 Plant defence mediated by resistosomes

5.8.1 Resistosomes function via calcium (Ca^{2+}) channel permeability

In the ZAR1 resistosome, a critical component is the N-terminal helix α1, which forms a distinct funnel-shaped structure. This structure, notably the only exposed portion of the CC domain, is thought to be crucial for the function of the ZAR1 resistosome. When two negatively charged residues (Glu11 and Glu18) located within this funnel-shaped structure are simultaneously mutated, it disrupts ZAR1-mediated immune responses, indicating a potential pore- or channel-related activity associated with the ZAR1 resistosome [134]. Additionally, protein fractionation experiments have shown that upon induction by AvrAC, ZAR1 associates with the plasma membrane. Similar plasma membrane localization has been observed for other CNLs (CC-NB-LRR proteins) like RESISTANT TO P. SYRINGAE 2 (RPS2) and RPM1 [172, 173]. Moreover, studies have demonstrated that the ZAR1 resistosome can form within *Arabidopsis* protoplasts and exhibit calcium-permeable cation-selective channel activity in lipid bilayers [174]. Activation of ZAR1 within plant cells triggers calcium influx, disrupts subcellular structures, and induces immune responses. These findings collectively suggest that the ZAR1 resistosome likely functions as a plasma membrane channel, playing a central role in mediating plant immune responses [174]. The available evidence suggests that the ability to function as Ca^{2+}-permeable channels may be conserved among CNLs across various plant species. For instance, the Sr35 resistosome in wheat shares a strikingly similar structure with the ZAR1 resistosome. Furthermore, when expressed in *Xenopus* oocytes, the Sr35 resistosome also exhibits Ca^{2+}-permeable channel activity. This indicates a potential evolutionary conservation of Ca^{2+}-permeable channel function among CNLs, highlighting their importance in plant immune responses across diverse plant species [175, 176].

5.8.2 CNL resistosome-mediated extracellular Ca^{2+} influx

Evidence from various studies strongly supports the essential role of extracellular Ca^{2+} influx in initiating ETI signalling [177, 178]. Early in the ETI process, there is a noticeable increase in intracellular Ca^{2+} levels. Mutations that enhance the activity of CNGC19/20 (CYCLIC NUCLEOTIDE GATED CHANNEL19/20), lead to a rise in Ca^{2+} influx, resulting in the constant activation of EDS1 and SA-dependent immunity in *Arabidopsis* [179, 180]. Additionally, resistosomes such as Sr35, NRG1, and ADR1 exhibit uncontrolled channel activity, capable of reproducing the cell death typically observed in plant CNL-mediated responses when introduced into eukaryotic cells [176, 181]. While Ca^{2+} release from internal cellular stores can contribute to ETI signalling. Studies indicate that blocking Ca^{2+} release from intracellular compartments with ruthenium red (RR) is less effective at inhibiting HR cell death compared to the Ca^{2+} influx blocker LaCl3. This suggests that extracellular Ca^{2+} influx is the primary trigger for ETI responses [182].

5.8.3 Cellular localization of CNLs

The necessity of nuclear localization for the disease-resistance function of many NLRs is well established, yet the precise mechanism connecting NLR cellular localization to their activity remains unclear [183]. Some CNLs have been observed to interact with TFs, suggesting a potential role in directly modulating transcriptional processes within the nucleus [184]. Despite both PTI and ETI signalling pathways activating transcription of similar gene sets, there are likely additional mechanisms involved in NLR-mediated transcriptional reprogramming, although these mechanisms remain poorly understood. NRG1A presents an intriguing scenario as it localizes to both the PM and the nucleus upon activation. However, only PM-resident NRG1A forms oligomers, suggesting that NRG1A function might be independent of oligomerization [185]. Various stresses have been found to cause increases in nuclear-free Ca^{2+} concentrations [186, 187]. The distribution of Ca^{2+} within a cell is not uniform, it is important to remember, that sharp gradients occur only a few nanometres away from Ca^{2+} channels [188]. Consequently, Ca^{2+} carried by PM-localized resistosomes or other Ca^{2+}-permeable channels may not reach the nucleus. However, it is still possible that CNLs form resistosomes inside the nucleus or at the junction of the nucleus and another organelle, like the endoplasmic reticulum (ER). This kind of cellular location may improve Ca^{2+} signalling within the nucleus through resistosome-mediated channels, similar to what has been shown for the Ca^{2+}-permeable channel CNGC15 during symbiotic Ca^{2+} oscillation [189].

5.8.4 ETI—an enhanced PTI response

PTI and ETI are two important layers of plant defence and involve different early signalling cascades and are triggered by various activation mechanisms. How they function is well explained in the 'zig-zag' model and was thought to function independently [7]. Immunity mediated by PRRs is well studied as compared to NLRs, but how these two pathways (PTI and ETI) interact is less studied [190]. In *Arabidopsis*, PTI is necessary for ETI to confer efficient pathogen resistance. PTI can restrict infections by limiting nutrition, strengthening cell walls, inhibiting bacterial type III secretion, and promoting the production of antimicrobial chemicals [191–193]. By enhancing the PTI signalling components and regulating protein turnover, translation, and transcription, ETI boosts PTI-induced defence responses. It is yet unknown how each PTI component is accomplishing this. It is further demonstrated that the mutual potentiation of these two systems results in a greater immune response when PTI and ETI are used together [190]. The new findings are consistent with a paradigm, that is, the main source of immunity in defences is triggered by PRR-dependent signalling. According to this concept, activated NLR receptors work to restore PRR signalling components and strengthen PRR-dependent signalling, which counteracts

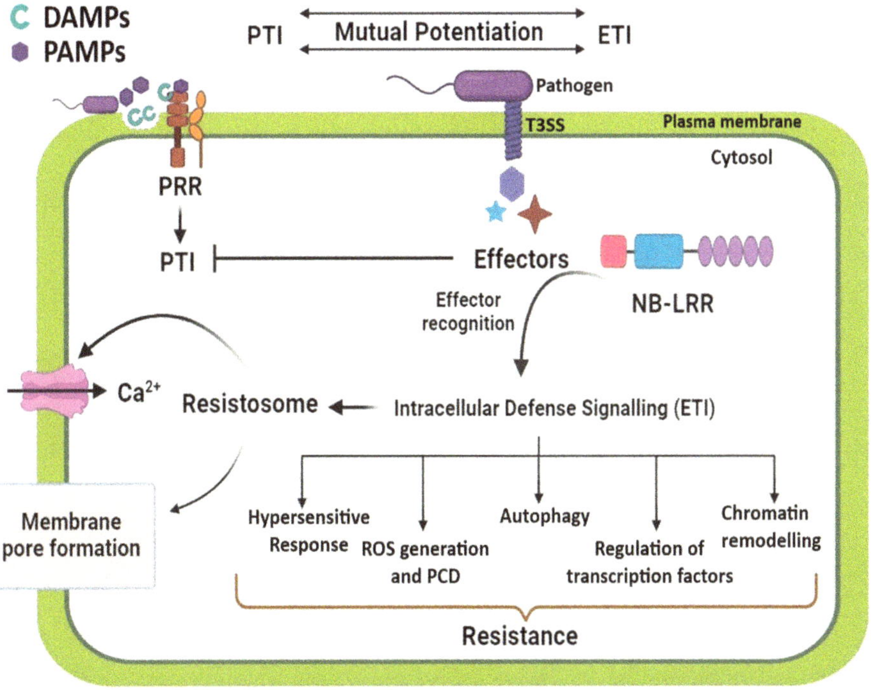

Figure 5.4. Diagrammatic representation of ETI. The first layer of defence is PTI, triggered by PAMPs/DAMPs perception by PRRs. To overcome this, pathogen secretes effectors that suppress the PTI. Plants contain NB-LRRs, that recognize these effectors and start ETI. ETI involves various intracellular signalling events such as resistosome formation, HR, ROS formation and PCD, autophagy, regulation of several TFs involved in defence signalling, and chromatin remodelling. On the other hand, resistosome initiates calcium (Ca^{2+}) signalling and it causes membrane pore formation. All of these provide resistance against pathogens. ETI and PTI work together and ETI is an enhanced PTI response.

the pathogen effector-induced reduction of PRR-associated proteins as well as the turnover of proteins following activation. Consequently, PRR-mediated immunity has the ability to enhance ETI outputs like HR in order to further limit the spread of infections [190, 194] (figure 5.4).

5.9 Conclusions

The field of plant immunity is one of the very important aspects to study. A lot of studies have been done in the last few decades, ranging from the identification of NLRs and NLR-mediated signalling pathways. Effector recognition by sensory and helper NLRs leads to resistosome formation and ETI. These helper NLRs and resistosomes are emerging concepts in the field of plant immunity. Further study on these areas will give new paths of research and may help in studying ETI pathways in a different perspective. Utilizing the knowledge from such studies in combination with modern molecular technologies will give better and highly resistant crops/plants that will fulfil the needs of the growing population.

References

[1] Teixeira P J P, Colaianni N R, Fitzpatrick C R and Dangl J L 2019 Beyond pathogens: microbiota interactions with the plant immune system *Curr. Opin. Microbiol.* **49** 7–17

[2] Dou D and Zhou J M 2012 Phytopathogen effectors subverting host immunity: different foes, similar battleground *Cell Host Microbe* **12** 484–95

[3] Dangl J L and Jones J D G 2001 Plant pathogens and integrated defence responses to infection *Nature* **411** 826–33

[4] Zipfel C 2014 Plant pattern-recognition receptors *Trends Immunol* **35** 345–51

[5] Karasov T L, Horton M W and Bergelson J 2014 Genomic variability as a driver of plant–pathogen coevolution? *Curr. Opin. Plant Biol.* **18** 24–30

[6] Dangl J L, Horvath D M and Staskawicz B J 2013 Pivoting the plant immune system from dissection to deployment *Science* **341** 746–51

[7] Jones J D G and Dangl J L 2006 The plant immune system *Nature* **444** 323–9

[8] Jacob F, Vernaldi S and Maekawa T 2013 Evolution and conservation of plant NLR functions *Front. Immunol* **4** 297

[9] Fu Z Q and Dong X 2013 Systemic acquired resistance: turning local infection into global defense *Annu. Rev. Plant. Biol.* **64** 839–63

[10] Takken F L, Albrecht M and Tameling W I 2006 Resistance proteins: molecular switches of plant defence *Curr. Opin. Plant Biol.* **9** 383–90

[11] Gao Y, Wang W, Zhang T, Gong Z, Zhao H and Han G Z 2018 Out of water: the origin and early diversification of plant R-genes *Plant Physiol.* **177** 82–9

[12] Baggs E L, Monroe J G, Thanki A S, O'Grady R, Schudoma C, Haerty W *et al* 2020 Convergent loss of an EDS1/PAD4 signaling pathway in several plant lineages reveals coevolved components of plant immunity and drought response *Plant Cell* **32** 2158–77

[13] Guo Y L, Fitz J, Schneeberger K, Ossowski S, Cao J and Weigel D 2011 Genome-wide comparison of nucleotide-binding site-leucine-rich repeat-encoding genes in Arabidopsis *Plant Physiol.* **157** 757–69

[14] Kanzaki 2012 Arms race co-evolution of *Magnaporthe oryzae* AVR-Pik and rice Pik genes driven by their physical interactions *Plant J.*

[15] Meyers B C, Kozik A, Griego A, Kuang H and Michelmore R W 2003 Genome-wide analysis of NBS-LRR–encoding genes in Arabidopsis *Plant Cell* **15** 809–34

[16] Li J, Ding J, Zhang W, Zhang Y, Tang P, Chen J Q *et al* 2010 Unique evolutionary pattern of numbers of gramineous NBS–LRR genes *Mol. Genet. Genomics* **283** 427–38

[17] Mun J H, Yu H J, Park S and Park B S 2009 Genome-wide identification of NBS-encoding resistance genes in *Brassica rapa Mol. Genet. Genomics* **282** 617–31

[18] Kang Y J, Kim K H, Shim S, Yoon M Y, Sun S, Kim M Y *et al* 2012 Genome-wide mapping of NBS-LRR genes and their association with disease resistance in soybean *BMC Plant Biol.* **12** 139

[19] Jupe F, Pritchard L, Etherington G J, MacKenzie K, Cock P J, Wright F *et al* 2012 Identification and localisation of the NB-LRR gene family within the potato genome *BMC Genomics* **13** 75

[20] Duxbury Z, Wu C-H and Ding P 2021 A comparative overview of the intracellular guardians of plants and animals: NLRs in innate immunity and beyond *Annu. Rev. Plant Biol.* **72** 155–84

[21] Sarris P F, Cevik V, Dagdas G, Jones J D G and Krasileva K V 2016 Comparative analysis of plant immune receptor architectures uncovers host proteins likely targeted by pathogens *BMC Biol.* **14** 8

[22] Yue J-X, Meyers B C, Chen J-Q, Tian D and Yang S 2012 Tracing the origin and evolutionary history of plant nucleotide-binding site–leucine-rich repeat (NBS-LRR) genes *New Phytol.* **193** 1049–63

[23] Wang J, Wang J, Hu M, Wu S, Qi J, Wang G *et al* 2019 Ligand-triggered allosteric ADP release primes a plant NLR complex *Science* **364** eaav5868

[24] Martin R, Qi T, Zhang H, Liu F, King M, Toth C *et al* 2020 Structure of the activated ROQ1 resistosome directly recognizing the pathogen effector XopQ *Science* **370** eabd9993

[25] Adachi H, Contreras M P, Harant A, hang Wu C, Derevnina L, Sakai T *et al* 2019 An N-terminal motif in NLR immune receptors is functionally conserved across distantly related plant species *eLife* **8** e49956

[26] Lukasik E and Takken F L 2009 STANDing strong, resistance proteins instigators of plant defence *Curr. Opin. Plant Biol.* **12** 427–36

[27] Horsefield S *et al* 2019 NAD+ cleavage activity by animal and plant TIR domains in cell death pathways *Science* **365** 793–9

[28] Wan L, Essuman K, Anderson R G, Sasaki Y, Monteiro F, Chung E H *et al* 2019 TIR domains of plant immune receptors are NAD^{+}-cleaving enzymes that promote cell death *Science* **365** 799–803

[29] Baudin M, Hassan J A, Schreiber K J and Lewis J D 2017 Analysis of the ZAR1 immune complex reveals determinants for immunity and molecular interactions *Plant Physiol.* **174** 2038–53

[30] Cesari S *et al* 2016 Cytosolic activation of cell death and stem rust resistance by cereal MLA-family CC–NLR proteins *Proc. Natl Acad. Sci.* **113** 36

[31] Grant M R, Godiard L, Straube E, Ashfield T, Lewald J, Sattler A *et al* 1995 Structure of the *Arabidopsis* RPM1 gene enabling dual specificity disease resistance *Science* **269** 843–6

[32] Kunkel B N, Bent A F, Dahlbeck D, Innes R W and Staskawicz B J 1993 RPS2, an Arabidopsis disease resistance locus specifying recognition of *Pseudomonas syringae* strains expressing the avirulence gene avrRpt2 *Plant Cell* **5** 865–75

[33] Hinsch M and Staskawicz B 1996 Identification of a new Arabidopsis disease resistance locus, RPs4, and cloning of the corresponding avirulence gene, avrRps4, from *Pseudomonas syringae* pv. pisi *Mol. Plant-Microbe Interact.* **9** 55–61

[34] Warren R F, Henk A, Mowery P, Holub E and Innes R W 1998 A mutation within the leucine-rich repeat domain of the Arabidopsis disease resistance gene RPS5 partially suppresses multiple bacterial and downy mildew resistance genes *Plant Cell* **10** 1439–52

[35] Bittner-Eddy P D, Crute I R, Holub E B and Beynon J L 2000 RPP13 is a simple locus in Arabidopsis thaliana for alleles that specify downy mildew resistance to different avirulence determinants in *Peronospora parasitica Plant* J. **21** 177–88

[36] Cooley M B, Pathirana S, Wu H J, Kachroo P and Klessig D F 2000 Members of the Arabidopsis HRT/RPP8 family of resistance genes confer resistance to both viral and Oomycete pathogens *Plant Cell* **12** 663–77

[37] Botella M A, Parker J E, Frost L N, Bittner-Eddy P D, Beynon J L, Daniels M J *et al* 1998 Three genes of the *Arabidopsis* RPP1 complex resistance locus recognize distinct *Peronospora parasitica* avirulence determinants *Plant Cell* **10** 1847

[38] Parker J E, Holub E B, Frost L N, Falk A, Gunn N D and Daniels M J 1996 Characterization of eds1, a mutation in *Arabidopsis* suppressing resistance to *Peronospora parasitica* specified by several different RPP genes *Plant Cell* **8** 2033–46

[39] Van Der Biezen E A, Freddie C T, Kahn K, Parker J E and Jones J D G 2002 Arabidopsis RPP4 is a member of the RPP5 multigene family of TIR-NB-LRR genes and confers downy mildew resistance through multiple signalling components *Plant* J. **29** 439–51

[40] Parker J E, Coleman M J, Szabò V, Frost L N, Schmidt R, van der Biezen E A *et al* 1997 The *Arabidopsis* downy mildew resistance gene RPP5 shares similarity to the toll and interleukin-1 receptors with N and L6 *Plant Cell* **9** 879–94

[41] Zhao B, Ardales E Y, Raymundo A, Bai J, Trick H N, Leach J E *et al* 2004 The avrRxo1 gene from the rice pathogen *Xanthomonas oryzae* pv. oryzicola confers a nonhost defense reaction on maize with resistance gene rxo1 *Mol. Plant-Microbe Interact.* **17** 771–9

[42] Tai T H, Dahlbeck D, Clark E T, Gajiwala P, Pasion R, Whalen M C *et al* 1999 Expression of the Bs2 pepper gene confers resistance to bacterial spot disease in tomato *Proc. Natl Acad. Sci.* **96** 14153–8

[43] Ballvora A, Ercolano M R, Weiß J, Meksem K, Bormann C A, Oberhagemann P *et al* 2002 The R1 gene for potato resistance to late blight (*Phytophthora infestans*) belongs to the leucine zipper/NBS/LRR class of plant resistance genes *Plant* J. **30** 361–71

[44] Huang S, Vleeshouwers V G A A, Werij J S, Hutten R C B, van Eck H J, Visser R G F *et al* 2004 The R3 resistance to *Phytophthora infestans* in potato is conferred by two closely linked R genes with distinct specificities *Mol. Plant-Microbe Interact.* **17** 428–35

[45] Karki H S, Abdullah S, Chen Y and Halterman D A 2021 Natural genetic diversity in the potato resistance gene RB confers suppression avoidance from Phytophthora effector IPI-O4 *Mol. Plant-Microbe Interact.* **34** 1048–56

[46] Champouret N, Bouwmeester K, Rietman H, van der Lee T, Maliepaard C, Heupink A *et al* 2009 Phytophthora infestans isolates lacking class I ipiO variants are virulent on rpi-blb1 potato *Mol. Plant-Microbe Interact.* **22** 1535–45

[47] Liu X, Lin F, Wang L and Pan Q 2007 The in silico map-based cloning of Pi36, a rice coiled-coil–nucleotide-binding site–leucine-rich repeat gene that confers race-specific resistance to the blast fungus *Genetics* **176** 2541–9

[48] Wang Z X, Yano M, Yamanouchi U, Iwamoto M, Monna L, Hayasaka H *et al* 1999 The Pib gene for rice blast resistance belongs to the nucleotide binding and leucine-rich repeat class of plant disease resistance genes *Plant* J. **19** 55–64

[49] Huang C L, Hwang S Y, Chiang Y C and Lin T P 2008 Molecular evolution of the Pi-ta gene resistant to rice blast in wild rice (*Oryza rufipogon*) *Genetics* **179** 1527–38

[50] Jiang N, Li Z, Wu J, Wang Y, Wu L, Wang S *et al* 2012 Molecular mapping of the Pi2/9 allelic gene Pi2-2 conferring broad-spectrum resistance to *Magnaporthe oryzae* in the rice cultivar Jefferson *Rice* **5** 29

[51] Ji C, Ji Z, Liu B, Cheng H, Liu H, Liu S *et al* 2020 Xa1 allelic R genes activate rice blight resistance suppressed by interfering TAL effectors *Plant Commun.* **1** 100087

[52] He Q, Li D, Zhu Y, Tan M, Zhang D and Lin X 2006 Fine mapping of Xa2, a bacterial blight resistance gene in rice *Mol. Breed* **17** 1–6

[53] Liu F, Zhang W, Schwessinger B, Wei T, Ruan D and Ronald P 2020 The rice Xa3 gene confers resistance to *Xanthomonas oryzae* pv. oryzae in the model rice kitaake genetic background *Front. Plant Sci.* **11** 49

[54] Wang W, Zhai W, Luo M, Jiang G, Chen X, Li X *et al* 2001 Chromosome landing at the bacterial blight resistance gene Xa4 locus using a deep coverage rice BAC library *Mol. Genet. Genomics MGG* **265** 118–25

[55] Flor H H 1971 Current status of the gene-for-gene concept *Annu. Rev. Phytopathol.* **9** 275–96

[56] Dodds P N and Rathjen J P 2010 Plant immunity: towards an integrated view of plant–pathogen interactions *Nat. Rev. Genet.* **11** 539–48

[57] Deslandes L, Olivier J, Peeters N, Feng D X, Khounlotham M, Boucher C *et al* 2003 Physical interaction between RRS1-R, a protein conferring resistance to bacterial wilt, and PopP2, a type III effector targeted to the plant nucleus *Proc. Natl Acad. Sci.* **100** 8024–9
[58] Dodds P N, Lawrence G J, Catanzariti A M, Teh T, Wang C I A, Ayliffe M A *et al* 2006 Direct protein interaction underlies gene-for-gene specificity and coevolution of the flax resistance genes and flax rust avirulence genes *Proc. Natl Acad. Sci.* **103** 8888–93
[59] Jia Y, McAdams S A, Bryan G T, Hershey H P and Valent B 2000 Direct interaction of resistance gene and avirulence gene products confers rice blast resistance *EMBO* J. **19** 4004-14
[60] Shahid Mukhtar M *et al* 2011 Independently evolved virulence effectors converge onto hubs in a plant immune system network *Science* **333** 596–601
[61] Kim M G, Cunha L D, McFall A J, Belkhadir Y, DebRoy S, Dangl J L *et al* 2005 Two *Pseudomonas syringae* type III effectors inhibit RIN4-regulated basal defense in *Arabidopsis Cell* **121** 749–59
[62] Wilton M, Subramaniam R, Elmore J, Felsensteiner C, Coaker G and Desveaux D 2010 The type III effector HopF2 Pto targets *Arabidopsis* RIN4 protein to promote *Pseudomonas syringae* virulence *Proc. Natl Acad. Sci.* **107** 2349–54
[63] van der Hoorn R A L and Kamoun S 2008 From guard to decoy: a new model for perception of plant pathogen effectors *Plant Cell* **20** 2009–17
[64] Ntoukakis V, Saur I M, Conlan B and Rathjen J P 2014 The changing of the guard: the Pto/Prf receptor complex of tomato and pathogen recognition *Curr. Opin. Plant Biol.* **20** 69–74
[65] Wu A J, Andriotis V M E, Durrant M C and Rathjen J P 2004 A patch of surface-exposed residues mediates negative regulation of immune signaling by tomato Pto kinase *Plant Cell* **16** 2809–21
[66] Eitas T K, Nimchuk Z L and Dangl J L 2008 Arabidopsis TAO1 is a TIR-NB-LRR protein that contributes to disease resistance induced by the *Pseudomonas syringae* effector AvrB *Proc. Natl Acad. Sci.* **105** 6475–80
[67] Ashfield T, Ong L E, Nobuta K, Schneider C M and Innes R W 2004 Convergent evolution of disease resistance gene specificity in two flowering plant families *Plant Cell* **16** 309–18
[68] Khan M, Subramaniam R and Desveaux D 2016 Of guards, decoys, baits and traps: pathogen perception in plants by type III effector sensors *Curr. Opin. Microbiol.* **29** 49–55
[69] Laflamme B, Dillon M M, Martel A, Almeida R N D, Desveaux D and Guttman D S 2020 The pan-genome effector-triggered immunity landscape of a host–pathogen interaction *Science* **367** 763–8
[70] Wang G *et al* 2015 The decoy substrate of a pathogen effector and a pseudokinase specify pathogen-Induced modified-self-recognition and immunity in plants *Cell Host Microbe* **18** 285–95
[71] Schultink A, Qi T, Bally J and Staskawicz B 2019 Using forward genetics in *Nicotiana benthamiana* to uncover the immune signaling pathway mediating recognition of the *Xanthomonas perforans* effector XopJ4 *New Phytol.* **221** 1001–9
[72] Deb S, Madhavan V N, Gokulan C G, Patel H K and Sonti R V 2021 Arms and ammunitions: effectors at the interface of rice and it's pathogens and pests *Rice* **14** 94
[73] Kay S, Hahn S, Marois E, Hause G and Bonas U 2007 A bacterial effector acts as a plant transcription factor and induces a cell size regulator *Science* **318** 648–51
[74] Römer P, Strauss T, Hahn S, Scholze H, Morbitzer R, Grau J *et al* 2009 Recognition of AvrBs3-like proteins is mediated by specific binding to promoters of matching pepper Bs3 alleles *Plant Physiol.* **150** 1697–712

[75] Boch J and Bonas U 2010 Xanthomonas AvrBs3 family-type III effectors: discovery and function *Annu. Rev. Phytopathol.* **48** 419–36
[76] Saur I M L *et al* 2019 Multiple pairs of allelic MLA immune receptor-powdery mildew AVRA effectors argue for a direct recognition mechanism *eLife* **8** e44471
[77] Lu X *et al* 2016 Allelic barley MLA immune receptors recognize sequence-unrelated avirulence effectors of the powdery mildew pathogen *Proc. Natl Acad. Sci.* **113** E6486–95
[78] Bettgenhaeuser J, Hernández-Pinzón I, Dawson A M, Gardiner M, Green P, Taylor J *et al* 2021 The barley immune receptor Mla recognizes multiple pathogens and contributes to host range dynamics *Nat. Commun.* **12** 6915
[79] Sweat T A, Lorang J M, Bakker E G and Wolpert T J 2008 Characterization of natural and induced variation in the LOV1 gene, a CC-NB-LRR gene conferring victorin sensitivity and disease susceptibility in *Arabidopsis Mol. Plant-Microbe Interact.* **21** 7–19
[80] Lorang J M, Hagerty C H, Lee R, McClean P E and Wolpert T J 2018 Genetic analysis of victorin sensitivity and identification of a causal nucleotide-binding site leucine-rich repeat gene in *Phaseolus vulgaris Mol. Plant-Microbe Interact.* **31** 1069–74
[81] Schmidt S M, Lukasiewicz J, Farrer R, van Dam P, Bertoldo C and Rep M 2016 Comparative genomics of *Fusarium oxysporum* f. sp. melonis reveals the secreted protein recognized by the Fom-2 resistance gene in melon *New Phytol.* **209** 307–18
[82] Dodds P and Thrall P 2009 Goldacre paper: recognition events and host–pathogen co-evolution in gene-for-gene resistance to flax rust *Funct. Plant Biol.* **36** 395–408
[83] Anderson C, Khan M A, Catanzariti A M, Jack C A, Nemri A, Lawrence G J *et al* 2016 Genome analysis and avirulence gene cloning using a high-density RADseq linkage map of the flax rust fungus, *Melampsora lini BMC Genomics* **17** 667
[84] Wu J *et al* 2015 Comparative genomics identifies the Magnaporthe oryzae avirulence effector AvrPi9 that triggers Pi9-mediated blast resistance in rice on JSTOR *New Phytol.* **206** 1463–75
[85] Vo K T X, Lee S K, Halane M K, Song M Y, Hoang T V, Kim C Y *et al* 2019 Pi5 and Pii Paired NLRs are functionally exchangeable and confer similar disease resistance specificity *Mol. Cells* **42** 637–45
[86] van Wersch R, Li X and Zhang Y 2016 Mighty dwarfs: *Arabidopsis* autoimmune mutants and their usages in genetic dissection of plant immunity *Front. Plant Sci.* **17** 1717
[87] Yi H and Richards E J 2009 Gene duplication and hypermutation of the pathogen resistance gene SNC1 in the *Arabidopsis* bal variant *Genetics* **183** 1227–34
[88] Shirano Y, Kachroo P, Shah J and Klessig D F 2002 A gain-of-function mutation in an *Arabidopsis* toll interleukin1 receptor–nucleotide binding site–leucine-rich repeat type R gene triggers defense responses and results in enhanced disease resistance *Plant Cell* **14** 3149–62
[89] Wang Y, Zhang Y, Wang Z, Zhang X and Yang S 2013 A missense mutation in CHS1, a TIR-NB protein, induces chilling sensitivity in *Arabidopsis Plant* J **75** 553–65
[90] van Wersch S, Tian L, Hoy R and Li X 2019 Plant NLRs: the whistleblowers of plant immunity *Plant Commun.* **1** 100016
[91] Lai Y and Eulgem T 2018 Transcript-level expression control of plant NLR genes *Mol. Plant Pathol.* **19** 1267–81
[92] Mohr T J, Mammarella N D, Hoff T, Woffenden B J, Jelesko J G and McDowell J M 2010 The *Arabidopsis* downy mildew resistance gene RPP8 is induced by pathogens and salicylic acid and is regulated by W box cis elements *Mol. Plant-Microbe Interact.* **23** 1303–15

[93] Halterman D A and Wise R P 2004 A single-amino acid substitution in the sixth leucine-rich repeat of barley MLA6 and MLA13 alleviates dependence on RAR1 for disease resistance signaling *Plant J.* **38** 215–26

[94] Vongs A, Kakutani T, Martienssen R A and Richards E J 1993 Arabidopsis thaliana DNA methylation mutants *Science* **260** 1926–8

[95] Li Y, Tessaro M J, Li X and Zhang Y 2010 Regulation of the expression of plant resistance gene SNC1 by a protein with a conserved BAT2 domain *Plant Physiol.* **153** 1425–34

[96] Jeddeloh J A, Stokes T L and Richards E J 1999 Maintenance of genomic methylation requires a SWI2/SNF2-like protein *Nat. Genet.* **22** 94–7

[97] Johnson K C M, Xia S, Feng X and Li X 2015 The chromatin remodeler SPLAYED negatively regulates SNC1-mediated immunity *Plant Cell Physiol.* **56** 1616–23

[98] Dowen R H, Pelizzola M, Schmitz R J, Lister R, Dowen J M, Nery J R *et al* 2012 Widespread dynamic DNA methylation in response to biotic stress *Proc. Natl Acad. Sci.* **109** E2183–91

[99] Le T N, Miyazaki Y, Takuno S and Saze H 2015 Epigenetic regulation of intragenic transposable elements impacts gene transcription in *Arabidopsisthaliana Nucl. Acids Res.* **43** 3911–21

[100] McDowell J M and Meyers B C 2013 A transposable element is domesticated for service in the plant immune system *Proc. Natl Acad. Sci.* **110** 14821–2

[101] Palma K, Zhao Q, Cheng Y T, Bi D, Monaghan J, Cheng W *et al* 2007 Regulation of plant innate immunity by three proteins in a complex conserved across the plant and animal kingdoms *Genes Dev.* **21** 1484–93

[102] Zhang S, Xie M, Ren G and Yu B 2013 CDC5, a DNA binding protein, positively regulates posttranscriptional processing and/or transcription of primary microRNA transcripts *Proc. Natl Acad. Sci.* **110** 17588–93

[103] Xu F, Xu S, Wiermer M, Zhang Y and Li X 2012 The cyclin L homolog MOS12 and the MOS4-associated complex are required for the proper splicing of plant resistance genes *Plant* J. **70** 916–28

[104] Pumplin N and Voinnet O 2013 RNA silencing suppression by plant pathogens: defence, counter-defence and counter-counter-defence *Nat. Rev. Microbiol.* **11** 745–60

[105] Wu L, Chen H, Curtis C and Fu Z Q 2014 Go in for the kill: how plants deploy effector-triggered immunity to combat pathogens *Virulence* **5** 710–21

[106] Shivaprasad P V, Chen H M, Patel K, Bond D M, Santos B A C M and Baulcombe D C 2012 A MicroRNA superfamily regulates nucleotide binding site–leucine-rich repeats and other mRNAs *Plant Cell* **24** 859–74

[107] Boccara M, Sarazin A, Thiébeauld O, Jay F, Voinnet O, Navarro L *et al* 2014 The Arabidopsis miR472-RDR6 silencing pathway modulates PAMP- and effector-triggered immunity through the post-transcriptional control of disease resistance genes *PLoS Pathog.* **10** e1003883

[108] Takahashi A, Casais C, Ichimura K and Shirasu K 2003 HSP90 interacts with RAR1 and SGT1 and is essential for RPS2-mediated disease resistance in Arabidopsis *Proc. Natl Acad. Sci.* **100** 11777–82

[109] Shirasu K 2009 The HSP90-SGT1 chaperone complex for NLR immune sensors *Annu. Rev. Plant Biol.* **60** 139–64

[110] Lopez V A *et al* 2019 A bacterial effector mimics a host HSP90 client to undermine immunity *Cell* **179** 205–8

[111] Goritschnig S, Zhang Y and Li X 2007 The ubiquitin pathway is required for innate immunity in Arabidopsis *Plant J.* **49** 540–51
[112] Duplan V and Rivas S 2014 E3 ubiquitin-ligases and their target proteins during the regulation of plant innate immunity *Front. Plant Sci.* **5** 42
[113] Shu K and Yang W 2017 E3 Ubiquitin ligases: ubiquitous actors in plant development and abiotic stress responses *Plant Cell Physiol.* **58** 1461–76
[114] Copeland C and Li X 2019 Regulation of plant immunity by the proteasome *International Review of Cell and Molecular Biology* **vol 343** ed L Galluzzi (New York: Academic) ch 2 pp 37–63
[115] Lee J *et al* 2007 Salicylic acid-mediated innate immunity in Arabidopsis is regulated by SIZ1 SUMO E3 ligase *Plant J.* **49** 79–90
[116] Cheng Y T, Li Y, Huang S, Huang Y, Dong X, Zhang Y *et al* 2011 Stability of plant immune-receptor resistance proteins is controlled by SKP1-CULLIN1-F-box (SCF)-mediated protein degradation *Proc. Natl Acad. Sci.* **108** 14694–9
[117] Gou M, Shi Z, Zhu Y, Bao Z, Wang G and Hua J 2012 The F-box protein CPR1/CPR30 negatively regulates R protein SNC1 accumulation *Plant* J **69** 411–20
[118] Dong O X, Ao K, Xu F, Johnson K C M, Wu Y, Li L *et al* 2018 Individual components of paired typical NLR immune receptors are regulated by distinct E3 ligases *Nat. Plants* **4** 699–710
[119] Zhang Y, Song G, Lal N K, Nagalakshmi U, Li Y, Zheng W *et al* 2019 TurboID-based proximity labeling reveals that UBR7 is a regulator of N NLR immune receptor-mediated immunity *Nat. Commun.* **10** 3252
[120] Kawasaki T, Nam J, Boyes D C, Holt B F III, Hubert D A, Wiig A *et al* 2005 A duplicated pair of Arabidopsis RING-finger E3 ligases contribute to the RPM1- and RPS2-mediated hypersensitive response *Plant* J. **44** 258–70
[121] Shirsekar G, Dai L, Hu Y, Wang X, Zeng L and Wang G L 2010 Role of ubiquitination in plant innate immunity and pathogen virulence *J. Plant Biol.* **53** 10–8
[122] Tong M *et al* 2017 E3 ligase SAUL1 serves as a positive regulator of PAMP-triggered immunity and its homeostasis is monitored by immune receptor SOC3 *New Phytol.* **215** 1516–32
[123] Lee C M, Feke A, Adamchek C, Webb K, Pruneda-Paz J, Bennett E J *et al* 2017 Decoys reveal the genetic and biochemical roles of redundant plant E3 ubiquitin ligases *bioRxiv* 115071
[124] Genoud T, Buchala A J, Chua N-H and Métraux J-P 2002 Phytochrome signalling modulates the SA-perceptive pathway in Arabidopsis *Plant J.* **31** 87–95
[125] Maekawa T, Kufer T A and Schulze-Lefert P 2011 NLR functions in plant and animal immune systems: so far and yet so close *Nat. Immunol.* **12** 817–26
[126] Zbierzak A M, Porfirova S, Griebel T, Melzer M, Parker J E and Dörmann P 2013 A TIR–NBS protein encoded by Arabidopsis Chilling Sensitive 1 (CHS1) limits chloroplast damage and cell death at low temperature *Plant* J. **75** 539–52
[127] Dinesh-Kumar S P and Baker B J 2000 Alternatively spliced N resistance gene transcripts: their possible role in tobacco mosaic virus resistance *Proc. Natl Acad. Sci.* **97** 1908–13
[128] Ade J, DeYoung B J, Golstein C and Innes R W 2007 Indirect activation of a plant nucleotide binding site–leucine-rich repeat protein by a bacterial protease *Proc. Natl Acad. Sci.* **104** 2531–6

[129] Mestre P and Baulcombe D C 2006 Elicitor-mediated oligomerization of the tobacco N disease resistance protein *Plant Cell* **18** 491–501
[130] Maekawa T, Cheng W, Spiridon L N, Töller A, Lukasik E, Saijo Y *et al* 2011 Coiled-coil domain-dependent homodimerization of intracellular barley immune receptors defines a minimal functional module for triggering cell death *Cell Host Microbe* **9** 187–99
[131] Césari S *et al* 2014 The NB-LRR proteins RGA4 and RGA5 interact functionally and physically to confer disease resistance *EMBO J.* **33** 1941–59
[132] Williams S J *et al* 2014 Structural basis for assembly and function of a heterodimeric plant immune receptor *Science* **344** 299–303
[133] Moltke J V, Ayres J S, Kofoed E M, Chavarría-Smith J and Vance R E 2013 Recognition of bacteria by inflammasomes *Annu. Rev. Immunol.* **31** 73–106
[134] Wang J *et al* 2019 Reconstitution and structure of a plant NLR resistosome conferring immunity *Science* **364** 6435
[135] Cai X, Chen J, Xu H, Liu S, Jiang Q X, Halfmann R *et al* 2014 Prion-like polymerization underlies signal transduction in antiviral immune defense and inflammasome activation *Cell* **156** 1207–22
[136] Seuring C, Greenwald J, Wasmer C, Wepf R, Saupe S J, Meier B H and Riek R 2012 The mechanism of toxicity in HET-S/HET-s prion incompatibility *PLoS Biol.* **10** e1001451
[137] Adachi H and Tsuda K 2019 Convergence of cell-surface and intracellular immune receptor signalling *New Phytol.* **221** 1676–8
[138] Torres M A, Dangl J L and Jones J D G 2002 *Arabidopsis* gp91phox homologues AtrbohD and AtrbohF are required for accumulation of reactive oxygen intermediates in the plant defense response *Proc. Natl Acad. Sci.* **99** 517–22
[139] Hackenberg T *et al* 2013 Catalase and *NO CATALASE ACTIVITY1* Promote autophagy-dependent cell death in *Arabidopsis Plant Cell* **25** 4616–26
[140] Liu Y, Schiff M, Czymmek K, Tallóczy Z, Levine B and Dinesh-Kumar S P 2005 Autophagy regulates programmed cell death during the plant innate immune response *Cell* **121** 567–77
[141] Patel S and Dinesh-Kumar S P 2008 *Arabidopsis* ATG6 is required to limit the pathogen-associated cell death response *Autophagy* **4** 20–7
[142] Hofius D, Schultz-Larsen T, Joensen J, Tsitsigiannis D I, Petersen N H T, Mattsson O *et al* 2009 Autophagic components contribute to hypersensitive cell death in *Arabidopsis Cell* **137** 773–83
[143] Yoshimoto K, Jikumaru Y, Kamiya Y, Kusano M, Consonni C, Panstruga R *et al* 2009 Autophagy negatively regulates cell death by controlling NPR1-dependent salicylic acid signaling during senescence and the innate immune response in *Arabidopsis Plant Cell* **21** 2914–27
[144] van Doorn W G 2011 Classes of programmed cell death in plants, compared to those in animals *J. Exp. Bot.* **62** 4749–61
[145] Muthamilarasan M and Prasad M 2013 Plant innate immunity: an updated insight into defense mechanism *J. Biosci.* **38** 433–49
[146] Hatsugai N, Iwasaki S, Tamura K, Kondo M, Fuji K, Ogasawara K *et al* 2009 A novel membrane fusion-mediated plant immunity against bacterial pathogens *Genes Dev.* **23** 2496–506
[147] Enyedi A J, Yalpani N and Raskin I 1992 Localization, conjugation, and function of salicylic acid in tobacco during the hypersensitive reaction to tobacco mosaic virus *Proc. Natl Acad. Sci.* **89** 2480–4

[148] Fu Z Q, Yan S, Saleh A, Wang W, Ruble J, Oka N *et al* 2012 NPR3 and NPR4 are receptors for the immune signal salicylic acid in plants *Nature* **486** 228–32
[149] Bhattacharjee S, Garner C M and Gassmann W 2013 New clues in the nucleus: transcriptional reprogramming in effector-triggered immunity *Front Plant Sci.* **4** 364
[150] Elmore J M, Lin Z J D and Coaker G 2011 Plant NB-LRR signaling: upstreams and downstreams *Curr. Opin. Plant Biol.* **14** 365–71
[151] Rivas S 2012 Nuclear dynamics during plant innate immunity *Plant Physiol.* **158** 87–94
[152] Pandey S P and Somssich I E 2009 The Role of WRKY transcription factors in plant immunity *Plant Physiol.* **150** 1648–55
[153] Lahaye T 2004 Illuminating the molecular basis of gene-for-gene resistance; *Arabidopsis thaliana RRS1-R* and its interaction with *Ralstonia solanacearum popP2 Trends Plant Sci.* **9** 1–4
[154] Tasset C, Bernoux M, Jauneau A, Pouzet C, Brière C, Kieffer-Jacquinod S *et al* 2010 Autoacetylation of the ralstonia solanacearum effector PopP2 targets a lysine residue essential for RRS1-R-mediated immunity in *Arabidopsis PLoS Pathog.* **6** e1001202
[155] Shen Q-H *et al* 2007 Nuclear activity of MLA immune receptors links isolate-specific and basal disease-resistance responses *Science* **315** 1098–103
[156] Inoue H, Hayashi N, Matsushita A, Xinqiong L, Nakayama A, Sugano S *et al* 2013 Blast resistance of CC-NB-LRR protein Pb1 is mediated by WRKY45 through protein–protein interaction *Proc. Natl Acad. Sci.* **110** 9577–82
[157] Matsushita A, Inoue H, Goto S, Nakayama A, Sugano S, Hayashi N *et al* 2013 Nuclear ubiquitin proteasome degradation affects WRKY45 function in the rice defense program *Plant* J. **73** 302–13
[158] Bhattarai K K, Atamian H S, Kaloshian I and Eulgem T 2010 WRKY72-type transcription factors contribute to basal immunity in tomato and *Arabidopsis* as well as gene-for-gene resistance mediated by the tomato R gene Mi-1 *Plant* J. **63** 229–40
[159] Mathur S, Vyas S, Kapoor S and Tyagi A K 2011 The mediator complex in plants: structure, phylogeny, and expression profiling of representative genes in a dicot (*Arabidopsis*) and a monocot (rice) during reproduction and abiotic stress *Plant Physiol.* **157** 1609–27
[160] Wang Y, An C, Zhang X, Yao J, Zhang Y, Sun Y *et al* 2013 The *Arabidopsis* elongator complex subunit2 epigenetically regulates plant immune responses *Plant Cell* **25** 762–76
[161] Caillaud M C, Asai S, Rallapalli G, Piquerez S, Fabro G and Jones J D G 2013 A downy mildew effector attenuates salicylic acid–triggered immunity in *Arabidopsis* by interacting with the host Mediator complex *PLoS Biol.* **11** e1001732
[162] An C and Mou Z 2012 The function of the mediator complex in plant immunity *Plant Signal. Behav.* **8** e23182
[163] Zhang X, Yao J, Zhang Y, Sun Y and Mou Z 2013 The *Arabidopsis* Mediator complex subunits MED14/SWP and MED16/SFR6/IEN1 differentially regulate defense gene expression in plant immune responses *Plant* J. **75** 484–97
[164] DeFraia C T, Zhang X and Mou Z 2010 Elongator subunit 2 is an accelerator of immune responses in *Arabidopsis thaliana Plant* J. **64** 511–23
[165] Alvarez-Venegas R, Abdallat A A, Guo M, Alfano J R and Avramova Z 2007 Epigenetic control of a transcription factor at the cross section of two antagonistic pathways *Epigenetics* **2** 106–13

[166] Palma K, Thorgrimsen S, Malinovsky F G, Fiil B K, Nielsen H B, Brodersen P *et al* 2010 Autoimmunity in *Arabidopsis* acd11 is mediated by epigenetic regulation of an immune receptor *PLoS Pathog.* **6** e1001137

[167] Kim S H, Kwon S I, Bhattacharjee S and Gassmann W 2009 Regulation of defense gene expression by *Arabidopsis* SRFR1 *Plant Signal Behav.* **4** 149–50

[168] Holt B F, Boyes D C, Ellerström M, Siefers N, Wiig A, Kauffman S *et al* 2002 An evolutionarily conserved Mediator of plant disease resistance gene function is required for normal *Arabidopsis* development *Dev. Cell* **2** 807–17

[169] Moissiard G, Cokus S J, Cary J, Feng S, Billi A C, Stroud H *et al* 2012 MORC family ATPases required for heterochromatin condensation and gene silencing *Science* **336** 1448–51

[170] Kang H G, Kuhl J C, Kachroo P and Klessig D F 2008 CRT1, an *Arabidopsis* ATPase that interacts with diverse resistance proteins and modulates disease resistance to Turnip crinkle virus *Cell Host Microbe* **3** 48–57

[171] Kang H G, Oh C S, Sato M, Katagiri F, Glazebrook J, Takahashi H *et al* 2010 Endosome-associated CRT1 functions early in resistance gene–mediated defense signaling in *Arabidopsis* and tobacco *Plant Cell* **22** 918–36

[172] Elmore J M, Liu J, Smith B, Phinney B and Coaker G 2012 Quantitative proteomics reveals dynamic changes in the plasma membrane during *Arabidopsis* immune signaling *Mol. Cell Proteomics* **11** M111.014555

[173] El Kasmi F *et al* 2017 Signaling from the plasma-membrane localized plant immune receptor RPM1 requires self-association of the full-length protein *Proc. Natl Acad. Sci.* **114** E7385–94

[174] Bi G, Su M, Li N, Liang Y, Dang S, Xu J *et al* 2021 The ZAR1 resistosome is a calcium-permeable channel triggering plant immune signaling *Cell* **184** 3528–3541.e12

[175] Zhao Y-B *et al* 2088 Pathogen effector AvrSr35 triggers Sr35 resistosome assembly via a direct recognition mechanism *Sci. Adv.* **8** eabq5108

[176] Förderer A, Li E, Lawson A W, nan Deng Y, Sun Y, Logemann E *et al* 2022 A wheat resistosome defines common principles of immune receptor channels *Nature* **610** 532–9

[177] Xu G, Moeder W, Yoshioka K and Shan L 2022 A tale of many families: calcium channels in plant immunity *Plant Cell* **34** 1551–67

[178] Kim N H, Jacob P and Dangl J L 2022 Con-Ca^{2+}-tenating plant immune responses via calcium-permeable cation channels *New Phytol.* **234** 813–8

[179] Yu X, Xu G, Li B, de Souza Vespoli L, Liu H, Moeder W *et al* 2019 The receptor kinases BAK1/SERK4 regulate Ca^{2+} channel-mediated cellular homeostasis for cell death containment *Curr. Biol.* **29** 3778–90.e8

[180] Zhao C *et al* 2021 A mis-regulated cyclic nucleotide-gated channel mediates cytosolic calcium elevation and activates immunity in Arabidopsis *New Phytol.* **230** 1078–94

[181] Jacob P *et al* 2021 Plant 'helper' immune receptors are Ca^{2+}-permeable nonselective cation channels *Science* **373** 420–5

[182] Gao F, Han X, Wu J, Zheng S, Shang Z, Sun D *et al* 2012 A heat-activated calcium-permeable channel—*Arabidopsis* cyclic nucleotide-gated ion channel 6—is involved in heat shock responses *Plant* J. **70** 1056–69

[183] Lüdke D, Yan Q, Rohmann P F W and Wiermer M 2022 NLR we there yet? Nucleocytoplasmic coordination of NLR-mediated immunity *New Phytol.* **236** 24–42

[184] Wang J, Han M and Liu Y 2021 Diversity, structure and function of the coiled-coil domains of plant NLR immune receptors *J. Integr. Plant Biol.* **63** 283–96

[185] Oligomerization of a plant helper NLR requires cell-surface and intracellular immune receptor activation *Proc. Natl Acad. Sci.*
[186] Pauly N, Knight M R, Thuleau P, Graziana A, Muto S, Ranjeva R *et al* 2001 The nucleus together with the cytosol generates patterns of specific cellular calcium signatures in tobacco suspension culture cells *Cell Calcium* **30** 413–21
[187] Xiong T C, Jauneau A, Ranjeva R and Mazars C 2004 Isolated plant nuclei as mechanical and thermal sensors involved in calcium signalling *Plant* J. **40** 12–21
[188] Pangršič T, Gabrielaitis M, Michanski S, Schwaller B, Wolf F, Strenzke N *et al* 2015 EF-hand protein Ca^{2+} buffers regulate Ca^{2+} influx and exocytosis in sensory hair cells *Proc. Natl Acad. Sci.* **112** E1028–37
[189] Charpentier M *et al* 2016 Nuclear-localized cyclic nucleotide–gated channels mediate symbiotic calcium oscillations *Science* **352** 1102–5
[190] Ngou B P M, Ahn H K, Ding P and Jones J D G 2021 Mutual potentiation of plant immunity by cell-surface and intracellular receptors *Nature* **592** 110–5
[191] Voigt C A 2014 Callose-mediated resistance to pathogenic intruders in plant defense-related papillae *Front. Plant Sci.* **5** 168
[192] Yamada K, Saijo Y, Nakagami H and Takano Y 2016 Regulation of sugar transporter activity for antibacterial defense in *Arabidopsis Science* **354** 1427–30
[193] Anderson J C, Wan Y, Kim Y M, Pasa-Tolic L, Metz T O and Peck S C 2014 Decreased abundance of type III secretion system-inducing signals in Arabidopsis mkp1 enhances resistance against *Pseudomonas syringae Proc. Natl Acad. Sci.* **111** 6846–51
[194] Yuan M, Jiang Z, Bi G, Nomura K, Liu M, Wang Y *et al* 2021 Pattern-recognition receptors are required for NLR-mediated plant immunity *Nature* **592** 105–9
[195] Adachi H, Kamoun S and Maqbool A 2019 A resistosome-activated 'death switch' *Nat. Plants* **5** 457–8

IOP Publishing

Advances in Biochemical and Molecular Mechanisms of Plant–Pathogen Interaction

Hitendra K Patel and Anirudh Kumar

Chapter 6

Silencing the enemy: role of RNAi-based immunity in plants

Somya Yadav, Priyanshi Porwal, Gargi Jauhari, Astha Singh and Alok Pandey

RNA interference (RNAi) or RNA silencing is one of the well-known strategies employed by plants to overcome plant pathogens. Dicer/dicer-like proteins, RNA-dependent RNA polymerase, and Argonaute proteins are the core components of the RNA silencing pathway. We summarize the function of core components and several cofactors of the RNAi pathway required in RNAi-based plant immunity against phytopathogens. A successfully adapted viral pathogen is generally capable of producing viral suppressor protein that can suppress the RNAi-based plant immunity eventually leading to plant pathogenesis. In this review, we focus mainly on RNAi-based plant immunity against viral pathogens.

6.1 Introduction

Plants are always under constant threat imposed by several abiotic and biotic stress conditions. These adverse conditions affect a plant's growth, development, and productivity. Crop plants are continuously threatened by phytopathogens such as viruses, bacteria, phytoplasmas, fungi, nematodes, insects, etc. Effective management of crop plant disease is very important in improving food security by reducing yield losses. Upon attack by pathogens, plants activate their defence mechanisms leading to enhanced resistance that controls pathogen spread or eliminates the phytopathogens. One of the important defence mechanisms employed by plants is RNA silencing or RNAi. RNA silencing is one of the prominent mechanisms by which plants display resistance against various kinds of pathogens [1]. RNA silencing was observed in *Caenorhabditis elegans* and the study indicated that gene expression is negatively controlled by non-coding RNAs [2].

RNA silencing or RNAi is an evolutionarily conserved pathway observed in eukaryotic organisms. In the RNA silencing events, regulation of gene expression is

doi:10.1088/978-0-7503-5673-2ch6

governed in a sequence-specific manner. Generally, RNA silencing or RNAi pathway is induced by dsRNA or hairpin RNA. This hairpin RNA or dsRNA is responsible for RNA degradation in a sequence-specific manner. The RNAi pathway involves critical factors like dicer and proteins belonging to the Argonaute family. It has been conclusively shown that dsRNAs are first processed into shorter small interfering RNAs. Thereafter, these small interfering RNAs target the homologous mRNAs [3, 4]. The small interfering RNAs direct the RNA-induced silencing complex (RISC) complex to destroy and disintegrate the target mRNAs [3]. The RNAi mechanism generally involves the processing of hairpin RNA or dsRNA by dicer/dicer-like proteins leading to the formation of 20–24 nucleotide-long small RNA. Then, this duplex small RNA is incorporated into an Argonaute protein to form the RISC. Thereafter, RISC is guided by a small RNA molecule to the target RNA which is cleaved by the Argonaute protein. In this way, the RNA silencing mechanism is responsible for the negative regulation of gene expression [5–7]. Involvement of the RNase III family enzymes has also been shown for the cleavage of dsRNAs leading to the production of guide RNA [8]. Many subsets of proteins such as dicer, dicer-like proteins, Argonaute proteins, and RNA-dependent RNA polymerase perform functions leading to the formation and magnification of the acting small RNA [9, 10]. The small RNA can be categorized into microRNAs (miRNAs) and small interfering RNAs as per their formation mechanism and origin [11]. The RNA-dependent RNA polymerase performs a very important function as they are involved in the synthesis of dsRNAs via single-stranded RNAs, which are then subjected to processing and leading to RNA silencing events [12].

The dicer/dicer-like proteins, Argonaute proteins, and RNA-dependent RNA polymerase have been shown to their prominent role in RNAi-mediated plant defence responses. The RNAi mechanism has therefore become the focus of the research-based study as it is one of the important mechanisms utilized by plants to counteract the plant pathogens. In this review, we focus on the role of the plant RNAi pathway against various viral pathogens.

6.2 Components of the RNAi pathway

One of the prominent components of the RNAi pathway is dicer-like proteins. Several studies have indicated that multiple numbers of dicer-like proteins are found in various plant species. In *Arabidopsis thaliana* and *Nicotiana benthamiana*, four dicer-like proteins are observed [13, 14]. In the rapeseed *Brassica napus*, thirteen dicer-like proteins are present [15, 16]. Seven dicer-like proteins are found in *Glycine max* and *Solanum lycopersicum* [17, 18] while eight dicer-like proteins have been observed in the rice plant [19]. Five dicer-like proteins have been observed in the *Zea mays* [20].

Dicer-like proteins display endoribonuclease activity belonging to the RNase III family. Dicer-like proteins contain a DExD helicase domain, a PAZ (Piwi-Argonaute-Zwille) domain, dsRNA binding domain (dsRBD), RNase III domain, and the DUF283 (domain of unknown function 283) [21–23]. Each domain of dicer-like proteins performs a highly specific function that is required for the initiation of the RNA silencing or RNAi pathway involving the cleavage of dsRNA into small RNAs [21]. It has been shown that the helicase domain plays a very important role in the dicer-like protein-dependent production of small RNA. In *A. thaliana*, a study has shown the function of the helicase domain in the formation of small RNA while the PAZ domain functions for the efficient processing of dsRNAs [24]. The PAZ domain contains a phosphate-binding pocket which has been shown for its important role in the formation of high-fidelity siRNAs [25, 26]. DUF283 facilitates hybridization between two complementary RNAs suggesting its role as a nucleic acid annealer [27]. The RNase III domain is involved in the cleavage of dsRNAs [28]. The central feature of dsRBD is the binding to the dsRNAs. Apart from this usual function, dsRBD also participates in the nucleocytoplasmic trafficking of the protein [29]. Various dicer-like proteins produce small RNAs of varying lengths and additionally display specificity for their substrates. Dicer-like 1 protein participates in the production of miRNAs [30]. In *the dcl2dcl3dcl4* triple mutant of *A. thaliana*, reduced production of small interfering RNAs was observed [31]. Taken together, various studies are suggesting towards function of the dicer-like proteins in the formation of small RNA representing the initiation stage of the RNAi pathway.

The second critical component of the RNAi pathway is argonaute proteins. Several studies involving various plant species reported the presence of multiple argonaute proteins in a single plant species. Ten argonaute proteins are found in *A. thaliana*, whereas nine argonaute proteins have been observed in *N. benthamiana* [13, 14]. A total of 28 argonaute proteins are present in the oilseed *B. napus* [15, 16]. *G. max* and *S. lycopersicum* have 21 and 25 argonaute proteins, respectively [17, 18] while 19 argonaute proteins have been observed in the rice plant [19]. Eighteen dicer-like proteins have been observed in the maize plant [20].

Argonaute proteins have small RNA-binding capacity and it is one of the important components of RISCs participating in the RNAi pathway [13, 32]. The small RNAs formed at the initiation phase of the RNAi pathway are loaded on argonaute protein to initiate the silencing event. Argonaute protein generally contains, PAZ (PIWI-Argonaute-Zwille/N-terminal), MID (Middle), and PIWI (P-element induced wimpy) domains [33]. The plant argonaute proteins perform various specific functions. Argonaute proteins are known to have various functions required for small interfering RNA accumulation, post-transcriptional gene silencing, DNA as well as histone methylation [34–36]. In *A. thaliana*, a heterochronic ZIPPY gene encodes ARGONAUTE7, one of the ten members of the ARGONAUTE family. Mutations in the ZIPPY gene cause the appearance of premature adult vegetative traits suggesting a critical function for Argonaute proteins in plant development [37]. Argonaute proteins have been demonstrated to enhance plant immunity and defence via their involvement in the RNA silencing pathway [38]. The assembly of RISCs is an important step during the gene silencing pathway. The small RNAs loading into argonaute protein have been conclusively shown to be dependent on the heat shock protein (Hsp90) chaperone machinery [39, 40]. Some argonaute proteins are responsible for transcriptional gene silencing while other argonaute proteins are responsible for post-transcriptional gene silencing. For example, ARGONAUTE4, ARGONAUTE6, and ARGONAUTE9 are involved in transcriptional gene silencing, whereas ARGONAUTE1, ARGONAUTE7, and ARGONAUTE10 are responsible for post-transcriptional gene silencing [32].

The third critical component of the RNAi pathway is RNA-dependent RNA polymerase proteins as these proteins are needed for the enhancement and amplification of the effect of the RNAi pathway mediated-silencing process. RNA-dependent RNA polymerase proteins convert ssRNAs into duplex RNAs. Now, these double-stranded RNAs are again subjected to dicer-like proteins leading to multiple new rounds of RNA silencing events [41]. In plants, multiple RNA-dependent RNA polymerase proteins have been observed. Six RNA-dependent RNA polymerase proteins (1, 2, 3a, 3b, 3c, and 6) are found in *A. thaliana*, whereas three RNA-dependent RNA polymerase proteins have been observed in *N. benthamiana* [13, 14]. A total of 16 RNA-dependent RNA polymerase proteins are encoded by the *B. napus* [15, 16]. *G. max* and *S. lycopersicum* have seven and six RNA-dependent RNA polymerase proteins, respectively [17, 18] while five RNA-dependent RNA polymerase proteins have been observed in the *Oryza sativa* and *Z. mays* [19, 20]. In the model plant *A. thaliana* and several other plant species, RNA-dependent RNA polymerase proteins have been demonstrated to be very important for the antiviral plant immunity and defence against RNA viruses [42–45]. Apart from their role for the antiviral defence, RNA-dependent RNA polymerase proteins also perform critical roles in post-transcriptional gene silencing, production of RNA templates, genome integrity maintenance, and defence against invading foreign RNA and DNA [46, 47]. Different RNA-dependent RNA polymerase proteins perform different functions and they have been categorized accordingly [17, 20]. The RNA-dependent RNA polymerase 1 provides resistance against abiotic and biotic stress conditions including insects [48, 49]. The RNA-dependent RNA polymerase 2 is essential for the small interfering RNA biogenesis. The Association of Jumonji (JmjC) domain-containing protein (JMJ24) with RNA-dependent RNA polymerase 2 ensures the basal level of transcription at the silenced loci [50, 51]. The RNA-dependent RNA polymerase 1/2/6-independent small RNA-producing loci have also been characterized adding a high level of complexity in the RNAi pathway [52].

6.3 Plant RNA silencing mechanism against viruses

Plant viruses are categorized as obligate intracellular parasites. Plant viruses require resources of plant host cells for multiplication and dissemination and they show dependency on the biotic factors such as insects for their survival and transmission. Aphids, leafhoppers, planthoppers, and whiteflies are the best-characterized plant viral insect vectors [53]. Plant viruses induce alteration in the physiology of plant cells to support their lifestyle. Metabolic alterations include phytohormone manipulation, reactive oxygen species accumulation, cell cycle disruptions, dysfunction of chloroplast processes, etc [54]. Virus-induced alteration in the cellular metabolism and physiology affects usual growth and development leading to characteristic viral symptoms in plants [55]. Various kinds of systemic symptoms are observed in plants infected with viruses. Symptoms such as yellowing or chlorosis, leaf deformation and discoloration, wilting, necrosis, mosaic patterns, stunted growth, and ringspots are observed in plants infected with viruses [55]. Generally, the genomes of the majority of plant viruses are made up of ss/dsRNA while the minority of plant viruses have ss/ds DNA genome [56].

The RNAi or RNA silencing pathway performs a critical role in the plant defence against pathogenic viral infections. RNA silencing pathway targets the plant RNA viruses as these classes of viruses produce double-stranded RNA intermediates during the replication process. The viral RNAs are processed by dicer-like proteins to generate small interfering RNAs which are loaded to the argonaute proteins leading to the cleavage of the viral genome and silencing-mediated defence [57]. Plant viruses having DNA genomes are processed by RNA-dependent RNA polymerase 6/SGS3 (suppressor of gene silencing 3) leading to RNAi-mediated defence responses [58]. RNA silencing pathway has been shown to provide protection against cabbage leaf curl begomovirus (a DNA geminivirus) in *Arabidopsis*. RNAi pathway is also operative against another DNA geminivirus, African cassava mosaic begomovirus, in *N. benthamiana* and cassava. Small interfering RNAs of varying lengths (21, 22, and 24 nt) have been observed that are derived from the intergenic as well as coding regions. The formation of small interfering RNAs in plants infected with pathogenic DNA viruses is also dependent on dicer-like proteins [59].

Infection with plant viruses enhances the accumulation of components of RNAi machinery as observed through various research studies involving different plant species. In the red sage plant, *Salvia miltiorrhiza*, the upregulated expression of RNA-dependent RNA polymerase 1, 2, and 3 is observed upon infection with cucumber mosaic virus [60]. Similarly, infection with tomato yellow leaf curl virus leads to enhanced expression and accumulation of RNA-dependent RNA polymerases, argonaute proteins, and dicer-like proteins [17] suggesting a functional RNAi pathway in the tomato plant. In *Capsicum annuum*, RNAi components such as argonaute proteins (ARGONAUTE2 and ARGONAUTE10b) and dicer-like protein 2/4 are induced in the response to the infections with potato virus Y, tobacco mosaic virus, and cucumber mosaic virus [61]. Enhanced production and accumulation of small interfering RNAs (siRNAs) have been observed in the *Arabidopsis* plants infected with the cucumber mosaic virus [62]. *Arabidopsis* lines lacking dicer-like proteins display a reduction in the cucumber mosaic virus-siRNAs indicating the importance of dicer-like proteins for the formation of siRNAs required for antiviral RNA silencing [63]. Wild-type *A. thaliana* displays resistance against silencing suppressor (TuMV P1/HC-Pro)-deficient turnip mosaic virus. However, *A. thaliana* mutants lacking dicer-like proteins 2 and 4 display susceptibility against silencing suppressor (TuMV P1/HC-Pro)-deficient turnip mosaic virus [64]. Systematic analysis revealed that viral-induced production of siRNAs is dependent on dicer-like protein 4 and RNA-dependent RNA polymerase 1 protein, but plants display complete antiviral defence when dicer-like protein 2 and RNA-dependent RNA polymerase 6 protein is also available along with dicer-like protein 4 and RNA-dependent RNA polymerase 1 protein [64]. Further, genetic analysis using *A. thaliana* plants revealed that different dicer-like proteins are needed for intracellular and systemic antiviral silencing [65]. Transgene-derived siRNAs (24 nt) are important for the immunity of transgenic plants against criniviruses (e.g. cucurbit yellow stunting disorder virus) [66].

A study involving *Arabidopsis* and cabbage leaf curl virus (a DNA virus) pathosystem, indicated that a total of four different types of dicer-like proteins take part in the production of viral siRNAs with varying lengths (21, 22, and 24 nt) and protection of plants via RNAi mechanism [67, 68]. SGS2/Silencing Defective 1 (SDE1) as well as SGS3 is required for endogenous gene silencing from DNA viruses indicating the role of SGS2/SDE1 in the reduction of geminivirus infections symptoms by targeting viral mRNAs. Increased viral symptoms and enhanced viral DNA accumulation have been observed in the *sgs2* and *sgs3* mutant lines of *Arabidopsis* infected with the cabbage leaf curl virus, whereas overexpression of SGS2/SDE1 leads to reduced symptoms [69]. *The HUA ENHANCER1* (*HEN1*) mutant of *Arabidopsis* displays multiple defects in plant development, and HEN1 has also been shown to be important for the stabilization of regulatory small RNAs [70]. HEN1 has been observed to be a key link molecule between miRNA-mediated developmental processes and siRNA-driven transgene silencing as well as resistance against plant viruses [71, 72]. Using *Arabidopsis* mutant lines lacking dicer-like 2, argonaute 2, and HEN1 proteins, it has been observed that the loss of function of these proteins makes *Arabidopsis* plants highly susceptible to turnip crinkle virus as compared with wild-type plants [73]. In the cabbage leaf curl virus, extensive genetic analysis has shown that all four dicer-like proteins are required for the production of viral small RNAs whereas the RNA-dependent RNA polymerases 1, 2, 6, and RNA polymerases IV, V are not required for the production of viral small RNAs [74]. Whitefly-transmitted tomato yellow leaf curl virus has a genome made up of single-stranded DNA and this virus infects tomato plants leading to typical viral symptoms including leaf curling and yellowing [75]. In tomato yellow leaf curl virus-infected tomato plants, upregulation of genes encoding RNAi components has been observed. In response to tomato yellow leaf curl virus, the tomato genes such as *SlAGO1*, *SlAGO4*, *SlDCL1*, *SlDCL2*, *SlDCL4*, *SlRDR2*, *SlRDR3a*, *SlRDR6a* are upregulated suggesting that alike RNA virus, RNA-dependent RNA polymerase proteins, argonaute proteins, and dicer-like proteins are also required for the RNA silencing event in the DNA viruses [76]. Inactivation or silencing of *SlDCL2* or *SlDCL4* enhances the susceptibility towards tomato yellow leaf curl virus. Production of RNAi components and salicylic acid together plays a very important role in the tomato defence against tomato yellow leaf curl virus [76].

In addition to core RNAi components, several double-stranded RNA-binding proteins (DRBs) perform important functions in the silencing pathways displayed by animals and plants [77]. In *Arabidopsis*, there are five DRBs and each one is composed of two double-stranded RNA-binding motifs. DRB proteins do not contain any catalytic domain [78]. In *Arabidopsis*, DRB2, DRB3, and DRB5 participate in the biogenesis of miRNA [79]. DRB4 protein is important for the *Arabidopsis* defence against turnip yellow mosaic virus [80]. The DAWDLE protein, a forkhead-associated domain protein, governs various dimensions of plant development. The *Arabidopsis dawdle* mutant plants produce defective flowers, roots, shoots, and reduced fertility [81]. Interaction of the DAWDLE protein with dicer-like proteins is required for the formation of miRNAs and siRNAs [82]. The DAWDLE protein positively regulates plant immunity against both adapted and non-adapted pathogens [83].

The argonaute proteins have been shown to be associated with the miRNAs and siRNAs. The ARGONAUTE2 protein is important for RNA-silencing-mediated protection against potato virus X in plants [84]. The *Arabidopsis ago2* mutant plants display hyper-susceptibility towards turnip crinkle virus and cucumber mosaic virus suggesting the role of ARGONAUTE2 in antiviral defence [85]. Further, ARGONAUTE1, 2, and 7 have been shown to perform an important role in RNAi-mediated antiviral defence mechanisms [86]. Using *Arabidopsis ago2ago5 double* mutant, the researcher observed that double mutants are more susceptible to potato virus X than a single mutant [87] indicating the role of ARGONAUTE5 in the RNAi-mediated antiviral defence mechanism. The ARGONAUTE1 and ARGONAUTE7 proteins are required in the antiviral RNA silencing pathway. ARGONAUTE1 protein targets viral RNAs having compact structures while ARGONAUTE7 protein targets viral RNAs displaying less compact structures [88]. Using CRISPR/Cas9 technology, inactivation of ARGONAUTE2 was performed in *N. benthamiana*. Mutational analysis indicated that ARGONAUTE2 is the critical component of the plant immune system and provides protection to *N. benthamiana* against potato virus X, turnip crinkle virus, and turnip mosaic virus [89]. ARGONAUTE1 protein has also been shown to be important for the temperature-dependent symptom recovery of *N. benthamiana* against tomato ringspot viral infection [90]. Tobacco rattle virus (TRV) is having a bipartite, positive-sense ssRNA genome [91]. Tobacco rattle virus infects several plant species including *A. thaliana, Solanum tuberosum*, and *N. benthamiana* [92]. Using *N. benthamiana ago4* mutants, it has been shown that plants lacking ARGONAUTE4 are highly susceptible to the tobacco rattle virus (TRV) [93]. ARGONAUTE4 is also involved in NB-LRR-mediated virus resistance and provides resistance against the cucumber mosaic virus [94, 95]. Multiple ARGONAUTE proteins have been located and identified in the various plant species. However, further research is needed to assign the specific function of each of the ARGONAUTE proteins.

The host RNA-dependent RNA polymerase proteins play a very important role in the RNA silencing processes. In *A. thaliana* plants, RNA-dependent RNA polymerase 1, 2, and 6 perform key functions during antiviral silencing events. In pepper (*C. annuum*) plants, RNA-dependent RNA polymerase 1 (*CaRDR1*) provides resistance against tobacco mosaic virus [96]. In pepper plants, the silencing of *CaRDR1* leads to enhanced susceptibility towards tobacco mosaic virus infection. Further, overexpression of *CaRDR1* in *N. benthamiana* leads to reduced symptoms as well as reduced accumulation of coat protein transcripts of tobacco mosaic virus [96]. Transgenic *N. benthamiana* plants expressing RNA-dependent RNA polymerase 1 of *Medicago truncatula* display reduced tobacco mosaic virus-induced symptoms [97]. Various works have shown enhanced viral susceptibility and increased viral load in plants associated with loss of function or down-regulation of *RDR1* gene encoding for RNA-dependent RNA polymerase 1 [97–100]. RNA-dependent RNA polymerase 2 primarily participates in RNA-directed DNA methylation. This protein is also required for the regulation of heterochromatin as well as plant development and governs transcriptional gene silencing [101]. Both RNA-dependent RNA polymerase 1/6 are required for restriction of viral replication in plants [102, 103]. The activity of RNA-dependent RNA polymerase 6 requires the association of SGS3 protein. The *N. benthamiana* plants lacking both *NbRDR6* and *NbSGS3 display* increased susceptibility toward geminivirus infection [104, 105]. Moreover, additional proteins such as ALA1 (a membrane-localized flippase Aminophospholipid ATPase 1) and ALA2 cooperate with RNA-dependent RNA polymerase 1 and 6 to promote RNAi-based immunity in plants [106]. In rice plants, the function of the RNA-dependent RNA polymerase 6 has been conclusively established in the defence against the rice stripe virus and rice dwarf phytoreo virus [107, 108]. Taken together, several studies have indicated the important role of each component and protein for efficient RNAi-based antiviral immunity in plants. Further studies might identify novel players participating in the RNAi pathway.

6.4 Plant antiviral RNA silencing blocked by viral suppressor proteins

Adapted pathogenic plant viruses are known to produce viral suppressors that interfere with the RNAi-based antiviral immunity leading to the establishment of pathogenesis and typical viral symptoms in the plants. Viral suppressors target any one of the components of RNAi pathways to suppress the plant host defence responses [109]. Cucumber mosaic virus produces 2b protein as a suppressor that decreases the accumulation of various siRNAs in *Arabidopsis* plants [63]. The P6 protein is encoded by the cauliflower mosaic virus that suppresses the RNA silencing factor DRB4 of *Arabidopsis* [110]. Similarly, the P6 protein encoded by strawberry vein banding virus also has been established as an RNA-silencing suppressor [111]. The cysteine-rich 16K protein of tobacco rattle virus targets ARGONAUTE4 to inhibit the formation of RISC [93]. The 2b protein of cucumber mosaic virus also impairs ARGONAUTE4 activity [95]. Polerovirus produces F box protein (P0) that acts as a silencing suppressor. The P0 protein targets the PAZ motif of ARGONAUTE1 to mediate its degradation eventually leading to severe infection [112, 113]. The P1 protein is produced by sweet potato mild mottle virus as a viral suppressor protein. The P1 protein inhibits the activity of RNA silencing machinery in the *Arabidopsis* plant to promote pathogenesis [114]. The coat protein of pelargonium flower break virus acts as a viral suppressor protein that sequesters siRNAs to inhibit the RNAi-mediated protection mechanism [115]. The NSs protein encoded by the tomato spotted wilt virus is one of the well-known RNA silencing suppressors [116]. Two RNA silencing suppressors p20 and p23 have been identified from Citrus tristeza virus. These proteins also interfere with the salicylic acid-mediated defence pathways in the orange plants [117, 118]. In olive mild mosaic virus, coat protein, as well as P6, display RNAi-suppressing activities [119]. More than one RNA silencing suppressor in plant pathogenic viruses provides a great advantage to these viruses as they can target multiple dimensions of RNAi-mediated plant defence mechanisms.

Pre-coat protein encoded by Geminivirus has been shown for its capability to modulate the RNA-dependent RNA polymerase 1-mediated antiviral RNA silencing, facilitating the onset of the disease symptoms [120]. The V2 protein produced by the tomato yellow leaf curl virus interacts with SGS3 to suppress the RNA-dependent RNA polymerase 6-mediated antiviral RNA silencing mechanism [121]. Similarly, the AC2 protein of the mungbean yellow mosaic Indian virus is capable of suppressing the antiviral silencing pathway by inhibiting the RNA-dependent RNA polymerase 6 and ARGONAUTE1 activities [122]. The cysteine-rich protein encoded by potato virus M suppresses the local as well as systemic silencing events, whereas another viral suppressor, triple gene block protein 1, only acts on systemic silencing processes [123]. The study indicates that the virus is capable of producing two distinct viral suppressor proteins governing viral multiplication and movement in the host plant. The 29K movement protein of the tobacco rattle virus also acts as an RNA silencing suppressor [124]. In another interesting study, it has been observed that RAV2 (an ethylene-inducible transcription factor) is needed for the inhibition of the antiviral RNAi pathway by the HC-Pro viral protein of potyvirus and the P38 protein of carmovirus. Both HC-Pro and P38 proteins are capable of blocking the activity of siRNAs [125]. Viral suppressor proteins target antiviral silencing by interfering with the components of the RNA decay pathway in plants [126]. Overall, these studies suggest that a successful viral pathogen produces diverse kinds of suppressor protein that targets multiple aspects of antiviral silencing-mediated host defence responses.

6.5 Conclusions

In this review, we have provided an overview of the RNAi mechanism. We have discussed the function of each component of the RNAi pathway and their indispensability in the defence pathways. We have focussed mainly on the antiviral silencing mechanisms, although the RNAi mechanism is also employed against fungal and bacterial plant pathogens. Bioinformatics analysis has indicated the presence of multiple dicer-like proteins, RNA-dependent RNA polymerase, and Argonaute proteins in a single plant species. It would be interesting to assign the function of each of these components of the RNAi pathway. RNAi mechanisms have been exploited to enhance the disease resistance in plants. This can become one of the important strategies for ensuring sustainable agricultural production

References

[1] Napoli C, Lemieux C and Jorgensen R 1990 Introduction of a chimeric chalcone synthase gene into petunia results in reversible co-suppression of homologous genes in trans *Plant Cell* **2** 279–89

[2] Guo S and Kemphues K J 1995 *par*-1, a gene required for establishing polarity in *C. elegans* embryos, encodes a putative Ser/Thr kinase that is asymmetrically distributed *Cell* **81** 611–20

[3] Hammond S M, Bernstein E, Beach D and Hannon G J 2000 An RNA-directed nuclease mediates post-transcriptional gene silencing in *Drosophila* cells *Nature* **404** 293–6

[4] Zamore P D, Tuschl T, Sharp P A and Bartel D P 2000 RNAi: double-stranded RNA directs the ATP-dependent cleavage of mRNA at 21 to 23 nucleotide intervals *Cell* **101** 25–33
[5] Hannon G J 2002 RNA interference *Nature* **418** 244–51
[6] Baulcombe D 2004 RNA silencing in plants *Nature* **431** 356–63
[7] Meister G and Tuschl T 2004 Mechanisms of gene silencing by double-stranded RNA *Nature* **431** 343–9
[8] Bernstein E, Caudy A A, Hammond S M and Hannon G J 2001 Role for bidentate ribnuclease in the initiation site of RNA interference *Nature* **409** 363–6
[9] Vaucheret H 2006 Post-transcriptional small RNA pathways in plants: mechanisms and regulations *Genes Dev.* **20** 759–71
[10] Chapman E J and Carrington J C 2007 Specialization and evolution of endogenous small RNA pathways *Nat. Rev. Genet.* **8** 884–96
[11] Voinnet O 2009 Origin, biogenesis, and activity of plant microRNAs *Cell* **136** 669–87
[12] Wassenegger M and Krczal G 2006 Nomenclature and functions of RNA-directed RNA polymerases *Trends Plant Sci.* **11** 142–51
[13] Vaucheret H 2008 Plant argonautes *Trends Plant Sci.* **13** 350–8
[14] Nakasugi K, Crowhurst R N, Bally J, Wood C C, Hellens R P and Waterhouse P M 2013 De novo transcriptome sequence assembly and analysis of RNA silencing genes of *Nicotiana benthamiana PLoS One* **8** e59534
[15] Cao J-Y, Xu Y-P, Li W, Li S-S, Rahman H and Cai X-Z 2016 Genome-wide identification of Dicer-like, Argonaute, and RNA-dependent RNA polymerase gene families in *Brassica* species and functional analyses of their Arabidopsis homologs in resistance to *Sclerotinia sclerotiorum Front. Plant Sci.* **7** 1614
[16] Zhao X, Zheng W, Zhong Z, Chen X, Wang A and Wang Z 2016 Genome-wide analysis of RNA-interference pathway in *Brassica napus*, and the expression profile of *BnAGOs* in response to *Sclerotinia sclerotiorum* infection *Eur. J. Plant Pathol.* **146** 565–79
[17] Bai M, Yang G S, Chen W T, Mao Z C, Kang H X, Chen G H, Yang Y H and Xie B Y 2012 Genome-wide identification of Dicer-like, Argonaute and RNA-dependent RNA polymerase gene families and their expression analyses in response to viral infection and abiotic stresses in *Solanum lycopersicum Gene* **501** 52–62
[18] Liu X, Lu T, Dou Y, Yu B and Zhang C 2014 Identification of RNA silencing components in soybean and sorghum *BMC Bioinform.* **15** 4
[19] Kapoor M, Arora R, Lama T, Nijhawan A, Khurana J P, Tyagi A K and Kapoor S 2008 Genome-wide identification, organization and phylogenetic analysis of Dicer-like, Argonaute and RNA-dependent RNA Polymerase gene families and their expression analysis during reproductive development and stress in rice *BMC Genom* **9** 451
[20] Qian Y, Cheng Y, Cheng X, Jiang H, Zhu S and Cheng B 2011 Identification and characterization of Dicer-like, Argonaute and RNA-dependent RNA polymerase gene families in maize *Plant Cell Rep.* **30** 1347–63
[21] Carmell M A and Hannon G J 2004 RNase III enzymes and the initiation of gene silencing *Nat. Struct. Mol. Biol.* **11** 214–8
[22] Margis R, Fusaro A F, Smith N A, Curtin S J, Watson J M, Finnegan E J and Waterhouse P M 2006 The evolution and diversification of Dicers in plants *FEBS Lett.* **580** 2442–50
[23] Court D L, Gan J, Liang Y H, Shaw G X, Tropea J E, Costantino N, Waugh D S and Ji X 2013 RNase III: genetics and function; structure and mechanism *Annu. Rev Genet.* **47** 405–31

[24] Montavon T, Kwon Y, Zimmermann A, Hammann P, Vincent T, Cognat V, Bergdoll M, Michel F and Dunoyer P 2018 Characterization of DCL4 missense alleles provides insights into its ability to process distinct classes of dsRNA substrates *Plant* J. **95** 204–18

[25] Tian Y, Simanshu D K, Ma J B, Park J E, Heo I, Kim V N and Patel D J 2014 A phosphate-binding pocket within the platform-PAZ-connector helix cassette of human Dicer *Mol. Cell* **53** 606–16

[26] Kandasamy S K and Fukunaga R 2016 Phosphate-binding pocket in Dicer-2 PAZ domain for high-fidelity siRNA production *Proc. Natl Acad. Sci. USA* **113** 14031–6

[27] Szczepanska A, Wojnicka M and Kurzynska-Kokorniak A 2021 The significance of the DUF283 domain for the activity of human ribonuclease dicer *Int. J. Mol. Sci.* **22** 8690

[28] Nicholson A W 2014 Ribonuclease III mechanisms of double-stranded RNA cleavage *Wiley Interdiscip. Rev. RNA* **5** 31–48

[29] Banerjee S and Barraud P 2014 Functions of double-stranded RNA-binding domains in nucleocytoplasmic transport *RNA Biol.* **11** 1226–32

[30] Kurihara Y and Watanabe Y 2004 *Arabidopsis* micro-RNA biogenesis through Dicer-like 1 protein functions *Proc. Natl Acad. Sci USA.* **101** 12753–8

[31] Henderson I R, Zhang X, Lu C, Johnson L, Meyers B C, Green P J and Jacobsen S E 2006 Dissecting *Arabidopsis thaliana* Dicer function in small RNA processing, gene silencing and DNA methylation patterning *Nat. Genet.* **38** 721–5

[32] Mallory A and Vaucheret H 2010 Form, function, and regulation of Argonaute proteins *Plant Cell* **22** 3879–89

[33] Wu J, Yang J, Cho W C and Zheng Y 2020 Argonaute proteins: structural features, functions and emerging roles *J. Adv. Res.* **24** 317–24

[34] Fagard M, Boutet S, Morel J B, Bellini C and Vaucheret H 2000 AGO1, QDE-2, and RDE-1 are related proteins required for post-transcriptional gene silencing in plants, quelling in fungi, and RNA interference in animals *Proc. Natl Acad. Sci. USA* **97** 11650–4

[35] Zilberman D, Cao X and Jacobsen S E 2003 Argonaute4 control of locus-specific siRNA accumulation and DNA and histone methylation *Science* **299** 716–9

[36] Havecker E R, Wallbridge L M, Hardcastle T J, Bush M S, Kelly K A, Dunn R M, Schwach F, Doonan J H and Baulcombe D C 2010 The Arabidopsis RNA-directed DNA methylation Argonautes functionally diverge based on their expression and interaction with target loci *Plant Cell* **22** 321–34

[37] Hunter C, Sun H and Poethig R S 2003 The *Arabidopsis* heterochronic gene *ZIPPY* is an Argonaute family member *Curr. Biol.* **13** 1734–9

[38] Kuo S Y, Hu C C, Huang Y W, Lee C W, Luo M J, Tu C W, Lee S C, Lin N S and Hsu Y H 2021 Argonaute 5 family proteins play crucial roles in the defence against *Cymbidium mosaic* virus and *Odontoglossum ringspot* virus in *Phalaenopsis aphrodite* subsp. *formosana Mol Plant Pathol.* **22** 627–43

[39] Iwasaki S, Kobayashi M, Yoda M, Sakaguchi Y, Katsuma S, Suzuki T and Tomari Y 2010 Hsc70/Hsp90 chaperone machinery mediates ATP-dependent RISC loading of small RNA duplexes *Mol Cell* **39** 292–9

[40] Rinaldi S, Colombo G and Paladino A 2020 Mechanistic model for the Hsp90-driven opening of human argonaute *J. Chem. Inf. Model.* **60** 1469–80

[41] Csorba T, Pantaleo V and Burgyan J 2009 RNA silencing: an antiviral mechanism *Adv. Virus Res.* **75** 35–71

[42] Dalmay T, Hamilton A, Rudd S, Angell S and Baulcombe D C 2000 An RNA-dependent RNA polymerase gene in *Arabidopsis* is required for posttranscriptional gene silencing mediated by a transgene but not by a virus *Cell* **101** 543–53
[43] Mourrain P *et al* 2000 Arabidopsis SGS2 and SGS3 genes are required for posttranscriptional gene silencing and natural virus resistance *Cell* **101** 533–42
[44] Xie Z, Fan B, Chen C and Chen Z 2001 An important role of an inducible RNA-dependent RNA polymerase in plant antiviral defense *Proc. Natl Acad. Sci. USA* **98** 6516–21
[45] Wang M B and Metzlaff M 2005 RNA silencing and antiviral defense in plants *Curr. Opin. Plant Biol.* **8** 216–22
[46] Zong J, Yao X, Yin J, Zhang D and Ma H 2009 Evolution of the RNA-dependent RNA polymerase (RdRP) genes: duplications and possible losses before and after the divergence of major eukaryotic groups *Gene* **447** 29–39
[47] Venkataraman S, Prasad B and Selvarajan R 2018 RNA dependent RNA polymerases: insights from structure, function and evolution *Viruses* **10** 76
[48] Pandey S P and Baldwin I T 2007 RNA-directed RNA polymerase 1 (RdR1) mediates the resistance of *Nicotiana attenuata* to herbivore attack in nature *Plant* J. **50** 40–53
[49] Liu Y, Gao Q, Wu B, Ai T and Guo X 2009 *NgRDR1*, an RNA-dependent RNA polymerase isolated from *Nicotiana glutinosa*, was involved in biotic and abiotic stresses *Plant Physiol. Biochem.* **47** 359–68
[50] Lu C, Kulkarni K, Muthuvalliappan R, Tej S S, Poethig R S, Henderson I R, Jacobsen S E, Wang W, Green P J and Meyers B C 2006 MicroRNAs and other small RNAs enriched in the *Arabidopsis* RNA-dependent RNA polymerase-2 mutant *Genome Res.* **16** 1276–88
[51] Deng S, Xu J, Liu J, Kim S H, Shi S and Chua N H 2015 JMJ24 binds to RDR2 and is required for the basal level transcription of silenced loci in *Arabidopsis Plant* J. **83** 770–82
[52] Polydore S and Axtell M J 2018 Analysis of RDR1/RDR2/RDR6-independent small RNAs in *Arabidopsis thaliana* improves miRNA annotations and reveals unexplained types of short interfering RNA loci *Plant* J. **94** 1051–63
[53] Whitfield A E, Falk B W and Rotenberg D 2015 Insect vector-mediated transmission of plant viruses *Virology* **479** 278–89
[54] Wang A 2015 Dissecting the molecular network of virus–plant interactions: the complex roles of host factors *Annu. Rev. Phytopathol.* **53** 45–66
[55] Jiang T and Zhou T 2023 Unraveling the mechanisms of virus-induced symptom development in plants *Plants* **12** 2830
[56] Mahmood M A, Naqvi R Z, Rahman S U, Amin I and Mansoor S 2023 Plant virus-derived vectors for plant genome engineering *Viruses* **15** 531
[57] Wang X-B, Jovel J, Udomporn P, Wang Y, Wu Q, Li W-X, Gasciolli V, Vaucheret H and Ding S-W 2011 The 21-nucleotide, but not 22-nucleotide, viral secondary small interfering RNAs direct potent antiviral defense by two cooperative Argonautes in *Arabidopsis thaliana Plant Cell* **23** 1625–38
[58] Bisaro D M 2006 Silencing suppression by geminivirus proteins *Virology* **344** 158–68
[59] Akbergenov R *et al* 2006 Molecular characterization of geminivirus-derived small RNAs in different plant species *Nucleic Acids Res.* **34** 462–71
[60] Shao F and Lu S 2014 Identification, molecular cloning and expression analysis of five RNA-dependent RNA polymerase genes in *Salvia miltiorrhiza PLoS One* **9** e95117
[61] Qin L, Mo N, Muhammad T and Liang Y 2018 Genome-wide analysis of DCL, AGO, and RDR gene families in pepper (*Capsicum annuum* L.) *Int. J. Mol. Sci.* **19** 1038

[62] Bouche N, Lauressergues D, Gasciolli V and Vaucheret H 2006 An antagonistic function for *Arabidopsis* DCL2 in development and a new function for DCL4 in generating viral siRNAs *EMBO J.* **25** 3347–56

[63] Diaz-Pendon J A, Li F, Li W-X and Ding S-W 2007 Suppression of antiviral silencing by cucumber mosaic virus 2b protein in *Arabidopsis* is associated with drastically reduced accumulation of three classes of viral small interfering RNAs *Plant Cell* **19** 2053–63

[64] Garcia-Ruiz H, Takeda A, Chapman E J, Sullivan C M, Fahlgren N, Brempelis K J and Carrington J C 2010 Arabidopsis RNA-dependent RNA polymerases and Dicer-like proteins in antiviral defense and small interfering RNA biogenesis during turnip mosaic virus infection *Plant Cell* **22** 481–96

[65] Andika I B, Maruyama K, Sun L, Kondo H, Tamada T and Suzuki N 2015 Different Dicer-like protein components required for intracellular and systemic antiviral silencing in *Arabidopsis thaliana Plant Signal. Behav.* **10** e1039214

[66] Qiao W, Zarzynska-Nowak A, Nerva L, Kuo Y-W and Falk B W 2018 Accumulation of 24 nucleotide transgene-derived siRNAs is associated with Crinivirus immunity in transgenic plants *Mol. Plant Pathol.* **19** 2236–47

[67] Blevins T *et al* 2006 Four plant dicers mediate viral small RNA biogenesis and DNA virus induced silencing *Nucleic Acids Res.* **34** 6233–46

[68] Aregger M *et al* 2012 Primary and secondary siRNAs in Geminivirus-induced gene silencing *PLoS Pathog.* **8** e1002941

[69] Muangsan N, Beclin C, Vaucheret H and Robertson D 2004 Geminivirus VIGS of endogenous genes requires SGS2/SDE1 and SGS3 and defines a new branch in the genetic pathway for silencing in plants *Plant* J. **38** 1004–14

[70] Tsai H L, Li Y H, Hsieh W P, Lin M C, Ahn J H and Wu S H 2014 HUA ENHANCER1 is involved in posttranscriptional regulation of positive and negative regulators in *Arabidopsis* photomorphogenesis *Plant Cell* **26** 2858–72

[71] Boutet S, Vazquez F, Jun L, Béclin C, Fagard M, Gratias A, Morel J-B, Crété P, Chen X and Vaucheret H 2003 *Arabidopsis* HEN1: a genetic link between endogenous miRNA controlling development and siRNA controlling transgene silencing and virus resistance *Curr. Biol.* **13** 843–8

[72] Li J, Yang Z, Yu B, Liu J and Chen X 2005 Methylation protects miRNAs and siRNAs from a 3′-end uridylation activity in *Arabidopsis Curr. Biol.* **15** 1501–7

[73] Zhang X, Zhang X, Singh J, Li D and Qu F 2012 Temperature-dependent survival of turnip crinkle virus-infected *Arabidopsis* plants relies on an RNA silencing-based defense that requires DCL2, AGO2, and HEN1 *J. Virol.* **86** 6847–54

[74] Blevins T, Rajeswaran R, Aregger M, Borah B K, Schepetilnikov M, Baerlocher L, Farinelli L, Meins F, Hohn T and Pooggin M M 2011 Massive production of small RNAs from a non-coding region of cauliflower mosaic virus in plant defense and viral counter-defense *Nucleic Acids Res.* **39** 5003–14

[75] Al-Hashash H, Akbar A and Al-Ali E 2023 Genome of tomato yellow leaf curl virus (TYLCV) obtained from tomato leaves in Kuwait *Microbiol. Resour. Announc* **12** e0115522

[76] Li Y, Muhammad T, Wang Y, Zhang D, Crabbe M J C and Liang Y 2018 Salicylic acid collaborates with gene silencing to tomato defense against tomato yellow leaf curl virus (TYLCV) *Pak. J. Bot.* **50** 2041–54

[77] Han M H, Goud S, Song L and Fedoroff N 2004 The Arabidopsis double-stranded RNA-binding protein HYL1 plays a role in microRNA-mediated gene regulation *Proc. Natl Acad. Sci. USA* **101** 1093–8
[78] Montavon T, Kwon Y, Zimmermann A, Hammann P, Vincent T, Cognat V, Michel F and Dunoyer P 2017 A specific dsRNA-binding protein complex selectively sequesters endogenous inverted-repeat siRNA precursors and inhibits their processing *Nucleic Acids Res.* **45** 1330–44
[79] Eamens A L and Wook Kim KPM DRB2 Waterhouse 2012 DRB3 and DRB5 function in a non-canonical microRNA pathway in *Arabidopsis thaliana Plant Signal. Behav.* **7** 1224–9
[80] Jakubiec A, Yang S W and Chua N H 2012 Arabidopsis DRB4 protein in antiviral defense against Turnip yellow mosaic virus infection *Plant* J. **69** 14–25
[81] Morris E R, Chevalier D and DAWDLE Walker J C 2006 a forkhead-associated domain gene, regulates multiple aspects of plant development *Plant Physiol.* **141** 932–41
[82] Zhang S, Dou Y, Li S, Ren G, Chevalier D, Zhang C and Yu B 2018 DAWDLE interacts with Dicer-like proteins to mediate small RNA biogenesis *Plant Physiol.* **177** 1142–51
[83] Feng B *et al* 2016 PARylation of the forkhead-associated domain protein DAWDLE regulates plant immunity *EMBO Rep.* **17** 1799–813
[84] Jaubert M, Bhattacharjee S, Mello A F S, Perry K L and Moffett P 2011 Argonaute 2 mediates RNA-silencing antiviral defenses against potato virus X in Arabidopsis *Plant Physiol.* **156** 1556–64
[85] Harvey J J W, Lewsey M G, Patel K, Westwood J, Heimstädt S, Carr J P and Baulcombe D C 2011 An antiviral defense role of AGO2 in plants *PLoS One* **6** e14639
[86] Garcia-Ruiz H *et al* 2015 Roles and programming of *Arabidopsis* Argonaute proteins during turnip mosaic virus infection *PLoS Pathog.* **11** e1004755
[87] Brosseau C and Moffett P 2015 Functional and genetic analysis identify a role for *Arabidopsis* Argonaute 5 in antiviral RNA silencing *Plant Cell* **27** 1742–54
[88] Qu F, Ye X and Morris T J 2008 Arabidopsis DRB4, AGO1, AGO7, and RDR6 participate in a DCL4-initiated antiviral RNA silencing pathway negatively regulated by DCL1 *Proc. Natl Acad. Sci. USA* **105** 14732–7
[89] Ludman M, Burgyan J and Fatyol K 2017 Crispr/Cas9 mediated inactivation of Argonaute 2 reveals its differential involvement in antiviral responses *Sci. Rep.* **7** 1010
[90] Paudel D B, Ghoshal B, Jossey S, Ludman M, Fatyol K and Sanfaçon H 2018 Expression and antiviral function of Argonaute 2 in *Nicotiana benthamiana* plants infected with two isolates of tomato ringspot virus with varying degrees of virulence *Virology* **524** 127–39
[91] Macfarlane S A 2010 Tobraviruses–plant pathogens and tools for biotechnology *Mol. Plant Pathol.* **11** 577–83
[92] Fernandez-Calvino L *et al* 2014 Virus-induced alterations in primary metabolism modulate susceptibility to *Tobacco rattle virus* in Arabidopsis *Plant Physiol.* **166** 1821–38
[93] Fernandez-Calvino L, Martínez-Priego L, Szabo E Z, Guzmán-Benito I, González I, Canto T, Lakatos L and Llave C 2016 Tobacco rattle virus 16K silencing suppressor binds Argonaute 4 and inhibits formation of RNA silencing complexes *J. Gen. Virol.* **97** 246–57
[94] Bhattacharjee S, Zamora A, Azhar M T, Sacco M A, Lambert L H and Moffett P 2009 Virus resistance induced by NB-LRR proteins involves Argonaute 4-dependent translational control *Plant* J. **58** 940–51
[95] Hamera S, Song X, Su L, Chen X and Fang R 2012 Cucumber mosaic virus suppressor 2b binds to AGO4-related small RNAs and impairs AGO4 activities *Plant* J. **69** 104–15

[96] Qin L, Mo N, Zhang Y, Muhammad T, Zhao G, Zhang Y and Liang Y 2017 *CaRDR1*, an RNA-dependent RNA polymerase plays a positive role in pepper resistance against TMV *Front. Plant Sci.* **8** 1068
[97] Lee W S, Fu S F, Li Z, Murphy A M, Dobson E A, Garland L, Chaluvadi S R, Lewsey M G, Nelson R S and Carr J P 2016 Salicylic acid treatment and expression of an RNA-dependent RNA polymerase 1 transgene inhibit lethal symptoms and meristem invasion during tobacco mosaic virus infection in *Nicotiana benthamiana BMC Plant Biol.* **16** 1–14
[98] Yu D, Fan B, MacFarlane S A and Chen Z 2003 Analysis of the involvement of an inducible Arabidopsis RNA-dependent RNA polymerase in antiviral defense *Mol. Plant. Microbe Interact.* **16** 206–16
[99] He J, Dong Z, Jia Z, Wang J and Wang G 2010 Isolation, expression and functional analysis of a putative RNA-dependent RNA polymerase gene from maize (*Zea mays* L.) *Mol. Biol. Rep.* **37** 865–74
[100] Liao Y W K, Liu Y R, Liang J Y, Wang W P, Zhou J, Xia X J, Zhou Y H, Yu J Q and Shi K 2014 The relationship between the plant-encoded RNA-dependent RNA polymerase 1 and alternative oxidase in tomato basal defense against tobacco mosaic virus *Planta* **241** 641–50
[101] Du X 2024 The cellular RNA-dependent RNA polymerases in plants *New Phytol.* **244** 2150–5
[102] Wang X-B, Wu Q, Ito T, Cillo F, Li W-X, Chen X, Yu J-L and Ding S-W 2010 RNAi-mediated viral immunity requires amplification of virus-derived siRNAs in *Arabidopsis thaliana Proc. Natl Acad. Sci. USA* **107** 484–9
[103] Qiu Y, Wu Y, Zhang Y, Xu W, Wang C and Zhu S 2018 Profiling of small RNAs derived from cucumber mosaic virus in infected *Nicotiana benthamiana* plants by deep sequencing *Virus Res.* **252** 1–7
[104] Li F, Wang Y and Zhou X 2017 SGS3 cooperates with RDR6 in triggering geminivirus-induced gene silencing and in suppressing geminivirus infection in *Nicotiana benthamiana Viruses* **9** 247
[105] Ludman M and Fatyol K 2019 The virological model plant, *Nicotiana benthamiana* expresses a single functional RDR6 homeolog *Virology* **537** 143–8
[106] Zhu B, Gao H, Xu G, Wu D, Song S, Jiang H, Zhu S, Qi T and Xie D 2017 Arabidopsis ALA1 and ALA2 mediate RNAi-based antiviral immunity *Front. Plant Sci.* **8** 422
[107] Jiang L, Qian D, Zheng H, Meng L Y, Chen J, Le W J, Zhou T, Zhou Y J, Wei C H and Li Y 2012 RNA-dependent RNA polymerase 6 of rice (*Oryza sativa*) plays role in host defense against negative-strand RNA virus, Rice stripe virus *Virus Res.* **163** 512–9
[108] Hong W, Qian D, Sun R, Jiang L, Wang Y, Wei C, Zhang Z and Li Y 2015 *OsRDR6* plays role in host defense against double-stranded RNA virus, rice dwarf phytoreovirus *Sci. Rep.* **5** 11324
[109] Burgyan J and Havelda Z 2011 Viral suppressors of RNA silencing *Trends Plant Sci.* **16** 265–72
[110] Haas G, Azevedo J, Moissiard G, Geldreich A, Himber C, Bureau M, Fukuhara T, Keller M and Voinnet O 2008 Nuclear import of CaMV P6 is required for infection and suppression of the RNA silencing factor DRB4 *EMBO J.* **27** 2102–12
[111] Feng M, Zuo D, Jiang X, Li S, Chen J, Jiang L, Zhou X and Jiang T 2018 Identification of Strawberry vein banding virus encoded P6 as an RNA silencing suppressor *Virology* **520** 103–10
[112] Baumberger N, Tsai C H, Lie M, Havecker E and Baulcombe D C 2007 The polerovirus silencing suppressor P0 targets Argonaute proteins for degradation *Curr. Biol.* **17** 1609–14

[113] Bortolamiol D, Pazhouhandeh M, Marrocco K, Genschik P and Ziegler-Graff V 2007 The polerovirus F box protein P0 targets Argonaute1 to suppress RNA silencing *Curr. Biol.* **17** 1615–21

[114] Kenesi E, Carbonell A, Lozsa R, Vertessy B and Lakatos L 2017 A viral suppressor of RNA silencing inhibits Argonaute 1 function by precluding target RNA binding to pre-assembled RISC *Nucleic Acids Res.* **45** 7736–50

[115] Martínez-Turiño S and Hernández C 2009 Inhibition of RNA silencing by the coat protein of Pelargonium flower break virus: distinctions from closely related suppressors *J. Gen. Virol.* **90** 519–25

[116] Garcia-Ruiz H, Gabriel Peralta S M and Harte-Maxwell P A 2018 Tomato spotted wilt virus NSs protein supports infection and systemic movement of a potyvirus and is a symptom determinant *Viruses* **10** 129

[117] Ruiz-Ruiz S, Soler N, Sanchez-Navarro J, Fagoaga C, Lopez C, Navarro L, Moreno P, Pena L and Flores R 2013 Citrus tristeza virus p23: determinants for nucleolar localization and their influence on suppression of RNA silencing and pathogenesis *Mol. Plant-Microbe Interact.* **26** 306–18

[118] Gomez-Munoz N, Velazquez K, Vives M C, Ruiz-Ruiz S, Pina J A, Flores R, Moreno P and Guerri J 2017 The resistance of sour orange to Citrus tristeza virus is mediated by both the salycilic acid and the RNA silencing defense pathways *Mol. Plant Pathol.* **18** 1253–66

[119] Varanda C M R, Materatski P, Campos M D, Clara M I E, Nolasco G and Félix M R 2018 Olive mild mosaic virus coat protein and P6 are suppressors of RNA silencing, and their silencing confers resistance against OMMV *Viruses* **10** 416

[120] Basu S, Kushwaha N K, Singh A K, Sahu P P, Vinoth Kumar R V and Chakraborty S 2018 Dynamics of a geminivirus-encoded pre-coat protein and host RNA-dependent RNA polymerase 1 in regulating symptom recovery in tobacco *J. Exp. Bot.* **69** 2085–102

[121] Glick E, Zrachya A, Levy Y, Mett A, Gidoni D, Belausov E, Citovsky V and Gafni Y 2008 Interaction with host SGS3 is required for suppression of RNA silencing by tomato yellow leaf curl virus V2 protein *Proc. Natl Acad. Sci. USA* **105** 157–61

[122] Kumar V, Mishra S K, Rahman J, Taneja J, Sundaresan G, Mishra N S and Mukherjee S K 2015 Mungbean yellow mosaic Indian virus encoded AC2 protein suppresses RNA silencing by inhibiting Arabidopsis RDR6 and AGO1 activities *Virology* **486** 158–72

[123] Senshu H, Yamaji Y, Minato N, Shiraishi T, Maejima K, Hashimoto M, Miura C, Neriya Y and Namba S 2011 A dual strategy for the suppression of host antiviral silencing: two distinct suppressors for viral replication and viral movement encoded by potato virus M *J. Virol.* **85** 10269–78

[124] Deng X, Kelloniemi J, Haikonen T, Vuorinen A L, Elomaa P, Teeri T H and Valkonen J P T 2013 Modification of Tobacco rattle virus RNA1 to serve as a VIGS vector reveals that the 29K movement protein is an RNA silencing suppressor of the virus *Mol. Plant-Microbe Interact.* **26** 503–14

[125] Endres M W, Gregory B D, Gao Z, Foreman A W, Mlotshwa S, Ge X, Pruss G J, Ecker J R, Bowman L H and Vance V 2010 Two plant viral suppressors of silencing require the ethylene-inducible host transcription factor RAV2 to block RNA silencing *PLoS Pathog.* **6** e1000729

[126] Li F and Wang A 2018 RNA decay is an antiviral defense in plants that is counteracted by viral RNA silencing suppressors *PLoS Pathog.* **14** e1007228

IOP Publishing

Advances in Biochemical and Molecular Mechanisms of Plant–Pathogen Interaction

Hitendra K Patel and Anirudh Kumar

Chapter 7

Remembering the attack: role of epigenetic modifications and their inheritance in plant defence

Niranjan Gattu, C G Gokulan and Hitendra K Patel

Organisms like bacteria, fungi, viruses, nematodes, and pests affect plant health, leading to a drastic decline in food production. Epigenetic processes including DNA methylation, histone modifications, chromatin assembly and remodelling play crucial functions in regulating the transcription of genes associated with plant defence mechanisms. Understanding these mechanisms holds great potential for revolutionizing breeding strategies aimed at developing crops with enhanced resistance against diseases and pests. Investigations in the model plant *Arabidopsis* have yielded significant findings on the mechanistic operations that govern epigenetic modifications. In this chapter, we compile and review the extant scientific literature pertaining to the interplay between crops and microbes, specifically in relation to the epigenetic phenomena. We discuss the impacts, obstacles, and approaches associated with harnessing epigenetic mechanisms to develop crops resilient to biotic stress factors.

7.1 Introduction

The practice of implementing crop rotation in agricultural systems has been observed to potentially enhance the nutrient content of the soil. However, it is worth noting that the adoption of monoculture, specifically with conventional crops, typically leads to an increase in the overall productivity of arable land in the field of agriculture [1]. Nonetheless, the vulnerability rate of the crops in these monocultures to the disease-causing pathogens is high. Hence, it became mandatory to develop disease-resistant crops to increase crop production and productivity for sustainable agriculture [2, 3]. For this, decoding the molecular mechanisms involved in plant–microbe interactions is essential.

doi:10.1088/978-0-7503-5673-2ch7

Co-evolution of plants and pathogens for a long time resulted in plants acquiring a sophisticated defence system to tackle pathogen infections [4]. This defence system is typically two-tiered. The first branch shows defence to a diverse collection of microbes including non-pathogens that possess common molecular signatures. The second tier responds to effector molecules of pathogens, either directly or by their effects on the host [5]. The first one uses transmembrane pattern recognition receptors (PRRs) to recognize pathogen-associated molecular patterns (PAMPs) or damage-associated molecular patterns (DAMPs) that activate pattern-triggered immunity (PTI). The second detects pathogen virulence factors by polymorphic nucleotide-binding-leucine rich receptor (NB-LRR) proteins encoded by the plant resistance (R) genes that initiate effector-triggered immunity (ETI) in the host cytoplasm [6–8]. Despite the variations observed in the magnitude and temporal dynamics of PTI and ETI in downstream defence responses, it is noteworthy that both mechanisms exhibit comparable potentiation in the context of plant immunity. Furthermore, it is evident that both PTI and ETI are intricately associated with a substantial transcriptional reprogramming of genes related to defence [9, 10]. The mounting body of evidence reveals that epigenetic modifications, including DNA methylation, modifications of histone proteins, chromatin assembly and remodelling, are responsible for orchestrating the transcriptional reprogramming associated with pathogen responses in plants. These epigenetic variations play a vital part in regulating the plant's ability to resist a diverse group of crop pathogens [11–14].

7.2 Interplay between DNA methylation and plant–pathogen interactions

DNA methylation is a heritable epigenetic modification that usually takes place at the C5 position of the cytosine base within the CG, CHG, and CHH contexts (where H can be A, C, or T), resulting in the formation of 5-methylcytosine (5 mC). DNA methylation plays a crucial role in plant development, stress adaptation, and evolution of the genome [15, 16]. A study conducted on the liverwort *Marchantia polymorpha*, which belongs to an ancient group of plants that adapted to terrestrial environments early on during evolution, has revealed that a significant amount of N4 cytosine methylation is necessary for the process of spermatogenesis. This expands our understanding of the role of functional DNA methylation in plants [17]. Typically, the DNA methylation profile is influenced by three mechanisms, namely *de novo* methylation, maintenance methylation, and active demethylation [18]. Small RNAs trigger the initiation of methylation through a process called RNA-dependent DNA methylation (RdDM), which is facilitated by domains rearranged methyltransferase 2 (DRM2) [19]. DNA methylation, once established, could potentially be upheld by enzymes like methyltransferase1 (MET1) and plant-specific chromomethylases (CMT2 and CMT3). Conversely, it can also be erased by DNA demethylases such as repressor of silencing 1 (ROS1), demeter (DME), demeter-Like 2 (DML2), and DML3 as reported in the model plant *Arabidopsis* [20].

The study of DNA (de)methylation (read as methylation and demethylation) in plant–microbe interactions has been thoroughly investigated using genetic mutants of DNA (de)methylation. Further, profiling DNA methylation has been performed using advanced techniques for DNA methylation profiling like methylation-sensitive amplified fragment length polymorphism (MSAP), whole-genome bisulfite sequencing (WGBS), methylated DNA immune-precipitation sequencing (MeDIPseq), and methyl CpG binding domain protein capture sequencing (MBDCap-seq) [16, 21]. Pathogen infections cause DNA hypomethylation in numerous plant species, including *Arabidopsis*, rice, tobacco, soybean, turnip, watermelon, and bread wheat [22–25]. In the context of crop and model plants, it has been observed that the application of flg22—the bacterial PAMP, the nematode PAMP 'NemaWater' and salicylic acid (SA) results in the occurrence of similar DNA hypomethylation [23, 26]. Several chromatin sites near defence-associated genes, such as promoters, coding sequence, and adjacent transposable elements (TEs), are affected by pathogen-induced DNA hypomethylation, according to high-resolution DNA methylation profiling [23]. DNA hypomethylation caused by pathogens typically results in the activation of nearby defence-related genes, as determined by a comprehensive analysis of gene expression and DNA methylation. However, the precise control of defence-related gene expression through DNA hypomethylation differs among plants and specific genes [23, 24]. An example of how DNA methylation can be chemically inhibited was shown by targeting the promoter of the rice resistance gene XA21G. This inhibition can activate the transcription of XA21G and lead to the development of resistance in rice against bacterial blight caused by *Xanthomonas oryzae pv. oryzae* (Xoo) [27]. Conversely, when the process of DNA methylation was blocked at the promoter of the Pib gene, which is responsible for rice blast disease resistance, it hindered the transcription of the Pib gene and consequently weakened the rice's ability to resist blast disease [28]. The process of DNA (de)methylation has the potential to impact defence genes in trans as well as exert influence on cis genes [16]. For example, a sophisticated examination of gene expression and DNA methylation in *Arabidopsis* mutants with altered levels of methylation indicated about 15% of defence-associated genes were activated in hypomethylated *nrpe1* mutants and suppressed in hypermethylated *ros1* mutants. These genes were found to be influenced by nearby transposable elements (TEs) and controlled by DNA methylation regulated by NRPE1 and/or ROS1. This suggests that the transcription of defence-related genes is regulated by both cis- and trans-regulatory mechanisms involving DNA (de)methylation [29].

Genetic studies, in addition to sequencing and *in silico* findings, have substantially implicated the role of DNA (de)methylation in pathogen responses of plants [30]. *Arabidopsis* plants with mutations in DNA hypomethylation genes, specifically *nrpe1*, *met1*–3, and *ddc* (*drm1–2, drm2–2, cmt3*–11), showed enhanced resistance to *Hyaloperonospora arabidopsidis* (Hpa), a biotrophic oomycete and *Pseudomonas syringae* pv. *tomato* DC3000, a hemi-biotrophic bacteria. However, resistance to the necrotrophic fungal pathogen *Plectosphaerella cucumerina* (Pc) was increased, whereas the DNA hypermethylation mutant *ros1-4* showed decreased resistance to Pst DC3000 and Hpa but increased resistance to Pc [29]. Using virus-induced gene silencing (VIGS) with barley stripe mosaic virus (BSMV), it was shown that the reduced expression of a DRM2 homologue in *Aegilops tauschii* resulted in enhanced resistance of the plant to *Blumeria graminis* f. sp. *tritici* (Bgt), the fungal pathogen responsible for wheat powdery mildew. Growing data suggests that DNA (de) methylation has a role in the regulation of plant defence responses to pests and nematodes [31]. For example, *Arabidopsis* mutant plants weak in DNA methylation and H3K9 (kyp) demonstrated resistance to the green peach aphid (*Myzus persicae*) [22]. In a similar vein, it was observed that mutants of rice RdDM and DDM1, namely *ago4*, *ddm*1, *drm*2, and *dcl3* exhibited a diminished susceptibility to infection by the nematode *Meloidogyne graminicola*. Further, it has been discovered that the process of DNA (de)methylation has a critical part in orchestrating the functioning of allelic defence-associated genes. The expression of WRKY45–1, which is an allele of the rice transcription factor gene WRKY45 that is susceptible to certain conditions, leads to the production of TE-siR815. This is a small interfering RNA (siRNA) derived from a TE that acts on other genes. TE-siR815 specifically targets and suppresses the expression of the defence-related gene known as siR815 Target 1 (ST1). This suppression occurs through RdDM. As a result, the resistance against rice blight, which is normally conferred by the presence of WRKY45-2, a resistant allele of the same gene, is abolished. Similarly, the process of DNA (de)methylation occurring at the promoter region of PigmS, a susceptible allele of the rice Pigm gene, exerts control over its transcription in a manner specific to certain tissues. This phenomenon plays a role in maintaining a delicate equilibrium between the resistance of rice plants to diseases and their overall yield [32]. These findings provide insight into the critical role of DNA (de)methylation in the control of plant response to pathogens (figure 7.1), as well as its tremendous potential for augmenting disease resistance in crops.

7.3 Multidimensional role of histone modifications in plant immunity

Histone post-translational modifications (PTMs) like acetylation, methylation, and ubiquitination, are fundamental epigenetic processes that transpire at the N-terminal tails of histones. These modifications are crucial for governing chromatin architecture and modulating various chromatin-associated activities [11–13]. Histone acetylation, a process that can be reversed, is subject to dynamic regulation by enzymes known as histone acetyltransferases (HATs) and histone deacetylases (HDACs) [33]. In a general context, it has been suggested that the process of histone acetylation facilitated by histone acetyltransferases (HATs) causes relaxation of the chromatin structure, thereby leading to an increase in the transcription of downstream genes. Conversely, the phenomenon of histone deacetylation, catalyzed by histone deacetylases (HDACs), has been found to potentially contribute to the repression of gene expression [34]. Extensive investigation pertaining to *Arabidopsis* HATs (AtELP2 and AtELP3) and HDACs (AtHDA6, AtHDA19, AtHD2B, and AtSRT2) has yielded compelling evidence supporting the involvement of histone (de)acetylation (read as acetylation and deacetylation) in plant–pathogen interactions, a subject that has been extensively reviewed [35–37]. Moreover, the investigation pertaining to the crop HATs (histone acetyltransferases) and HDACs (histone deacetylases) substantiated the notion that the process of histone (de)

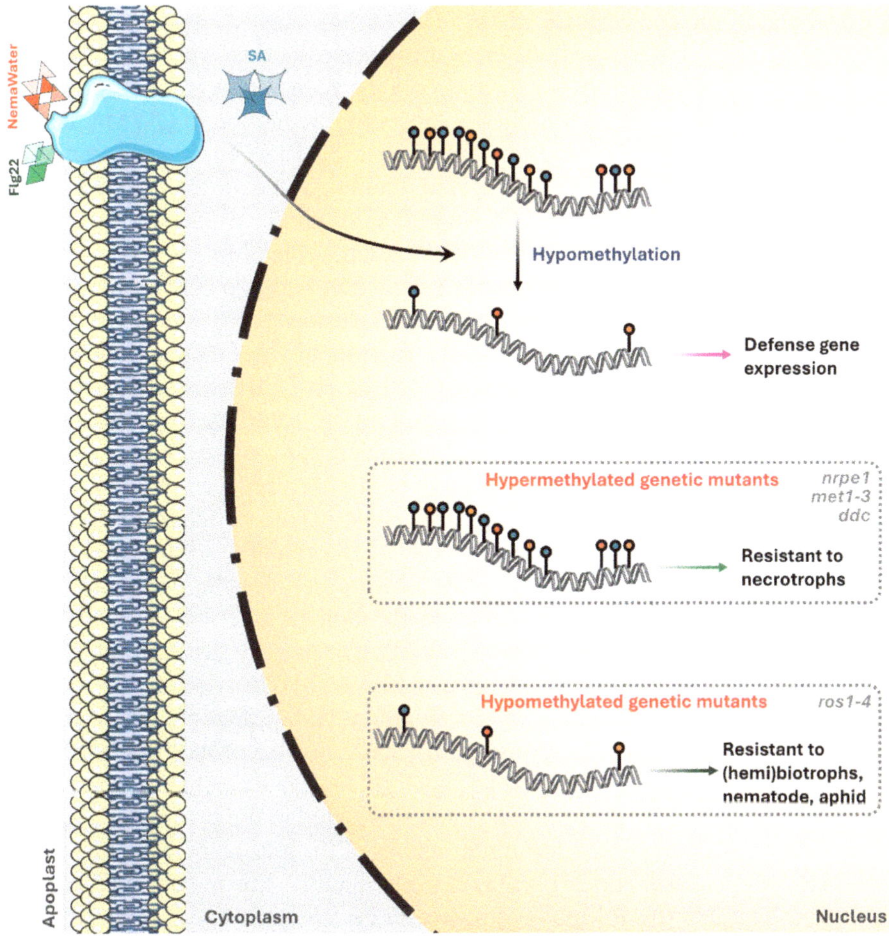

Figure 7.1. DNA (de)methylation is involved in determining response to pathogen attack. The schematic shows how the extent of DNA methylation is controlled by various defence-eliciting molecules like Flg22 (bacterial flagellin epitope), SA, and the nematode elicitor NemaWater. Further, the relation between the DNA methylation level and the pathogen lifestyle is depicted.

acetylation plays a pivotal role in governing the interactions between crops and pathogens [38, 39]. The wheat HAT complex, consisting of TaGCN5 and TaADA2, plays a crucial role in facilitating histone acetylation near the promoters of wheat wax biosynthesis genes. These genes are associated with the production of wax, which in turn provides signals necessary for the conidial growth of the fungal pathogen Bgt. By means of histone deacetylation occurring at the promoter regions of PRR and defence genes, the rice histone deacetylase (HDAC) OsHDT701 operates in conjunction with the rice ribonuclease P subunit Rpp30 to exert a negative regulatory effect on the defence responses of rice against the pathogenic fungus *Magnaporthe oryzae* and the bacterial pathogen *Xoo* [40]. In addition, a recent analysis of gene expression showed that nearly all TaHDACs were upregulated in response to infection with BSMV, Chinese wheat mosaic virus (CWMV), and wheat yellow mosaic virus (WYMV). This finding suggests that TaHDACs play a widespread role in the wheat defence mechanism against viral infections [41]. Remarkably, the utilization of virus-induced gene silencing (VIGS) employing the Foxtail mosaic virus (FoMV) exhibited a substantial reduction in CWMV infection in bread wheat through the targeted suppression of TaSRT1, a gene encoding a histone deacetylase (HDAC) in wheat. The findings indicate that the presence of TaSRT1 is a contributing factor to the vulnerability of wheat plants to infection by CWMV [41]. Interestingly, studies have identified the role of histone (de)acetylation in the pathogens that in turn regulate the interaction between crops and pathogens [42, 43]. In the case of the soybean oomycete pathogen *Phytophthora sojae*, the effector PsAvh23 interacts with the ADA2 subunit of the HAT complex SAGA, thereby inducing disruption in its assembly. Consequently, this disruption suppresses the initiation of defence-related genes by impeding the acetylation of H3K9 through the HAT complex. This in turn elevates the vulnerability of soybean plants to the infection caused by *P. sojae* [42]. Plants have been observed to employ methylation-mediated transcriptional gene silencing as a robust defence mechanism in response to gemini virus infections [44]. The potential of the tomato yellow leaf curl virus (TYLCV) lies in its ability to impede the process of methylation-mediated transcriptional gene silencing. This is achieved through its interaction with the host protein known as histone deacetylase 6 (NbHDA6), thereby disrupting the recruitment of MET1 by HDA6 [44]. The process of histone (de)acetylation plays a pivotal role in the regulation of pathogen development and infection, as well as the modulation of crop defence responses. For instance, the protein MoSNT2 from the fungus *M. oryzae* is hypothesized to play a role in the recruitment of the histone deacetylase complex. This recruitment is believed to aid in the process of histone H3 deacetylation specifically at the promoter regions of the autophagy genes MoATG6, MoATG15, MoATG16, and MoATG22. Consequently, this molecular mechanism is thought to regulate the autophagy process that is associated with infection caused by *M. oryzae* [45]. A separate investigation has revealed that the chemical compound phenazine1-carboxamide, which is secreted by the bacteria, *Pseudomonas piscium* residing in the wheat microbiome, possesses the capability to directly impede the functioning of the fungal histone acetyltransferase FgGcn5. Consequently, this alteration in histone acetylation leads to a decrease in the virulence of the wheat fungal pathogen *Fusarium graminearum* [46]. These investigations demonstrated that histone (de)acetylation plays a crucial role in governing a diverse array of biological processes, including pathogen growth and virulence, plant immune responses, and thus plant–pathogen interactions (figure 7.2). These findings open new avenues for the management of agricultural diseases.

Histone methylation is another highly dynamic process that is intricately regulated by the coordinated actions of histone methyltransferases and demethylases. In stark contrast to the phenomenon of histone acetylation, which is commonly associated with the activation of genes, histone methylation exhibits a dual role in both gene activation and repression. H3K4 and H3K36 methylation play a crucial role in facilitating the process of active transcription, while H3K9 and

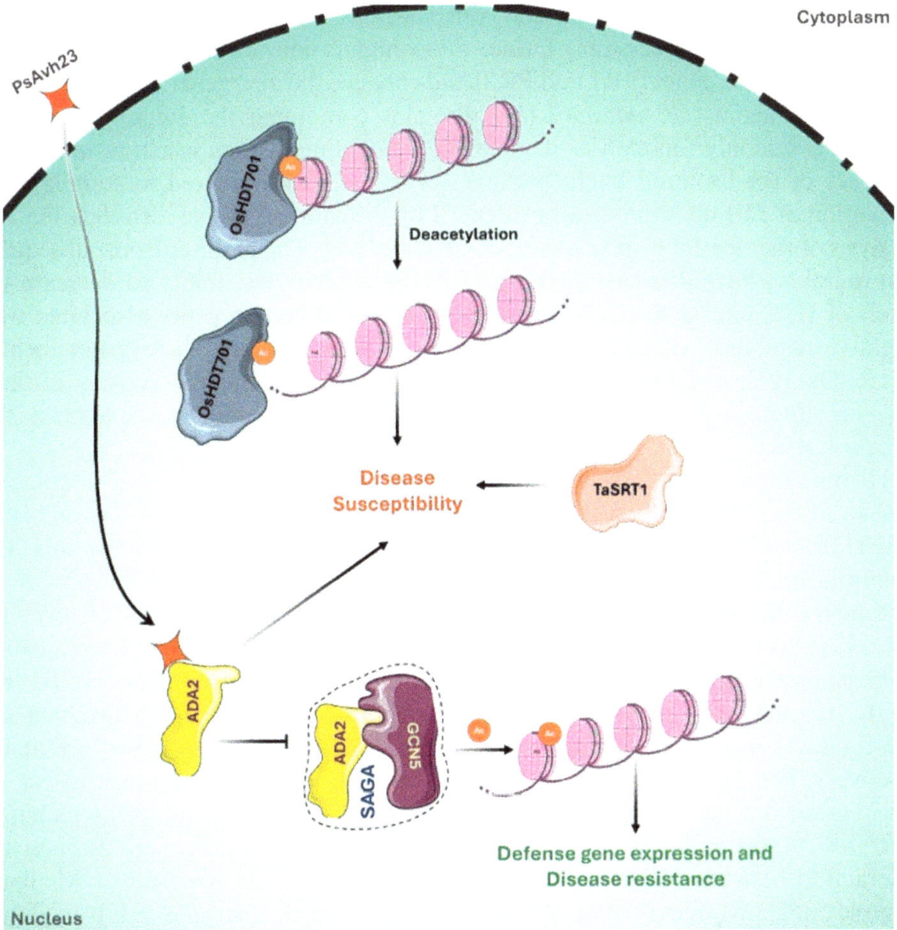

Figure 7.2. Histone acetylation controls the outcome of plant–pathogen interaction. Histone deacetylation appears to be a potent determinant of disease susceptibility as shown by studies in rice, wheat, and *Arabidopsis*. On the other hand, pathogen-mediated disruption of a histone acetylation complex (SAGA) results in disease susceptibility. OsHDT701 and TaSRT are histone deacetylases whereas the subcomplex ADA2 and GCN5 are part of the histone acyltransferase complex SAGA.

H3K27 methylation causes the repression of gene expression [12, 13]. According to the available reports, the regulation of plant–pathogen interactions is purportedly facilitated by the involvement of two specific histone demethylases, namely AtJMJ27 and AtIBM1, as well as three distinct *Arabidopsis* histone methyltransferases, namely AtATX1, AtSDG8, and AtSDG25 [47–49]. Furthermore, there is mounting proof that crop–pathogen interactions are significantly controlled by histone (de)methylation [50, 51]. One potential avenue for enhancing plant disease resistance involves the application of β-aminobutyric acid (BABA), a non-protein amino acid, onto potato leaves via spraying. This phenomenon may result in an elevation of the histone modifications H3K4me2 and H3K27me3 on the genes that contribute to systemic acquired resistance (SAR) like NPR1 and SNI1. Consequently, this process may induce a reconfiguration of the expression patterns of defence-related genes (PR1 and PR2) and enhance the intergenerational resistance of potatoes against *P. infestans* [51]. Proteins containing the Jumonji C (JmjC) domain commonly function as demethylases targeting histone lysine residues. The presence of the bacterial blight pathogen *Xoo* has been observed to stimulate the activation of 15 JmjCs in rice, as evidenced by the comprehensive identification and analysis of the rice JmjC gene at a genome-wide level. The findings from subsequent investigations have demonstrated that JMJ704 exhibits the ability to decrease the levels of H3K4me2/3 specifically at the promoters of certain genes associated with negative regulation of defence mechanisms in rice. Notably, these genes include NRR, Os11N3, and OsWRKY62. Therefore, the transcriptional activity of these genes is suppressed, leading to an augmented defence response in rice against *Xoo* infection [50].

Histone mono-ubiquitination, although comparatively less explored in comparison to histone methylation and acetylation predominantly occurs on histone H2A and H2B, exerting a substantial influence on the regulation of transcription. The ubiquitin ligases HUB1 and HUB2 are responsible for facilitating the process of histone H2B mono-ubiquitination (H2Bub1) in the plant species *Arabidopsis* [52–54]. Through alterations to cortical microtubule dynamics and amplification of resistance gene transcription (SNC1 and RPP4), investigations on HUB1 and HUB2 revealed that H2Bub1 improves *Arabidopsis* resistance to Pst DC3000 and the fungus *Verticillium dahliae* (Vd) [54, 55]. It is intriguing to observe that the discovery has been made that HUB1 exhibits a positive regulatory effect on *Arabidopsis* defence mechanisms against necrotrophic fungal pathogens including *Botrytis cinerea* and *Alternaria brassicicola*. This effect is achieved through the interaction between HUB1 and MED21, a subunit of the *Arabidopsis* Mediator complex [52]. Moreover, the study has demonstrated that the HUB-mediated H2Bub1 mechanism exerts a positive regulatory effect on the expression of NADPH oxidase RbohD, a pivotal modulator involved in defence responses. This regulation is achieved by enhancing the enrichment of H3K4me3, a histone modification associated with active gene expression. The findings indicate that there exists a complex interplay between histone mono-ubiquitination, methylation, and the mediator complex in governing the regulatory mechanisms of defence responses in plants [53]. In addition to HUBs, it has been shown that binding proteins associated with mono-ubiquitinated histones are implicated in the regulation of plant defence responses [56]. For instance, in the absence of pathogen infection, rice SWI/SNF2 ATPase BRHIS1-containing complex exhibited a specific interaction with mono-ubiquitinated histone variants H2B.7 and H2A.Xa/H2A.Xb/H2A.3. This interaction resulted in the repression of transcription for a select few disease defence-associated genes, namely OsPBZc and OsSIRK1, as well as the resistance to rice blast disease [56]. To enhance our comprehension of the diverse functions of histone mono-ubiquitination in the regulation of crop–pathogen interactions, it is imperative to undertake the characterization of additional histone ubiquitin ligases (HUBs) and proteins that bind to mono-ubiquitinated histones in crop species.

7.4 Regulation of pathogen response by chromatin assembly and remodelling

The organization of the plant genome is predominantly dependent on the structural arrangement of chromatin, a complex of DNA and proteins. This arrangement is subject to dynamic regulation by various modulators and histones. The gene transcription regulation is dependent on the control of DNA accessibility to the transcription machinery, which is governed by the structure of chromatin [57]. Growing data shows that the regulation of plant defensive responses is influenced by chromatin shape [58, 59]. Utilizing the tobacco rattle virus (TRV)-based VIGS, the suppression of histone H2B in *Nicotiana benthamiana* resulted in the activation of genes associated with SA biosynthesis and signalling pathways, such as EDS1 and NPR1. The observed phenomenon involves an augmented endogenous accumulation of SA, leading to an enhanced ability to withstand infection caused by potato virus X (PVX) [59]. For histones to be assembled into nucleosomes after replication, a chaperone called CHROMATIN ASSEMBLY FACTOR 1 (CAF-1) must be present. This chaperone has remained unchanged throughout evolution [58, 60]. In *Arabidopsis*, the absence of the chromatin assembly factor 1 (CAF-1) has been observed to have a suppressive effect on the transcription of defence-related genes. This effect is further substantiated by the findings that the absence of CAF-1 leads to a decrease in nucleosome occupancy and an increase in the presence of histone H3K4me3 at the defence response genes PR1 and WRKY53. The observed outcome led to the unauthorized initiation of a defence response dependent on SA in plants cultivated under standard aseptic growth conditions [58, 60]. The regulation of plant defence responses is governed by chromatin remodelling factors. These factors, along with histone H2B and histone chaperon CAF-1, possess the ability to utilize the energy derived from ATP hydrolysis to disrupt the association between DNA and histone molecules [61–63]. Furthermore, SWP73A may inhibit the RNA splicing regulator gene CDC5's transcription and have an impact on RPS2 and RPS4's alternative splicing, which would reduce defence responses in *Arabidopsis* [62]. Remarkably, a recent revelation has unveiled the distinct functionalities of H2A. *Z* and SWR1c subunits in the realm of plant immunity and gene regulation. It is worth noting that the SWR1 chromatin remodelling complex (SWR1c) is responsible for the catalysis of histone H2A substitution with the histone variant H2AZ within nucleosomes during the gene regulation processes of eukaryotes [61]. Simultaneously, new data supported the idea that chromatin remodelling proteins control plant disease resistance [64]. According to reports, the wheat CHD3-type chromatin remodelling protein TaCHR729 has been observed to exhibit binding affinity towards the promoter of the wheat wax biosynthetic genes, specifically 3-ketoacyl-CoA synthase (*TaKCS6*). The transcription of TaKCS6 is subsequently facilitated by the positive regulation exerted by TaCHR729, which enhances the presence of the flexible epigenetic mark H3K4me3 at the promoter region of TaKCS6. The silencing of TaCHR729 through VIGS appears to have the potential to impede the biosynthesis of wheat wax and the germination of Bgt conidia. This suggests that TaCHR729, a factor involved in chromatin remodelling, could potentially epigenetically activate the biosynthesis of cuticular wax in wheat and play a role in the interaction between wheat and powdery mildew [64].

7.5 The role of epigenetic memory in defence priming

Priming agents or crop pathogens, such as azelaic acid (AzA), methyl jasmonic acid (MeJA), SA, β aminobutyric acid (BABA), pipecolic acid (Pip) when applied to crop plants, can prime the plants' defence mechanisms and develop SAR to forthcoming pathogen encounters [65]. This SAR and defence priming could last for several generations (figure 7.3) [66–68]. For example, applying BABA as a priming agent to common beans may increase the plants' resistance against bacterial pathogen *P. syringae* pv. phaseolicocca [69]. Interestingly, for a minimum of two generations, the defence-related PvPR1 gene expression (ortholog of the *Arabidopsis* PR1-1, *Phaseolus vulgaris* PR1, common bean) demonstrated a priming reaction against pathogen attack [69]. Furthermore, epigenetic profiling showed that histone modifications and the DNA methylation are typically linked to defence priming [70, 71]. For example, treating leaves of *Arabidopsis* with bacterial pathogen *P. syringae* pv. maculicola or the synthetic analogue of SA, BTH may increase the epigenetic mark H3K4me2/3 on the defensive genes, that is intended to prime defence genes activation in the event of pathogen re-infection [72]. Comparatively, MeJA-induced defence priming in rice may result in an increase in the levels of H3K9ac and H3K4me3 at the promoters of the defence gene OsBBPI, as well as a modification of

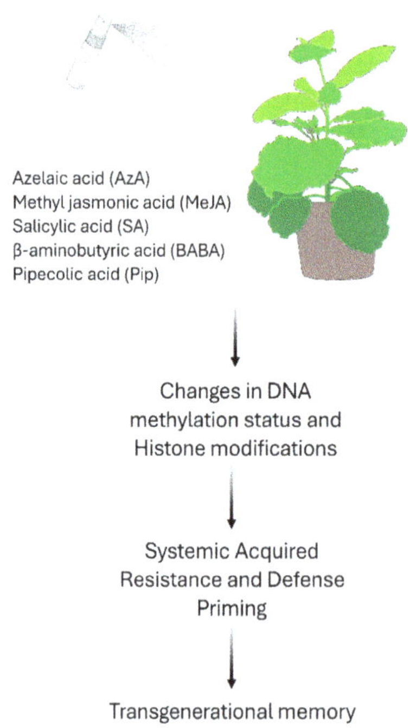

Figure 7.3. Defence priming using epigenetics-modifying chemicals as an approach to control plant disease. Flowchart showing the steps involved in the chemical-mediated defence priming in plants.

the levels of DNA methylation (5mC) throughout the genome. This ultimately leads to a chromatin-based wounding stress memory [73]. There is growing evidence that histone modifications and the DNA methylation play a role in the transgenerational defence priming [68, 70, 74]. For example, Progeny (P1) with enhanced resistance to the hemi-biotrophic bacterial pathogen *Pst* DC3000 and the biotrophic oomycete pathogen *Hpa*, as well as increased gene expression linked to SA-inducible defence, such as PR1, and WRKY53, can be produced through inheritance of SAR induced by the bacterial pathogen Pst DC3000 [68]. This P1 progeny also showed decreased resistance to the necrotrophic fungal pathogen *A. brassicicola*, as well as decreased expression of JA inducible defence-related genes like PDF1.2 and VSP2 [72]. Subsequent research indicated that the variation in the expression of defence-related genes related to SA and JA in P1 progeny was linked to changes in histone modifications including H3K27me3 and H3K9ac at the promoters of these defence-associated genes rather than hormone-level variations [72].

7.6 Employing epigenetic modifications to improve biotic stress resilience in crops

The frequency and prevalence of crop diseases are increased by the rigorous monoculture of cultivated crops on fertilized land. These diseases become more severe during extreme weather events brought on by climate change, such as heat waves and droughts [1, 75]. Despite crop-protection approaches such as regular chemical control and environment-friendly biocontrol, disease outbreaks account for around 20% of worldwide yield loss in notable crops [2, 3]. The traditional breeding methods for improving crop biotic stress resistance, which primarily depend on the use of resistance (R) genes, have turned out to be less effective because of the faster co-evolution of crop pathogens [76]. Furthermore, selective breeding has reduced the genetic diversity of R genes [76]. Creating genetically modified organisms (GMOs) and breeding crop varieties with their wild relatives are new innovative and promising methods for crop biotic stress breeding [68]. Alternatively, given the role of epigenetic processes like DNA methylation, histone modifications, chromatin remodelling, and assembly in plant–pathogen interactions, breeding for selective epigenetic modifications may enhance crop resistance to biotic stress. This process termed 'epi-breeding' opens new avenues in plant breeding [77–79].

Together, naturally occurring and artificially produced epigenetic modifications have the potential to impact crop biotic stress resistance and hold great promise for improving crop resistance through epi-breeding. For example, certain model plants and crops, including *Arabidopsis*, rice, tomato, oilseed rape, and melon, have shown to carry heritable natural epialleles linked to plant development and stress acclimatization [78, 80–86]. DNA hypomethylation was typically triggered by abiotic and biotic stresses, as was covered in detail in DNA (de)methylation section. This suggests that these types of stress conditions could be used to induce disease resistance-related epigenetic modifications in crops. Moreover, chemical treatments, gene specific DNA methylation, epigenome editing, and mutations in epigenetic marks could all be employed in experiments to produce epigenetic variants [16, 77–79, 87]. For example, two non-methylable cytosine analogues, zebularine and 5-azacytidine, are frequently employed as chemical inducers of DNA demethylation by inhibiting DNA methyltransferases [88, 89]. Even though the epigenetic changes brought about by these chemical treatments and stressors are temporary, changes in the epigenetic regulation may cause TEs to be activated and heritable epialleles to form, which may be used for breeding crops [90]. Through mutagenization of DNA (de)methylation machinery and subsequent crosses between genetically identical plant with varying levels of DNA methylation, epigenetic recombinant inbred line (epiRIL) populations were successfully generated in *Arabidopsis*. These populations had similar genomic backgrounds, but they had different levels of DNA methylation. These epiRIL populations exhibited remarkable phenotypic variations, such as modified disease resistance [91]. The generation of epiRILs in crops such as maize and bread wheat on the other hand, has proven more difficult due to a lack of mutants in DNA (de)methylation, minimal genome stability, and modest rates of mutagenesis transformation, and regeneration in crops [92].

Epigenome editing and gene specific DNA methylation represent emerging and promising techniques for inducing allelic diversity in both crop and model plant species. Indeed, it has been observed that the generation of 24-nt short-interfering RNAs (siRNAs) through the processing of double-stranded RNAs (dsRNAs) can exert regulatory control over DNA methylation processes. This, in turn, leads to the induction of gene silencing within the endogenous genetic material of plants. Notably, this silencing phenomenon has been observed to be inherited across successive generations of plants [93, 94]. Significantly, the exogenous administration of double-stranded RNA (dsRNA) with the purpose of selectively targeting gene promoter regions resulted in the induction of promoter RNA-directed DNA methylation (RdDM), thereby potentially circumventing obstacles encountered in crop breeding such as limited transformation efficiency and regrowth rate. The present proposition embodies an innovative methodology for the manipulation of the epigenome in crops [93, 95]. Within mammalian cells, a modified form of Cas9 (CRISPR associated protein 9), which serves as a prevalent sequence-specific nuclease (SSN) in the CRISPR system (clustered regularly interspaced short palindromic repeat), has been combined with histone-modifying enzymes such as histone demethylase LSD1, DNA methyltransferase DNMT3A, and the catalytic core of human histone acetyltransferase p300. This fusion has been devised to facilitate the manipulation of the epigenome [96–100]. In this study, the authors employed a fusion strategy by combining a zinc finger (ZF) peptide with the RdDM component SUDH9 to achieve targeted DNA methylation in specific regions of the *Arabidopsis* genome. The discovery presents an opportunity to engage in locus-specific epigenome editing within the realm of crop plants [101, 102]. However, to achieve a similar level of epigenome editing in crop plants using CRISPR or ZF technology, it is imperative to either introduce preassembled CRISPR or ZF complexes into crop protoplasts through transfection, or to incorporate the appropriate CRISPR or ZF elements into the genome of the crop plants to ensure stability. The feasibility of the latter is impeded by suboptimal rates of transformation and regeneration observed in certain crop species, such as *Triticum aestivum* (common wheat) [103]. Moreover, the introduction of epigenome modification techniques to enhance agricultural products necessitates the revision and elucidation of the existing regulatory framework employed in certain countries for GMOs and genome editing using engineered endonucleases (GEENs) [104].

Transmission characteristics of epigenetic variations, like inheritability and stability, should be explored as crucial issues in epi-breeding. It has been demonstrated that epialleles contain differentially methylated regions (DMRs), which are transmitted in Mendelian-fashion during meiosis. This finding implies that DNA methylation could potentially function as a heritable and persistent epigenetic marker that persists for multiple generations [105, 106]. The stable transmission of histone modifications via mitosis has been demonstrated by meticulous investigations into the process of plant vernalization [107]. For instance, when *Arabidopsis* plants are at rest, the repressive epigenetic mark H3K27me3 could be increased at the flowering locus C (FLC), and cold exposure could cause FLC to be silenced [108]. While some studies have indicated that histone modifications play a role in transgenerational stress memory, there is not much evidence regarding the inheritance of histone modifications across generations. For example, every generation resets the vernalization process involving H3K27 methylation [107]. Consequently, it is unlikely that histone modifications could be passed down steadily during sexual reproduction.

Transmission characteristics of epigenetic variations and crop plant propagation types, influence epi-breeding strategies for improving crop resistance [77, 78]. Given the enduring transmission of these epigenetic markers across successive plant generations, it is plausible to harness DNA methylation as a valuable tool in crop plant breeding methods encompassing both vegetative and seed propagation techniques. In contrast, it should be noted that histone modifications are exclusively applicable to the cultivation of clonally propagated crops and do not possess the ability to be inherited across successive generations. Clonal propagation, a highly efficient method of reproduction, has been observed to be a viable strategy for the proliferation of numerous crops, encompassing more than 60% of agricultural species. Notable examples include *Solanum tuberosum*, *Sorghum bicolor*, and *Manihot esculenta*. Additionally, it possesses the advantageous attribute of rectifying advantageous epigenetic traits [78, 109]. Furthermore, the utilization of molecular epigenomic markers, such as DMRs, holds promise for the initial assessment of epigenetic conditions in meristem/seed screening. This approach also plays a significant role in marker-assisted selection within crop epi-breeding procedures, specifically aimed at enhancing disease resistance [77, 78].

The utilization of epigenetic modelling holds promise in the realm of predicting the impact of epigenetic variations on the observable characteristics of plants, known as the plant phenotype. This approach aims to establish a connection between epigenetic variations and the practice of epi-breeding, which focuses on enhancing disease resistance in plants. By employing epigenetic modelling, scientists aspire to narrow the current distance between epigenetic variations and the improvement of disease resistance through epi-breeding techniques [77]. A refined process-oriented model was formulated to forecast the impacts of epigenetic fluctuations on gene expression and plant characteristics through the utilization of equations that delineate the underlying principles of extensively investigated biological mechanisms. As previously elucidated, the utilization of statistical models presents a viable approach for delving into the intricate interplay between DNA methylomes and transcriptomes, devoid of any prior comprehension regarding the fundamental mechanisms governing biological processes [77, 110, 111]. In this study, statistical models were employed to identify novel associations between DNA methylation patterns and gene expression in the species *Mimulus guttatus*. Additionally, a correlation was established between DNA methylation data and variations in plant height in the model plant *Arabidopsis thaliana* [112, 113]. In a parallel vein, process-oriented models were devised to establish a causal nexus between histone modifications, gene expression, and phenotypic traits in plants, such as the metabolic pathways governing lycopene biosynthesis and the vernalization response in *A. thaliana* and *Solanum lycopersicum* [77, 114]. Therefore, the decision of whether to induce or repress particular epigenetic variants for the purpose of improving resistance in crop epi-breeding can be led by the utilization of epigenetic modelling, namely the process-based model, in order to investigate the influence that epigenetic variations have on the resistance of plants to disease.

7.7 Conclusions

Decades of research has made it certain that epigenetic variations influence the response of plants to various environmental factors including pathogens. Such variations include changes at the level of DNA methylation, histone modifications, and chromatin remodelling. The change in the epigenome directly affects the transcriptional status of several genes that are involved in defence responses. Further, accumulating evidence is unravelling the molecular underpinnings of defence priming which can be exploited to generate translatable products that could be deployed in the fields. From the breeding point of view, DNA methylation is an attractive epigenetic modification due to its inheritable nature and involvement in defence processes. Despite the available research, epigenomics of plant–pathogen interaction is not fully unveiled. For instance, the evolutionary trajectory of epialleles that are involved in shaping the co-evolution of plants and pathogens is not known. On a similar note, how different pathogens harness the plant epigenetic machinery to successfully promote disease is yet to be elucidated. Another relatively poorly studied avenue is the relation between plant epigenetics and the plant microbiota. Future studies on these areas will deepen our understanding of epigenetics in plant–microbe interaction, thereby informing us of novel strategies of crop breeding for disease resistance.

This chapter has been reproduced with permission from [115].

References

[1] Bruce T J A 2012 GM as a route for delivery of sustainable crop protection *J. Exp. Bot.* **63** 537–41

[2] Nicaise V 2017 Boosting innate immunity to sustainably control diseases in crops *Curr. Opin. Virol* **26** 112–9

[3] Silva M S, Arraes F B M, Campos M D A, Grossi-de-Sa M, Fernandez D, Cândido E D S *et al* 2018 Potential biotechnological assets related to plant immunity modulation applicable in engineering disease-resistant crops *Plant Sci.* **270** 72–84
[4] van der Burgh A M and Joosten M H A J 2019 Plant immunity: thinking outside and inside the box *Trends Plant Sci.* **24** 587–601
[5] Jones J D G and Dangl J L 2006 The plant immune system *Nature* **444** 323–9
[6] Jones J D G, Vance R E and Dangl J L 2016 Intracellular innate immune surveillance devices in plants and animals *Science* **354** aaf6395
[7] Saijo Y, Loo E P and Yasuda S 2018 Pattern recognition receptors and signaling in plant–microbe interactions *Plant* J. **93** 592–613
[8] Yu X, Feng B, He P and Shan L 2017 From chaos to harmony: responses and signaling upon microbial pattern recognition *Annu. Rev. Phytopathol.* **55** 109–37
[9] Adachi H and Tsuda K 2019 Convergence of cell-surface and intracellular immune receptor signalling *New Phytol.* **221** 1676–8
[10] Bjornson M, Pimprikar P, Nürnberger T and Zipfel C 2021 The transcriptional landscape of *Arabidopsis thaliana* pattern-triggered immunity *Nat. Plants* **7** 579–86
[11] Alonso C, Ramos-Cruz D and Becker C 2019 The role of plant epigenetics in biotic interactions *New Phytol.* **221** 731–7
[12] Ding B and Wang G-L 2015 Chromatin versus pathogens: the function of epigenetics in plant immunity *Front. Plant Sci.* **6** 675
[13] Wang C, Wang C, Zou J, Yang Y, Li Z and Zhu S 2019 Epigenetics in the plant–virus interaction *Plant Cell Rep.* **38** 1031–8
[14] Zhu Q H, Shan W X, Ayliffe M A and Wang M B 2016 Epigenetic mechanisms: an emerging player in plant–microbe interactions *Mol. Plant Microbe Interact.* **29** 187–96
[15] Colot V and Rossignol J L 1999 Eukaryotic DNA methylation as an evolutionary device *Bioessays* **21** 402–11
[16] Tirnaz S and Batley J 2019 DNA methylation: toward crop disease resistance improvement *Trends. Plant Sci.* **24** 1137–50
[17] Walker J, Zhang J, Liu Y, Vickers M, Dolan L, Nakajima K *et al* 2021 Extensive N4 cytosine methylation is essential for marchantia sperm function *bioRxiv* 428880
[18] Henderson I R and Jacobsen S E 2007 Epigenetic inheritance in plants *Nature* **447** 418–24
[19] Erdmann R M and Picard C L 2020 RNA-directed DNA methylation *PLoS Genet.* **16** e1009034
[20] Elhamamsy A R 2016 DNA methylation dynamics in plants and mammals: overview of regulation and dysregulation *Cell Biochem. Funct.* **34** 289–98
[21] Hsu H K, Weng Y I, Hsu P Y, Huang T H M and Huang Y W 2020 Detection of DNA methylation by MeDIP and MBDCap assays: an overview of techniques *Molecular Toxicology Protocols Methods in Molecular Biology* vol 2102 ed P Keohavong, K P Singh and W Gao (New York: Springer) pp 225–34
[22] Annacondia M L, Markovic D, Reig-Valiente J L, Scaltsoyiannes V, Pieterse C M J, Ninkovic V *et al* 2021 Aphid feeding induces the relaxation of epigenetic control and the associated regulation of the defense response in *Arabidopsis New Phytol.* **230** 1185–200
[23] Atighi M R, Verstraeten B, De Meyer T and Kyndt T 2020 Genome-wide DNA hypomethylation shapes nematode pattern-triggered immunity in plants *New Phytol.* **227** 545–58

[24] Geng S, Kong X, Song G, Jia M, Guan J, Wang F *et al* 2019 DNA methylation dynamics during the interaction of wheat progenitor *Aegilops tauschii* with the obligate biotrophic fungus *Blumeria graminis* f. sp. *tritici New Phytol.* **221** 1023–35
[25] Sun Y, Fan M and He Y 2019 DNA methylation analysis of the *Citrullus lanatus* response to cucumber green mottle mosaic virus infection by whole-genome bisulfite sequencing *Genes* **10** 344
[26] Ngom B, Sarr I, Kimatu J, Mamati E and Kane N A 2017 Genome-wide analysis of cytosine DNA methylation revealed salicylic acid promotes defense pathways over seedling development in pearl millet *Plant Signal. Behav.* **12** e1356967
[27] Akimoto K, Katakami H, Kim H J, Ogawa E, Sano C M, Wada Y *et al* 2007 Epigenetic inheritance in rice plants *Ann. Bot.* **100** 205–17
[28] Li Y, Xia Q, Kou H, Wang D, Lin X, Wu Y *et al* 2011 Induced pib expression and resistance to *Magnaporthe grisea* are compromised by cytosine demethylation at critical promoter regions in rice *J. Integr. Plant Biol.* **53** 814–23
[29] López Sánchez A, Stassen J H M, Furci L, Smith L M and Ton J 2016 The role of DNA (de)methylation in immune responsiveness of Arabidopsis *Plant* J. **88** 361–74
[30] Annacondia M L, Magerøy M H and Martinez G 2018 Stress response regulation by epigenetic mechanisms: changing of the guards *Physiol. Plant.* **162** 239–50
[31] Leonetti P and Molinari S 2020 Epigenetic and metabolic changes in root-knot nematode–plant interactions *IJMS* **21** 7759
[32] Deng Y, Zhai K, Xie Z, Yang D, Zhu X, Liu J *et al* 2017 Epigenetic regulation of antagonistic receptors confers rice blast resistance with yield balance *Science* **355** 962–5
[33] Imhof A and Wolffe A P 1998 Transcription: gene control by targeted histone acetylation *CurrBiol.* **8** R422–4
[34] Song G and Walley J W 2016 Dynamic protein acetylation in plant–pathogen interactions *Front. Plant Sci.* **7** 421
[35] DeFraia C T, Zhang X and Mou Z 2010 Elongator subunit 2 is an accelerator of immune responses in *Arabidopsis thaliana*: function of AtELP2 in plant immunity *Plant* J. **64** 511–23
[36] Latrasse D, Jégu T, Li H, De Zelicourt A, Raynaud C, Legras S *et al* 2017 MAPK-triggered chromatin reprogramming by histone deacetylase in plant innate immunity *Genome Biol.* **18** 131
[37] Wang C, Gao F, Wu J, Dai J, Wei C and Li Y 2010 Arabidopsis putative deacetylase AtSRT2 regulates basal defense by suppressing PAD4, EDS5 and SID2 expression *Plant Cell Physiol.* **51** 1291–9
[38] Kong L, Liu Y, Wang X and Chang C 2020 Insight into the role of epigenetic processes in abiotic and biotic stress response in wheat and barley *Int. J. Mol. Sci.* **21** 1480
[39] Liu J, Zhi P, Wang X, Fan Q and Chang C 2019 Wheat WD40-repeat protein TaHOS15 functions in a histone deacetylase complex to fine-tune defense responses to *Blumeria graminis* f.sp. *tritici J. Exp. Bot.* **70** 255–68
[40] Li W, Xiong Y, Lai L B, Zhang K, Li Z, Kang H *et al* 2021 The rice RNase P protein subunit Rpp30 confers broad-spectrum resistance to fungal and bacterial pathogens *Plant Biotechnol. J.* **19** 1988–99
[41] Jin P, Gao S, He L, Xu M, Zhang T, Zhang F *et al* 2020 Genome-wide identification and expression analysis of the histone deacetylase gene family in wheat (*Triticum aestivum* L.) *Plants* **10** 19

[42] Kong L, Qiu X, Kang J, Wang Y, Chen H, Huang J *et al* 2017 A phytophthora effector manipulates host histone acetylation and reprograms defense gene expression to promote infection *Curr. Biol.* **27** 981–91

[43] Walley J W, Shen Z, McReynolds M R, Schmelz E A and Briggs S P 2018 Fungal-induced protein hyperacetylation in maize identified by acetylome profiling *Proc. Natl Acad. Sci. USA* **115** 210–5

[44] Wang B, Yang X, Wang Y, Xie Y and Zhou X 2018 Tomato yellow leaf curl virus V2 Interacts with host histone deacetylase 6 to suppress methylation-mediated transcriptional gene silencing in plants *J. Virol.* **92** e00036–18

[45] He M, Xu Y, Chen J, Luo Y, Lv Y, Su J *et al* 2018 MoSnt2-dependent deacetylation of histone H3 mediates MoTor-dependent autophagy and plant infection by the rice blast fungus *Magnaporthe oryzae Autophagy* **14** 1543–61

[46] Chen Y, Wang J, Yang N, Wen Z, Sun X, Chai Y *et al* 2018 Wheat microbiome bacteria can reduce virulence of a plant pathogenic fungus by altering histone acetylation *Nat. Commun.* **9** 3429

[47] Alvarez-Venegas R, Abdallat A A, Guo M, Alfano J R and Avramova Z 2007 Epigenetic control of a transcription factor at the cross section of two antagonistic pathways *Epigenetics* **2** 106–13

[48] Chan C and Zimmerli L 2019 The histone demethylase IBM1 positively regulates arabidopsis immunity by control of defense gene expression *Front Plant Sci.* **10** 1587

[49] Zhang X, Ménard R, Li Y, Coruzzi G M, Heitz T, Shen W H *et al* 2020 Arabidopsis SDG8 potentiates the sustainable transcriptional induction of the pathogenesis-related genes PR1 and PR2 during plant defense response *Front Plant Sci.* **11** 277

[50] Hou Y, Wang L, Wang L, Liu L, Li L, Sun L *et al* 2015 JMJ704 positively regulates rice defense response against *Xanthomonas oryzae* pv. oryzae infection via reducing H3K4me2/3 associated with negative disease resistance regulators *BMC Plant Biol.* **15** 286

[51] Meller B, Kuźnicki D, Arasimowicz-Jelonek M, Deckert J and Floryszak-Wieczorek J 2018 BABA-primed histone modifications in potato for intergenerational resistance to phytophthora infestans *Front. Plant Sci.* **9** 1228

[52] Dhawan R, Luo H, Foerster A M, AbuQamar S, Du H N, Briggs S D *et al* 2009 HISTONE MONOUBIQUITINATION1 interacts with a subunit of the mediator complex and regulates defense against necrotrophic fungal pathogens in *Arabidopsis Plant Cell* **21** 1000–19

[53] Zhao J, Chen Q, Zhou S, Sun Y, Li X and Li Y 2020 H2Bub1 regulates *RbohD*-dependent hydrogen peroxide signal pathway in the defense responses to *Verticillium dahliae* toxins *Plant Physiol.* **182** 640–57

[54] Zou B, Yang D L, Shi Z, Dong H and Hua J 2014 Monoubiquitination of histone 2B at the disease resistance gene locus regulates its expression and impacts immune responses in *Arabidopsis Plant Physiol.* **165** 309–18

[55] Hu M, Pei B L, Zhang L F and Li Y Z 2014 Histone H2B monoubiquitination is involved in regulating the dynamics of microtubules during the defense response to *Verticillium dahliae* toxins in *Arabidopsis Plant Physiol.* **164** 1857–65

[56] Li X, Jiang Y, Ji Z, Liu Y and Zhang Q 2015 BRHIS 1 suppresses rice innate immunity through binding to monoubiquitinated H2A and H2B variants *EMBO Rep.* **16** 1192–202

[57] Li B, Carey M and Workman J L 2007 The role of chromatin during transcription *Cell* **128** 707–19

[58] Muñoz-Viana R, Wildhaber T, Trejo-Arellano M S, Mozgová I and Hennig L 2017 Arabidopsis chromatin assembly factor 1 is required for occupancy and position of a subset of nucleosomes *Plant* J. **92** 363–74
[59] Yang X, Lu Y, Zhao X, Jiang L, Xu S, Peng J *et al* 2019 Downregulation of nuclear protein H2B induces salicylic acid mediated defense against PVX infection in *Nicotiana benthamiana Front. Microbiol.* **10** 1000
[60] Mozgová I, Wildhaber T, Liu Q, Abou-Mansour E, L'H F, Métraux J P *et al* 2015 Chromatin assembly factor CAF-1 represses priming of plant defence response genes *Nat. Plants* **1** 15127
[61] Berriri S, Gangappa S N and Kumar S V 2016 SWR1 chromatin-remodeling complex subunits and H2A.*Z* have non-overlapping functions in immunity and gene regulation in Arabidopsis *Mol. Plant* **9** 1051–65
[62] Huang C Y, Rangel D S, Qin X, Bui C, Li R, Jia Z *et al* 2021 The chromatin-remodeling protein BAF60/SWP73A regulates the plant immune receptor NLRs *Cell Host Microbe* **29** 425–34
[63] Walley J W, Rowe H C, Xiao Y, Chehab E W, Kliebenstein D J, Wagner D *et al* 2008 The chromatin remodeler SPLAYED regulates specific stress signaling pathways *PLoS Pathog.* **4** e1000237
[64] Wang X, Zhi P, Fan Q, Zhang M and Chang C 2019 Wheat CHD3 protein TaCHR729 regulates the cuticular wax biosynthesis required for stimulating germination of *Blumeria graminis* f.sp. *tritici J. Exp. Bot.* **70** 701–13
[65] Reimer-Michalski E M and Conrath U 2016 Innate immune memory in plants *Semin. Immunol.* **28** 319–27
[66] Iwasaki M and Paszkowski J 2014 Epigenetic memory in plants *EMBO J.* **33** 1987–98
[67] Jaskiewicz M, Conrath U and Peterhänsel C 2011 Chromatin modification acts as a memory for systemic acquired resistance in the plant stress response *EMBO Rep.* **12** 50–5
[68] Slaughter A, Daniel X, Flors V, Luna E, Hohn B and Mauch-Mani B 2012 Descendants of primed *Arabidopsis* plants exhibit resistance to biotic stress *Plant Physiol.* **158** 835–43
[69] Ramírez-Carrasco G, Martínez-Aguilar K and Alvarez-Venegas R 2017 Transgenerational defense priming for crop protection against plant pathogens: a hypothesis *Front. Plant Sci.* **8** 696
[70] He Y and Li Z 2018 Epigenetic environmental memories in plants: establishment, maintenance, and reprogramming *Trends Genet* **34** 856–66
[71] Lämke J and Bäurle I 2017 Epigenetic and chromatin-based mechanisms in environmental stress adaptation and stress memory in plants *Genome Biol.* **18** 124
[72] Luna E, Bruce T J A, Roberts M R, Flors V and Ton J 2012 Next-generation systemic acquired resistance *Plant Physiol.* **158** 844–53
[73] Laura B, Silvia P, Francesca F, Benedetta S and Carla C 2018 Epigenetic control of defense genes following MeJA-induced priming in rice (*O. sativa*) *J. Plant Physiol.* **228** 166–77
[74] Sharrock J and Sun J C 2020 Innate immunological memory: from plants to animals *Curr. Opin. Immunol.* **62** 69–78
[75] Zytynska S E, Eicher M, Rothballer M and Weisser W W 2020 Microbial-mediated plant growth promotion and pest suppression varies under climate change *Front. Plant Sci.* **11** 573578
[76] Rodriguez-Moreno L, Song Y and Thomma B P 2017 Transfer and engineering of immune receptors to improve recognition capacities in crops *Curr. Opin. Plant Biol.* **38** 42–9
[77] Gallusci P, Dai Z, Génard M, Gauffretau A, Leblanc-Fournier N, Richard-Molard C *et al* 2017 Epigenetics for plant improvement: current knowledge and modeling avenues *Trends Plant Sci.* **22** 610–23

[78] Latutrie M, Gourcilleau D and Pujol B 2019 Epigenetic variation for agronomic improvement: an opportunity for vegetatively propagated crops *Am. J. Bot* **106** 1281–4
[79] Varotto S, Tani E, Abraham E, Krugman T, Kapazoglou A, Melzer R *et al* 2020 Epigenetics: possible applications in climate-smart crop breeding *J. Exp. Bot.* **71** 5223–36
[80] Chen X and Zhou D X 2013 Rice epigenomics and epigenetics: challenges and opportunities *Curr. Opin. Plant Biol.* **16** 164–9
[81] Cubas P, Vincent C and Coen E 1999 An epigenetic mutation responsible for natural variation in floral symmetry *Nature* **401** 157–61
[82] Liu R, How-Kit A, Stammitti L, Teyssier E, Rolin D, Mortain-Bertrand A *et al* 2015 A DEMETER-like DNA demethylase governs tomato fruit ripening *Proc. Natl Acad. Sci. USA* **112** 10804–9
[83] Long Y, Xia W, Li R, Wang J, Shao M, Feng J *et al* 2011 Epigenetic QTL mapping in *Brassica napus Genetics* **189** 1093–102
[84] Manning K, Tör M, Poole M, Hong Y, Thompson A J, King G J *et al* 2006 A naturally occurring epigenetic mutation in a gene encoding an SBP-box transcription factor inhibits tomato fruit ripening *Nat. Genet.* **38** 948–52
[85] Martin A, Troadec C, Boualem A, Rajab M, Fernandez R, Morin H *et al* 2009 A transposon-induced epigenetic change leads to sex determination in melon *Nature* **461** 1135–8
[86] Telias A, Lin-Wang K, Stevenson D E, Cooney J M, Hellens R P, Allan A C *et al* 2011 Apple skin patterning is associated with differential expression of MYB10 *BMC Plant Biol.* **11** 93
[87] Springer N M and Schmitz R J 2017 Exploiting induced and natural epigenetic variation for crop improvement *Nat. Rev. Genet.* **18** 563–75
[88] Baubec T, Pecinka A, Rozhon W and Mittelsten Scheid O 2009 Effective, homogeneous and transient interference with cytosine methylation in plant genomic DNA by zebularine *Plant* J. **57** 542–54
[89] Griffin P T, Niederhuth C E and Schmitz R J 2016 A comparative analysis of 5-azacytidine- and zebularine-induced DNA demethylation *G3 Genes|Genomes|Genetics* **6** 2773–80
[90] Mirouze M and Paszkowski J 2011 Epigenetic contribution to stress adaptation in plants *Curr. Opin. Plant Biol.* **14** 267–74
[91] Zhang Y Y, Latzel V, Fischer M and Bossdorf O 2018 Understanding the evolutionary potential of epigenetic variation: a comparison of heritable phenotypic variation in epiRILs, RILs, and natural ecotypes of *Arabidopsis thaliana Heredity* **121** 257–65
[92] Kapazoglou A, Ganopoulos I, Tani E and Tsaftaris A 2018 Epigenetics, epigenomics and crop improvement *Advances in Botanical Research* (Amsterdam: Elsevier) pp 287–324
[93] Dalakouras A and Papadopoulou K K 2020 Epigenetic modifications: an unexplored facet of exogenous RNA application in plants *Plants* **9** 673
[94] Kasai M and Kanazawa A 2013 Induction of RNA-directed DNA methylation and heritable transcriptional gene silencing as a tool to engineer novel traits in plants *Plant Biotechnol.* **30** 233–41
[95] Dalakouras A and Ganopoulos I 2021 Induction of promoter DNA methylation upon high-pressure spraying of double-stranded RNA in plants *Agronomy* **11** 789
[96] Hilton I B, D'Ippolito A M, Vockley C M, Thakore P I, Crawford G E, Reddy T E *et al* 2015 Epigenome editing by a CRISPR-Cas9-based acetyltransferase activates genes from promoters and enhancers *Nat. Biotechnol.* **33** 510–7

[97] Kearns N A, Pham H, Tabak B, Genga R M, Silverstein N J, Garber M *et al* 2015 Functional annotation of native enhancers with a Cas9–histone demethylase fusion *Nat. Methods* **12** 401–3

[98] Kungulovski G and Jeltsch A 2016 Epigenome editing: state of the art, concepts, and perspectives *Trends Genet.* **32** 101–13

[99] Liu C and Moschou P N 2018 Phenotypic novelty by CRISPR in plants *Dev. Biol.* **435** 170–5

[100] Vojta A, Dobrinić P, Tadić V, Bočkor L, Korać P, Julg B *et al* 2016 Repurposing the CRISPR-Cas9 system for targeted DNA methylation *Nucleic Acids Res.* **44** 5615–28

[101] Johnson L M, Du J, Hale C J, Bischof S, Feng S, Chodavarapu R K *et al* 2014 SRA- and SET-domain-containing proteins link RNA polymerase V occupancy to DNA methylation *Nature* **507** 124–8

[102] Kolb A F, Coates C J, Kaminski J M, Summers J B, Miller A D and Segal D J 2005 Site-directed genome modification: nucleic acid and protein modules for targeted integration and gene correction *Trends Biotechnol.* **23** 399–406

[103] Woo J W, Kim J, Kwon S I, Corvalán C, Cho S W, Kim H *et al* 2015 DNA-free genome editing in plants with preassembled CRISPR-Cas9 ribonucleoproteins *Nat. Biotechnol.* **33** 1162–4

[104] Metje-Sprink J, Menz J, Modrzejewski D and Sprink T 2019 DNA-Free genome editing: past, present and future *Front. Plant Sci.* **9** 1957

[105] Li Q, Song J, West P T, Zynda G, Eichten S R, Vaughn M W *et al* 2015 Examining the causes and consequences of context-specific differential DNA methylation in maize *Plant Physiol.* **168** 1262–74

[106] Schmitz R J, He Y, Valdés-López O, Khan S M, Joshi T, Urich M A *et al* 2013 Epigenome-wide inheritance of cytosine methylation variants in a recombinant inbred population *Genome Res.* **23** 1663–74

[107] Baulcombe D C and Dean C 2014 Epigenetic regulation in plant responses to the environment *Cold Spring Harb. Perspect. Biol.* **6** a019471

[108] Coustham V, Li P, Strange A, Lister C, Song J and Dean C 2012 Quantitative modulation of polycomb silencing underlies natural variation in vernalization *Science* **337** 584–7

[109] Meyer R S, DuVal A E and Jensen H R 2012 Patterns and processes in crop domestication: an historical review and quantitative analysis of 203 global food crops *New Phytol.* **196** 29–48

[110] Buck-Sorlin G 2013 Process-based model *Encyclopedia of Systems Biology* ed W Dubitzky, O Wolkenhauer, K H Cho and H Yokota (New York: Springer) pp 1755–5

[111] Richards D, Berry S and Howard M 2012 Illustrations of mathematical modeling in biology: epigenetics, meiosis, and an outlook *Cold Spring Harbor Symp. Quant. Biol.* **77** 175–81

[112] Colicchio J M, Miura F, Kelly J K, Ito T and Hileman L C 2015 DNA methylation and gene expression in *Mimulus guttatus BMC Genom.* **16** 507

[113] Hu Y, Morota G, Rosa G J M and Gianola D 2015 Prediction of plant height in *Arabidopsis thaliana* using DNA methylation data *Genetics* **201** 779–93

[114] Angel A, Song J, Dean C and Howard M 2011 A polycomb-based switch underlying quantitative epigenetic memory *Nature* **476** 105–8

[115] Zhi P and Chang C 2021 Exploiting epigenetic variations for crop disease resistance improvement *Front. Plant Sci.* **12** 692328

IOP Publishing

Advances in Biochemical and Molecular Mechanisms of Plant–Pathogen Interaction

Hitendra K Patel and Anirudh Kumar

Chapter 8

Players in action: role of metabolites in plant defence

Lavanya Tayi

Plant growth and development is adversely affected by several biotic and abiotic environmental stress factors. It is estimated that abiotic stresses such as salinity, drought, temperature, harmful UV radiations and trace metals cause the loss of 50% of the crop yield worldwide. Plants respond to these stresses by accumulating metabolites. Primary metabolites, such as carbohydrates, lipids, amino acids and polyamines have a direct role in plant growth and development. Primary metabolites majorly contribute to abiotic stress tolerance by functioning as the osmolytes and osmoprotectants. Secondary metabolites include phenolics, terpenes, nitrogen and sulfur-containing compounds. Secondary metabolites play a significant role in tolerance to biotic stresses such as attacks by pathogens, herbivores and insects. This review provides insights into the roles of primary and secondary metabolites in tolerating various biotic and abiotic stress factors and in plant defence.

8.1 Introduction

Life is a constant struggle. This struggle for existence applies to every living organism on earth from tiny microbes to humans. Even sessile plants are not exempted. They have to struggle and deal with various kinds of abiotic and biotic stress factors. Abiotic stress factors include extreme sunlight, intense wind, high salinity, drought, flood, UV radiation, nutritional imbalances etc, which can drastically affect the plants in the area subjected to stress, while living organisms like bacteria, fungi, insects, parasites etc are responsible for biotic stress in plants (figure 8.1). Because of their sessile lifestyle and absence of an adaptive immune system, plants rely on the innate immunity of each cell to protect themselves.

The plant immune system can be described by a classical four-phased zig-zag model [1]. In phase 1, pathogen/microbe associated molecular patterns (PAMPs/

doi:10.1088/978-0-7503-5673-2ch8

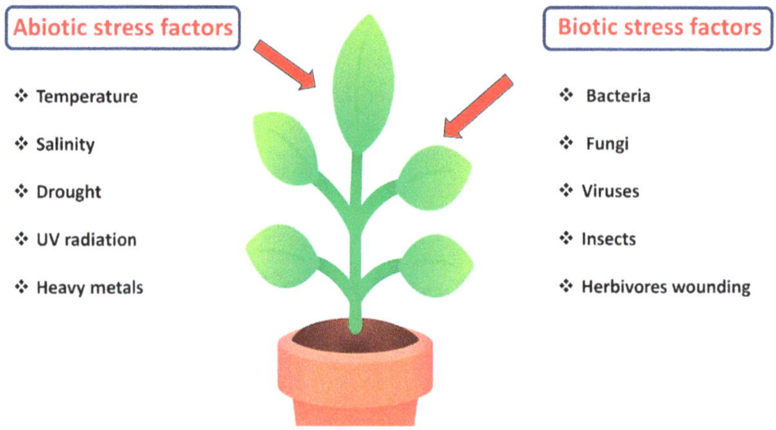

Figure 8.1. Various abiotic and biotic stress factors that affect plant health.

MAMPs) and damage associated molecular patterns (DAMPs) are recognized by host transmembrane pattern recognition receptors (PRRs) to mount PTI (PAMP-triggered immunity) and DTI (DAMP-triggered immunity) which can prevent further colonization by microbes. In phase 2, successful pathogens interfere with PTI by deploying effectors, leading to ETS (effector-triggered susceptibility) in the host. In phase 3, plants mount ETI (effector-triggered immunity) using nucleotide binding leucine rich receptor (NB-LRR) proteins. These are coded by the R genes that specifically recognize the pathogen-deployed effectors either directly or indirectly. The fourth phase is characterized by natural selection. Natural selection drives pathogens to modify the recognized effector gene, or to acquire additional effectors to avoid ETI. Natural selection can also lead to new *R* specificities in plants so that ETI can be triggered again. Thus, the key players in plant immunity are the transmembrane PRRs and the cytoplasmic R gene encoded NB-LRR proteins [1].

Plants have evolved to use various metabolite molecules to protect themselves from the environmental stresses to which they are exposed. Metabolites are broadly classified into two types: primary and secondary metabolites that are different in structure, function and their distribution in plants. Primary metabolites are important for normal plant growth and development [2]. These include carbohydrates, amino acids and proteins, lipids, etc that are essential for the proper functioning of the cell. Secondary metabolites, although they play no role in basic processes of the cell, are important in adaptation to biotic stress factors such as herbivores, insects and pathogens besides enhancing the resistance against abiotic stress factors. The secondary metabolites include alkaloids, terpenes, quinones, phenolics, nitrogen-containing compounds etc [2, 3]. Primary metabolites serve as the precursors for the production of the secondary metabolites suggesting that despite having different functional roles both primary and secondary metabolites are linked to each other [4] (figure 8.2).

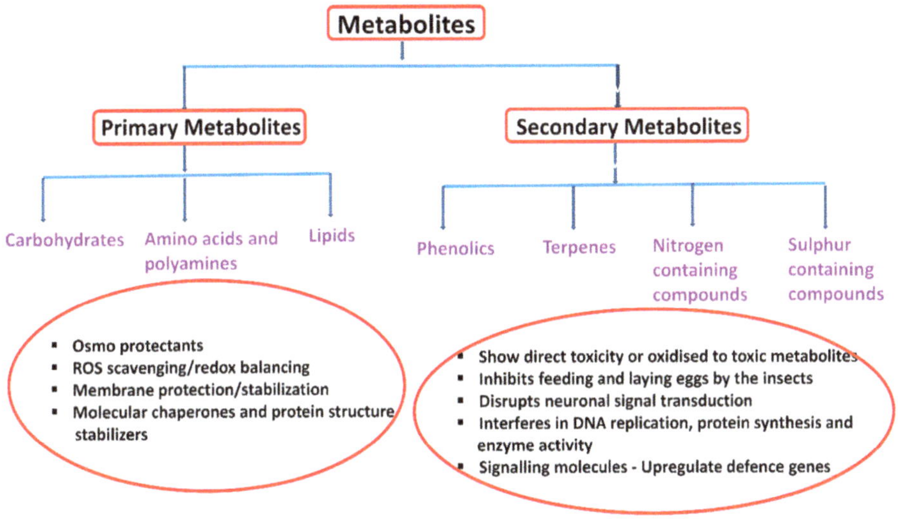

Figure 8.2. Role of different types of plant metabolites in mitigating abiotic and biotic stress factors.

8.2 Primary metabolism during defence

Primary metabolic pathways during resistance responses to avirulent pathogens are excellently reviewed in [5]. Plant defence responses are energetically costly. For example, in *Arabidopsis*, it was found that within 10 min of inoculation of avirulent strain of *P. syrigae* pv. *tomato* hundreds of genes were differentially expressed to clear the infection, showing increased demand for energy and biosynthetic capacity [6]. Plants during defence responses display compromised growth and fitness [5]. *Arabidopsis* mutants that show constitutive production of salicylic acid, a defence hormone or the mutants in which there is a continuous expression of defence genes show a decrease in growth and reproduction, while the mutants that are suppressed for induced resistance signalling show increased fitness [7]. Similarly, infection of avirulent strains of powdery mildew on barley led to reduced protein content and seed weight [8].

8.2.1 Photosynthesis

Studies on photosynthesis and plant defence in several plants such as tomato, maize, barley, *Arabidopsis*, tobacco have shown decreased rates of photosynthesis after infection with virulent or avirulent pathogens or application of elicitins [9–17]. Wounding or herbivore attack and the hormone treatment causes reprogramming of the primary metabolism including the photosynthesis [18–21].

8.2.2 Carbohydrates metabolism and respiration

During a defence response there is an increased cellular demand of sugars to meet the energy requirements. Upon infection, there is an increase in the apoplastic sucrose levels and an increase in the cell wall invertase gene expression and activity [22]. The invertase functions to cleave sucrose into hexoses that are immediately transported into the plant cell via the monosaccharide transporters to meet the energy requirements [23]. It was observed in tobacco leaves that in response to avirulent *Phytophthora nicotianae* infection the callose depositions blocks the sucrose export [15]. It was speculated that this callose deposition along with increased invertase activity, may help in retaining the carbohydrates in plant cells that are actively utilized in mounting the defence response [15]. The carbohydrate increase, signals the expression of defence genes and suppression of photosynthesis [9, 24–26].

Glycolysis, TCA cycle, and the mitochondrial electron transport are the three main energy-yielding pathways of plant respiration [27]. These pathways also supply the carbon skeletons for the biosynthesis of various metabolites. Respiration rates are known to be increased during defence responses [8]. For example, the PFK (phospho fructo kinase), the main regulatory enzyme of glycolysis was found to be upregulated in wheat in response to *Puccinia triticina* [28]. Similarly, upregulation of rate-limiting enzymes citrate synthase or α-ketoglutarate dehydrogenase of TCA cycle (both considered to be rate-limiting enzymes) was observed during resistance responses [29, 30], Increased levels of metabolites of oxidative phase of pentose phosphate pathway were demonstrated at the infection site during *P. nicotianae* infection of tobacco [15].

Besides these key metabolic pathways, the energy-yielding glyoxalate cycle and malate metabolism were also upregulated during the defence response [6, 31, 32].

8.2.3 Nitrogen, amino acid and fatty acid metabolism

During infections, amino acids are shuttled into energy-yielding pathways. All the 20 amino acids can be converted to intermediates of TCA cycle. It was observed that plants deprive pathogens of nitrogen by mobilizing nitrogen away from the infection sites [33, 34]. Also, there is upregulation of several genes involved in nitrogen mobilization during infection [33, 35, 36]. Nitrogen in the form of nitric oxide (NO) a reactive nitrogen species is directly toxic to invading microbes and together with reactive oxygen species (ROS), induce various defence-related processes [37, 38]. The β-oxidation of fatty acids pathway is upregulated during the resistance response to several pathogens and thus contributes energy during defence [28, 39].

Thus, during resistance responses in plants, the primary metabolic pathways are highly activated to meet the increased cellular energy demands.

8.3 Role of primary metabolites in abiotic stress tolerance

Both biotic and abiotic factors badly affect the growth and development of crops leading to reduced yield. It is estimated that abiotic factors such as salinity, drought, temperature, ultraviolet radiation, trace metals, and soil pH are the reason for the loss of 50% of crop production [2]. Metabolites such as sugars and sugar alcohols (sucrose, trehalose, sorbitol, and mannitol), amino acids (proline, ectoine) and lipids have been proved to be the contributing factors for combating abiotic stresses. They are effective in reducing the effects of salinity, drought and water stresses by mainly functioning as the osmolytes [40].

8.3.1 Amino acids and polyamines

Amino acids play an active role upon plant exposure to several abiotic stress factors [41, 42]. Amino acids function as an osmolyte to adjust ions passage, reduce stomatal opening, alter gene expression and synthesis and activity of several enzymes [42] Amino acid proline and ectoine act as osmoprotectants to protect plants from water, temperature, salinity, and UV radiation stresses [43, 44]. Besides functioning as an osmolyte, proline performs several defence-related roles as an ROS scavenger, redox balancer, as a molecular chaperon and as a protein structure stabilizer [45]. Overexpression of the P5CS (pyrroline-5-carboxylate synthetase) the proline synthesis gene of soybean in transgenic plants increased proline content and conferred salinity stress tolerance [46].

Polyamines are positively charged organic compounds with more than two amino groups that interact with negatively charged nucleic acids, phospholipids, and protein molecules [47]. Triamine tetraamine, spermine, spermidine, and putrescine are the common polyamines [48]. Polyamines are not only essential for plant growth and development but are also required for abiotic stress tolerance in higher plants [49, 50]. Overexpression of spermidine synthase in transgenic *Arabidopsis* improved waterlogging and salinity stresses tolerance [51]. Overexpression of arginine decarboxylase gene of polyamine biosynthesis pathway provides tolerance to drought in transgenic *Arabidopsis* and rice, respectively [52, 53]. Under stress conditions polyamines prevent membrane damage and regulate the antioxidant system and suppress ROS production [54, 55].

8.3.2 Carbohydrates

Sugars play a critical role in tolerance to salt, cold and drought stress [56]. The concentration of sugars like glucose, fructose and sucrose increases under high salt conditions and they play a variety of roles in carbon storage, osmoprotection and in scavenging of free radicals [57]. It is reported in rice that glucose and fructose function as *osmoprotectants* and free radical scavengers under salinity [58]. Hu *et al* have shown that, exogenous application of glucose to wheat seedlings under salt stress prevents water loss, causes proline accumulation, maintains ion homeostasis, and activates antioxidant enzyme activity [59]. Under drought stress, photosynthesis is inhibited significantly [60]. Drought stress induces stomatal closure mediated by abscisic acid (ABA) and enhances the plant's adaptability [61]. In several plant species, during seed desiccation accumulation of sugars like raffinose, stachyose and verbascose was observed [62, 63]. It is reported that transgenic rice overexpressing trehalose are tolerant to abiotic stresses like salt and drought [64]. *Cold stress* causes membranes disruption, ROS accumulation, *protein denaturation*, etc [65, 66]. Soluble sugars, are shown to provide freezing tolerance in higher plants [65]. Similarly, water soluble *fructans* provides resistance to drought and chilling stress through membrane stabilization by inserting the polysaccharide chain into the lipid bilayer and thus preventing water leakage[67].

8.3.3 Lipids

Lipids of the membrane serve as signalling molecules. Fatty acids, sphingolipids, lysophospholipids, triacylglycerol, etc are the major signalling lipid molecules [68]. Lipid mediated signalling occurs in plants in response to a variety of abiotic stresses such as low-temperature, drought and salinity and during biotic stress as well [69–72].

Fatty acids are identified as the stress-responsive lipids in plants and are shown to regulate salt, drought, heavy metals tolerance and also wound-induced responses of pathogens or herbivores in plants [73]. Oleic acids were demonstrated to promote tolerance in *Arabidopsis* through NO [74]. In plant tissues under stressed conditions such as freezing, the concentrations of phosphatidic acid and lysophospholipids increase, which are generally present in very low levels [75]. The responsive roles lipids such as phosphatidic acid (PA), inositol polyphosphates, oxylipins, sphingolipids, and others in mitigating stress tolerance are also studied in various plant species [76, 77].

8.4 Role of primary metabolites in biotic stress tolerance

Primary metabolites are involved in mounting defence responses to bacterial, fungal and viral pathogens. PR proteins form the first line of defence in plants. In healthy plant tissues, PR proteins are generally present in low concentrations but upon infection their levels increase significantly. PR proteins are the best weapons against bacterial pathogens and are therefore used to develop bacterial resistant plants [78]. PR proteins as antimicrobial agents inhibit pathogen growth and spore germination by functioning as hydrolases and proteinase inhibitors [79, 80]. So far multiple families (about 20) of PR proteins have been identified, including peroxidases (PR-9), chitinases (PR-3, PR-8, PR-11), β-1,3-glucanases (PR-2), ribonucleases(PR-10), defensins and so on [80–84].

Degradation of cell walls of microorganisms by plant β-1,3 glucanases and chitinases releases oligomers that are sensed as elicitors leading to induction of defence responses. Upon detection of pathogen elicitors or abiotic stress, there is a production of ROS leading to oxidative burst. Plant peroxidases function in cellular detoxification and ROS removal and are therefore important in maintaining the redox homeostasis of plant cells [85]. Plant peroxidases also participate in lignification, cell elongation, promote cell wall strengthening through macromolecules polymerization and their deposition at cell surface, wound healing and oxidation of polyphenols [86]. PR10 family proteins are necessary to reduce the damage caused to plants by biotic and abiotic stress. PR10 proteins exhibit ribonuclease activity. The ribonucleolytic activity of PR10 proteins causes cleavage of pathogens RNA and thus inhibits or destroys pathogens. The ribonuclease activity of PR-10 proteins on bacteria, fungi and viruses is well established [87, 88]. Besides ribonuclease activity, PR-10 protein showed protease inhibitory activity against the root-knot nematode *Meloidogyne incognita* [89]. PR10 proteins also participate in plant programmed cell death or apoptotic processes [90, 91].

Defence-related proteins were found to play an effective role in defence against viral infections by inhibiting viral replication [92]. The beetin 27 protein is effective against virus attack in sugar beet besides being effective against several pathogens [93]. Several such defensive-related proteins were reported to be effective against tobacco mosaic virus (TMV), papaya ring spot virus (PRSV) etc. The *CCP25* and *27* proteins isolated from the *Celosia cristata* extracts are effective against TMV [94] and extract of *Clerodendrum aculeatum* with *CAP-34* defence-related protein was found to inhibit PRSV [95].

8.5 Secondary metabolites and response to abiotic and biotic stresses

Secondary metabolites serve as a protective barrier to shield plants from biotic stresses, particularly those that are caused by phytophagous herbivores, in contrast to primary metabolites, which are directly engaged in plant growth, development, and reproduction [96]. Secondary metabolites belong to the following major groups: phenolics, terpenes, sulfur-containing compounds and nitrogen comprising compounds [2, 97]. Biosynthesis of secondary metabolites is a very complicated but a coordinated process wherein primary metabolites serve as precursors for the secondary metabolites [97]. The phenolic compounds are synthesized in the shikimate pathway. The nitrogen and sulfur-containing compounds are produced from the tricarboxylic acid cycle pathway. Terpenes are synthesized via the mevalonic phosphate pathway (MVP) or methylerythritol phosphate pathway (MEP) [97].

8.5.1 Phenolics

Phenols are the most widely distributed secondary metabolites that include simple phenols, flavonoids, isoflavonoids, anthocyanins, lignins, stilbenes, tannins, coumarins etc. Phenolic compounds are produced by the phenylpropanoid pathway which is regulated by both abiotic and biotic stress factors. Plants cope with the oxidative stress by either preventing the production of ROS or by producing antioxidant enzymes through the activities of enzymes like PAL (phenyl ammonia lyase), CHS (chalcone synthase) and other enzymes [98, 99]. Under drought conditions, fluctuations in the phenolics of cell wall inhibit wall extension and root growth enabling maize roots to acclimatize to drought [100]. The reduced ferulic acid and increased p-coumaric acid and caffeic acid contents in maize xylem sap, were speculated to be responsible for lignification and thus hardening of the cell wall under drought conditions [101].

Flavonoids with their antioxidant properties improve resistance and protect plants from water-deficit situations. Nakabayashi *et al* have reported that flavonoids significantly improved resistance to water scare conditions in *Arabidopsis* [102, 103]. In *Hypericum brasiliense* a herbaceous plant, drought and waterlogging conditions increased the content of quercetin and rutin flavonoids [104]. In citrus under water-stressed conditions the flavonoid content was shown to be critical in maintaining the antioxidant activity [105].

Studies have shown that different polyphenols were associated with drought tolerance in potato plants [106]. Polyphenols aid in drought stress tolerance by maintaining osmotic potential and seizing free radicals in cells. The levels of anthocyanins were found to be increased in plant tissues under drought and cold stress conditions to provide protection to plant cells through their antioxidant and ROS scavenging properties [107, 108]. Under low-temperature stress, phenolics accumulate as suberin or lignin in the cell walls to increase the resistance to cold stress [109, 110].

Phenolics are also actively involved in defending plants against herbivores. They can inhibit insects growth and developmental processes by functioning as direct toxins to herbivores or can be converted to toxic metabolites by peroxidases or polyphenol oxidases [111].

8.5.2 Terpenes

Terpenes form the largest group of secondary metabolites in plants and are produced from isoprene units [112]. As phytohormones, terpenoids play an important role in dealing with biotic and abiotic stresses. The phytohormone ABA, which is a sesquiterpenoid, functions in heavy metal stress mitigation [113]. In response to abiotic and biotic factors, terpenoids also function as phytoalexins during plant defence. UV radiation induced the accumulation of phytoalexins in rice leaves [114]. Phytoalexins are also accumulated in maize roots in response to drought stress [115]. It was reported in *Brassica juncea* that the total tocopherol levels increased under salt, heavy metal and osmotic stress situations and the γ-TMT (γ-tocopherol methyl transferase) overexpressing transgenic *Brassica juncea* plants showed enhanced tolerance [115]. Saponins in soybean plants were identified to be crucial for salt stress resistance [116].

Terpenes as phytoalexins majorly help in plant defence, especially against herbivores either by their direct toxicity or by inhibiting feeding by the herbivores. Multiple diterpene phytoalexins with herbicidal and antifungal activities are reported in rice [117]. Sesquiterpenoid phytoalexins functions against pathogens and herbivores in cotton plants [118]. However, some herbivores utilize these plant terpenoids for locating their host plants. Some terpenoids serve as feeding stimulants for herbivores and as attractants for insects and help in pollination [119].

8.5.3 Nitrogen-containing secondary metabolites

Alkaloids are secondary metabolites with one or more nitrogen atoms [120]. Alkaloids display a defensive role under abiotic and biotic stress (herbivores) conditions. It was observed that poppy plants make more alkaloids under drought and salinity stress [121]. There are reports of alkaloids from primitive plant genera, angiosperms, and gymnosperms [122]. Alkaloids are further divided into true alkaloids, pseudo-alkaloids and proto-alkaloids based on their mode of synthesis. Pseudo-alkaloids are not made from amino acids, in contrast to true alkaloids and proto-alkaloids [123]. Alkaloid toxicity to herbivores is because of their interference in DNA replication, protein synthesis, enzyme activity and disruption of neuronal signal transduction [124].

8.5.4 Sulfur-containing secondary metabolites

Glucosides are the sulfur-containing secondary metabolites that are derived from glucose and amino acids. Four groups of glucosides that include aliphatic glucosinolates, indole glucosinolates, aromatic glucosinolates and glucosinolates are reported [125]. Glucosides play a significant role against different biotic and abiotic stresses. Increased level of glucosinolates was observed in *A. thaliana* and in *B. juncea* under waterlogging and water-deficit conditions, respectively [126, 127].

Upon injury by herbivores, glucosinolates present in plant cell vacuoles are broken down by myrosinases (enzymes that cover the glucosinolates) into toxic metabolites that function as effective synthetic insecticides. They are extremely toxic to herbivorous insects and repel them from feeding [128, 129].

8.5.5 Secondary metabolites as regulators of plant defence

Multiple mutational studies have provided evidence that secondary metabolites are key regulators of plant defence. *Arabidopsis* mutants defective in indole glucosinolate biosynthesis and benzoxazinoid-deficient *bx1* maize mutants could not mount callose, a hallmark defence response [130–132]. Benzoxazinoids-mediated callose response is conserved in wheat and maize, but not in *Arabidopsis*, whereas glucosinolates show callose deposition in *Arabidopsis* but not in maize [133], indicating a highly specific regulation of callose by secondary metabolites. Glucosinolates and benzoxazinoids may promote callose production by either regulating hormonal pathways through transcriptional regulation or by directly initiating callose formation post-translationally [134–136]. Glucosinolates and benzoxazinoids also regulate the production of other secondary metabolites [133, 136, 137]. *Arabidopsis* mutants deficient in indole glucosinolate production showed lower accumulation of the phenylpropanoid sinapoylmalate [136], whereas myrosinase defective mutants release lower amounts of camalexin (a tryptophan-derived metabolite) upon flg22 treatment and infection by *Pseudomonas syringae* [138, 139].

Besides glucosinolates and benzoxazinoids, other secondary metabolites such as terpenoids, aromatic compounds and green-leaf volatiles, also regulate plant defences [140–143]. Many of these metabolites that are released upon herbivore or pathogen attack are capable of priming or directly inducing hormonal signalling pathways and resistance in plants. Maize mutants defective in producing volatile indole are unable to release terpenes in systemic tissues upon herbivore attack. This phenotype is restored by indole addition [142]. Similarly, *Arabidopsis* mutants defective in systemic acquired resistance against *P. syringae* were restored by addition of pathogen-induced volatiles α- and β-pinene to the headspace of the mutant [144]. Transgenic rice plants deficient in the expression of map kinase *OsMPK3* are not responsive to indole [145]. Evidence for the volatiles functioning as suppressors of defence responses comes from *LOX2*-silenced *Nicotiana attenuata* plants. These plants are deficient in the production of herbivory-induced, green-leaf volatiles induced expression of defence-related genes more in neighbours than in the wild-type plants, suggesting that volatiles can also suppress defence [146].

Thus, taken together various secondary metabolites such as glucosinolates, benzoxazinoids, terpenes, aromatic compounds, and green-leaf volatiles function as potential regulators of defence in plants.

8.6 Conclusions

Plants being sessile depend upon metabolites for protecting themselves from various environmental stresses. In the current review, we have provided an overview of the role of primary metabolites and secondary metabolites against multiple abiotic and biotic factors. The primary metabolites such as carbohydrates, amino acids and lipids greatly protect plants from abiotic stresses by functioning as osmolytes and by maintaining redox homeostasis of the cell. Overexpression of pathogenesis-related proteins (primary metabolite) in plants is a useful tool to combat biotic stress by bacterial and fungal pathogens. Secondary metabolites that include alkaloids, terpenes, phenolics and glycosides are key players of defence in plant–herbivore interaction. Some of the secondary metabolites can be utilized as an alternative to pesticides as they repel insects and impair their digestion. For using secondary metabolites in pest management in an eco-friendly manner it is important to understand their biosynthesis processes. Large-scale production of secondary metabolites can be achieved by using efficient biotechnological tools such as metabolic engineering and *in vitro* plant cell cultures. Overall, a clear understanding of the role of primary and secondary metabolites in mitigating various types of environmental stresses is necessary for improving the stress tolerance of plants and thus eventually in preventing huge yield losses of crop plants.

Acknowledgements

Lavanya Tayi acknowledges Inspire Faculty Award (DST/INSPIRE/04/2017/002275) from Department of Science and Technology, Ministry of Science and Technology, Government of India.

References

[1] Jones J D G and Dangl J L 2006 The plant immune system *Nature* **444** 323–9

[2] Salam U, Ullah S, Tang Z-H, Elateeq A A, Khan Y, Khan J, Khan A and Ali S 2023 Plant metabolomics: an overview of the role of primary and secondary metabolites against different environmental stress factors *Life* **13** 706

[3] Jamwal K, Bhattacharya S and Puri S 2018 Plant growth regulator mediated consequences of secondary metabolites in medicinal plants *J. Appl. Res. Med. Aromat. Plants* **9** 26–38

[4] War A R, Buhroo A A, Hussain B, Ahmad T, Nair R M and Sharma H C 2020 Plant defense and insect adaptation with reference to secondary metabolites *Co-Evolution of Secondary Metabolites* (Springer) pp 795–822

[5] Bolton M D 2009 Primary metabolism and plant defense—fuel for the fire *Mol. Plant–Microbe Interact.* **22** 487–97

[6] Scheideler M, Schlaich N L, Fellenberg K, Beissbarth T, Hauser N C, Vingron M, Slusarenko A J and Hoheisel J D 2002 Monitoring the switch from housekeeping to

pathogen defense metabolism in *Arabidopsis thaliana* using cDNA arrays *J. Biol. Chem.* **277** 10555–61
[7] Heil M and Baldwin I T 2002 Fitness costs of induced resistance: emerging experimental support for a slippery concept *Trends Plant Sci.* **7** 61–7
[8] Smedegaard-Petersen V and Stolen O 1981 Effect of energy-requiring defense reactions on yield and grain quality in a powdery mildew-resistant barley cultivar *Phytopathology* **71** 396–9
[9] Berger S, Papadopoulos M, Schreiber U, Kaiser W and Roitsch T 2004 Complex regulation of gene expression, photosynthesis and sugar levels by pathogen infection in tomato *Physiol. Plant.* **122** 419–28
[10] Berger S, Benediktyová Z, Matouš K, Bonfig K, Mueller M J, Nedbal L and Roitsch T 2007 Visualization of dynamics of plant–pathogen interaction by novel combination of chlorophyll fluorescence imaging and statistical analysis: differential effects of virulent and avirulent strains of *P. syringae* and of oxylipins on *A. thaliana J. Exp. Bot.* **58** 797–806
[11] Bonfig K B, Schreiber U, Gabler A, Roitsch T and Berger S 2006 Infection with virulent and avirulent *P. syringae* strains differentially affects photosynthesis and sink metabolism in Arabidopsis leaves *Planta* **225** 1–12
[12] Doehlemann G, Wahl R, Horst R J, Voll L M, Usadel B, Poree F, Stitt M, Pons-Kühnemann J, Sonnewald U and Kahmann R 2008 Reprogramming a maize plant: transcriptional and metabolic changes induced by the fungal biotroph Ustilago maydis *Plant J.* **56** 181–95
[13] Horst R J, Engelsdorf T, Sonnewald U and Voll L M 2008 Infection of maize leaves with *Ustilago maydis* prevents establishment of C4 photosynthesis *J. Plant Physiol.* **165** 19–28
[14] Manter D K, Kelsey R G and Karchesy J J 2007 Photosynthetic declines in Phytophthora ramorum-infected plants develop prior to water stress and in response to exogenous application of elicitins *Phytopathology* **97** 850–6
[15] Scharte J, Schön H and Weis E 2005 Photosynthesis and carbohydrate metabolism in tobacco leaves during an incompatible interaction with *Phytophthora nicotianae Plant Cell Environ.* **28** 1421–35
[16] Swarbrick P J, Schulze-Lefert P and Scholes J D 2006 Metabolic consequences of susceptibility and resistance (race-specific and broad-spectrum) in barley leaves challenged with powdery mildew *Plant Cell Environ.* **29** 1061–76
[17] Zou J, Rodriguez-Zas S, Aldea M, Li M, Zhu J, Gonzalez D O, Vodkin L O, DeLucia E and Clough S J 2005 Expression profiling soybean response to *Pseudomonas syringae* reveals new defense-related genes and rapid HR-specific downregulation of photosynthesis *Mol. Plant–Microbe Interact.* **18** 1161–74
[18] Hermsmeier D, Schittko U and Baldwin I T 2001 Molecular interactions between the specialist herbivore *Manduca sexta* (Lepidoptera, Sphingidae) and its natural host *Nicotiana attenuata*. I. Large-scale changes in the accumulation of growth- and defense-related plant mRNAs *Plant Physiol.* **125** 683–700
[19] Schröder R, Forstreuter M and Hilker M 2005 A plant notices insect egg deposition and changes its rate of photosynthesis *Plant Physiol.* **138** 470–7
[20] Schwachtje J and Baldwin I T 2008 Why does herbivore attack reconfigure primary metabolism? *Plant Physiol.* **146** 845–51
[21] Reinbothe S, Mollenhauer B and Reinbothe C 1994 JIPs and RIPs: the regulation of plant gene expression by jasmonates in response to environmental cues and pathogens *Plant Cell* **6** 1197

[22] Roitsch T, Balibrea M, Hofmann M, Proels R and Sinha A 2003 Extracellular invertase: key metabolic enzyme and PR protein *J. Exp. Bot.* **54** 513–24
[23] Truernit E, Schmid J, Epple P, Illig J and Sauer N 1996 The sink-specific and stress-regulated Arabidopsis STP4 gene: enhanced expression of a gene encoding a monosaccharide transporter by wounding, elicitors, and pathogen challenge *Plant Cell* **8** 2169–82
[24] Chou H M, Bundock N, Rolfe S A and Scholes J D 2000 Infection of *Arabidopsis thaliana* leaves with *Albugo candida* (white blister rust) causes a reprogramming of host metabolism *Mol. Plant Pathol.* **1** 99–113
[25] Kocal N, Sonnewald U and Sonnewald S 2008 Cell wall-bound invertase limits sucrose export and is involved in symptom development and inhibition of photosynthesis during compatible interaction between tomato and *Xanthomonas campestris* pv vesicatoria *Plant Physiol.* **148** 1523–36
[26] Sinha A K, Hofmann M G, Römer U, Köckenberger W, Elling L and Roitsch T 2002 Metabolizable and non-metabolizable sugars activate different signal transduction pathways in tomato *Plant Physiol.* **128** 1480–9
[27] Fernie A R, Carrari F and Sweetlove L J 2004 Respiratory metabolism: glycolysis, the TCA cycle and mitochondrial electron transport *Curr. Opin. Plant Biol.* **7** 254–61
[28] Bolton M D, Kolmer J A, Xu W W and Garvin D F 2008 Lr34-mediated leaf rust resistance in wheat: transcript profiling reveals a high energetic demand supported by transient recruitment of multiple metabolic pathways *Mol. Plant–Microbe Interact.* **21** 1515–27
[29] Strumilo S 2005 Short-term regulation of the α-ketoglutarate dehydrogenase complex by energy-linked and some other effectors *Biochemistry* **70** 726–9
[30] Wiegand G and Remington S J 1986 Citrate synthase: structure, control, and mechanism *Ann. Rev. Biophys. Biophys. Chem.* **15** 97–117
[31] Casati P, Drincovich M F, Edwards G E and Andreo C S 1999 Malate metabolism by NADP-malic enzyme in plant defense *Photosynth. Res.* **61** 99–105
[32] Zulak K G, Khan M F, Alcantara J, Schriemer D C and Facchini P J 2009 Plant defense responses in opium poppy cell cultures revealed by liquid chromatography-tandem mass spectrometry proteomics *Mol. Cell. Proteomics* **8** 86–98
[33] Tavernier V, Cadiou S, Pageau K, Laugé R, Reisdorf-Cren M, Langin T and Masclaux-Daubresse C 2007 The plant nitrogen mobilization promoted by *Colletotrichum lindemuthianum* in Phaseolus leaves depends on fungus pathogenicity *J. Exp. Bot.* **58** 3351–60
[34] Newingham B A, Callaway R M and BassiriRad H 2007 Allocating nitrogen away from a herbivore: a novel compensatory response to root herbivory *Oecologia* **153** 913–20
[35] Pageau K, Reisdorf-Cren M, Morot-Gaudry J-F and Masclaux-Daubresse C 2006 The two senescence-related markers, GS1 (cytosolic glutamine synthetase) and GDH (glutamate dehydrogenase), involved in nitrogen mobilization, are differentially regulated during pathogen attack and by stress hormones and reactive oxygen species in *Nicotiana tabacum* L. leaves *J. Exp. Bot.* **57** 547–57
[36] Stephenson S-A, Green J R, Manners J M and Maclean D J 1997 Cloning and characterisation of glutamine synthetase from *Colletotrichum gloeosporioides* and demonstration of elevated expression during pathogenesis on *Stylosanthes guianensis Curr. Genet.* **31** 447–54
[37] Zaninotto F, Camera S L, Polverari A and Delledonne M 2006 Cross talk between reactive nitrogen and oxygen species during the hypersensitive disease resistance response *Plant Physiol.* **141** 379–83

[38] Romero-Puertas M C, Perazzolli M, Zago E D and Delledonne M 2004 Nitric oxide signalling functions in plant–pathogen interactions *Cell. Microbiol.* **6** 795–803

[39] Schenk P M, Kazan K, Manners J M, Anderson J P, Simpson R S, Wilson I W, Somerville S C and Maclean D J 2003 Systemic gene expression in Arabidopsis during an incompatible interaction with *Alternaria brassicic*ola *Plant Physiol.* **132** 999–1010

[40] Teklić T, Parađiković N, Špoljarević M, Zeljković S, Lončarić Z and Lisjak M 2021 Linking abiotic stress, plant metabolites, biostimulants and functional food *Ann. Appl. Biol.* **178** 169–91

[41] Hildebrandt T M 2018 Synthesis versus degradation: directions of amino acid metabolism during Arabidopsis abiotic stress response *Plant Mol. Biol.* **98** 121–35

[42] Khan N, Ali S, Zandi P, Mehmood A, Ullah S, Ikram M, Ismail I and Babar M 2020 Role of sugars, amino acids and organic acids in improving plant abiotic stress tolerance *Pak. J. Bot.* **52** 355–63

[43] Dikilitas M, Simsek E and Roychoudhury A 2020 Role of proline and glycine betaine in overcoming abiotic stresses, protective chemical agents in the amelioration of plant abiotic stress: biochemical and molecular perspectives *Protective Chemical Agents in the Amelioration of Plant Abiotic Stress: Biochemical and Molecular Perspectives* (Wiley) ch 1 pp 1–23

[44] Verbruggen N and Hermans C 2008 Proline accumulation in plants: a review *Amino Acids* **35** 753–9

[45] Meena M, Divyanshu K, Kumar S, Swapnil P, Zehra A, Shukla V, Yadav M and Upadhyay R 2019 Regulation of L-proline biosynthesis, signal transduction, transport, accumulation and its vital role in plants during variable environmental conditions *Heliyon* **5** e02952

[46] Kishor P K, Hong Z, Miao G-H, Hu C-A A and Verma D P S 1995 Overexpression of [delta]-pyrroline-5-carboxylate synthetase increases proline production and confers osmotolerance in transgenic plants *Plant Physiol.* **108** 1387–94

[47] Fischer W, Calderón M and Haag R 2010 Hyperbranched polyamines for transfection *Nucleic Acid Transfection* (Springer) pp 95–129

[48] Chen D, Shao Q, Yin L, Younis A and Zheng B 2019 Polyamine function in plants: metabolism, regulation on development, and roles in abiotic stress responses *Front. Plant Sci.* **9** 1945

[49] Liu J-H, Wang W, Wu H, Gong X and Moriguchi T 2015 Polyamines function in stress tolerance: from synthesis to regulation *Front. Plant Sci.* **6** 827

[50] Alcázar R, Bueno M and Tiburcio A F 2020 Polyamines: small amines with large effects on plant abiotic stress tolerance *Cells* **9** 2373

[51] Kasukabe Y, He L, Nada K, Misawa S, Ihara I and Tachibana S 2004 Overexpression of spermidine synthase enhances tolerance to multiple environmental stresses and up-regulates the expression of various stress-regulated genes in transgenic *Arabidopsis thaliana Plant Cell Physiol.* **45** 712–22

[52] Ran J, Shang C, Mei L, Li S, Tian T and Qiao G 2022 Overexpression of CpADC from Chinese cherry (*Cerasus pseudocerasus* Lindl.'Manaohong') promotes the ability of response to drought in *Arabidopsis thaliana Int. J. Mol. Sci.* **23** 14943

[53] Capell T, Bassie L and Christou P 2004 Modulation of the polyamine biosynthetic pathway in transgenic rice confers tolerance to drought stress *Proc. Natl. Acad. Sci.* **101** 9909–14

[54] Zhao J, Wang X, Pan X, Jiang Q and Xi Z 2021 Exogenous putrescine alleviates drought stress by altering reactive oxygen species scavenging and biosynthesis of polyamines in the seedlings of *Cabernet Sauvignon Front. Plant Sci.* **12** 767992

[55] Gill S S and Tuteja N 2010 Polyamines and abiotic stress tolerance in plants *Plant Signal. Behav.* **5** 26–33

[56] Sami F, Yusuf M, Faizan M, Faraz A and Hayat S 2016 Role of sugars under abiotic stress *Plant Physiol. Biochem.* **109** 54–61

[57] Rosa M, Prado C, Podazza G, Interdonato R, González J A, Hilal M and Prado F E 2009 Soluble sugars: metabolism, sensing and abiotic stress: a complex network in the life of plants *Plant Signal. Behav.* **4** 388–93

[58] Pattanagul W and Thitisaksakul M 2008 Effect of salinity stress on growth and carbohydrate metabolism in three rice (*Oryza sativa* L.) cultivars differing in salinity tolerance *Indian J. Exp. Biol.* **46** 736–42

[59] Hu M, Shi Z, Zhang Z, Zhang Y and Li H 2012 Effects of exogenous glucose on seed germination and antioxidant capacity in wheat seedlings under salt stress *Plant Growth Regul.* **68** 177–88

[60] Ashraf M and Harris P J 2013 Photosynthesis under stressful environments: an overview *Photosynthetica* **51** 163–90

[61] Osakabe Y, Yamaguchi-Shinozaki K, Shinozaki K and Tran L S P 2014 ABA control of plant macroelement membrane transport systems in response to water deficit and high salinity *New Phytol.* **202** 35–49

[62] Peterbauer T and Richter A 1998 Galactosylononitol and stachyose synthesis in seeds of adzuki bean: purification and characterization of stachyose synthase *Plant Physiol.* **117** 165–72

[63] Mohammadkhani N and Heidari R 2008 Drought-induced accumulation of soluble sugars and proline in two maize varieties *World Appl. Sci. J.* **3** 448–53

[64] Garg A K, Kim J-K, Owens T G, Ranwala A P, Choi Y D, Kochian L V and Wu R J 2002 Trehalose accumulation in rice plants confers high tolerance levels to different abiotic stresses *Proc. Natl Acad. Sci.* **99** 15898–903

[65] Yuanyuan M, Yali Z, Jiang L and Hongbo S 2009 Roles of plant soluble sugars and their responses to plant cold stress *Afr. J. Biotechnol.* **8** 2004–10

[66] Mahajan S and Tuteja N 2005 Cold, salinity and drought stresses: an overview *Arch. Biochem. Biophys.* **444** 139–58

[67] Livingston D P, Hincha D K and Heyer A G 2009 Fructan and its relationship to abiotic stress tolerance in plants *Cell. Mol. Life Sci.* **66** 2007–23

[68] Hou Q, Ufer G and Bartels D 2016 Lipid signalling in plant responses to abiotic stress *Plant Cell Environ.* **39** 1029–48

[69] Pinosa F, Buhot N, Kwaaitaal M, Fahlberg P, Thordal-Christensen H, Ellerström M and Andersson M X 2013 *Arabidopsis* phospholipase Dδ is involved in basal defense and nonhost resistance to powdery mildew fungi *Plant Physiol.* **163** 896–906

[70] Sun Y, Li Y, Sun X, Wu Q, Yang C and Wang L 2022 Overexpression of a phosphatidylinositol-specific phospholipase C gene from *Populus simonii* × *P. nigra* improves salt tolerance in transgenic tobacco *J. Plant Biol.* **65** 365–76

[71] Arisz S A, van Wijk R v, Roels W, Zhu J-K, Haring M A and Munnik T 2013 Rapid phosphatidic acid accumulation in response to low temperature stress in Arabidopsis is generated through diacylglycerol kinase *Front. Plant Sci.* **4** 20590

[72] McLoughlin F, Arisz S A, Dekker H L, Kramer G, De Koster C G, Haring M A, Munnik T and Testerink C 2013 Identification of novel candidate phosphatidic acid-binding proteins involved in the salt-stress response of *Arabidopsis thaliana* roots *Biochem. J.* **450** 573–81

[73] Upchurch R G 2008 Fatty acid unsaturation, mobilization, and regulation in the response of plants to stress *Biotechnol. Lett.* **30** 967–77

[74] Mandal M K, Chandra-Shekara A, Jeong R-D, Yu K, Zhu S, Chanda B, Navarre D, Kachroo A and Kachroo P 2012 Oleic acid–dependent modulation of NITRIC OXIDE ASSOCIATED1 protein levels regulates nitric oxide–mediated defense signaling in Arabidopsis *Plant Cell* **24** 1654–74

[75] Welti R, Li W, Li M, Sang Y, Biesiada H, Zhou H-E, Rajashekar C, Williams T D and Wang X 2002 Profiling membrane lipids in plant stress responses: role of phospholipase Dα in freezing-induced lipid changes in Arabidopsis *J. Biol. Chem.* **277** 31994–2002

[76] Kang L, Wang Y S, Uppalapati S R, Wang K, Tang Y, Vadapalli V, Venables B J, Chapman K D, Blancaflor E B and Mysore K S 2008 Overexpression of a fatty acid amide hydrolase compromises innate immunity in Arabidopsis *Plant J.* **56** 336–49

[77] Markham J E, Lynch D V, Napier J A, Dunn T M and Cahoon E B 2013 Plant sphingolipids: function follows form *Curr. Opin. Plant Biol.* **16** 350–7

[78] Dos Santos C and Franco O L 2023 Pathogenesis-related proteins (PRs) with enzyme activity activating plant defense responses *Plants* **12** 2226

[79] Jain D and Khurana J P 2018 Role of pathogenesis-related (PR) proteins in plant defense mechanism *Molecular Aspects of Plant–Pathogen Interaction* (Springer) pp 265–81

[80] Zribi I, Ghorbel M and Brini F 2021 Pathogenesis related proteins (PRs): from cellular mechanisms to plant defense *Curr. Protein Pep. Sci.* **22** 396–412

[81] Stintzi A, Heitz T, Prasad V, Wiedemann-Merdinoglu S, Kauffmann S, Geoffroy P, Legrand M and Fritig B 1993 Plant 'pathogenesis-related'proteins and their role in defense against pathogens *Biochimie* **75** 687–706

[82] Finkina E I, Melnikova D N, Bogdanov I V and Ovchinnikova T V 2017 Plant pathogenesis-related proteins PR-10 and PR-14 as components of innate immunity system and ubiquitous allergens *Curr. Med. Chem.* **24** 1772–87

[83] Sudisha J, Sharathchandra R, Amruthesh K, Kumar A and Shetty H S 2012 Pathogenesis related proteins in plant defense response *Plant Defence: Biological Control* (Springer) pp 379–403

[84] Kaur A, Pati P K, Pati A M and Nagpal A K 2020 Physico-chemical characterization and topological analysis of pathogenesis-related proteins from *Arabidopsis thaliana* and Oryza sativa using in-silico approaches *PLoS One* **15** e0239836

[85] Sellami K, Couvert A, Nasrallah N, Maachi R, Abouseoud M and Amrane A 2022 Peroxidase enzymes as green catalysts for bioremediation and biotechnological applications: a review *Sci. Total Environ.* **806** 150500

[86] Kaur S, Samota M K, Choudhary M, Choudhary M, Pandey A K, Sharma A and Thakur J 2022 How do plants defend themselves against pathogens—biochemical mechanisms and genetic interventions *Physiol. Mol. Biol. Plants* **28** 485–504

[87] Aglas L, Soh W T, Kraiem A, Wenger M, Brandstetter H and Ferreira F 2020 Ligand binding of PR-10 proteins with a particular focus on the Bet v 1 allergen family *Curr. Allergy Asthma Rep.* **20** 1–11

[88] McBride J K, Cheng H, Maleki S J and Hurlburt B K 2019 Purification and characterization of pathogenesis related class 10 panallergens *Foods* **8** 609

[89] Andrade L B D S, Oliveira A S, Ribeiro J K, Kiyota S, Vasconcelos I M, de Oliveira J T A and de Sales M P 2010 Effects of a novel pathogenesis-related class 10 (PR-10) protein from *Crotalaria pallida* roots with papain inhibitory activity against root-knot nematode *Meloidogyne incognita J. Agric. Food Chem.* **58** 4145–52

[90] Choi D S, Hwang I S and Hwang B K 2012 Requirement of the cytosolic interaction between Pathogenesis-Related Protein10 and Leucine-Rich Repeat Protein1 for cell death and defense signaling in pepper *Plant Cell* **24** 1675–90

[91] Lee O R, Pulla R K, Kim Y-J, Balusamy S R D and Yang D-C 2012 Expression and stress tolerance of PR10 genes from *Panax ginseng* CA Meyer *Mol. Biol. Rep.* **39** 2365–74

[92] Zaynab M, Fatima M, Abbas S, Sharif Y, Umair M, Zafar M H and Bahadar K 2018 Role of secondary metabolites in plant defense against pathogens *Microb. Pathog.* **124** 198–202

[93] Iglesias R, Citores L, Di Maro A and Ferreras J M 2015 Biological activities of the antiviral protein BE27 from sugar beet (*Beta vulgaris* L.) *Planta* **241** 421–33

[94] Balasubrahmanyam A, Baranwal V, Lodha M, Varma A and Kapoor H 2000 Purification and properties of growth stage-dependent antiviral proteins from the leaves of *Celosia cristata Plant Sci.* **154** 13–21

[95] Fitch M M 2010 Papaya Ringspot Virus (PRSV) coat protein gene virus resistance in papaya: update on progress worldwide *Transgenic Plant J.* **4** 16–28

[96] Khare S, Singh N, Singh A, Hussain I, Niharika K, Yadav V, Bano C, Yadav R K and Amist N 2020 Plant secondary metabolites synthesis and their regulations under biotic and abiotic constraints *J. Plant Biol.* **63** 203–16

[97] Divekar P A, Narayana S, Divekar B A, Kumar R, Gadratagi B G, Ray A, Singh A K, Rani V, Singh V and Singh A K 2022 Plant secondary metabolites as defense tools against herbivores for sustainable crop protection *Int. J. Mol. Sci.* **23** 2690

[98] Signorelli S, Coitinño E L, Borsani O and Monza J 2014 Molecular mechanisms for the reaction between ·OH radicals and proline: insights on the role as reactive oxygen species scavenger in plant stress *J. Phys. Chem.* B **118** 37–47

[99] Dehghan S, Sadeghi M, Pöppel A, Fischer R, Lakes-Harlan R, Kavousi H R, Vilcinskas A and Rahnamaeian M 2014 Differential inductions of phenylalanine ammonia-lyase and chalcone synthase during wounding, salicylic acid treatment, and salinity stress in safflower, *Carthamus tinctorius Biosci. Rep.* **34** e00114

[100] Fan L, Linker R, Gepstein S, Tanimoto E, Yamamoto R and Neumann P M 2006 Progressive inhibition by water deficit of cell wall extensibility and growth along the elongation zone of maize roots is related to increased lignin metabolism and progressive stelar accumulation of wall phenolics *Plant Physiol.* **140** 603–12

[101] Alvarez S, Marsh E L, Schroeder S G and Schachtman D P 2008 Metabolomic and proteomic changes in the xylem sap of maize under drought *Plant Cell Environ.* **31** 325–40

[102] Li B, Fan R, Sun G, Sun T, Fan Y, Bai S, Guo S, Huang S, Liu J and Zhang H 2021 Flavonoids improve drought tolerance of maize seedlings by regulating the homeostasis of reactive oxygen species *Plant Soil.* **461** 389–405

[103] Nakabayashi R, Yonekura-Sakakibara K, Urano K, Suzuki M, Yamada Y, Nishizawa T, Matsuda F, Kojima M, Sakakibara H and Shinozaki K 2014 Enhancement of oxidative and drought tolerance in Arabidopsis by overaccumulation of antioxidant flavonoids *Plant J.* **77** 367–79

[104] de Abreu I N and Mazzafera P 2005 Effect of water and temperature stress on the content of active constituents of *Hypericum brasiliense* Choisy *Plant Physiol. Biochem.* **43** 241–8

[105] Djoukeng J D, Arbona V, Argamasilla R and Gomez-Cadenas A 2008 Flavonoid profiling in leaves of citrus genotypes under different environmental situations *J. Agric. Food Chem.* **56** 11087–97

[106] André C M, Schafleitner R, Legay S, Lefèvre I, Aliaga C A A, Nomberto G, Hoffmann L, Hausman J-F, Larondelle Y and Evers D 2009 Gene expression changes related to the production of phenolic compounds in potato tubers grown under drought stress *Phytochemistry* **70** 1107–16

[107] Xiang M, Ding W, Wu C, Wang W, Ye S, Cai C, Hu X, Wang N, Bai W and Tang X 2021 Production of purple Ma bamboo (*Dendrocalamus latiflorus* Munro) with enhanced drought and cold stress tolerance by engineering anthocyanin biosynthesis *Planta* **254** 50

[108] Wang N, Zhang Z, Jiang S, Xu H, Wang Y, Feng S and Chen X 2016 Synergistic effects of light and temperature on anthocyanin biosynthesis in callus cultures of red-fleshed apple (*Malus sieversii* f. *niedzwetzkyana*) *Plant Cell Tiss. Organ Cult.* **127** 217–27

[109] Naikoo M I, Dar M I, Raghib F, Jaleel H, Ahmad B, Raina A, Khan F A and Naushin F 2019 Role and regulation of plants phenolics in abiotic stress tolerance: an overview *Plant Signal. Mol.* 157–68

[110] Cabane M, Afif D and Hawkins S 2012 Lignins and abiotic stresses *Advances in Botanical Research* (Amsterdam: Elsevier) pp 219–62

[111] Bhonwong A, Stout M J, Attajarusit J and Tantasawat P 2009 Defensive role of tomato polyphenol oxidases against cotton bollworm (*Helicoverpa armigera*) and beet armyworm (*Spodoptera exigua*) *J. Chem. Ecol.* **35** 28–38

[112] Toffolatti S L, Maddalena G, Passera A, Casati P, Bianco P A and Quaglino F 2021 Role of terpenes in plant defense to biotic stress *Biocontrol Agents and Secondary Metabolites* (Amsterdam: Elsevier) pp 401–17

[113] Kumar S, Shah S H, Vimala Y, Jatav H S, Ahmad P, Chen Y and Siddique K H 2022 Abscisic acid: metabolism, transport, crosstalk with other plant growth regulators, and its role in heavy metal stress mitigation *Front. Plant Sci.* **13** 972856

[114] Kodama O, Suzuki T, Miyakawa J and Akatsuka T 1988 Ultraviolet-induced accumulation of phytoalexins in rice leaves *Agric.Biol.Chem.* **52** 2469–73

[115] Vaughan M M, Christensen S, Schmelz E A, Huffaker A, Mcauslane H J, Alborn H T, Romero M, Allen L H and Teal P E 2015 Accumulation of terpenoid phytoalexins in maize roots is associated with drought tolerance *Plant Cell Environ.* **38** 2195–207

[116] Wu W, Zhang Q, Zhu Y, Lam H-M, Cai Z and Guo D 2008 Comparative metabolic profiling reveals secondary metabolites correlated with soybean salt tolerance *J. Agric. Food Chem.* **56** 11132–8

[117] Gu C-Z, Xia X-M, Lv J, Tan J-W, Baerson S R, Pan Z-q, Song Y-Y and Zeng R-S 2019 Diterpenoids with herbicidal and antifungal activities from hulls of rice (*Oryza sativa*) *Fitoterapia* **136** 104183

[118] Yang C-Q, Wu X-M, Ruan J-X, Hu W-L, Mao Y-B, Chen X-Y and Wang L-J 2013 Isolation and characterization of terpene synthases in cotton (*Gossypium hirsutum*) *Phytochemistry* **96** 46–56

[119] Aharoni A, Jongsma M A and Bouwmeester H J 2005 Volatile science? Metabolic engineering of terpenoids in plants *Trends Plant Sci.* **10** 594–602

[120] Miyagawa H 2009 Studies on nitrogen-containing secondary metabolites playing a defensive role in plants *J. Pestic. Sci.* **34** 110–2

[121] Davodnia B, Ahmahdi J and Ourang S F 2017 Evaluation of drought and salinity stresses on morphological and biochemical characteristics in four species of papaver, eco-phyto-chemical *J. Med. Plants* **5** 24–36

[122] Wink M 2003 Evolution of secondary metabolites from an ecological and molecular phylogenetic perspective *Phytochemistry* **64** 3–19
[123] Dey P, Kundu A, Kumar A, Gupta M, Lee B M, Bhakta T, Dash S and Kim H S 2020 Analysis of alkaloids (indole alkaloids, isoquinoline alkaloids, tropane alkaloids) *Recent Advances in Natural Products Analysis* (Amsterdam: Elsevier) pp 505–67
[124] Züst T and Agrawal A A 2016 Mechanisms and evolution of plant resistance to aphids *Nat. Plants* **2** 1–9
[125] Hopkins R J, van Dam N M and van Loon J J 2009 Role of glucosinolates in insect-plant relationships and multitrophic interactions *Annu. Rev. Entomol.* **54** 57–83
[126] Mewis I, Khan M, Glawischnig E, Schreiner M and Ulrichs C 2012 Water stress and aphid feeding differentially influence metabolite composition in *Arabidopsis thaliana* **7** e48661
[127] Jensen C, Mogensen V, Mortensen G, Fieldsend J, Milford G, Andersen M N and Thage J 1996 Seed glucosinolate, oil and protein contents of field-grown rape (*Brassica napus* L.) affected by soil drying and evaporative demand *Field Crops Res.* **47** 93–105
[128] Li Q, Eigenbrode S D, Stringam G and Thiagarajah M 2000 Feeding and growth of *Plutella xylostella* and *Spodoptera eridania* on *Brassica juncea* with varying glucosinolate concentrations and myrosinase activities *J. Chem. Ecol.* **26** 2401–19
[129] Bennett R N and Wallsgrove R M 1994 Secondary metabolites in plant defence mechanisms *New Phytol.* **127** 617–33
[130] Clay N K, Adio A M, Denoux C, Jander G and Ausubel F M 2009 Glucosinolate metabolites required for an *Arabidopsis* innate immune response *Science* **323** 95–101
[131] Ahmad S, Veyrat N, Gordon-Weeks R, Zhang Y, Martin J, Smart L, Glauser G, Erb M, Flors V and Frey M 2011 Benzoxazinoid metabolites regulate innate immunity against aphids and fungi in maize *Plant Physiol.* **157** 317–27
[132] Meihls L N, Handrick V, Glauser G, Barbier H, Kaur H, Haribal M M, Lipka A E, Gershenzon J, Buckler E S and Erb M 2013 Natural variation in maize aphid resistance is associated with 2,4-dihydroxy-7-methoxy-1,4-benzoxazin-3-one glucoside methyltransferase activity *Plant Cell* **25** 2341–55
[133] Li B, Förster C, Robert C A M, Züst T, Hu L, Machado R, Berset J-D, Handrick V, Knauer T and Hensel G 2018 Convergent evolution of a metabolic switch between aphid and caterpillar resistance in cereals *Sci. Adv.* **4** eaat6797
[134] Burow M, Atwell S, Francisco M, Kerwin R E, Halkier B A and Kliebenstein D J 2015 The glucosinolate biosynthetic gene AOP2 mediates feed-back regulation of jasmonic acid signaling in *Arabidopsis Mol. Plant* **8** 1201–12
[135] Katz E, Nisani S, Yadav B S, Woldemariam M G, Shai B, Obolski U, Ehrlich M, Shani E, Jander G and Chamovitz D A 2015 The glucosinolate breakdown product indole-3-carbinol acts as an auxin antagonist in roots of *Arabidopsis thaliana Plant J.* **82** 547–55
[136] Kim J I, Dolan W L, Anderson N A and Chapple C 2015 Indole glucosinolate biosynthesis limits phenylpropanoid accumulation in *Arabidopsis thaliana Plant Cell* **27** 1529–46
[137] Hemm M R, Ruegger M O and Chapple C 2003 The Arabidopsis ref2 mutant is defective in the gene encoding CYP83A1 and shows both phenylpropanoid and glucosinolate phenotypes *Plant Cell* **15** 179–94
[138] Frerigmann H, Piślewska-Bednarek M, Sánchez-Vallet A, Molina A, Glawischnig E, Gigolashvili T and Bednarek P 2016 Regulation of pathogen-triggered tryptophan metabolism in *Arabidopsis thaliana* by MYB transcription factors and indole glucosinolate conversion products *Mol. Plant* **9** 682–95

[139] Stahl E, Bellwon P, Huber S, Schlaeppi K, Bernsdorff F, Vallat-Michel A, Mauch F and Zeier J 2016 Regulatory and functional aspects of indolic metabolism in plant systemic acquired resistance *Mol. Plant* **9** 662–81
[140] Baldwin I T, Halitschke R, Paschold A, Von Dahl C C and Preston C A 2006 Volatile signaling in plant–plant interactions: 'talking trees' in the genomics era *Science* **311** 812–5
[141] Godard K-A, White R and Bohlmann J 2008 Monoterpene-induced molecular responses in *Arabidopsis thaliana Phytochemistry* **69** 1838–49
[142] Erb M 2018 Volatiles as inducers and suppressors of plant defense and immunity—origins, specificity, perception and signaling *Curr. Opin. Plant Biol.* **44** 117–21
[143] Bouwmeester H, Schuurink R C, Bleeker P M and Schiestl F 2019 The role of volatiles in plant communication *Plant J.* **100** 892–907
[144] Riedlmeier M, Ghirardo A, Wenig M, Knappe C, Koch K, Georgii E, Dey S, Parker J E, Schnitzler J-P and Vlot A C 2017 Monoterpenes support systemic acquired resistance within and between plants *Plant Cell* **29** 1440–59
[145] Ye M, Glauser G, Lou Y, Erb M and Hu L 2019 Molecular dissection of early defense signaling underlying volatile-mediated defense regulation and herbivore resistance in rice *Plant Cell* **31** 687–98
[146] Paschold A, Halitschke R and Baldwin I T 2006 Using 'mute' plants to translate volatile signals *Plant J.* **45** 275–91

IOP Publishing

Advances in Biochemical and Molecular Mechanisms of Plant–Pathogen Interaction

Hitendra K Patel and Anirudh Kumar

Chapter 9

Fast-forward to molecular and mutation breeding for developing disease-resistant varieties

Monika Naik, Rekha Balodi and A S Kotasthane

Challenges of evolving pathogens, climate change, and demographic shifts make developing disease-resistant crop varieties crucial for food security. This chapter explores the blending of fast-forward molecular and mutation breeding approaches to combat plant diseases effectively. Synergy between molecular and mutation breeding holds significant potential for accelerating the generation of disease-resistant cultivars for a broad spectrum of crops. Molecular breeding, using genomic tools like genomic selection (GS) and marker-assisted selection (MAS), accelerates trait selection and breeding. Mutation breeding, enhanced by CRISPR-Cas9, induces genetic diversity and targets disease-resistance genes. Combining these methods can rapidly produce disease-resistant cultivars, improving global food resilience and addressing evolving agricultural challenges.

9.1 Introduction

Taking into consideration the ever-expanding global population, changing demographic patterns, dwindling availability of agricultural land, and the looming spectre of climate change, securing food availability and safety has surfaced as a major concern. Traditional breeding methods have long been instrumental in fortifying crop plants against the ravages of pests and diseases, thereby bolstering agricultural productivity. However, the escalating pace of environmental degradation, exacerbated by climate change-induced shifts in pest and disease dynamics, demands a paradigm shift in agricultural practices. Fast resistance breeding is a cutting-edge approach that harnesses the power of advanced technologies to expedite the generation of resilient crop varieties. This ground-breaking approach may play a

doi:10.1088/978-0-7503-5673-2ch9

crucial role in ensuring global food security along with its potential to promote sustainability and resilience amidst growing environmental challenges. Traditional plant breeding progresses slowly attributable to multiple factors. Firstly, the generation time of crops significantly influences the pace of breeding. Plants undergo growth, flowering, and seed production stages, which cannot be rushed beyond a certain extent. Secondly, achieving stable lines necessitates growing plants over multiple generations. With each new generation, undesirable traits are bred out while desirable ones are retained, an inherently time-consuming process. Additionally, field selection processes span entire seasons, further contributing to the slow advancement of each generation in traditional breeding methods. Lastly, traditional breeding often relies on phenotypic characteristics to identify superior cultivars, lacking the efficiency of modern molecular methods that segregate gametes based on genetic markers. These factors collectively contribute to the perceived sluggishness of traditional plant breeding practices. The integration of molecular biology techniques with mutation breeding, known as 'molecular mutation breeding', is paving the way for the development of 'designer' crop varieties tailored to specific needs [1].

Development of disease-resistant crop varieties through traditional breeding methods can be quite time-consuming due to the long cycles of crossing and selection. To accelerate this process, fast-forward molecular breeding and mutation breeding are increasingly being used. This chapter delves into the multifaceted implications of fast resistance breeding, exploring its pivotal role in enhancing food security, promoting sustainability, and confronting the formidable challenges of climate change. Fast-forward molecular breeding involves the use of advanced genomic tools and techniques to quickly identify and incorporate disease-resistance traits into crops. This method relies on genomic selection, utilizing genome-wide markers to predict the performance of plant genotypes, along with MAS, which employs molecular markers for selection of linked disease-resistance genes and traits. Furthermore, gene editing methods such as CRISPR/Cas9 are used to accurately alter plant DNA, accelerating resistance development to diseases. In contrast, mutation breeding utilizes physical or chemical mutagens to induce genetic variations that can lead to disease resistance. This approach, known as induced mutagenesis, has been successful in creating new crop varieties with improved traits by exposing seeds or plant tissues to mutagens like radiation or chemicals. TILLING (Targeting Induced Local Lesions in Genomes) is a type of mutation breeding that uses a reverse genetics approach, blending traditional mutagenesis with contemporary genomics to identify mutations in specific genes, thereby enhancing the creation of disease-resistant varieties or cultivars.

9.2 Disease resistance in plants

Disease resistance is an innate ability of plants to combat diseases and ranges between the extremities of immunity and high susceptibility. Disease resistance is categorized into different classes based on various parameters (figure 9.1). For instance, based on the genetics/genes involved, disease resistance is commonly categorized into two types: complete resistance, synonymized as qualitative resistance; and partial resistance, also referred to as quantitative resistance. The former is mainly governed by resistance

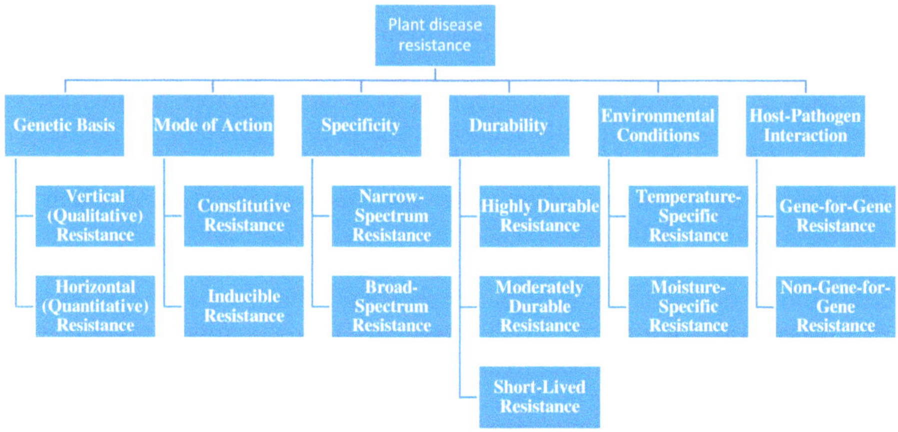

Figure 9.1. Classification of plant disease resistance based on different parameters.

genes (R genes), usually encoding surface or intracellular immune receptors, for example receptor-like kinases (RLKs) or nucleotide-binding leucine-rich repeat proteins (NLRs), respectively. These receptors can directly or indirectly detect conserved pathogenic molecules or specific effector/avirulence (Avr) proteins. This type of resistance is 'race-specific' following the 'gene-for-gene model', and is prone to breakdown in the field as pathogens can evolve to evade detection by the host by developing the corresponding Avr gene. This necessitates a constant search for new R genes by plant breeders to combat crop diseases effectively.

In contrast, partial resistance is regulated by multiple genes or quantitative resistance loci (popularly known as quantitative trait loci (QTLs) for resistance), which offer wide-ranging resistance to multiple pathogens. However, utilizing a single QTL in crop breeding is challenging due to their limited effects. Hence, identifying genes responsible for broad-spectrum resistance in crops has become a significant objective in breeding programmes.

Resistant crop varieties are developed through a combination of traditional breeding techniques and modern biotechnological methods. Here's an overview of the process:

1. **Identification of resistance genes:** Identify genes that confer resistance to specific diseases in plants. This can be done through genetic studies of plants that naturally exhibit resistance.
2. **Cross breeding**: Traditional breeding involves crossing plants with desirable traits, such as disease resistance, with commercial cultivars to combine the best characteristics of both.
3. **Selection**: Breeders then select the offspring that show the desired resistance traits and continue to breed these over several generations to stabilize the trait in the new variety
4. **Molecular breeding**: This approach uses molecular markers to select plants that carry the resistance genes, speeding up the breeding process.
5. **Genetic engineering**: In some cases, resistant varieties are developed by transferring resistance genes from one organism to another using biotechnological techniques, even between unrelated species.

6. **Field testing**: The new varieties are tested in the field to ensure that they maintain resistance under natural conditions and do not have any adverse effects on yield or quality.
7. **Release of new varieties**: Once the new varieties have passed field testing and regulatory approval, they are released for commercial cultivation.

Creating disease-resistant varieties is an ongoing effort, as pathogens can evolve (arms race between host and pathogens) and bypass existing resistance measures. Therefore, plant breeders and scientists are always working to identify new sources of resistance for developing new resistant varieties.

9.3 Mutation breeding: unleashing nature's potential

Experiments performed by Lweis John Stadler on induction of genetic alterations by x-rays in plants laid the foundation of mutation breeding during the 1920s and early 1930s. 'Mutation breeding' a term coined by Freisleben and Lein outline the 'intentional creation and advancement of mutant lines to enhance agricultural practices' [2]. It involves inducing genetic changes/mutations in crops by exposing them to physical/radiation or chemical mutagens. These genetic alterations can result in genetic variation, induction of novel traits, such as disease resistance and are utilized for breeding programmes. Concisely, it has several advantages over other breeding approaches (figure 9.2). Some commonly employed physical and chemical mutagens in plant mutation breeding are listed below [3].

9.3.1 Physical mutagens

- Ionizing radiation: Sources like gamma rays, x-rays, and neutron radiation are commonly used to induce mutations in plant genomes. These radiations induce DNA damage by ionizing atoms and molecules within the DNA, leading to mutations.
- Ultraviolet (UV) radiation: UV radiation is used to induce mutations primarily in the form of UV lamps. It can cause thymine dimers and other DNA lesions, resulting in mutations.
- Gamma rays are highly penetrating electromagnetic radiation capable of causing a range of mutations in plant DNA.

9.3.2 Chemical mutagens

These mutagens are applied in controlled environments to treat seeds, cuttings, or tissues with the goal of introducing genetic variation into plant populations. The resulting mutated individuals are then screened for desirable traits, such as disease resistance, and selected for further breeding programmes. Mutation breeding serves as a valuable complement to traditional breeding techniques by broadening the genetic diversity available for selection and crop improvement. Fast-forward mutation techniques accelerate this process by rapidly generating genetic diversity. These methods allow scientists to create a large pool of diverse plant variants in a shorter time. For disease resistance, researchers focus on manipulating specific traits

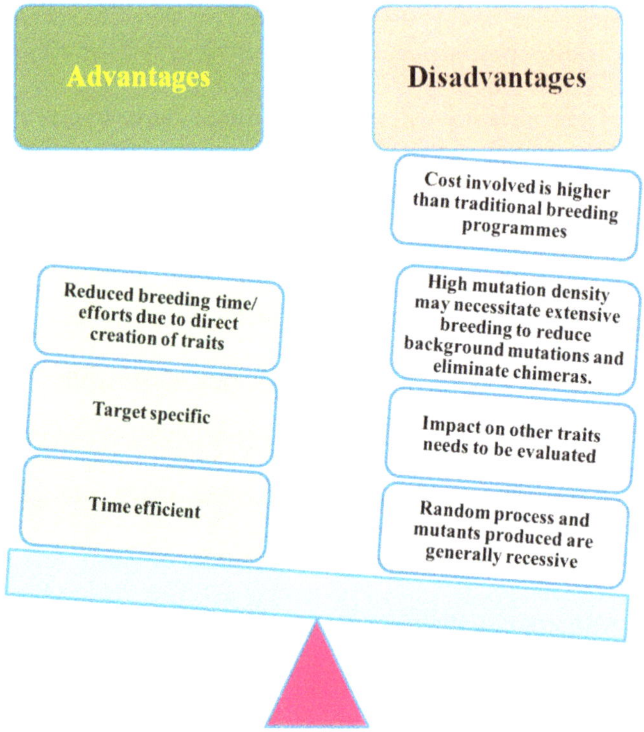

Figure 9.2. Advantages and disadvantages of mutation breeding.

related to flowering, reproductive capacity, and biomass allocation under varying climatic conditions. By identifying and selecting mutants with improved disease resistance, breeders can develop resilient crop varieties [4].

- Ethyl methane sulfonate (EMS): EMS is an alkylating agent that methylates DNA bases, particularly guanine, which results in point mutations such as GC to AT transitions.
- Sodium azide: This chemical mutagen primarily induces point mutations by causing base substitutions or frameshift mutations.
- Nitrous oxide (N_2O): It can induce mutations by deaminating adenine or cytosine bases, resulting in transitions or transversions.
- Hydroxylamine: This chemical mutagen induces mutations by causing base substitutions, frameshift mutations, or DNA strand breaks.
- Ethyleneimine: It is a highly reactive alkylating agent that can cause a broad spectrum of mutations in plant genomes.

9.4 Molecular breeding: unveiling genetic secrets

At the heart of molecular breeding lies the intricate unravelling of plant genomes—a task made feasible by ground-breaking advancements in genomic sequencing and molecular marker technologies. By pinpointing genes associated with desirable traits such as disease resistance, molecular breeding enables breeders to expedite the selection of superior plant genotypes with unprecedented accuracy. Through techniques such as MAS and GS, breeders can swiftly identify and introgress beneficial alleles into elite germplasm, thus accelerating the development of disease-resistant varieties.

9.5 Recessive class analysis for disease resistance

For analysis of disease resistance using F2 segregating material, the process begins with evaluating the material under field conditions through challenge inoculation with the pathogen. Phenotyping is then conducted to identify resistant and susceptible lines. Following this, DNA from the respective lines is used for marker analysis. In this context, the F2 segregating material can be classified into two groups: the dominant class, which represents resistance, and the recessive class, consisting of susceptible lines. The utility of the recessive class is significant; these susceptible lines exhibit clear phenotypic expressions of disease, allowing for quantitative assessments that establish strong linkage relationships between resistance genes and genetic markers. Conversely, while the dominant class is crucial for confirming resistance, its lack of disease symptoms complicates the genetic assessment, as this absence may arise from factors (disease escape) like incomplete dominance or interactions with other genes.

9.6 Sequencing technology-based genotyping

Sequencing technology-based genotyping is becoming increasingly useful for genetic marker development and mapping. One of the most promising options for a quick breeding programme is inclusion of perennial crops in this technique. Compared to a single marker association, the genotyping-by-sequencing (GBS) pipeline allowed for the simultaneous extraction of a wide array of markers from a population exhibiting various extreme phenotypes, which reduced the time required to identify markers linked to disease resistance [5].

9.6.1 Pan-genomics

Pan-genomics involves studying the entire genomic content of a species, including core genes and accessory genes. Pan-genomics studies provide insights into genetic diversity beyond the reference genome enabling the identification of genes related to disease resistance and other adaptive traits, facilitating the development of resilient crop variety. Pan-genomics considers a wide range of genetic variations, ensuring robustness.

9.6.2 Breeding for desirable traits: phenotyping derived genotyping

To achieve genetic progress in terms of disease resistance, it is essential to leverage the genetic loci associated with this trait and understand their inheritance patterns [6]. This has been extensively explored through the identification of QTLs to ascertain the adaptive characteristics of the relevant traits. Through the integration of refined methods that combine modified phenotype-based genotyping, there exists the potential to enhance disease resistance.

9.6.3 Speed breeding

Speed breeding is based on NASA's pioneering technology designed for crop cultivation within enclosed spaces, coupled with strategic manipulation of photoperiods. The technology has found applications across a broad spectrum of crops, encompassing legumes and cereals. This innovative approach necessitates the meticulous maintenance of specific environmental parameters, including a seamless light cycle with minimal intervals of darkness, moderate humidity levels ranging between 60%–70%, and tightly controlled temperatures between 22 °C/17 °C. The integration of advanced LED and halogen lighting systems within controlled environments such as greenhouses or growth chambers has yielded remarkable reductions in generation times [7]. The targeted application of crop-specific LED lighting alongside far-red light has showcased accelerated flowering, obviating the necessity for embryo rescue and other tissue culture methodologies [8]. Notably, within confined greenhouse settings, generation times have been dramatically reduced to six for wheat, barley, durum wheat, and peas, and to four for canola, in stark contrast to the conventional two to three generations typically observed in traditional greenhouse setups [9]. Moreover, in a study focusing on chickpeas, the strategic manipulation of light, temperature, and the germination process of immature seeds has yielded generation times of 7 for early, 6.2 for medium and 6.0 for late-maturing genotypes per annum, representing a diverse array of varieties from various agro-ecological contexts [10].

9.6.4 Rapid generation advancement (RGA)

Disease-resistant plant breeding requires a swift approach to develop outcomes and target varieties, aiming to shorten long crop cycles. To achieve this, various techniques have been evolving. For example, cultivating crops in winter nurseries and using the double-haploid (DH) technique have long been established methods to accelerate crop cycles. Accelerating breeding cycles in response to rapid climate changes involves modifying allelic frequencies, selecting commercial cultivars, and proactively distributing them for adaptation in farmers' fields [11]. These modified varieties serve as parent stock for future breeding efforts. Utilizing speed breeding, especially through the single-seed descent (SSD) method, shortens breeding cycles, enhances the development of favourable breeding lines, and lowers programme costs [12]. SSD has been effectively used for isolating desirable transgressive segregants for many years [13]. Another expedited breeding approach, the Biotron Breeding System (BBS), manipulates CO_2 levels, temperature, and day length to artificially hasten crop growth rates. This system, coupled with embryo rescue techniques, minimizes seed maturation and dormancy periods, enabling generation times as short as two months, as demonstrated with the Nipponbare rice variety [14]. BBS proves more effective with photoperiod-sensitive, long-duration cultivars.

9.6.5 Double-haploid and shuttle breeding

Shuttle breeding is devised as a means to expedite variety development, enhancing adaptability across various environments and facilitating widespread adoption [15]. Initially applied in rice, this approach has garnered attention in the private sector due to limitations on cross-border material movement for off-season trials [16]. DH techniques have long been pivotal in rapid breeding strategies, swiftly yielding homozygous lines [17]. DH methodologies are extensively employed in developing various rice and cereal varieties, constituting a crucial element in forward breeding [18]. This technology accelerates line evaluation, enhancing genetic gains [19]. Additionally, DH techniques bolster genomic selections for traits with low heritability, especially in regions with limited off-season nursery opportunities [20]. By allocating resources away from repetitive inbreeding toward climate resilience evaluations, DH technologies streamline varietal development processes [21].

9.6.6 Prediction-based early phenotyping

Advancements in genotyping technology, progressing from basic mapping to genomic selection, heavily rely on robust phenotyping data. However, the absence of recent field-level phenotyping platforms poses challenges for traditional plant breeding programmes, as genomic predictions are built upon statistical models reliant on such data [22]. Efficient phenotyping is paramount for future genomic plant breeding endeavours [23]. Early stress detection, employing methods like chlorophyll fluorescence, infrared spectroscopy, and RGB imaging [24, 25], shows potential but faces complexities when applied at canopy levels [26]. It is imperative to develop robust phenotyping protocols tailored to specific traits for continuous cultivar enhancement. Recent technological innovations, such as RGB imaging, LiDAR for measuring plant height, and yield estimation using line sensors and fat-based scanners on conveyor belts [27], enable rigorous phenotyping, thus boosting genetic gain crucial for climate resilience [28, 29]. Bridging the genotype–phenotype gap amidst dynamic environmental changes entails studying the plasticity of plant traits and responses within the Soil–Plant–Atmosphere Continuum (SPAC) framework [30]. Computational models predicting spatiotemporal phenotypes and analyzing complex traits are increasingly prevalent, utilizing NIR-based phenotyping to correlate plant physiological status with wild and mutant studies across diverse climates, aiding in the prediction of biochemical content for model-based phenotype analysis [31].

9.7 Molecular genotyping strategies: advances and prospects for quick breeding

Advancements in molecular biology techniques, particularly marker-based genotyping, have enhanced genetic understanding and facilitated the linking of markers to target genes or QTL, thus speeding up product delivery [32, 33]. This rapidity is increasingly important given climate change and other human-induced alterations. Linking markers to traits like disease resistance has significantly benefited breeding programmes. MAS can predict the genetic potential of genomes and help select among varieties already identified through conventional pedigree methods in crops such as cereals, legumes, and forest species [34, 35], and it can also improve existing varieties [36].

9.7.1 Genotyping by sequencing

GBS is a sophisticated MAS tool used to speed up plant breeding. GBS leverages next-generation sequencing (NGS) technologies to identify and genotype single-nucleotide polymorphisms (SNPs) in crop genomes and populations. GBS has applications in identifying genetic variants associated with specific traits within or across populations; mapping genes and elucidating inheritance patterns of the genes, identifying markers linked to desirable traits and speeding up of the breeding procedure.

However, the application of GBS technologies faces limitations due to difficulties in associating genetic markers with polymorphic traits, accounting for genetic interactions such as epistatic effects, and navigating complex genetic and biochemical pathways (figure 9.3). To address these challenges, integrating multi-omics data —including metabolomics, transcriptomics, and epigenomics—into GBS may be beneficial [37]. For example, analyzing the transcriptomes of individual genotypes under various environmental conditions can aid in candidate gene identification and link environmental responses to specific traits [38].

9.7.2 Genotyping arrays

These arrays allow simultaneous genotyping of multiple markers across the genome. They are particularly useful for high-throughput genotyping in large-scale breeding programmes. By analyzing specific genetic variations, breeders can make informed decisions about selecting desirable traits. Genotyping arrays includes use of allele-specific oligonucleotide (ASO) probes or other hybridization probes which help in

Figure 9.3. Methodology of genotyping by sequencing.

detection of specific variants, detection of SNPs or simultaneous high-throughput genotyping of multiple loci.

9.7.3 Extreme trait GWAS (Et-GWAS)

This is a powerful approach for unravelling rare genetic variants in crop genomes. Et-GWAS aims to identify rare genetic variants with significant impacts linked to specific traits, which is essential for enhancing crops [39]. Et-GWAS employs bulk pooling and allele frequency analysis to effectively identify rare variants from large datasets. It concentrates on quantifying rare alleles, allowing for precise association analysis.

9.7.4 RNA-sequence analysis

RNA-Seq is a contemporary technique that employs NGS to analyze the transcriptome, enabling the study of unknown genes or transcripts. This method involves generating cDNA molecules through reverse transcription of RNA, followed by library preparation and high-throughput deep sequencing.

RNA sequencing can be conducted on various platforms, including the following.

9.7.4.1 Illumina

This is one of the most widely utilized platforms for RNA-Seq, offering a range of kits with differing read lengths, coverage depths, and other parameters.

9.7.4.2 PacBio

Known for its long-read sequencing capabilities, PacBio provides high accuracy and extended read lengths, which are beneficial for identifying alternative splicing and other complex features.

9.7.4.3 Oxford Nanopore

Another long-read option, Oxford Nanopore is notable for its portability and ability to deliver real-time sequencing data.

9.7.4.4 Ion torrent

This semiconductor-based platform offers rapid turnaround times and is suitable for targeted sequencing applications.

9.7.4.5 BGISEQ

A newer entrant in the RNA-Seq arena, BGISEQ utilizes DNA nanoballs for sequencing RNA, providing high accuracy and throughput.

9.8 Genome editing

Genome editing enables the targeted manipulation of an organism's genome for mutagenesis, gene transfer, and gene expression modulation [40]. The main techniques are as follows.

9.8.1 Oligonucleotide-directed mutagenesis (ODM)

This uses oligonucleotides (20–100 bp) to induce single base pair changes.

9.8.2 Site-directed nucleases (SDNs)

This uses nucleases to cut DNA at specific locations, introducing mutations during DNA repair.

SDNs include **transcription activator-like effector nucleases (TALENs)** and **zinc-finger nucleases (ZFNs)**, designed to target specific DNA sequences and facilitate cleavage. The resulting double-strand breaks (DSBs) are repaired via pathways like **non-homologous end joining (NHEJ)** or **homology-directed repair (HDR)** [41].

CRISPR-Cas9 is another powerful tool, consisting of clustered regularly interspaced short palindromic repeats and the Cas9 protein, originating from archaea. It acts as molecular scissors, with guide RNA directing Cas9 to create DSBs that are repaired through NHEJ or other mechanisms [42–44]. CRISPR-Cas9 is cost-effective, user-friendly, and can edit multiple target sites, making it suitable for gene pyramiding in new cultivars [45–47] (table 9.1). However, its effective application requires thorough knowledge of target genes, which may limit its use across different crops, though it holds great promise for advancing genetic engineering in agriculture.

Overall, genome editing offers a rapid and precise approach for biotechnology-driven crop breeding, and is particularly beneficial for fast breeding programmes [53].

Table 9.1. Enhanced disease resistance in various crops using CRISPR

Gene	Crop	Disease
OsERF922	Rice	Blast disease [48]
TaMLO	Wheat	Powdery mildew [49, 50]
eIF4E	Cucumber	Broad virus resistance [51]
SlDMR6-1	Tomato	Multiple diseases [52]

9.9 Conclusions

Resistance to diseases is a complex trait. Phenotypic manifestation of resistance is dependent upon the pathogen population and other environmental factors in addition to the genotype of the host. Development of resistant varieties is a demanding task amid climate change and shrinking resources of agriculture. The challenge is to develop climate smart varieties with higher resilience to pathogens in a shorter duration of time. This can be achieved by the integration of advanced high-throughput phenotyping, rapid generation advancement and fast-forwarding genotyping tools. With the advancements in the genome editing tools like CRISPR-Cas technology, accurate changes in the targeted genome are possible helping breeders to develop sturdier varieties.

References

[1] Raina A, Laskar R A, Khursheed S, Amin R, Tantray Y R, Parveen K and Khan S 2016 Role of mutation breeding in crop improvement- past, present and future *Asian Res. J. Agric.* **2** 1–13

[2] Lamo K, Bhat D J, Kour K and Solanki S P S 2017 Mutation studies in fruit crops: a review *Int. J. Curr. Microbiol. Appl. Sci.* **6** 3620–33

[3] Yousaf S, Rehman T and Qaisar U 2021 Mutagenesis and plant breeding *Mutagenesis, Cytotoxicity and Crop Improvement: Revolutionizing Food Science* (Cambridge Scholars Publishing) p 446

[4] Shu Q Y, Forster B P and Nakagawa H (ed) 2012 *Plant Mutation Breeding and Biotechnology* (CABI)

[5] Gardiner S E, Volz R K, Chagne D *et al* 2014 Tools to breed better cultivars faster at plant and food research *Proc. of the 1st Int. Rapid Cycle Crop Breeding Conf.(7–9 January 2014) (Leesburg, VA)*

[6] Kole C *et al* 2015 Application of genomicsassisted breeding for generation of climate resilient crops: progress and prospects *Front. Plant. Sci.* **6** 563

[7] Ghosh S *et al* 2018 Speed breeding in growth chambers and glasshouses for crop breeding and model plant research *Nat. Protoc.* **13** 2944–63

[8] Jähne F, Volker H, Würschum T and Leiser W 2020 Speed breeding short-day crops by LED-controlled light schemes *Theor. Appl. Genet.* **133** 2335–42

[9] Watson A *et al* 2018 Speed breeding is a powerful tool to accelerate crop research and breeding *Nat. Plants* **4** 23–9

[10] Samineni S, Sen M, Sajja S and Gaur P 2019 Rapid generation advance (RGA) in chickpea to produce up to seven generations per year and enable speed breeding *Crop J.* **8** 164–9

[11] Atlin G, Cairns J and Das B 2017 Rapid breeding and varietal replacement are critical to adaptation of cropping systems in the developing world to climate change *Glob. Food Sec.* **12** 31–7

[12] Janwan M, Sreewongch T and Sripichitt P 2013 Rice breeding for high yield by advanced single seed descent method of selection *J. Plant Sci.* **8** 24–30

[13] Moon H P, Kang K H, Choi I S, Jeong O Y, Hong H C, Choi S H and Choi H C 2003 Comparing agronomic performance of breeding populations derived from anther culture and single seed sescent in rice *Advances in Rice Genetics* ed G S Khush, D S Brar and B Hardy (International Rice Research Institute) pp 3–5

[14] Tanaka J, Hayashi T and Iwata H 2016 A practical, rapid generation-advancement system for rice breeding using simplifed biotron breeding system *Breed. Sci.* **66** 542–51

[15] Ortiz R, Trethowan G O, Ferrara M, Iwanaga J H, Dodds J H, Crouch J H, Crossa J and Braun H J 2007 High yield potential, shuttle breeding, genetic diversity, and a new international wheat improvement strategy *Euphytica* **157** 365–84

[16] Lenaerts B, Collard B C Y and Demont M 2019 Review: improving global food security through accelerated plant breeding *Plant Sci.* **287** 110207

[17] Mishra R and Rao G J N 2016 In-vitro androgenesis in rice: advantages, constraints and future prospects *Rice Sci.* **23** 57–68

[18] Pauk J, Jancsó M and Simon-Kiss I 2009 Rice doubled haploids and breeding *Advances in Haploid Production in Higher Plants.* (Dordrecht: Springer) pp 189–97

[19] Chang M T and Coe E H 2009 Doubled haploids *Biotechnology in Agriculture and Forestry Molecular genetic approaches to maize improvement* **vol 63** ed A L Kriz and B A Larkins (Berlin: Springer) pp 127–42

[20] Bouchez A and Gallais A 2000 Efficiency of the use of doubled haploids in recurrent selection for combining ability *Crop Sci.* **40** 23–9
[21] Prasanna B M, Cairns J and Xu Y 2013 Genomic tools and strategies for breeding climate resilient cereals *Genomics and Breeding for Climate-Resilient Crops* ed C Kole (Berlin: Springer)
[22] Desta Z A and Ortiz R 2014 Genomic selection: genome-wide prediction in plant improvement *Trends Plant Sci.* **19** 592–601
[23] Ghanem M E, Marrou H and Sinclair T R 2015 Physiological phenotyping of plants for crop improve ment *Trends Plant Sci.* **20** 139–44
[24] Fang Y and Ramasamy R P 2015 Current and prospective methods for plant disease detection *Biosensors (Basel)* **5** 537–61
[25] Mutka A M and Bart R S 2015 Image-based phenotyping of plant disease symptoms *Front. Plant Sci.* **5** 734
[26] Pauli D, Chapman S C, Bart R, Topp C N, Lawrence-Dill C J, Poland J and Gore M A 2016 The quest for understanding phenotypic variation via integrated approaches in the field environment *Plant Physiol.* **172** 622–34
[27] Tanger P *et al* 2017 Field-based high throughput phenotyping rapidly identifies genomic regions controlling yield components in rice *Sci. Rep.* **7** 42839
[28] Liang X, Wang K, Huang C, Zhang X, Yan J and Yang W 2016 A high-throughput maize kernel traits scorer based on line-scan imaging *Measurement* **90** 453–60
[29] Miller N D, Haase N J, Lee J, Kaeppler S M, de Leon N and Spalding E P 2017 A robust, high throughput method for computing maize ear, cob, and kernel attributes automatically from images *Plant J.* **89** 169–78
[30] Negin B and Moshelion M 2017 The advantages of functional phenotyping in pre-field screening for drought-tolerant crops *Funct. Plant Biol.* **44** 107
[31] Anderssen R and Edwards M 2012 Mathematical modelling in the science and technology of plant breeding *Int. J. Numer. Anal. Model. Ser.* B **3** 242–58
[32] Kumar J, Choudhary A K, Solanki R and Pratap A 2011 A towards marker-assisted selection in pulses: a review *Plant Breed.* **130** 297–313
[33] Lenaerts B, Collard B C Y and Demont M 2019 Review: improving global food security through accel erated plant breeding *Plant Sci.* **287** 110207
[34] Harfouche A, Meilan R, Kirst M, Morgante M, Boerjan W, Sabatti M and Scarascia Mugnozza G 2012 Accelerating the domestication of forest trees in a changing world *Trends Plant Sci.* **17** 64–72
[35] Wallace J G, Rodgers-Melnick E and Buckler E S 2018 On the road to breeding 4.0: unraveling the good, the bad, and the boring of crop quantitative genomics *Ann. Rev. Gene.* **52** 421–44
[36] Collard B C Y and Mackill D J 2008 Marker-assisted selection: an approach for precision plant breeding in the twenty-first century *Philos. Trans R. Soc. B. Biol. Sci.* **363** 557–72
[37] Harfouche A, Jacobson D, Kainer D, Romero C J, Harfouche A, Mugnozza G, Moshelion M, Tuskan G A, Keurentjes J and Altman A 2019 Accelerating climate resilient plant breeding by applying next-generation artificial intelligence *Trends Biotechnol.* **37** 1217–35
[38] Urano K, Kurihara Y, Seki M and Shinozaki K 2010 Omics' analyses of regulatory networks in plant abiotic stress responses *Curr. Opin. Plant Biol.* **13** 132–8
[39] Gnanapragasam N *et al* 2024 Extreme trait GWAS (Et-GWAS): unraveling rare variants in the 3,000-rice genome *Life Sci. Alliance* **7** e202302352

[40] Cardi T, Batelli G and Nicolia A 2017 Opportunities for genome editing in vegetable crops *Emerg. Top Life Sci.* **1** 193–207
[41] Gaj T, Gersbach C A and Barbas C F 2013 ZFN, TALEN, and CRISPR/Cas-based methods for genome engineering *Trends Biotechnol.* **31** 397–405
[42] Jinek M, Chylinski K, Fonfara I, Hauer M, Doudna J A and Charpentier E 2012 A programmable dual-RNA–guided DNA endonuclease in adaptive bacterial immunity *Science* **337** 816–21
[43] Kumar R, Nizampatnam N R, Alam M, Thakur T K and Kumar A 2021 Genome editing technologies for plant improvement: advances, applications and challenges *Omics Technologies for Sustainable Agriculture and Global Food Security* **vol 1** 1st edn A Kumar, R Kumar, P Shukla and M K Pandey (Singapore: Springer) pp 213–40
[44] Xiong J S, Ding J and Li Y 2015 Genome-editing technologies and their potential application in horticultural crop breeding *Hortic. Res.* **2** 15019
[45] Shen B *et al* 2014 Efficient genome modification by CRISPR-Cas9 nickase with minimal off-target effects *Nat. Methods* **11** 399–402
[46] Gasiunas G, Barrangou R, Horvath P and SikSnys V 2012 Cas9-crRNA ribonucleoprotein complex mediates specific DNA cleavage for adaptive immunity in bacteria *Proc. Natl Acad. Sci. USA* **109** E2579–86
[47] Ran F A *et al* 2013 Double nicking by RNA-guided CRISPR Cas9 for enhanced genome editing specificity *Cell* **154** 1380–9
[48] Wang F, Wang C, Liu P, Lei C, Hao W, Gao Y, Liu Y G and Zhao K 2016 Enhanced rice blast resis tance by CRISPR/Cas9-targeted mutagenesis of the ERF transcription factor gene OsERF922 *PLoS One* **11** e0154027
[49] Appiano M, Catalano D, Santillán Martínez M, Lotti C, Zheng Z, Visser R G F, Ricciardi L, Bai Y and Pavan S 2015 Monocot and dicot mlo powdery mildew susceptibility factors are functionally conserved in spite of the evolution of class-specific molecular features *BMC Plant Biol.* **15** 257
[50] Wang Y, Cheng X, Shan Q, Zhang Y, Liu J, Gao C and Qiu J L 2014 Simultaneous editing of three homoeoalleles in hexaploid bread wheat confers heritable resistance to powdery mildew *Nat. Biotechnol.* **32** 947–51
[51] Zhang J *et al* 2016 LIN28 regulates stem cell metabolism and conversion to primed pluripotency *Cell Stem Cell* **19** 66–80
[52] de Toledo Thomazella D P, Brail Q, Dahlbeck D and Staskawicz B 2016 Loss of function of a DMR6 ortholog in tomato confers broad-spectrum disease resistance *PNAS* **118** e2026152118
[53] Taranto F, Nicolia A, Pavan S, De Vita P and D'Agostino N 2018 Biotechnological and digital revolution for climate-smart plant breeding *Agronomy* **8** 277

IOP Publishing

Advances in Biochemical and Molecular Mechanisms of Plant–Pathogen Interaction

Hitendra K Patel and Anirudh Kumar

Chapter 10

Impact of plant disease management in mitigating global food security

S J S Rama Devi and Ranjith Pamirelli

Phytopathogens (bacteria, fungi, oomycetes, viruses, insects and pests) are responsible for significant yield losses to agriculture every year. The present chapter aims to summarize the advancements accomplished for sustainable plant disease management by utilizing living organisms as biostimulants and biocontrol agents. Further, we discuss the efficiency of nanotechnology, bioinformatics, and molecular breeding techniques along with genome editing with relevance to disease management among the major food crops rice, wheat and maize. Also, we discuss proposed strategies to bridge the existing research gaps. The presented information will be extremely valuable to plant biologists to design future crop protection management strategies.

10.1 Introduction

The global population is projected to reach 9.7 billion by 2050. In order to ensure food security for the growing population, food production is estimated to increase by 56% [1]. Food security is defined as 'all the people, at all times have access to secure, nutritious and adequate food to meet the dietary necessities leading to a healthy life' [2]. According to a Food and Agriculture Organization (FAO) estimate, a total of 768 million people suffered malnutrition and one in every three globally faced hunger in 2020 [3]. Correspondingly, in India 189.2 million children (accounting for 14% of the total population) are undernourished, ranking it 101 among the Global Hunger Index 2021, which is a serious concern that needs to be addressed to ensure national food security (https://www.fao.org/3/ca9692en/online/ca9692en.html;https://www.globalhungerindex.org/india.html).

Plant diseases caused by phytopathogens are a severe threat to agriculture. Phytopathogens are responsible for reducing agricultural productivity qualitatively and quantitatively. Historically, plant diseases have been accountable for famines

doi:10.1088/978-0-7503-5673-2ch10

leading to millions of deaths. For instance, the late blight of potato caused by *Phytophthora infestans* led to the great Irish famine (1845) and brown spot-on rice caused by *Cochliobolus miyabeanus* resulted in the Indian Bengal famine. Additionally, global climate change also contributes to the increase in crop losses. Recent studies have demonstrated the impact of climate change on phytopathogen distribution, altered lifecycle and evolution of new strains/races [4]. It is estimated that every degree Celsius increase in global temperature may increase crop losses by 10%–25% in staple food crops like wheat, rice, and maize [5]. Keeping the above-cited bottlenecks in view, the sustainable development goal SDG 2 (zero hunger) was framed to increase food production. According to the FAO and IPPC 2021, plant health management is directly involved in meeting the seven sustainable development goals. Also, the United Nations declared the year 2020 as the International Year of Plant Health (IYPH) so as to raise awareness for plant disease protection. This signifies the importance of plant disease management in increasing food production.

Phytopathogens are responsible for significant yield losses to agricultural productivity both at pre and post-harvest stages. It is estimated that approximately 100 000 diseases are caused by diverse phytopathogens (like bacteria; viruses; fungi and other microbes) [6]. The definition and consequences of plant diseases in relevance to food security have previously been discussed with examples [7]. Statistically, insects are responsible for 5%–20% reduction in crop yields for major food crops [5]. Similarly, plant pests and parasitic nematodes are accountable for 40% and 12.3% of global food loss worth 220 and 173 billion dollars each year [8, 9]. Correspondingly, India accounts for 21.3% of major crop losses worth 1.58 billion dollars by parasitic nematodes [10]. Savary *et al* conducted a first-of-its-kind survey for crop health assessment across 67 countries covering 137 pathogens and pests among the five staple crops (wheat, rice, maize, potato and soybean) [11]. The percentage yield losses caused by various phytopathogens in staple food crops such as rice, wheat and maize are represented in figure 10.1.

Disease management includes maintenance as well as improvement in plant health and production. Hence, it is essential to have advanced disease detection methods and prevention strategies so as to reduce the impact of plant diseases over agricultural productivity. This chapter aims to summarize the recent scientific advancements for controlling plant diseases. In this direction, the roles of biostimulants along with

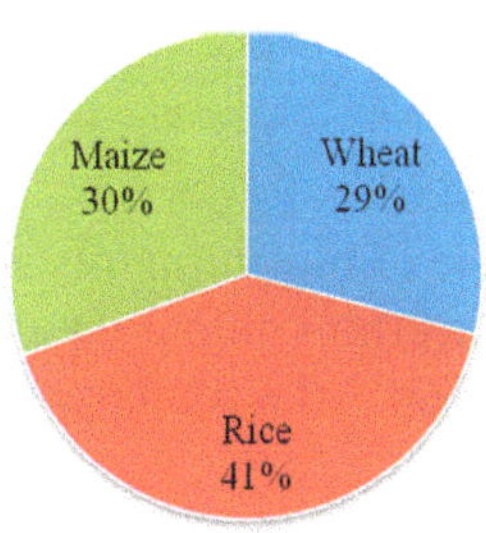

Figure10.1. Global yield loss estimate caused by diseases and pests (source: adopted and modified from [11]).

biocontrol agents were discussed as a part of sustainable plant disease management strategies. Further, the utilization of molecular breeding, nanotechnology, bioinformatics and genome editing technologies are emphasized in the context of plant disease management. The chapter can serve as an updated multidimensional resource for plant biologists to investigate and frame new approaches for plant disease management.

10.2 Sustainable plant disease management strategies

Sustainable plant disease management includes the utilization of beneficial microorganisms for efficient and sustainable crop production. Numerous soil bacteria and fungi stimulate plant growth and development as well as counteracting pathogen attack either directly or indirectly [12]. These include biostimulants along with biocontrol agents. Biostimulants are the formulations of biological origin which have the capability to stimulate biological processes like nutrient uptake, improving plant health, crop quality and developing resistance towards biotic as well as abiotic stresses [13, 14]. The biostimulants include varied types of biomolecules or substances of biological origin like: humic acid [15]; fulvic acid [16]; microalgae [17]; arbuscular mycorrhiza fungi (AMF) [18]; and plant growth promoting rhizobacteria (PGPRs) [14]. Biostimulants are being used in agricultural crops [19] and horticultural crops [20, 21], developing biotic and abiotic stress tolerance [22, 23], thus aiding in sustainable plant production [24, 25]. The biostimulants are effective in protecting the plants from pathogens. For example, the application of biostimulants on zucchini plants infected with tomato leaf curl New Delhi virus increased the plant fitness towards the disease [26]. The endophytes belonging to the genus, *Bacillus* had shown excellent antifungal activity, and treatment of unhealthy rice seeds with endophytic bacteria could recover seed fertility and ensure protection from fungal pathogens during the lifecycle [27]. Likewise, chitosan is a biocompatible, biodegradable and cost-effective polymer which acts as a natural elicitor and induces defence responses by downstream signalling mechanisms [28]. The fungal chitosan produced by growing *Aspergillus niger* on rice straw was demonstrated as a sustainable approach to improve disease resistance in rice against bacterial leaf blight and blast [29]. FytoSol (a combination of chitosan and pectine derived oligo galacturonide) had proved to be effective and a promising alternative to fungicide against tomato early blight [30]. The application of biostimulants has emerged as a promising alternative for the reduced utilization of pesticides and fertilizers [31, 32]. There are numerous commercial biostimulants are available today for agricultural and horticultural important crops [14]. The market value of biostimulants was 3.2 billion US dollars and is expected to reach 5.6 billion US dollars by 2026 (https://www.marketsandmarkets.com/Market-Reports/biostimulant-market-1081.html). The main challenges for the usage of biostimulants are the commercialization of low quality commercial products; uncertain global regulatory framework. Recently, the European Union, Regulation (EU) 2019/1009, recognized the use of biostimulants within the scope of organic fertilizers [33].

Biocontrol agents can be defined as the utilization of natural or modified microbial strains that can reduce the incidence or severity of plant diseases caused by phytopathogens [34]. Biocontrol agents can be used as antifungal, nematicidal, antibacterial, biopesticides and bioinsecticides. For instance, the genus *Streptomyces* has proved to be effective against plant fungal pathogens [35]. Similarly, three rhizobacterial strains *Bacillus subtilis*, *Bacillus thuringiensis* and *Enterobacter cloacae* were proved to be antifungal and inhibit the growth of *Rhizoctonia solani* causing fungal root rot [36]. Likewise, many endophytic bacteria were proven to be effective against phytopathogenic fungi and nematodes [37]. Additionally, bacteriophages were also proved efficient for controlling phytobacteria [38, 39]. Use of fungal derived biopesticide through integrated pest management (IPM) has proved to be an effective alternative to minimize the use of chemical insecticides [40]. *Metarhizium anisopliae* is an entomo-pathogenic fungal strain identified as a potential biocontrol agent for three major rice insect pests (i.e. *Nilaparvatalugens*—rice planthopper, *Cnaphalocrocismedinalis*—causes leaf rolling and *Chilosuppressalis*—rice stem-borer) [41]. The potential biocontrol properties of *Trichoderma* against several diseases associated with agricultural crops was also reviewed earlier [42]. Likewise, Tariq *et al* has summarized the bacterial and fungal strains which are reported as biocontrol agents against diverse phytopathogens [43]. Biocontrol agents interact with plants either directly through nitrogen fixation [44]; phosphate solubilization [45]; by inducing phytoharmones [46]; and siderophore production [47]; or indirectly by producing the HCN [48]; antibiotics [49]; antimicrobial compounds and enhancing induced systemic resistance (ISR) [50]. The commercial manufacturing of biocontrol agents and their mode of action were earlier reviewed [51]. The green revolution has significantly increased the global food production. However, the indiscriminate utilization of pesticides and herbicides in agriculture has created serious environmental and human health concerns. Thus, exploiting the biocontrol agents in combating the phytopathogens and insect pests has proved to be an effective alternative strategy. In addition, it is an eco-friendly approach for sustainable plant disease management. However, the commercialization of biocontrol agents has many constraints from field to reach the market. This includes proper identification and assessment of biocontrol agents followed by government registration to use in sustainable agriculture production [52].

10.3 Scope of new technologies for plant disease management

The following section aims in summarizing the multi-disciplinary progress achieved through new scientific technologies or advancements to understand plant diseases and their control mechanisms in relevance to major food crops rice, wheat and maize.

10.3.1 Role of nanotechnology in plant disease management

Nanoparticles are tiny molecules of size ranging between 1 and 100 nanometres (nm). Nanotechnology has gained importance in the recent past because of its specificity, efficacy and low toxicity towards soil and water, and further ensures resistance against varied phytopathogens [53]. They are deployed in agriculture for detection and control of phytopathogens and pests [54, 55]. Nanomaterials can be exploited as carriers to carry actives like pesticides, insecticides or small interfering RNA (siRNA) targeting a wide range of phytopathogens and pests [56]. Chitosan-based nanoparticles were proved to be promising in plant health management [57]. Green nanotechnology is a novel approach of combination of plant extracts or microorganisms coated with metal nanoparticles which have antibacterial and antifungal activities [58]. Additionally, researchers have developed nanobiosensors which are useful for the early recognition of plant diseases and their management [59, 60]. Nanobiosensors allow rapid, sensitive detection [61] and monitoring of phytopathogens [62]. They have several applications in plant stress management [63]; pesticide detection [64] and also for sustainable agricultural practices [65]. For instance, nanotechnology has been successfully deployed for rice fungal diseases [66], for insect pest management [67] and plant viruses' management [68, 69]. The potential of nanotechnology in wheat and barley disease management was reviewed earlier [70]. Additionally, nanotechnology is being employed to increase the shelf-life of perishable agricultural products through post-harvest technology [71]. The scope of nanotechnology in agriculture has been critically reviewed recently [72]. In spite of numerous advantages of nanotechnology in agriculture, several factors like long-term usage in the field for environmental safety and off target affects need to be addressed by generating data followed by critical analysis. Hence it is necessary to establish collaborative and integrated research between plant biologists, ecologists and material scientists. This information is essential to make harmonized policies for effective utilization of nanoparticles in plant health management.

10.3.2 Role of bioinformatics in plant disease management

Bioinformatics is the study of large biological information with the help of computers. In the recent past this area has gained much attention in biological studies to understand the complex molecular data. Thus, it has become a novel tool for plant researchers to achieve sustainable agriculture [73]. In this context, in order to understand the complex plant–pathogen interactions and develop different disease management strategies, bioinformatics-based tools were developed for comparative analysis of genomes, to decipher plant as well as pathogen evolutionary analysis etc. For example, *in silico* tools like PATRIC (https://www.patricbrc.org) and Phytopath (http://www.phytopathdb.org) were developed to understand pathogen genomes [74, 75]. Also, plant resistance proteins can be predicted using RRGPredictor (https://github.com/RanerJSS/RRG_predictor) and RFPDR—Random Forest (RF) for Plant Disease Resistance (PDR) protein prediction) (https://github.com/cvfilippi/rfpdr) [76, 77]. Likewise, the plant–pathogen interactions can be studied using PHI-base (http://www.phi-base.org) [78]. Recently the utilization, advantages and limitations of various bioinformatics tools in plant disease management were critically reviewed [79].

10.3.3 Molecular breeding for plant disease management

Genomics-assisted breeding (GAB) and genome-wide association studies (GWAS) were identified as alternative approaches for novel gene(s) identification, for development of disease resistance crops, as well as for the development of climate-resilient crop varieties so as to ensure food and nutritional security [80]. GWAS in rice has helped in identifying the resistance loci from wild species *Oryza rufipogon* against sheath blight [81, 82] and bacterial blight [83]. Likewise, in wheat the resistance loci against powdery mildew [84]; Karnal bunt disease [85]; stripe, stem, and leaf rust [86–88], *Septoria tritici* blotch [89], black point disease [90] were identified using GWAS. Similarly, GWAS was successfully utilized to decipher the disease resistance in maize [91]. For example, resistance genes/loci were identified against maize ear rot fungus *Fusarium verticillioides* [92], corn leaf blight [93, 94] and rust [95]. In contrast, GWAS is also successfully deployed from a pathogen perspective too. For example, the pathogenicity loci were identified in rice neck blast pathogen *Magnaporthe oryzae* [96], in wheat *Septoria nodorum* leaf blotch (SNB) and glume blotch causing fungus *Parastagonospora nodorum* [97]. However, not many GWAS were carried out among the varied phytopathogens [98]. Bridging this research gap may certainly give more clues about the evolutionary dynamics of phytopathogens and plant–pathogen interactions.

Mining new resistance gene(s) and utilizing in breeding programs was proposed to be the effective strategy for plant disease management [99, 100]. Gene pyramiding (combination of two or more resistance genes) proved effective for the development of durable disease resistance [101]. In rice, gene pyramiding is highly successful in the development of resistance against major rice diseases. The recent advances in gene pyramiding for biotic stress resistance in rice was reviewed in the past [102]. For instance, the disease resistance in rice was developed against bacterial leaf blight (BLB) [103], blast [104], BLB and gall midge [105], BLB and blast [106–108], blast, BLB and BPH [109], blast and sheath blight [110], BLB, blast and sheath blight [111]. Similarly, marker-assisted selection (MAS) was successfully deployed for wheat rust management [112–116] and powdery mildew management [117–119]. The pyramided lines with leaf rust and fusarium head blight (FHB) in wheat have shown more resistance than controls [120]. Similarly, gene pyramiding with resistance genes, *Cre1* and *Cre3* and *Cre8* is proposed to be effective to in controlling cereal cyst nematode in wheat [121, 122]. Gene pyramiding in maize is successful in achieving resistance against necrotrophic pathogens [123]. The pyramiding of *cry* genes (*Cry1Ie2*, *Cry1Ab* or *Cry1Fa*) in corn was proposed to be affective against the Europeon corn borer (*Ostrinia nubilalis*) [124].

10.3.4 Genome editing technologies for plant disease management

Genome editing technology has potential implications in mitigating pre- and post-harvest losses without compromising the yield, quality and nutritional qualities in agriculturally important crops like cereals [125], in improving the shelf-life of fruit and vegetables [126] and also for plant disease management [127, 128]. For example, the genome editing advancements for rice blast disease resistance [129], for wheat improvement [130] and for maize banded leaf and sheath blight (BLSB) disease resistance [131] were collated lucidly earlier. Similarly, harnessing the genome editing tools for development of virus resistance in cereals and other plants were

Table10.1. List of genes which are targeted for CRISPR/Cas genome editing in staple food crops rice, wheat and maize for disease resistance crop improvement.

Pathogen type	Host plant	Disease	*Pathogen name*	*Target gene*	References
Bacteria	Rice	Bacterial leaf blight	*Xanthomonas oryzaepv. oryzae.*	*OsSWEET13*	[139, 140]
				Os09g29100	[141]
				OsSWEET11	[142]
				OsSWEET14	[143, 144]
Fungus	Rice	Blast resistance	*Magnaporthe grisea*	*Bsr-dl; Pi21; OsERF922*	[145, 146]
				OsSEC3A	[147]
				Pita	[148]
		Sheath blight	*Rhizoctonia solani*	*OsPFT1*	[149]
	Maize	Banded leaf and sheath blight	*Rhizoctonia solani*	*GRMZM2G700188 (ZmCAD)*	[150]
	Wheat	stripe rust	*Puccinia striiformis f. sp. tritici*	*TaPsIPK1*	[151, 152]
		Fusarium head blight	*Fusarium graminearum*	*TaHRC*	[153]
		Powdery mildew	*Blumeriagraminisf. sp. tritici*	*TaMLO-A1*	[154]
				Ta EDR 1	[155]
Virus	Rice		*Rice tungro virus*	*eIF4G*	[156, 157]

also compiled [132, 133]. The genes targeted for CRISPR/Cas editing in the three staple food crops rice, wheat and maize against varied phytopathogens are compiled in table 10.1. Additionally, CRISPR/Cas9 has also been utilized for precisely editing the fungal and oomycetes genomes as a new strategy for plant disease management. For instance, the rice blast fungus, *Magnaporthe oryzae* has been successfully edited to understand its functional genomics [134]. Paul *et al* reviewed the recent advancements of CRISPR/Cas technology adopted for editing the genomes of phytopathogens, fungi and oomycetes [135].

Furthermore, CRISPR-based detection tools are also gaining importance in relevance to plant disease management. For example, the CRISPR/Cas12a-based DNA detection method was proved to be efficient in rice pathogen detection [136]. Likewise, CRISPR-based genotyping and pathogen detection tool named BIOSCAN was recently developed and validated in elite wheat germplasm [137]. To increase the precision in onsite detection of viruses, CRISPR/Cas technology was coupled with the existing assays for plant viruses' visual detection [138]. Thus, genome editing has wide and immense scope for plant disease resistance improvement programs. The precision in genome editing is highly useful for researchers to gain in-depth insights towards plant–pathogen interactions. In addition, CRISPR/Cas technology can efficiently overcome the limitations of conventional breeding

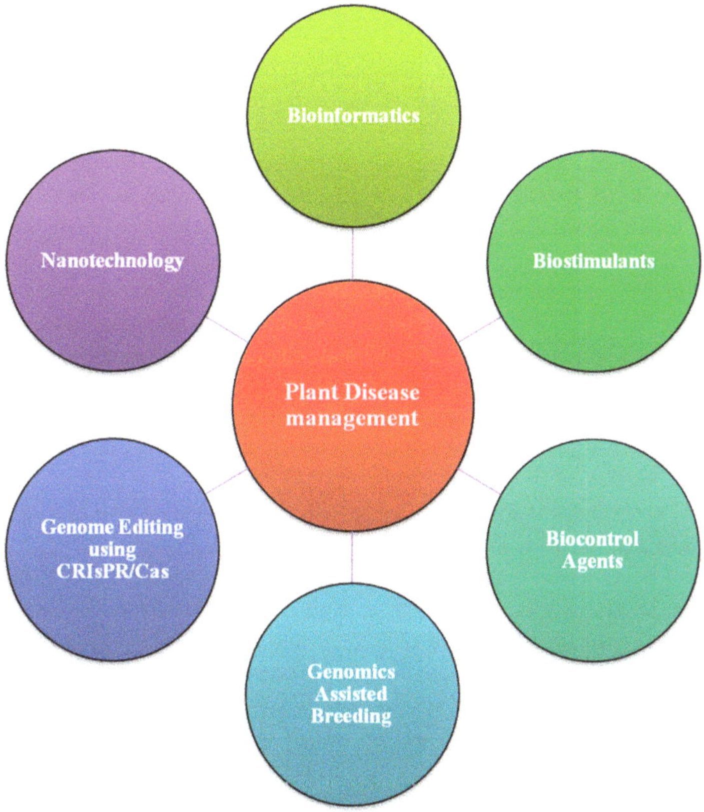

Figure10.2. Summary of different strategies used for plant disease management.

and can accelerate disease resistance crop improvement programs. The different strategies used for plant disease management are summarized in figure 10.2.

10.4 Conclusions

In the recent past, several efforts have been made by international researchers to increase food production and reduce the yield losses caused by diseases and pests. Conversely, the emerging new phytopathogen variants pose a serious threat to sustainable food production. Hence, it is essential to have scientifically valid continuous tracking systems to assess the timely variations among the pathogens. Although these kind of systemic studies were conducted earlier, they were limited to few crops and pathogens. These kinds of monitoring systems need to be established worldwide for all major food crops. Further, the asynchronized policies in nations are a major constraint to utilize the new technologies like CRISPR/Cas in crop improvement programs. This needs to be addressed, so as to adopt the new technology and develop new varieties for staple crops in a faster way. Finally, most of the bioinformatics and advanced research has been focused on model plant–pathogen systems, but they need to be extended for others too.

References

[1] Van Dijk M, Morley T, Rau M L and Saghai Y 2021 A meta-analysis of projected global food demand and population at risk of hunger for the period 2010–2050 *Nat. Food* **2** 494–501

[2] Food and Agriculture Organization of the United Nations 2019 *The State of Food and Agriculture 2019: Moving Forward on Food Loss and Waste Reduction* (Rome: Food and Agricultural Organization)

[3] FAO, IFAD, UNICEF, WFP and WHO 2021 The state of food security and nutrition in the world 2021 *Transforming Food Systems for Food Security, Improved Nutrition and Affordable Healthy Diets for All* (Rome: Food and Agriculture Organization)

[4] Chakraborty S and Newton A C 2011 Climate change, plant diseases and food security: an overview *Plant Pathol.* **60** 2–14

[5] Deutsch C A, Tewksbury J J, Tigchelaar M, Battisti D S, Merrill S C, Huey R B and Naylor R L 2018 Increase in crop losses to insect pests in a warming climate *Science* **361** 916–9

[6] Dhaliwal G S, Jindal V and Mohindru B 2015 Crop losses due to insect pests: global and Indian scenario *Indian J. Entomol.* **77** 165–8

[7] Ristaino J B, Anderson P K, Bebber D P, Brauman K A, Cunniffe N J, Fedoroff N V and Wei Q 2021 The persistent threat of emerging plant disease pandemics to global food security *Proc. Natl Acad. Sci.* **118** e2022239118

[8] IPPC Secretariat 2021 International year of plant health—final report. Protecting plants, protecting life *FAO on Behalf of the Secretariat of the International Plant Protection Convention* FAO 10.4060/cb7056en

[9] Elling A A 2013 Major emerging problems with minor Meloidogyne species *Phytopathology* **103** 1092–102

[10] Kumar V, Khan M R and Walia R K 2020 Crop loss estimations due to plant-parasitic nematodes in major crops in India *Natl. Acad. Sci. Lett.* **43** 409–12

[11] Savary S, Willocquet L, Pethybridge S J, Esker P, McRoberts N and Nelson A 2019 The global burden of pathogens and pests on major food crops *Nat. Ecol. Evol.* **3** 430–9

[12] Harish S, Parthasarathy S, Durgadevi D, Anandhi K and Raguchander T 2019 Plant growth-promoting rhizobacteria: harnessing its potential for sustainable plant disease management *Plant Growth Promoting Rhizobacteria for Agricultural Sustainability* (Singapore: Springer) pp 151–87

[13] Du Jardin P 2015 Plant biostimulants: definition, concept, main categories and regulation *Sci. Hortic.* **196** 3–14

[14] Hamid B, Zaman M, Farooq S, Fatima S, Sayyed R Z, Baba Z A and Suriani N L 2021 Bacterial plant biostimulants: a sustainable way towards improving growth, productivity, and health of crops *Sustainability* **13** 2856

[15] Pačuta V, Rašovský M, Michalska-Klimczak B and Wyszyňski Z 2021 Grain yield and quality traits of durum wheat (Triticum durum Desf.) treated with seaweed-and humic acid-based biostimulants *Agronomy* **11** 1270

[16] Canellas L P, Olivares F L, Aguiar N O, Jones D L, Nebbioso A, Mazzei P and Piccolo A 2015 Humic and fulvic acids as biostimulants in horticulture *Sci. Hortic.* **196** 15–27

[17] Ronga D, Biazzi E, Parati K, Carminati D, Carminati E and Tava A 2019 Microalgal biostimulants and biofertilisers in crop productions *Agronomy* **9** 192

[18] Bitterlich M, Rouphael Y, Graefe J and Franken P 2018 Arbuscular mycorrhizas: a promising component of plant production systems provided favorable conditions for their growth *Front. Plant Sci.* **9** 1329
[19] Calvo P, Nelson L and Kloepper J W 2014 Agricultural uses of plant biostimulants *Plant Soil.* **383** 3–41
[20] Colla G and Rouphael Y 2015 Biostimulants in horticulture *Sci. Hortic.* **196** 1–134
[21] Parađiković N, Teklić T, Zeljković S, Lisjak M and Špoljarević M 2019 Biostimulants research in some horticultural plant species—a review *Food Energy Secur.* **8** e00162
[22] Sharma H S, Fleming C, Selby C, Rao J R and Martin T 2014 Plant biostimulants: a review on the processing of macroalgae and use of extracts for crop management to reduce abiotic and biotic stresses *J. Appl. Phycol.* **26** 465–90
[23] Dubey A, Kumar A and Khan M L 2020 Role of biostimulants for enhancing abiotic stress tolerance in Fabaceae plants *The Plant Family Fabaceae* (Singapore: Springer) pp 223–36
[24] Caradonia F, Ronga D, Tava A and Francia E 2021 Plant biostimulants in sustainable potato production: an overview *Potato Res.* **65** 83–104
[25] Ganugi P, Martinelli E and Lucini L 2021 Microbial biostimulants as a sustainable approach to improve the functional quality in plant-based foods: a review *Curr. Opin. Food Sci.* **41** 217–23
[26] Donati L, Bertin S, Gentili A, Luigi M, Taglienti A, Manglli A and Ferretti L 2022 Effects of organic biostimulants added with Zeolite on Zucchini squash plants infected by tomato leaf curl New Delhi virus *Viruses* **14** 607
[27] Rangjaroen C, Lumyong S, Sloan W T and Sungthong R 2019 Herbicide-tolerant endophytic bacteria of rice plants as the biopriming agents for fertility recovery and disease suppression of unhealthy rice seeds *BMC Plant Biol.* **19** 1–16
[28] Riseh R S, Tamanadar E, Hajabdollahi N, Vatankhah M, Thakur V K and Skorik Y A 2022 Chitosan microencapsulation of rhizobacteria for biological control of plant pests and diseases: recent advances and applications *Rhizosphere* **23** 100565
[29] Stanley-Raja V, Senthil-Nathan S, Chanthini K M P, Sivanesh H, Ramasubramanian R, Karthi S and Kalaivani K 2021 Biological activity of chitosan inducing resistance efficiency of rice (Oryza sativa L.) after treatment with fungal based chitosan *Sci. Rep.* **11** 20488
[30] Bektas Y 2022 FytoSol, a promising plant defense elicitor, controls early blight (Alternaria solani) disease in the tomato by inducing host resistance-associated gene expression *Horticulturae* **8** 484
[31] Rouphael Y and Colla G 2018 Synergistic biostimulatory action: designing the next generation of plant biostimulants for sustainable agriculture *Front. Plant Sci.* **9** 1655
[32] Rouphael Y and Colla G 2020 Toward a sustainable agriculture through plant biostimulants: from experimental data to practical applications *Agronomy* **10** 1461
[33] Barros-Rodríguez A, Rangseekaew P, Lasudee K, Pathom-Aree W and Manzanera M 2020 Regulatory risks associated with bacteria as biostimulants and biofertilizers in the frame of the European Regulation (EU) 2019/1009 *Sci. Total Environ.* **740** 140239
[34] Gupta M, Topgyal T, Zahoor A and Gupta S 2021 Rhizobium: eco-friendly microbes for global food security *Microbial Management of Plant Stresses* (Cambridge: Woodhead Publishing) pp 221–33
[35] LeBlanc N 2022 Bacteria in the genus Streptomyces are effective biological control agents for management of fungal plant pathogens: a meta-analysis *BioControl* **67** 111–21

[36] Abdeljalil N O B, Vallance J, Gerbore J, Yacoub A, Daami-Remadi M and Rey P 2021 Combining potential oomycete and bacterial biocontrol agents as a tool to fight tomato Rhizoctonia root rot *BiolControl.* **155** 104521
[37] Daulagala W H 2021 Chitinolytic endophytic bacteria as biocontrol agents for phytopathogenic fungi and nematode pests: a review *Asian J. Biotechnol.* **5** 14–24
[38] Sabri M, Benkirane R, Habbadi K, Sadik S, Ou-Zine M, Diouri M and Achbani E H 2021 Phages as a potential biocontrol of phytobacteria *Arch. Phytopathol. Plant Protect.* **54** 1277–91
[39] Holtappels D, Fortuna K, Lavigne R and Wagemans J 2021 The future of phage biocontrol in integrated plant protection for sustainable crop production *Curr. Opin. Biotechnol.* **68** 60–71
[40] Jin S F, Feng M G, Ying S H, Mu W J and Chen J Q 2011 Evaluation of alternative rice planthopper control by the combined action of oil-formulated Metarhizium anisopliae and low-rate buprofezin *Pest Manag. Sci.* **67** 36–43
[41] Peng G, Xie J, Guo R, Keyhani N O, Zeng D, Yang P and Xia Y 2021 Long-term field evaluation and large-scale application of a Metarhizium anisopliae strain for controlling major rice pests *J. Pest Sci.* **94** 969–80
[42] Ferreira F V and Musumeci M A 2021 Trichoderma as biological control agent: scope and prospects to improve efficacy *World J. Microbiol. Biotechnol.* **37** 1–17
[43] Tariq M, Khan A, Asif M, Khan F, Ansari T, Shariq M and Siddiqui M A 2020 Biological control: a sustainable and practical approach for plant disease management *Acta Agric. Scand. B Soil Plant Sci.* **70** 507–24
[44] Das K, Prasanna R and Saxena A K 2017 Rhizobia: a potential biocontrol agent for soilborne fungal pathogens *Folia Microbiol.* **62** 425–35
[45] Mitra D, Anđelković S, Panneerselvam P, Senapati A, Vasić T, Ganeshamurthy A N and Radha T K 2020 Phosphate-solubilizing microbes and biocontrol agent for plant nutrition and protection: current perspective *Commun. Soil Sci. Plant Anal.* **51** 645–57
[46] Kawaguchi A and Noutoshi Y 2022 Insight into inducing disease resistance with Allorhizobium vitis strain ARK-1, a biological control agent against grapevine crown gall disease *Eur. J. Plant Pathol.* **162** 981–7
[47] Chaiharn M, Chunhaleuchanon S and Lumyong S 2009 Screening siderophore producing bacteria as potential biological control agent for fungal rice pathogens in Thailand *World J. Microbiol. Biotechnol.* **25** 1919–28
[48] Azeem S, Agha S I, Jamil N, Tabassum B, Ahmed S, Raheem A and Khan A 2022 Characterization and survival of broad-spectrum biocontrol agents against phytopathogenic fungi *Rev. Argent. Microbiol.* **54** 233–42
[49] Junaid J M, Dar N A, Bhat T A, Bhat A H and Bhat M A 2013 Commercial biocontrol agents and their mechanism of action in the management of plant pathogens *Int. J. Mod. Plant Anim. Sci.* **1** 39–57
[50] Ngalimat M S, Yahaya R S R, Baharudin M M A A, Yaminudin S M, Karim M, Ahmad S A and Sabri S 2021 A review on the biotechnological applications of the operational group Bacillus amyloliquefaciens *Microorganisms* **9** 614
[51] Teixidó N, Usall J and Torres R 2022 Insight into a successful development of biocontrol agents: production, formulation, packaging, and self life as key aspects *Horticulturae* **8** 305
[52] Velivelli S L, De Vos P, Kromann P, Declerck S and Prestwich B D 2014 Biological control agents: from field to market, problems, and challenges *Trends Biotechnol.* **32** 493–6

[53] Singh A, Rajput V D, Rawat S, Sharma R, Singh A K, Kumar P and Singh S 2022 Geoinformatics and nanotechnological approaches for coping up abiotic and biotic stress in crop plants *Sustainable Agriculture Systems and Technologies* (Wiley) ch 17 pp 337–59
[54] Younas A, Yousaf Z, Rashid M, Riaz N, Fiaz S, Aftab A and Haung S 2020 Nanotechnology and plant disease diagnosis and management *Nanoagronomy* (Cham: Springer) pp 101–23
[55] Sahoo B, Rath S K, Mahanta S K and Arakha M 2022 Nanotechnology mediated detection and control of phytopathogens *Bio-Nano Interface* (Singapore: Springer) pp 109–25
[56] Worrall E A, Hamid A, Mody K T, Mitter N and Pappu H R 2018 Nanotechnology for plant disease management *Agronomy* **8** 285
[57] Hoang N H, Le Thanh T, Sangpueak R, Treekoon J, Saengchan C, Thepbandit W and Buensanteai N 2022 Chitosan nanoparticles-based ionic gelation method: a promising candidate for plant disease management *Polymers* **14** 662
[58] Nargund V B, Vinay J U, Basavesha K N, Chikkanna S, Jahagirdar S and Patil R R 2021 Green nanotechnology and its application in plant disease management *Emerging Trends in Plant Pathology* (Singapore: Springer) pp 591–609
[59] Sellappan L, Manoharan S, Sanmugam A and Anh N T 2022 Role of nanobiosensors and biosensors for plant virus detection *Nanosensors for Smart Agriculture* (Amsterdam: Elsevier) pp 493–506
[60] Marinello F, Scaramuzzo F A, Dinarelli S, Passeri D and Rossi M 2022 Nanoscale characterization methods in plant disease management *Nanotechnology-Based Sustainable Alternatives for the Management of Plant Diseases* (Amsterdam: Elsevier) pp 149–77
[61] Avila-Quezada G D, Golinska P and Rai M 2022 Erratum: Engineered nanomaterials in plant diseases: can we combat phytopathogens? *Appl. Microbiol. Biotechnol.* **106** 1771
[62] John S A, Chattree A, Ramteke P W, Shanthy P, Nguyen T A and Rajendran S 2022 Nanosensors for plant health monitoring *Nanosensors for Smart Agriculture* (Amsterdam: Elsevier) pp 449–61
[63] Kordrostami M, Mafakheri M and Al-Khayri J M 2021 Contributions of nano biosensors in managing rnvironmental plant stress under climatic changing era *Nanobiotechnology* (Cham: Springer) pp 117–37
[64] Zia R, Taj A, Younis S, Bukhari S Z, Latif F, Feroz Y and Bajwa S Z 2022 Application of nanosensors for pesticide detection *Nanosensors for Smart Agriculture* (Amsterdam: Elsevier) pp 259–302
[65] Zhu L, Chen L, Gu J, Ma H and Wu H 2022 Carbon-based nanomaterials for sustainable agriculture: their application as light converters, nanosensors, and delivery tools *Plants* **11** 511
[66] Ahmad S, Husnain M G, Iqbal Z, Ghazanfar M U, ur Rehman F, Ahmad I and Ahmad S 2022 Nanotechnology for rice fungal diseases *Modern Techniques of Rice Crop Production* (Singapore: Springer) pp 493–515
[67] Rai M and Ingle A 2012 Role of nanotechnology in agriculture with special reference to management of insect pests *Appl. Microbiol. Biotechnol.* **94** 287–93
[68] Farooq T, Adeel M, He Z, Umar M, Shakoor N, da Silva W and Rui Y 2021 Nanotechnology and plant viruses: an emerging disease management approach for resistant pathogens *ACS Nano.* **15** 6030–7
[69] Shidore T, Zuverza-Mena N, White J C and da Silva W 2021 Nanoenabled delivery of RNA molecules for prolonged antiviral protection in crop plants: a review *ACS Appl. Nano Mater.* **4** 12891–904

[70] Kashyap P L, Kumar S, Kaul N, Aggarwal S K, Jasrotia P, Bhardwaj A K and Singh G P 2022 Nanotechnology for wheat and barley health management: current scenario and future prospectus *New Horizons in Wheat and Barley Research* (Singapore: Springer) pp 337–63
[71] Jena B, Ningthoujam R, Pattanayak S, Dash S, Panda M K, Jit B P and Singh Y D 2022 Nanotechnology and its potential application in postharvest technology *Bio-Nano Interface* (Singapore: Springer) pp 93–107
[72] Reddy S S and Chhabra V 2022 Nanotechnology: its scope in agriculture *J. Phys. Conf. Ser.* **2267** 012112
[73] Mishra D P, Badajena J C, Nayak S K and Baliyarsingh B 2022 Bioinformatics: a tool for sustainable agriculture *Advances in Agricultural and Industrial Microbiology* (Singapore: Springer) pp 233–46
[74] Wattam A R, Abraham D, Dalay O, Disz T L, Driscoll T, Gabbard J L and Sobral B W 2014 PATRIC, the bacterial bioinformatics database and analysis resource *Nucleic Acids Res.* **42** D581–91
[75] Pedro H, Maheswari U, Urban M, Irvine A G, Cuzick A, McDowall M D and Kersey P J 2016 PhytoPath: an integrative resource for plant pathogen genomics *Nucleic Acids Res.* **44** D688–93
[76] Silva R J S and Micheli F 2020 RRGPredictor, a set-theory-based tool for predicting pathogen-associated molecular pattern receptors (PRRs) and resistance (R) proteins from plants *Genomics* **112** 2666–76
[77] Simón D, Borsani O and Filippi C V 2022 RFPDR: a random forest approach for plant disease resistance protein prediction *PeerJ* **10** e11683
[78] Urban M, Cuzick A, Seager J, Wood V, Rutherford K, Venkatesh S Y and Hammond-Kosack K E 2020 PHI-base: the pathogen–host interactions database *Nucleic Acids Res.* **48** D613–20
[79] Dong A Y, Wang Z, Huang J J, Song B A and Hao G F 2021 Bioinformatic tools support decision-making in plant disease management *Trends Plant Sci.* **26** 953–67
[80] Mir R R, Rustgi S, Zhang Y M and Xu C 2022 Multi-faceted approaches for breeding nutrient-dense, disease-resistant, and climate-resilient crop varieties for food and nutritional security *Heredity* **128** 387–90
[81] Aggarwal S K, Malik P, Neelam K, Kumar K, Kaur R, Lore J S and Singh K 2022 Genome-wide association mapping for identification of sheath blight resistance loci from wild rice Oryza rufipogon *Euphytica* **218** 1–20
[82] Li D, Zhang F, Pinson S R, Edwards J D, Jackson A K, Xia X and Eizenga G C 2022 Assessment of rice sheath blight resistance including associations with plant architecture, as revealed by genome-wide association studies *Rice* **15** 1–30
[83] Kim S M and Reinke R F 2019 A novel resistance gene for bacterial blight in rice, Xa43(t) identified by GWAS, confirmed by QTL mapping using a bi-parental population *PLoS One* **14** e0211775
[84] Li G, Xu X, Tan C, Carver B F, Bai G, Wang X and Cowger C 2019 Identification of powdery mildew resistance loci in wheat by integrating genome-wide association study (GWAS) and linkage mapping *Crop J.* **7** 294–306
[85] Singh S, Sehgal D, Kumar S, Arif M A R, Vikram P, Sansaloni C P and Ortiz C 2020 GWAS revealed a novel resistance locus on chromosome 4D for the quarantine disease Karnal bunt in diverse wheat pre-breeding germplasm *Sci. Rep.* **10** 1–11

[86] Kumar D, Kumar A, Chhokar V, Gangwar O P, Bhardwaj S C, Sivasamy M and Tiwari R 2020 Genome-wide association studies in diverse spring wheat panel for stripe, stem, and leaf rust resistance *Front. Plant Sci.* **11** 748

[87] Saleem K, Shokat S, Waheed M Q, Arshad H M I and Arif M A R 2022 A GBS-based GWAS analysis of leaf and stripe rust resistance in diverse pre-breeding germplasm of bread wheat (Triticum aestivum L.) *Plants* **11** 2363

[88] Aoun M, Rouse M N, Kolmer J A, Kumar A and Elias E M 2021 Genome-wide association studies reveal all-stage rust resistance loci in elite durum wheat genotypes *Front. Plant Sci.* **12** 640739

[89] Odilbekov F, Armonienė R, Koc A, Svensson J and Chawade A 2019 GWAS-assisted genomic prediction to predict resistance to Septoria tritici blotch in Nordic winter wheat at seedling stage *Front. Genet.* **10** 1224

[90] Li Q, Niu H, Xu K, Xu Q, Wang S, Liang X and Niu J 2020 GWAS for resistance against black point caused by Bipolaris sorokiniana in wheat *J. Cereal Sci.* **91** 102859

[91] Shrestha V, Awale M and Karn A 2019 Genome wide association study (GWAS) on disease resistance in maize *Disease Resistance in Crop Plants* (Cham: Springer) pp 113–30

[92] Stagnati L, Lanubile A, Samayoa L F, Bragalanti M, Giorni P, Busconi M and Marocco A 2019 A genome wide association study reveals markers and genes associated with resistance to Fusarium verticillioides infection of seedlings in a maize diversity panel *G3-Genes Genom Genet.* **9** 571–9

[93] Rashid Z, Sofi M, Harlapur S I, Kachapur R M, Dar Z A, Singh P K and Nair S K 2020 Genome-wide association studies in tropical maize germplasm reveal novel and known genomic regions for resistance to Northern corn leaf blight *Sci. Rep.* **10** 1–16

[94] Rizzardi D A, Peterlini E, Scapim C A, Pinto R J B, Faria M V and Contreras-Soto R I 2022 Genome wide association study identifies SNPs associated with northern corn leaf blight caused by Exserohilum turcicum in tropical maize germplasm (Zea mays L.) *Euphytica* **218** 1–12

[95] Kibe M, Nyaga C, Nair S K, Beyene Y, Das B, Bright J M and Gowda M 2020 Combination of linkage mapping, GWAS, and GP to dissect the genetic basis of common rust resistance in tropical maize germplasm *Int. J. Mol. Sci.* **21** 6518

[96] Myint N N A, Korinsak S, Chutteang C, Laosatit K, Thunnom B, Toojinda T and Siangliw J L 2022 Identification of pathogenicity loci in Magnaporthe oryzae using GWAS with neck blast phenotypic data *Genes* **13** 916

[97] Phan H T, Furuki E, Hunziker L, Rybak K and Tan K C 2021 GWAS analysis reveals distinct pathogenicity profiles of Australian Parastagonospora nodorum isolates and identification of marker-trait-associations to septoria nodorum blotch *Sci. Rep.* **11** 1–14

[98] Bartoli C and Roux F 2017 Genome-wide association studies in plant pathosystems: toward an ecological genomics approach *Front. Plant Sci.* **8** 763

[99] Lv H, Fang Z, Yang L, Zhang Y and Wang Y 2020 An update on the arsenal: mining resistance genes for disease management of Brassica crops in the genomic era *Hortic. Res.* **7** 34

[100] Wani S H, Samantara K, Razzaq A, Kakani G and Kumar P 2022 Back to the wild: mining maize (Zea mays L.) disease resistance using advanced breeding tools *Mol. Biol. Rep.* **49** 5787–803

[101] Raj K J and Sanghamitra N 2010 Gene pyramiding—a broad spectrum technique for developing durable stress resistance in crops *Biotechnol. Mol. Biol. Rev.* **5** 51–60

[102] Haque M A, Rafii M Y, Yusoff M M, Ali N S, Yusuff O, Datta D R and Ikbal M F 2021 Recent advances in rice varietal development for durable resistance to biotic and abiotic stresses through marker-assisted gene pyramiding *Sustainability* **13** 10806
[103] Pradhan M and Bastia D N 2022 Validation of linked markers of bacterial leaf blight resistance genes in rice variety of Odisha (Oryza sativa) *J Pharm. Innov.* **11** 446–8
[104] Fukuoka S, Saka N, Mizukami Y, Koga H, Yamanouchi U, Yoshioka Y and Yano M 2015 Gene pyramiding enhances durable blast disease resistance in rice *Sci. Rep.* **5** 7773
[105] Chandrasekar A, Kumari M, Navaneetha Krishnan J, Suresh S, Gnanam R, Sundaram R M and Kumaravadivel N 2022 Marker-assisted introgression of bacterial blight resistance gene xa13 into improved CO43 *Euphytica* **218** 118
[106] Jamaloddin M, Durga Rani C V, Swathi G, Anuradha C, Vanisri S, Rajan C P D and Madhav M S 2020 Marker assisted gene pyramiding (MAGP) for bacterial blight and blast resistance into mega rice variety 'Tellahamsa' *PLoS One* **15** e0234088
[107] Singh U M, Dixit S, Alam S, Yadav S, Prasanth V V, Singh A K and Kumar A 2022 Marker-assisted forward breeding to develop a drought-, bacterial-leaf-blight-, and blast-resistant rice cultivar *Plant Genome* **15** e20170
[108] Badri J, Lakshmidevi G, JaiVidhya L R K, Prasad M S, Laha G S, Lakshmi V J and Sundaram R M 2022 Multiparent-derived, marker-assisted introgression lines of the Elite Indian Rice Cultivar,'Krishna Hamsa' show resistance against bacterial blight and blast and tolerance to drought *Plants* **11** 622
[109] Yang D, Xiong L, Mou T and Mi J 2022 Improving the resistance of the rice PTGMS line Feng39S by pyramiding blast, bacterial blight, and brown plant hopper resistance genes *Crop J.* **10** 1187–97
[110] Senthilvel V, Chockalingam V, Raman R, Rangasamy S, Sundararajan T and Jegadeesan R 2021 Performance of gene pyramided rice lines for blast and sheath blight resistance *In Biol. Forum Int. J.* **13** 913–7
[111] Ramalingam J, Raveendra C, Savitha P, Vidya V, Chaithra T L, Velprabakaran S and Vanniarajan C 2020 Gene pyramiding for achieving enhanced resistance to bacterial blight, blast, and sheath blight diseases in rice *Front. Plant Sci.* **11** 591457
[112] Mallick N, Agarwal P, Jha S K and Niranjana M 2021 Marker-assisted breeding for rust management in wheat *Indian Phytopathol* **74** 365–70
[113] Rana M, Kaldate R, Nabi S U, Wani S H and Khan H 2021 Marker-assisted breeding for resistance against wheat rusts *Physiological, Molecular, and Genetic Perspectives of Wheat Improvement* (Cham: Springer) pp 229–62
[114] Liu R, Lu J, Zhou M, Zheng S, Liu Z, Zhang C and Zhang L 2020 Developing stripe rust resistant wheat (Triticum aestivum L.) lines with gene pyramiding strategy and marker-assisted selection *Genet. Resour. Crop Evol.* **67** 381–91
[115] Randhawa M S, Bains N S, Sohu V S, Chhuneja P, Trethowan R M, Bariana H S and Bansal U 2019 Marker assisted transfer of stripe rust and stem rust resistance genes into four wheat cultivars *Agronomy* **9** 497
[116] Zheng X, Zhou J, Zhang M, Tan W, Ma C, Tian R and Yang S 2022 Transfer of durable stripe rust resistance gene Yr39 into four chinese elite wheat cultivars using marker-assisted selection *Agronomy* **12** 1791
[117] Shah L, Rehman S, Ali A, Yahya M, Riaz M W, Si H and Lu J 2018 Genes responsible for powdery mildew resistance and improvement in wheat using molecular marker-assisted selection *J. Plant Dis. Prot.* **125** 145–58

[118] Ye X, Zhang S, Li S, Wang J, Chen H, Wang K and Yan Y 2019 Improvement of three commercial spring wheat varieties for powdery mildew resistance by marker-assisted selection *Crop Prot.* **125** 104889

[119] Zhang X, Wang W, Liu C, Zhu S, Gao H, Xu H and Ma P 2021 Diagnostic kompetitive allele-specific PCR markers of wheat broad-spectrum powdery mildew resistance genes Pm21, PmV, and Pm12 developed for high-throughput marker-assisted selection *Plant Dis.* **105** 2844–50

[120] Zhang B, Chi D, Hiebert C, Fetch T, McCallum B, Xue A and Fedak G 2019 Pyramiding stem rust resistance genes to race TTKSK (Ug99) in wheat *Can. J. Plant. Pathol.* **41** 443–9

[121] Ogbonnaya F C, Eastwood R F and Lagudah E 2009 Identification and utilisation of genes for cereal cyst nematode resistance (Heterodera avenae) resistance in wheat: the Australian experience *Proc. of the First Workshop of the International Cereal Cyst Nematode Initiative: Cereal Cyst Nematode: Status, Research and Outlook* 166

[122] Ali M A, Shahzadi M, Zahoor A, Dababat A A, Toktay H, Bakhsh A and Li H 2019 Resistance to cereal cyst nematodes in wheat and barley: an emphasis on classical and modern approaches *Int. J. Mol. Sci.* **20** 432

[123] Zhu X, Zhao J, Abbas H M K, Liu Y, Cheng M, Huang J and Dong W 2018 Pyramiding of nine transgenes in maize generates high-level resistance against necrotrophic maize pathogens *Theor. Appl. Genet.* **131** 2145–56

[124] Zhao C, Jurat-Fuentes J L, Abdelgaffar H M, Pan H, Song F and Zhang J 2015 Identification of a new cry1I-type gene as a candidate for gene pyramiding in corn to control Ostrinia species larvae *Appl. Environ. Microbiol.* **81** 3699–705

[125] Krishna T P A, Maharajan T and Ceasar S A 2022 Application of CRISPR/Cas9 genome editing system to reduce the pre-and post-harvest yield losses in cereals *Open Biotechnol. J.* **16** e187407072205190

[126] Kumari C, Sharma M, Kumar V, Sharma R, Kumar V, Sharma P and Irfan M 2022 Genome editing technology for genetic amelioration of fruits and vegetables for alleviating post-harvest loss *Bioengineering* **9** 176

[127] Yin K and Qiu J L 2019 Genome editing for plant disease resistance: applications and perspectives *Philos. Trans. Royal Soc.* B **374** 20180322

[128] Singh R 2021 Genome Editing for Plant Disease Resistance *Emerging Trends in Plant Pathology* (Singapore: Springer) pp 577–90

[129] Khanale V, Bhattacharya A, Prashar M and Char B 2022 Genome editing interventions to combat rice blast disease *Plant Biotechnol. Rep.* **17** 1–13

[130] Ye X, Wang K, Liu H, Tang H, Qiu Y and Gong Q 2022 Genome editing toward wheat improvement *Genome Editing Technologies for Crop Improvement* (Singapore: Springer) pp 241–69

[131] Ajayo B S, Huang Y and Huang H 2022 Utilizing transcription factors for improving banded leaf and sheath blight disease resistance in maize: a review *J. Plant Interact.* **17** 911–26

[132] Talakayala A, Ankanagari S and Garladinne M 2022 Harnessing CRISPR/Cas tools for installing virus resistance in cereals: an overview *Next-Generation Plant Breeding Approaches for Stress Resilience in Cereal Crops* (Springer) pp 433–52

[133] Khan Z A, Kumar R and Dasgupta I 2022 CRISPR/Cas-mediated resistance against viruses in plants *Int. J. Mol. Sci.* **23** 2303

[134] Arazoe T 2021 Genome editing using CRISPR/Cas9 system in the rice blast fungus *Magnaporthe Oryzae* (New York: Humana) pp 149–60

[135] Paul N C, Park S W, Liu H, Choi S, Ma J, MacCready J S and Sang H 2021 Plant and fungal genome editing to enhance plant disease resistance using the CRISPR/Cas9 system *Front. Plant Sci.* **12** 700925

[136] Zhang Y M, Zhang Y and Xie K 2020 Evaluation of CRISPR/Cas12a-based DNA detection for fast pathogen diagnosis and GMO test in rice *Mol. Breed.* **40** 11

[137] Sánchez E, Ali Z, Islam T and Mahfouz M 2022 A CRISPR-based lateral flow assay for plant genotyping and pathogen diagnostics *Plant Biotechnol. J.* **20** 2418–29

[138] Bhat A I, Aman R and Mahfouz M 2022 Onsite detection of plant viruses using isothermal amplification assays *Plant Biotechnol. J.* **20** 1859–73

[139] Li T, Liu B, Spalding M H, Weeks D P and Yang B 2012 High-efficiency TALEN-based gene editing produces disease-resistant rice *Nat. Biotechnol.* **30** 390–2

[140] Zhou J, Peng Z, Long J, Sosso D, Liu B O, Eom J S and Yang B 2015 Gene targeting by the TAL effector PthXo2 reveals cryptic resistance gene for bacterial blight of rice *Plant J.* **82** 632–43

[141] Cai L, Cao Y, Xu Z, Ma W, Zakria M, Zou L and Chen G 2017 A transcription activator-like effector Tal7 of *Xanthomonas oryzae pv. oryzicola* activates rice gene *Os09g29100* to suppress rice immunity *Sci. Rep.* **7** 5089

[142] Kim Y A, Moon H and Park C J 2019 CRISPR/Cas9-targeted mutagenesis of *Os8N3* in rice to confer resistance to *Xanthomonas oryzae pv. oryzae Rice* **12** 67

[143] Oliva R, Ji C, Atienza-Grande G, Huguet-Tapia J C, Perez-Quintero A, Li T and Yang B 2019 Broad-spectrum resistance to bacterial blight in rice using genome editing *Nat. Biotechnol.* **37** 1344–50

[144] Duy P N, Lan D T, Pham Thu H, Thi Thu H P, Nguyen Thanh H, Pham N P and Pham X H 2021 Improved bacterial leaf blight disease resistance in the major elite Vietnamese rice cultivar TBR225 via editing of the OsSWEET14 promoter *PLoS One* **16** e0255470

[145] Wang F, Wang C, Liu P, Lei C, Hao W, Gao Y and Zhao K 2016 Enhanced rice blast resistance by CRISPR/Cas9-targeted mutagenesis of the ERF transcription factor gene OsERF922 *PLoS One* **11** e0154027

[146] Zhou Y, Xu S, Jiang N, Zhao X, Bai Z, Liu J and Yang Y 2022 Engineering of rice varieties with enhanced resistances to both blast and bacterial blight diseases via CRISPR/Cas9 *Plant Biotechnol. J.* **20** 876–85

[147] Ma J, Chen J, Wang M, Ren Y, Wang S, Lei C and Cheng Z 2018 Disruption of OsSEC3A increases the content of salicylic acid and induces plant defense responses in rice *J. Exp. Bot.* **69** 1051–64

[148] Xu Y, Wang F, Chen Z, Wang J, Li W Q, Fan F and Yang J 2020 Intron-targeted gene insertion in rice using CRISPR/Cas9: a case study of the Pi-ta gene *Crop J.* **8** 424–31

[149] Shah P R, Varanavasiappan S, Kokiladevi E, Ramanathan A and Kumar K K 2019 Genome editing of rice PFT1 gene to study its role in rice sheath blight disease resistance *Int. J. Curr. Microbiol. Appl. Sci.* **8** 2356–64

[150] Li N, Lin B, Wang H, Li X, Yang F, Ding X and Chu Z 2019 Natural variation in ZmFBL41 confers banded leaf and sheath blight resistance in maize *Nat. Genet.* **51** 1540–8

[151] Wang N, Tang C, Fan X, He M, Gan P, Zhang S and Wang X 2022 Inactivation of a wheat protein kinase gene confers broad-spectrum resistance to rust fungi *Cell* **185** 2961–74

[152] Macho A, Wang P and Zhu J K 2022 Modification of the susceptibility gene TaPsIPK1-a win-win for wheat disease resistance and yield *Stress Biology* **2** 40

[153] Karmacharya A 2022 Molecular mapping of QTL for genetic transformation-related trait and CRISPR/Cas9-mediated gene editing in wheat *Doctoral Dissertation* North Dakota State University

[154] Wang Y, Cheng X, Shan Q, Zhang Y, Liu J, Gao C and Qiu J L 2014 Simultaneous editing of three homoeoalleles in hexaploid bread wheat confers heritable resistance to powdery mildew *Nat. Biotechnol.* **32** 947–51

[155] Zhang Y, Bai Y, Wu G, Zou S, Chen Y, Gao C and Tang D 2017 Simultaneous modification of three homoeologs of Ta EDR 1 by genome editing enhances powdery mildew resistance in wheat *Plant J.* **91** 714–24

[156] Macovei A, Sevilla N R, Cantos C, Jonson G B, Slamet-Loedin I, Čermák T and Chadha-Mohanty P 2018 Novel alleles of rice eIF4G generated by CRISPR/Cas9-targeted mutagenesis confer resistance to rice tungro spherical virus *Plant Biotechnol. J.* **16** 1918–27

[157] Kumam Y, Rajadurai G, Kumar K K, Varanavasiappan S, Reddy M K, Krishnaveni D and Sudhakar D 2022 Genome editing of indica rice ASD16 for imparting resistance against rice tungro disease *J. Plant Biochem. Biotechnol.* **31** 880–93

Chapter 11

Major protein families involved in plant defence

Apoorva Masade, Anirudh Kumar and Hitendra K Patel

Plants face constant threats from pathogens and herbivorous insects, necessitating a robust defence system composed of various protein families. Central to this immune response are pattern recognition receptors (PRRs), which detect pathogen-associated molecular patterns (PAMPs) and damage-associated molecular patterns (DAMPs), initiating defensive mechanisms. Nucleotide-binding site-leucine-rich repeat (NLR) proteins play a crucial role in effector-triggered immunity, activating strong responses upon recognizing specific pathogen effectors. Additionally, pathogenesis-related (PR) proteins contribute to inhibiting pathogen growth through diverse mechanisms. This chapter explores the major protein families integral to plant defence mechanisms against biotic stresses, such as pathogens and herbivorous insects. By understanding these protein families and their interactions, the chapter sheds light on the complexity of plant immunity and its implications for developing resilient crops.

11.1 Introduction

In this chapter, we explore the multi-layered defence systems in plants, emphasizing the diverse protein families integral to pathogen detection, signalling, and defence response. Plants face ongoing threats from bacteria, fungi, viruses, and insects, relying on protein PRRs to detect pathogens through PAMPs/DAMPs. PRRs, including receptor-like kinases (RLKs) and receptor-like proteins (RLPs), initiate downstream signalling involving calcium flux, MAPKs, and hormonal signalling (SA, JA, ET), leading to transcriptional activation of defence genes. NLRs activate robust, often localized responses against specific pathogen effectors, contributing to both hypersensitive responses and systemic acquired resistance.

PR proteins, crucial for systemic resistance, function in antifungal, antibacterial, and cell-wall-degrading activities, including chitinases and β-1,3-glucanases

doi:10.1088/978-0-7503-5673-2ch11

targeting fungal components. Complementary defence proteins, such as peroxidases and lipid transfer proteins, contribute to reactive oxygen species (ROS) generation, detoxification, and membrane repair. In the signalling domain, salicylic acid (SA) mediates long-lasting resistance mechanisms, jasmonic acid (JA) and ethylene (ET) address necrotrophic pathogens and herbivory, and abscisic acid (ABA) modulates stomatal defence in tandem with drought responses. Other significant contributors include calcium-dependent protein kinases (CDPKs), which phosphorylate targets to reinforce defence, and the ubiquitin–proteasome system, controlling immune receptor turnover.

Conserved hubs in these hormone networks, such as nonexpressor of pathogenesis-related genes 1 (NPR1) in SA signalling, JAZ–MYC in JA pathways, and EIN2 in ethylene networks, function as integrative points that balance plant growth-defence trade-offs. The chapter highlights practical applications of these findings, showcasing how genetic studies on PR proteins, PRRs, and signalling molecules pave the way for enhancing crop resilience through biotechnological approaches, presenting a detailed framework for advancing sustainable agricultural practices.

11.2 Pattern recognition receptors in plants: receptor-like kinases (RLKs) and receptor-like proteins (RLPs)

Plants, unlike animals, lack adaptive immunity; however, they possess a highly evolved and sophisticated innate immune system to combat the constant threat posed by various pathogens, including bacteria, fungi, viruses, and insects. Central to the plant's innate immune system is the PRRs, which are cell-surface proteins that detect PAMPs or DAMPs. These molecular patterns are highly conserved across broad classes of pathogens or are released from host cells upon damage. Detecting PAMPs by PRRs initiates pattern-triggered immunity (PTI), a primary defence mechanism to prevent pathogen entry, colonization, and spread [1–3].

PRRs are specialized transmembrane proteins located primarily on the surface of plant cells. Their main role is to detect external signals, such as PAMPs or DAMPs, and initiate intracellular signalling pathways that lead to immune responses. Unlike animals, which rely on mobile immune cells to fight infection, plants depend entirely on these surface receptors to recognize external threats.

PAMPs include various molecular structures essential for pathogen survival, such as bacterial flagellin (flg22), elongation factor Tu (EF-Tu), and fungal chitin. These PAMPs are conserved across different pathogens, making them ideal targets for recognizing PRRs. DAMPs, on the other hand, are molecules released from plant cells during pathogen attack or mechanical damage. Examples include fragments of plant cell walls, such as oligogalacturonides, which serve as danger signals and alert the plant immune system to an ongoing attack [3, 4].

Plant PRRs can be classified into two major families based on their structural domains: RLKs and RLPs. Both families share common structural features [5–7]:

1. Extracellular domain: This domain is crucial for PAMP/DAMP recognition. It often contains leucine-rich repeats (LRRs), which enhance binding specificity to ligands.

2. Transmembrane domain: A single alpha-helix or a glycophosphatidylinositol (GPI) anchors the receptor in the plasma membrane.
3. Cytoplasmic kinase domain (only in the case of RLKs): This domain contains the catalytic site necessary for phosphorylating target proteins, thereby initiating intracellular signalling cascades.

11.2.1 Distribution of different cell surface receptors

The sizes of RLPs and RLKs families differ among plant species and are thought to increase over evolutionary periods in reaction to pathogenic pressure. Plant cell surface receptors can be characterized based on their ectodomains. Variability in the ectodomains is pivotal for diverse pattern/ligand recognition [8–10].

Recent studies have found that immune induction in plants generally leads to higher expression of certain RLKs than RLPs, indicating that the RLPs might be playing a more subdued role in plant immunity.

In *Arabidopsis*, ~610 RLKs and ~170 RLPs have been reported so far. In rice, on the other hand, ~1,100 RLKs and ~90 RLPs have been reported [5, 11–13].

11.2.2 Ligand perception by PRRs

In plants, PRRs are pivotal components in detecting pathogenic signals and activating immune responses.

The ectodomains of PRRs, which facilitate pathogen recognition, may display various motifs, including LRRs, lysine motifs (LysMs), lectins, and epidermal growth factor (EGF)-like motifs. LRR-type PRRs bind to protein ligands, such as bacterial flagellin or elongation factor Tu (EF-Tu), and plant-derived Pep peptides, while other motifs enable recognition of carbohydrates, such as fungal chitin or bacterial peptidoglycans. This structural diversity reflects the broad spectrum of potential PRR ligands encoded in plant genomes.

Although many RLKs and RLPs are hypothesized to act as PRRs, their specific ligands remain unknown. However, recent advances in molecular, genetic, and structural studies have begun to clarify the details of how PRRs recognize pathogenic molecules and activate immune responses within plants, offering valuable insights into the mechanics of plant innate immunity.

11.2.2.1 Chitin perception in Arabidopsis

Chitin, a major component of fungal cell walls, acts as a PAMP and triggers immune responses in plants. In *Arabidopsis thaliana*, chitin perception is mediated by CERK1 (chitin elicitor receptor kinase 1), an LysM-type receptor-like kinase. CERK1 interacts with the extracellular chitin oligomers and forms a complex with LYK5, another LysM receptor-like kinase, to initiate downstream signalling events [14–16].

Upon ligand binding, CERK1 undergoes autophosphorylation, which activates mitogen-activated protein kinases (MAPKs) and triggers the production of ROS. The activated pathway ultimately leads to the expression of defence-related genes, strengthening cell walls through callose deposition and inducing the secretion of antimicrobial compounds. The importance of CERK1 in chitin perception is highlighted by mutant *Arabidopsis* plants lacking this receptor, which exhibit increased susceptibility to fungal pathogens [17, 18].

11.2.2.2 Flagellin perception in Arabidopsis

Flagellin, a bacterial PAMP, is perceived by the PRR FLS2 (flagellin-sensitive 2) in *Arabidopsis*. FLS2 recognizes a conserved 22-amino-acid peptide within the N-terminal region of bacterial flagellin, known as flg22. Upon binding to flg22, FLS2 interacts with the co-receptor BAK1 (BRI1-associated receptor kinase 1), which is essential for the full activation of immune signalling [19–21].

This interaction triggers a phosphorylation cascade that activates MAPKs and other downstream effectors, leading to the production of ROS and the upregulation of defence genes. The perception of flagellin also induces stomatal closure, a defence mechanism that restricts bacterial entry into the plant. *Arabidopsis* mutants deficient in FLS2 are more vulnerable to bacterial pathogens, underscoring the critical role of this receptor in bacterial immunity [22, 23].

11.2.2.3 Chitin perception in rice

Like *Arabidopsis*, rice (*Oryza sativa*) also recognizes chitin as a PAMP to initiate defence responses. In rice, chitin perception involves OsCERK1, which is homologous to *Arabidopsis* CERK1, and OsCEBiP (chitin elicitor binding protein), a LysM receptor-like protein. Unlike CERK1, OsCEBiP lacks a kinase domain and thus relies on OsCERK1 for downstream signalling [16, 24].

Upon chitin recognition, OsCEBiP binds to chitin oligomers and forms a complex with OsCERK1, leading to the activation of immune responses. This interaction triggers MAPK cascades, ROS production, and the expression of defence genes that inhibit fungal growth. Rice plants with impaired OsCERK1 or OsCEBiP function show increased susceptibility to fungal pathogens like *Magnaporthe oryzae*, the causal agent of rice blast disease [25–29].

11.2.2.4 Glucan receptor of soybean

In soybean (glycine max), β-glucans derived from fungal cell walls serve as elicitors of immune responses. The perception of β-glucans is mediated by a β-glucan receptor, which triggers immune signalling. One such receptor is GmGRM (glucan receptor for β-glucan elicitor in soybean), which plays a role in the recognition of fungal β-glucans and the activation of defence responses.

Upon recognition of β-glucans, GmGRM activates downstream signalling pathways, including ROS production and the activation of defence-related genes. This response limits fungal colonization and protects the plant from disease. The identification and characterization of GmGRM have provided insights into how legumes such as soybeans detect and respond to fungal pathogens through β-glucan perception [11, 30].

11.2.2.5 Xylanase receptor of tomato

Tomato (*Solanum lycopersicum*) perceives xylanase, a cell wall-degrading enzyme produced by fungal pathogens, through the receptor LeEIX2 (*Lycopersicon esculentum* EIX Receptor 2). Xylanase acts as an elicitor in tomatoes, inducing defence responses when recognized by LeEIX2. This receptor belongs to the class of RLPs that lack a kinase domain and instead rely on co-receptors for signal transduction.

The perception of xylanase by LeEIX2 triggers the rapid production of ROS, callose deposition, and the activation of defence genes. Tomato plants with compromised LeEIX2 function show increased susceptibility to fungal pathogens, demonstrating the importance of this receptor in fungal immunity [11, 31, 32].

11.2.2.6 MAMP receptors

MAMPs are conserved microbial molecules that are recognized by PRRs to trigger immune responses in plants. MAMPs include molecules such as flagellin, EF-Tu (elongation factor thermo-unstable), lipopolysaccharides (LPS), and chitin. Each MAMP is recognized by a specific PRR, leading to the activation of PTI.

For example, EFR (EF-Tu Receptor) in *Arabidopsis* recognizes the N-terminal region of bacterial EF-Tu, a protein involved in protein synthesis. Like FLS2, EFR forms a complex with BAK1 to activate immune signalling. The activation of MAPKs and other downstream components leads to the induction of defence genes and the reinforcement of cell walls [11, 33, 34].

11.2.2.7 DAMP receptors

DAMPs are endogenous molecules released from damaged plant cells during a pathogen attack or mechanical injury. DAMPs are recognized by PRRs, triggering immune responses similar to those activated by MAMPs. Common DAMPs include oligogalacturonides (OGs), which are derived from the breakdown of plant cell wall pectin [35].

In *Arabidopsis*, the perception of OGs is mediated by the receptor WAK1 (wall-associated kinase 1). WAK1 recognizes OGs and activates immune signalling pathways that lead to the expression of defence-related genes. The detection of DAMPs allows plants to respond not only to PAMPs but also to signals generated by host tissue damage, providing an additional layer of immune protection [12, 35].

Ligand perception by PRRs is a critical aspect of plant innate immunity, enabling plants to detect and respond to pathogen invasion rapidly. The examples discussed in this chapter—ranging from chitin and flagellin perception in *Arabidopsis* to glucan and xylanase receptors in soybean and tomato—highlight the diversity of PRRs and their roles in plant defence. By recognizing both PAMPs and DAMPs, PRRs initiate a complex network of signalling events that culminate in the activation of immune responses, helping plants fend off a wide range of pathogens [11].

11.2.3 Events following PRR activation

Following the perception of immunogens by plant PRRs, a sequence of events unfolds that can be categorized into three phases: very early response, early response, and late response. Each phase is characterized by distinct biochemical and physiological changes, ultimately leading to enhanced plant immunity [11, 36, 37].

11.2.3.1 Very early response

The very early response occurs within seconds to minutes (1–5 min) after the perception of a PAMP or DAMP. This phase is marked by:

- **Ligand binding:** PRRs on the plant cell surface bind to specific ligands, initiating the immune signalling cascade.
- **Receptor dimerization:** Ligand binding often leads to the dimerization of PRRs, which is crucial for downstream signalling.
- **Ion fluxes:** The binding event triggers rapid changes in ion fluxes across the plasma membrane, particularly calcium ions (Ca^{2+}). An influx of Ca^{2+} can result in a rapid increase in cytosolic calcium levels, which is a key secondary messenger in plant signalling.

11.2.3.2 Early response

The early response phase spans from a few minutes (5–30 min) to several hours post-perception and involves more complex signalling events:

- **MAPK activation:** MAPKs are activated, leading to phosphorylation cascades that propagate the immune signal. This activation results in the modulation of various transcription factors.
- **ROS production:** The early response is characterized by the production of ROS, which can act as signalling molecules and have direct antimicrobial properties.
- **defence gene expression:** Activation of transcription factors leads to the upregulation of defence-related genes, resulting in the production of antimicrobial compounds, pathogenesis-related proteins, and other defensive metabolites.
- **Stomatal closure:** In response to some PAMPs, plants may close their stomata as a defensive measure to limit pathogen entry.

11.2.3.3 Late response

The late response phase can occur hours to days after pathogen perception and is associated with longer-term defence mechanisms:

- **Callose deposition:** Callose, a polysaccharide, is deposited at the site of infection, reinforcing the plant cell wall and acting as a physical barrier to pathogen entry.
- **Systemic acquired resistance (SAR):** The early local responses can lead to systemic changes in the plant, conferring broad-spectrum resistance to subsequent pathogen attacks throughout the plant.
- **Programmed cell death (PCD):** In some cases, PCD may be initiated in localized areas to prevent the spread of pathogens, creating a hypersensitive response that restricts pathogen growth.
- **Induction of secondary metabolites:** The late response phase also sees an increase in the synthesis of secondary metabolites that can have antimicrobial properties or play roles in signalling and defence.

The time course of events following the perception of immunogens by plant PRRs illustrates a highly coordinated immune response. Understanding these phases is crucial for unravelling the complexities of plant immunity and enhancing disease resistance through breeding or biotechnological approaches.

11.3 Nucleotide-binding leucine-rich repeat (NLR) receptor proteins

Pathogens employ a strategy of host manipulation by secreting effector proteins directly into plant cells. These effectors can alter host cellular processes to facilitate infection. To counteract this, plants utilize surface-localized PRRs that detect various pathogen components, initiating the immune response. Additionally, intracellular NLR receptor proteins play a crucial role in recognizing these pathogenic effectors once they have infiltrated the host.

The immune responses triggered by PRRs and NLRs—termed PTI and NLR-triggered immunity or effector-triggered immunity (ETI), respectively—both initiate a cascade of defence mechanisms. These include the production of ROS, the influx of extracellular calcium ions, activation of protein kinases, and a comprehensive reprogramming of gene expression geared towards defence [2, 29].

While PTI and ETI share many similarities, they exhibit notable differences in their timing, intensity, and duration. NLR activation typically results in a stronger, more sustained response, often leading to a hypersensitive reaction characterized by programmed cell death at the infection site. This robust defence strategy serves to limit pathogen spread and reinforces the overall resilience of the plant.

NLR proteins are pivotal components of the plant immune response, responsible for recognizing pathogen-derived effectors and initiating downstream signalling to trigger ETI. NLR-mediated immune signalling is complex and involves additional molecular components that facilitate signalling beyond effector recognition [38].

11.3.1 Overview of NLR receptors

NLR receptors are characterized by two main functional domains: the nucleotide-binding domain (NBD) and the LRR domain. These receptors are generally classified into two groups based on their structural features:

- **TIR-NLRs** (toll/interleukin-1 receptor-NLRs) contain a TIR domain at the N-terminus.
- **CC-NLRs** (coiled-coil NLRs) have a coiled-coil domain at their N-terminus.

Both types share the NBD and LRR domains, crucial for pathogen detection and signalling.

NLRs, along with pathogen effectors, localize to various cellular compartments, such as the cytoplasm, nucleus, and plasma membrane. For instance, the barley CNL MLA10 and *Arabidopsis* TNL RPS4 localize to both the nucleus and cytoplasm, while the CNL RPM1 primarily associates with the plasma membrane, recognizing specific effector proteins from *Pseudomonas* species.

NLRs exhibit varied mechanisms for recognizing pathogen effectors, with some binding directly to effectors, while others detect disruptions in host proteins. For example, *Nicotiana* TNL Roq1 interacts with the Xanthomonas effector XopQ, whereas RPM1 identifies the phosphorylation of the host protein RIN4.

Additionally, some NLRs can function as pairs for coordinated expression, typically consisting of a signalling NLR and a sensor NLR, such as RPS4 and RRS1-R. Certain NLRs also require helper NLRs to form a functional resistance unit, highlighting the complex interactions in plant immunity [39, 40].

11.3.2 Structural features of NLR receptors

11.3.2.1 Nucleotide binding domain

The NBD is a highly conserved region that is critical for the function of NLRs. It is responsible for binding nucleotides (ATP or ADP) and is involved in the energy-dependent activation of downstream signalling pathways. The NBD comprises several subdomains, including:

- **Walker A Motif:** This motif is essential for ATP binding and hydrolysis. It contains a conserved lysine residue that interacts with the phosphate groups of ATPs.
- **Walker B Motif:** This region is involved in coordinating the magnesium ions necessary for nucleotide binding and hydrolysis.
- **Sensor 1 and Sensor 2:** These subdomains play crucial roles in sensing conformational changes that occur upon nucleotide binding. The conformational changes are critical for the activation of the NLR.

The NBD often exists in two conformations: an active form when bound to ATP and an inactive form when bound to ADP or lacking nucleotide. These conformational changes are integral to the receptor's signalling function [39, 41].

11.3.2.2 Leucine-rich repeat domain

The LRR domain is characterized by repeated motifs rich in leucine residues, typically 20–30 amino acids in length. This domain plays a significant role in protein–protein interactions and ligand recognition. The LRR structure can be described as follows:

- **Helical and loop structures:** The LRRs adopt a horseshoe-like shape, composed of β-strands and loops. The loops between the repeats are often involved in specific interactions with pathogen effectors or other signalling proteins.
- **Diversity in LRRs:** The LRR domain can vary significantly among different NLRs, allowing for the recognition of diverse pathogen effectors. Some LRRs can also exhibit variations in length and composition, contributing to specificity in pathogen recognition.

The LRR domain is vital for the receptor's ability to recognize and bind to various PAMPs and DAMPs, thereby initiating downstream immune responses [39, 40].

11.3.2.3 N-terminal domains: TIR and CC domains

The N-terminal regions of NLRs can differ between TIR-NLRs and CC-NLRs:

- **TIR domain:** This domain is homologous to the cytoplasmic domains of the toll-like receptors in animals. It plays a role in downstream signalling by

interacting with other immune signalling components. The TIR domain facilitates the formation of multi-protein complexes that propagate the immune signal.
- **Coiled-coil (CC) domain:** In CC-NLRs, the coiled-coil domain mediates dimerization and interaction with other signalling proteins. The CC domain is crucial for the formation of higher-order complexes necessary for effective immune signalling.

Recent advancements in structural biology have provided significant insights into the architecture of NLR proteins. Techniques such as x-ray crystallography and cryo-electron microscopy have revealed detailed structures of both the NBD and LRR domains in various NLRs. These studies highlight the dynamic nature of NLR receptors and their ability to adapt to different pathogen effectors [13, 42].

For instance, the crystal structure of the TIR-NLR protein, RPS4 from *Arabidopsis*, has been elucidated, revealing how the LRR domain engages with specific effectors. Similarly, the structural analysis of CC-NLRs has shed light on their oligomerization and dimerization mechanisms.

11.3.3 Activation of NLRs

Plant immunity is a complex, multifaceted system that allows plants to detect and respond to a wide range of pathogens. At the heart of this immune response are NLRs, a class of proteins that play a critical role in recognizing pathogen-derived molecules and activating defence mechanisms. NLRs are classified into two main groups based on their structural domains: TIR-NLRs (TNLs) and CC-NLRs (CNLs). This report explores the immune signalling mechanisms mediated by plant NLRs, focussing on their activation processes, structural dynamics, downstream signalling pathways, and the evolutionary context of these immune receptors [29, 43].

The processes involved in activating sensor NLRs and their associated helper proteins are expected to vary significantly. However, the presence of similar structural domains and a shared evolutionary lineage suggests that multimerization via the NB-ARC domain is essential for NLR activation. This multimerization occurs after ADP is replaced by ATP, a phenomenon that underscores why NLRs are frequently referred to as molecular switches in immune signalling.

11.3.3.1 NLR states: ADP-bound versus ATP-bound

NLRs can exist in two primary states: the inactive ADP-bound state and the active ATP-bound state. In the ADP-bound state, the NLR is in an 'off' configuration, wherein the LRR domain interacts with the NB-ARC domain to maintain this inactive status. This interaction stabilizes the receptor, preventing it from initiating an immune response.

Conversely, the ATP-bound state signifies an active 'on' configuration. Recent studies using cryo-electron microscopy on *A. thaliana*'s HOPZ-ACTIVATED RESISTANCE 1 (ZAR1) have revealed both the inactive and active forms, suggesting the presence of an intermediate state. In its ADP-bound form, ZAR1 operates as a monomer, characterized by multiple intramolecular interactions between the LRR and the NB-ARC domain, specifically between helical domain 1 (HD1) and the winged helix domain (WHD). Recognition of effector-induced changes by the LRR alters the protein's conformation, thereby relieving the inhibitory constraints and allowing for activation [4, 29].

11.3.3.2 Oligomerization and formation of the resistosome

When ATP binds to the NLR, it transitions into an oligomeric state, forming a pentameric structure known as the resistosome. This structure is reminiscent of the mammalian apoptosome, which signifies the active form of Apaf-1, as well as the inflammasomes activated in mammalian NLRs. The CC domain of ZAR1 plays a crucial role in this oligomerization process, resulting in the formation of an α-helical barrel.

The active resistosome subsequently associates with the plasma membrane, where specific charged residues within its funnel are critical for triggering cell death and mediating disease resistance. This interaction highlights the importance of the resistosome in establishing a localized immune response [29, 42, 43].

11.3.3.3 Mechanisms of NLR activation

Further investigation is needed to understand how ligand binding or modifications influence the nucleotide state of the NB-ARC domain. One prevailing hypothesis suggests that effector binding induces conformational changes that expose the NB-ARC domain, facilitating the transition from ADP to ATP. This phenomenon has been observed in ZAR1, where binding with the uridylated host protein kinase PBS1-LIKE 2 (PBL2) causes a rotation in the NB-ARC domain, decreasing its affinity for ADP and promoting activation [39].

11.3.3.4 Equilibrium-based switch model

In contrast, the equilibrium-based switch model posits that pathogen effectors do not exhibit strong binding to the ADP-bound state of the NLR. In this framework, NLRs are thought to oscillate continuously between on and off states, with effectors stabilizing the on state and thereby enhancing defensive responses. This model suggests that both activation paradigms may coexist and be employed by various NLRs, offering different mechanisms for activation. The diversity inherent in these activation modes strengthens immune responses against pathogen strategies and enables adaptation to varying effector binding affinities [40, 44].

11.3.3.5 TNLs and their signalling mechanisms

Although no active TNL complex structures have been documented, it is plausible that they adopt a similar oligomeric configuration to that of the NB-ARC domain. TNLs are known for their oligomerization capabilities, and TIR domains can self-associate across multiple surfaces. While the active ZAR1 resistosome engages with membranes through its N-terminal α-helix, TNL complexes likely transmit signals by enzymatically degrading NAD + via their TIR domains [38, 39, 41, 45].

11.3.3.6 TIR domains and NAD⁺ cleavage

Various plant TIRs, alongside a mammalian TIR from SARM1, have been shown to cleave NAD^+ into nicotinamide and cyclic ADP-ribose, with some plant TIRs yielding an additional product, v-cADPR. This enzymatic activity requires self-association and is contingent on a conserved catalytic glutamate, highlighting the functional significance of TIR domains in plant immunity [38].

11.3.4 Downstream signalling pathways

The activation of NLRs initiates downstream responses that can lead to the hypersensitive response (HR), a localized form of cell death that serves as a defence mechanism against pathogens. All identified TNLs rely on ENHANCED DISEASE SUSCEPTIBILITY 1 (EDS1) and PHYTOALEXIN DEFICIENT 4 (PAD4), genes that code for lipase-like proteins essential for signalling. These proteins play critical roles in downstream signalling, facilitating the transfer of information from NLR activation to effector response [46].

11.3.4.1 Role of EDS1 and PAD4

The precise effects of TIR domain enzymatic products on downstream signalling components remain to be fully understood. Recent research has significantly enhanced our understanding of RNLs functioning downstream of sensor NLRs, indicating that NRG1 interacts with EDS1 and SENESCENCE-ASSOCIATED GENE 101 (SAG101) to form a complex essential for initiating HR in *A. thaliana*. This complex may directly facilitate HR, as the RPW8 domain of NRG1 resembles pore-forming toxins, potentially disrupting membrane integrity and enhancing cell death in response to pathogen attack [13, 43, 46].

11.3.4.2 NDR1 and pore-forming activities

Certain CNLs depend on NON-RACE SPECIFIC DISEASE RESISTANCE 1 (NDR1), a plasma membrane-anchored protein involved in the immune response. Given that the active resistosome of ZAR1 associates with membranes, it is plausible that NDR1 contributes to pore-forming activities, furthering the defence against pathogens. Together, these findings deepen our understanding of HR initiation and function, which is crucial for unravelling the complexities of NLR-mediated immunity [13, 43, 46].

11.4 Plant disease resistance (R) proteins

In the 1940s, H H Flor formulated the gene-for-gene hypothesis by working on flax and *Melampsora lini* (flax rust). This concept suggests that for each resistance (R) gene in a plant, there is a corresponding avirulence (*avr*) gene in the pathogen that, when recognized, can initiate the plant's defence. C G Oort observed similar principles in wheat's interaction with *Ustilago tritici*.

Expanding this in the 1960s, researchers illustrated that the R protein-effector interaction largely dictates infection success. The zig-zag model (2006) by Jones and

Dangl further refined this concept, showing an evolutionary 'arms race' between pathogen effectors and plant R genes [2, 47]. This model highlights two immune layers: PTI, activated by pathogen markers like bacterial flagellin, and ETI, driven by intracellular NLR proteins that recognize specific pathogen effectors to activate stronger defences.

In essence, the continuous adaptation of pathogens to bypass plant defences and plants' evolution of new R proteins in response represents a dynamic co-evolutionary cycle in plant immunity.

11.4.1 Classification of R proteins

R proteins can be categorized into classes based on their different structural variations. Different classes of R proteins are listed in table 11.1 (the table has been adapted from Martin *et al* 2003) [29, 48–50].

11.5 Mitogen-activated protein kinases

MAPK cascades are pivotal signalling pathways in eukaryotic cells, crucial for translating both stress and developmental cues into appropriate responses. These cascades involve a series of kinases that undergo reversible phosphorylation. The activation of these kinases allows for the relay and amplification of signals, ultimately resulting in the phosphorylation of various substrate proteins. Such phosphorylation alters the activities of these proteins, which, in turn, leads to a diverse range of cellular responses, including significant changes in gene expression [51].

The specificity of MAPK signalling pathways is maintained even when certain components are shared among different cascades. This is achieved through several factors, such as the precise timing and location of signalling events (spatiotemporal regulation) and the dynamic interactions that occur between proteins. Additionally, regulatory mechanisms play a critical role in maintaining signalling specificity. Cross-inhibition, where one signalling pathway can inhibit another, feedback control that adjusts the intensity of the response, and scaffolding proteins that help organize and stabilize signalling complexes all contribute to the complexity of these pathways.

In the context of plants, MAPK cascades are fundamental to a variety of physiological processes, including responses to both abiotic stress (like drought and salinity) and biotic stress (such as pathogen attacks), hormonal signalling, innate immune responses, and various developmental programs. Genetic studies, particularly in the model organism *A. thaliana*, have identified several key components essential to these signalling networks. Prominent MAPKs involved include MPK3, MPK4, and MPK6, along with mitogen-activated protein kinase kinases (MAP2Ks) such as MKK1, MKK2, MKK4, and MKK5. Together, these proteins interact in various configurations to coordinate the intricate signalling networks that manage plant responses to environmental challenges and developmental signals, thereby ensuring their survival and growth [51, 52].

Table 11.1. Different classes of R proteins, structural motifs, examples and their corresponding effectors, pathogens and pests.

Class of R proteins	Basis of differentiation (structural motif)	Examples of R proteins	Corresponding effectors/pathogens/pest
Class 1	Contains an N-terminal myristylation motif and a serine/threonine kinase catalytic domain.	**Pto** in tomato	AvrPto, AvrPtoB in *Pseudomonas syringae*
Class 2	Contains leucine-rich repeats (LRRs), a putative nucleotide-binding site (NBS), and an N-terminal leucine zipper (LZ) or coiled-coil (CC) sequence.	**HRT**[b] in *Arabidopsis*	Coat protein of turnip crinkle virus
		RPM1 in *Arabidopsis*	AvrRpm1, AvrB in *Pseudomonas syringae*
		RPP8[b] in *Arabidopsis*	Effectors in *Peronospora parasitica*
		RPP13 in *Arabidopsis*	Effectors in *Peronospora parasitica*
		RPS2 in *Arabidopsis*	AvrRpt2 in *Pseudomonas syringae*
		RPS5 in *Arabidopsis*	AvrPphB in *Pseudomonas syringae*
		Gpa2[a] in potato	Effectors in *Globodera pallida*
		Hero in potato	Effectors in *Globodera pallida* and *Globodera rostochiensis*
		R1 in potato	Effectors in *Phytophthora infestans*
		Rx1[a] in potato	Coat proteins of potato virus *X*
		Rx2 in potato	Coat proteins of potato virus *X*
		I2 in tomato	Effectors in *Fusarium oxysporum*
		Mi in tomato	Effectors in *Meloidogyne incognita* and *Macrosiphum euphorbiae*
		Sw-5 in tomato	Effectors in tomato spotted wilt virus
		Pib in rice	Effectors in *Magnaporthe grisea*
		Pi-ta in rice	AVR-Pi-ta in *Magnaporthe grisea*
		Xa1 in rice	Effectors in *Xanthomonas oryzae*
		Bs2 in pepper	AvrBs2 in *Xanthomonas campestris*
		Dm3 in lettuce	Effectors in *Bremia lactucae*
		Mla in barley	Effectors in *Blumeria graminis*
		Rp1 in maize	Effectors in *Puccinia sorghi*

(Continued)

Table 11.1. *(Continued)*

Class of R proteins	Basis of differentiation (structural motif)	Examples of R proteins	Corresponding effectors/pathogens/pest
Class 3	It resembles Class 2 but has a TIR region, similar to the N-terminus of Toll and IL-1R proteins, instead of a CC sequence.	**RPP1** in *Arabidopsis*	Effectors in *Peronospora parasitica*
		RPP4 in *Arabidopsis*	Effectors in *Peronospora parasitica*
		RPP5 in *Arabidopsis*	Effectors in *Peronospora parasitica*
		RPS4 in *Arabidopsis*	Effectors in *Pseudomonas syringae*
		L in flax	Effectors in *Melampsora lini*
		M in flax	Effectors in *Melampsora lini*
		P in flax	Effectors in *Melampsora lini*
		N in tobacco	Helicase in tobacco mosaic virus
Class 4	They lack an NBS but contain a TM, an extracellular LRR, and a small cytoplasmic tail without distinct motifs.	**Cf-2**[c] in tomato	Avr2 in *Cladosporium fulvum*
		Cf-4[d] in tomato	Avr4 in *Cladosporium fulvum*
		Cf-5[c] in tomato	Effectors in *Cladosporium fulvum*
		Cf-9[d] in tomato	Avr9 in *Cladosporium fulvum*
Class 5	Contains an extracellular LRR, a TM, and a cytoplasmic serine/threonine kinase region.	**Xa21** in rice	Effectors in *Xanthomonas oryzae*
Class 6	Others	Jacalin-like **RTM1** in *Arabidopsis*	Effectors in tobacco etch virus
		HSP-like **RTM2** in *Arabidopsis*	Effectors in tobacco etch virus
		TIR-NBS-LRR protein, **RRS1-R** in *Arabidopsis*	Effectors in *Ralstonia solanacearum*
		RPW8 in *Arabidopsis*	Effectors in *Erisyphe chicoracearum*
		Membrane protein **Mlo** in barley	Effectors in *Blumeria graminis*
		A pair of adjacent protein kinase domains and a likely weak transmembrane domain containing **Rpg1** in barley	Effectors in *Puccinia graminis*
		Toxin reductase, **Hm1** in maize	Effectors in *Cochiobolus carbonum*
		HS1[pro-1], without clear domains for protein interaction, in beet	Effectors in *Heterodera schachtii*
		Predicted cell-surface glycoproteins with receptor-like endocytosis signals, **Ve1**[e], **Ve2**[e] in tomato	Effectors in *Verticillium alboatrum*

11.5.1 Mitogen-activated protein kinase kinases (MAP3Ks)

MAP3Ks represent a vital group of enzymes in eukaryotic cells that play a crucial role in cellular signalling. These enzymes function at the top of the MAPK cascade, activating the downstream MAP2Ks, which then activate MAPKs. This signalling pathway is fundamental in relaying various external stimuli—such as environmental stress, hormones, and developmental signals—into the cell's adaptive responses.

MAP3Ks are essential for numerous biological processes, including stress responses, growth regulation, and immune responses. Understanding the mechanisms of action and regulation of MAP3Ks can provide insights into their roles in both plant and animal systems, ultimately leading to applications in agriculture and medicine [53].

MAP3Ks typically contain a well-conserved serine/threonine kinase catalytic domain responsible for their enzymatic activity. In addition to the catalytic domain, many MAP3Ks possess various regulatory domains that can modulate their activity, such as LRRs, which are thought to play a role in protein–protein interactions and substrate specificity.

The structural diversity of MAP3Ks is noteworthy, with different homologs found in various organisms, including plants, animals, and fungi. This diversity reflects the unique adaptations of signalling pathways in different biological contexts [54, 55].

MAP3Ks initiate MAPK signalling cascades by responding to various extracellular signals. Upon activation, they phosphorylate and activate MAP2Ks, thereby propagating the signal downstream to MAPKs. The activation mechanisms can involve multiple pathways, including direct phosphorylation, protein interactions, and the binding of specific signalling molecules.

One of the significant roles of MAP3Ks is their involvement in stress responses. For instance, in plants, MAP3Ks are critical in mediating responses to drought, salinity, and pathogen infections. When plants encounter abiotic stress, MAP3Ks can activate downstream pathways that lead to the expression of stress-related genes, helping the plant adapt to challenging conditions.

Beyond stress responses, MAP3Ks are integral in regulating plant growth and development. They influence processes such as embryogenesis, root and shoot development, and flowering time. MAP3Ks interact with various hormonal signalling pathways, including auxins, cytokinins, and gibberellins, to coordinate growth responses.

For example, in response to auxin, MAP3Ks can modulate the expression of genes that promote cell elongation and division, thereby influencing overall plant morphology. This hormonal interplay underscores the importance of MAP3Ks in maintaining proper developmental processes.

One of the fascinating aspects of MAP3K signalling is its ability to integrate signals from other pathways. MAP3Ks can interact with calcium signalling pathways, where calcium ions act as secondary messengers that can activate MAP3Ks in response to stress signals. Additionally, ROS signalling can influence MAP3K activity, highlighting the complexity of these signalling networks.

Feedback mechanisms are also crucial in these pathways, allowing for the fine-tuning of responses to ensure that the plant maintains homeostasis. For instance, upon activation, a MAP3K may upregulate proteins that inhibit its pathway, thus preventing overactivation and potential cellular damage.

In the realm of plant immunity, MAP3Ks are fundamental components of both PTI and ETI. Upon detecting PAMPs, plants can activate MAP3Ks, which leads to a series of downstream signalling events that bolster the plant's defence mechanisms.

Key MAP3Ks have been identified that specifically respond to pathogen attacks, mediating the expression of defence-related genes and the production of antimicrobial compounds. By understanding these pathways, researchers can develop strategies to enhance disease resistance in crops, leading to improved agricultural yields.

Genetic approaches have been pivotal in elucidating the roles of MAP3Ks. Researchers have employed techniques such as reverse genetics to identify MAP3K genes and examine their functions. Studies involving loss-of-function mutants have provided insights into the physiological roles of specific MAP3Ks, revealing their contributions to stress tolerance, growth regulation, and immune responses [53, 56, 57].

Functional characterization through biochemical assays and genetic interactions further clarifies how MAP3Ks operate within larger signalling networks. Such studies underscore the intricate relationships between different MAP3K family members and their distinct roles in various signalling pathways.

The insights gained from studying MAP3Ks have significant implications for agriculture and human health. In agriculture, enhancing MAP3K signalling can lead to crops that are more resilient to environmental stresses and diseases, ultimately supporting food security in the face of climate change.

In the medical field, MAP3Ks have been implicated in various diseases, including cancer. Understanding their signalling pathways can lead to the identification of potential therapeutic targets for disease intervention. Thus, MAP3Ks hold promise not only for advancing agricultural practices but also for developing novel medical treatments.

MAP3Ks are critical components of cellular signalling networks that mediate responses to a myriad of environmental and developmental signals. Their roles in stress responses, growth regulation, and immunity highlight their importance in both plant and animal systems. As research continues to unravel the complexities of MAP3K signalling pathways, the potential applications in agriculture and medicine are becoming increasingly evident. Future studies aimed at understanding the nuances of MAP3K function and regulation will undoubtedly contribute to advances in biotechnology and therapeutic development [57].

11.5.2 Mitogen-activated protein kinase kinases

MAP2Ks are integral components of the MAPK signalling pathways in plants, playing crucial roles in mediating responses to a variety of environmental stimuli and developmental signals. This report provides an in-depth examination of the structure, activation mechanisms, and functional roles of MAP2Ks in plant biology. The involvement of MAP2Ks in stress responses, developmental processes, and their interactions with hormonal signalling pathways is discussed. The report concludes with insights into the potential applications of MAP2K research in agriculture and biotechnology [58, 59].

MAPKs are part of a highly conserved signalling pathway that transduces extracellular signals into appropriate cellular responses. MAP2Ks, positioned between MAP3Ks (MAPK Kinase Kinases) and MAPKs, are vital for the phosphorylation cascade that modulates various cellular functions, including growth, differentiation, and stress adaptation. In plants, the MAP2K family is especially important due to their roles in regulating responses to environmental stressors and hormonal signals.

The significance of MAP2Ks extends beyond mere signal transduction; they serve as crucial integrators of multiple signalling pathways, allowing plants to adapt to changing environmental conditions effectively. This report aims to provide a comprehensive overview of MAP2Ks in plants, focussing on their structure, functional roles, and implications for agriculture.

MAP2Ks are characterized by a conserved serine/threonine kinase domain that is essential for their catalytic activity. This domain is responsible for the transfer of phosphate groups from ATP to target substrates, a key process that regulates various cellular functions. The general structure of MAP2Ks consists of [60, 61]:

- **Catalytic domain:** This domain is crucial for the kinase activity and is responsible for the phosphorylation of downstream MAPKs. It contains the conserved motifs necessary for ATP binding and substrate recognition.
- **Regulatory regions:** MAP2Ks may possess additional regulatory regions that modulate their activity. These regions can influence how MAP2Ks interact with upstream MAP3Ks and downstream MAPKs.

The plant kingdom exhibits significant diversity in MAP2K isoforms, which reflects adaptations to specific environmental conditions and developmental processes. For instance, in *A. thaliana*, several MAP2K genes have been identified, including MKK1, MKK2, MKK4, and MKK5. Each of these isoforms may be differentially expressed in response to specific signals or stressors, contributing to the functional specialization of MAP2Ks in various plant species.

Apart from the catalytic domain, some MAP2Ks possess additional functional domains that enhance their regulatory capabilities. These may include:

- **LRRs:** In some MAP2Ks, LRRs are involved in protein–protein interactions, facilitating communication with other signalling components.
- **CC domains:** These structures can mediate oligomerization or interactions with other proteins, playing a role in the assembly of signalling complexes.

MAP2Ks are activated by upstream MAP3Ks through phosphorylation. Upon receiving an external signal, MAP3Ks undergo conformational changes that lead to their activation. Activated MAP3Ks then phosphorylate specific residues on MAP2Ks, triggering their enzymatic activity. This process is crucial for propagating the signal downstream to MAPKs.

Activation of MAP2Ks is tightly regulated by various mechanisms, ensuring specificity and preventing overactivation. Key regulatory mechanisms include [53, 59]:

- **Feedback loops:** Activated MAP2Ks may initiate feedback regulation that limits their activity, helping to prevent excessive signalling.
- **Inhibitory proteins:** The presence of specific inhibitors can modulate MAP2K activity, further contributing to the regulation of MAPK signalling.
- **Scaffold proteins:** Scaffolding proteins can organize MAP2K and MAP3K interactions, enhancing the efficiency and specificity of the signalling cascade.

MAP2Ks function as essential mediators in the MAPK signalling cascade, acting as the bridge between MAP3Ks and MAPKs. Upon activation, MAP2Ks phosphorylate downstream MAPKs, leading to a series of signalling events that result in various physiological responses. This signal relay is crucial for maintaining cellular homeostasis and facilitating adaptive responses to environmental changes.

Despite the shared components in different MAPK cascades, the specificity of MAP2K signalling is maintained through spatial and temporal regulation. MAP2Ks exhibit distinct expression patterns in different tissues and developmental stages, allowing for tailored responses to specific stimuli. Additionally, interactions with other signalling pathways, such as calcium and hormonal signals, further modulate MAP2K activity, enhancing the specificity of the responses.

MAP2Ks are integral to the integration of multiple signalling pathways. They can respond to various stimuli, including abiotic stress (e.g., drought, salinity) and biotic stress (e.g., pathogen attacks). The ability of MAP2Ks to integrate these signals allows plants to mount effective and coordinated responses, balancing growth and defence.

MAP2Ks play a significant role in regulating plant growth and development. They are involved in key processes such as cell division, elongation, and differentiation. For example, specific MAP2Ks have been shown to influence root development by regulating cell division in root meristems [62].

11.5.3 Mitogen-activated protein kinase

MAPKs are integral components of cellular signalling pathways in plants, functioning as key mediators of diverse physiological responses to environmental stimuli, developmental cues, and biotic and abiotic stresses. The MAPK signalling pathway is characterized by a highly conserved three-tiered phosphorylation cascade that involves MAPKKKs (MAPK kinase kinases), MAPKKs (MAPK kinases), and MAPKs themselves, forming a signalling network that transduces signals from the cell surface to the nucleus. The structural foundation of MAPKs consists of a serine/threonine kinase domain, which is essential for their enzymatic activity, as well as a conserved phosphorylation motif located in the activation loop that includes a threonine and a tyrosine residue, both of which must be phosphorylated for full activation. This dual phosphorylation mechanism is a defining feature of MAPKs, differentiating them from other protein kinases and allowing for precise regulation of their activity in response to various stimuli [53, 55, 63, 64]. In plants, a notable example of a well-characterized MAPK is MPK6, which is part of a signalling cascade activated by

PAMPs and is crucial for defence responses against biotic stresses. The activation of MPK6 occurs upon the recognition of PAMPs by PRRs, leading to a cascade involving MAPKKK, such as MEKK1, which phosphorylates and activates MAPKK, such as MKK4 or MKK5, resulting in the activation of MPK6. Upon activation, MPK6 translocates to the nucleus, where it phosphorylates transcription factors and other substrate proteins, leading to the expression of defence-related genes, thereby enhancing the plant's resistance to pathogens. Beyond biotic stress, MAPKs also play significant roles in abiotic stress responses, such as drought and salinity. For instance, the MPK3 and MPK6 kinases are activated in response to osmotic stress, triggering downstream signalling pathways that modulate gene expression and protein activity to enhance tolerance. This involves the interaction with ABA signalling pathways, where MAPKs phosphorylate and activate factors that regulate stomatal closure and root growth, critical for maintaining water homeostasis under drought conditions. Furthermore, the structural variability among plant MAPKs is noteworthy; for example, while some MAPKs like MPK3 and MPK6 are involved in stress responses, others, such as MPK4, play a role in regulating developmental processes and immunity, indicating the specialized functions that different MAPK isoforms can have. The integration of MAPK signalling with other pathways is another crucial aspect; MAPKs often interact with various hormonal signalling pathways, including those mediated by auxins, gibberellins, and cytokinins. This cross-talk allows for a coordinated response to environmental conditions and developmental signals. For example, in *Arabidopsis*, the interaction between the auxin signalling pathway and MAPK cascades has been shown to be vital for regulating root development, with MAPK activation influencing the expression of auxin-responsive genes. Additionally, the scaffolding proteins play a significant role in the specificity of MAPK signalling, as they help organize the signalling components to form distinct complexes that can regulate specific cellular responses. A well-studied example is the protein MPKBP, which interacts with both MKK4 and MPK6, facilitating the formation of a signalling module that enhances the efficiency and specificity of the MAPK cascade during stress responses. The dynamics of MAPK signalling also include the role of phosphatases, which provide a mechanism for deactivating MAPKs and thus fine-tuning the signalling output. For instance, the dual-specificity phosphatases (DUSPs) can dephosphorylate the active MAPKs, thereby reversing their activation and allowing for a reset of the signalling pathway. This dynamic regulation ensures that plants can adapt to fluctuating environmental conditions effectively. The implications of MAPK signalling extend into the realm of biotechnology and agricultural applications. Given their pivotal roles in stress tolerance and developmental processes, manipulating MAPK pathways holds potential for enhancing crop resilience to abiotic stresses such as drought and salinity. For instance, transgenic approaches that overexpress specific MAPKs or their upstream regulators have been shown to confer increased stress tolerance in various plant species, suggesting that targeted engineering of MAPK pathways could be a promising strategy for developing climate-resilient crops [51, 57, 65, 66]. Moreover, the detailed understanding of MAPK signalling pathways opens avenues for precision agriculture, where tailored interventions could be designed to modulate specific signalling pathways in response to real-time environmental changes.

In conclusion, the structure and signalling dynamics of MAPKs represent a critical aspect of plant biology, underlining their significance in mediating responses to a wide array of internal and external stimuli. The interplay between MAPKs, their upstream activators, downstream targets, and the integration with other signalling pathways exemplifies a complex network that is essential for plant adaptability and survival. As research continues to unravel the intricacies of MAPK signalling, the potential for leveraging this knowledge in agricultural biotechnology and crop improvement becomes increasingly promising, offering pathways to enhance food security in the face of global climate challenges [53, 66].

11.6 Calmodulin binding proteins

Plant innate immunity is a critical response system allowing plants to detect and defend against pathogens and environmental stresses. Calcium signalling, an ancient and versatile intracellular signalling mechanism, plays a central role in regulating these responses. Calcium (Ca^{2+}) ions act as secondary messengers, translating external signals into cellular responses through precise changes in cytosolic calcium concentrations. This section explores the roles of calcium signalling and calmodulin-binding proteins in plant immunity, focussing on molecular mechanisms, major components, and applications in plant biology and agriculture.

Plants, as immobile organisms, face constant environmental challenges, particularly from biotic stressors like herbivores and pathogens. Such biological threats significantly hinder plant growth, with substantial impacts on agricultural yields and food quality. For example, the rice blast disease caused by *M. oryzae* is responsible for around 6% of annual global rice yield losses [67].

To counteract these threats, plants have developed complex signalling systems, with calcium/calmodulin (CaM/CML) signalling playing a central role in defence responses. Calcium ions (Ca^{2+}), serving as essential secondary messengers, are integral in plant immune responses, triggering various physiological activities such as enzyme adjustments, gene expression changes, membrane transport, and cytoskeletal reorganization in response to stress cues.

A significant focus has been on understanding Ca^{2+} signals generated in response to external stimuli, as well as their spatial and temporal dynamics, amplitude, and oscillation frequency. These properties collectively enable precise cellular responses. Ca^{2+} signals are regulated by a network of permeable channels, antiporters, and pumps across membranes, including the ER, mitochondria, chloroplasts, and vacuoles, which respond to a range of second messengers to maintain cellular balance [68].

In addition to Ca^{2+} signalling, ROS play a pivotal role in plant immunity. The enzyme respiratory burst oxidase homolog D (RbohD) is mainly responsible for ROS production, a critical component in defence signalling. Interactions between Ca^{2+} and ROS are essential; Ca^{2+} fluctuations influence RbohD activity, thereby modulating immune responses.

Despite advancements in understanding CaM and CML functions in biotic stress, many questions remain. For instance, the mechanisms by which Ca^{2+} oscillations differ among stresses and how CaMs/CMLs interpret these signals remain largely unresolved. Plant immunity mechanisms, like PTI and ETI, rely on complex Ca^{2+} signatures coordinated by Ca^{2+}-binding channels and pumps, including cyclic nucleotide-gated channels (CNGCs) and mildew resistance locus O (MLO) proteins. However, the molecular pathways through which CaMs/CMLs regulate these components are not yet fully understood [69, 70].

While multiple CaM and CML isoforms exist, their interaction specificity with downstream calcium-binding proteins (CBPs) under biotic stress is still under exploration. Many CML targets in this context remain unidentified, highlighting the need for further research. High-throughput analyses, multi-omics approaches, and advanced imaging technologies are expected to aid in identifying these targets and clarifying their roles in stress responses.

Most current knowledge on CaM/CML-mediated signalling stems from studies in model species, such as *A. thaliana*. However, the role of these pathways in crop species remains less understood. Modern genetic transformation and genome editing tools, like CRISPR/Cas9, offer promising opportunities for enhancing biotic stress tolerance in crops by modifying CaMs/CMLs expression or adjusting CBP interactions, possibly without affecting growth [67, 68].

11.6.1 Structure of CaM/CML

CaMs and CaM-like proteins (CMLs) are essential for calcium-based signalling, integral to many plants cellular processes. CaMs are well-known for their distinctive structure—a dumbbell shape featuring two globular domains connected by a flexible linker. Each domain contains two EF-hand motifs, which are specific helix-loop-helix structures crucial for binding Ca^{2+} ions CaMs can bind up to four Ca^{2+} ions, with two high-affinity sites in the C-terminal and two lower-affinity sites in the N-terminal. When Ca^{2+} ions attach, CaMs undergo a major conformational change, shifting from a compact, calcium-free (apo-CaM) form to an extended, calcium-loaded state. This change not only alters the protein's shape but also adjusts its surface characteristics from hydrophilic to primarily hydrophobic, enabling interactions with target proteins that are key to triggering cellular responses [71–73].

In comparison, CMLs display greater structural variability, possessing anywhere from one to six EF-hand motifs. However, not all of these motifs bind Ca^{2+} effectively. For instance, *Arabidopsis* CML14 binds calcium at only one of its three EF-hand motifs, which enables CMLs to respond to diverse calcium concentrations and signals. This variability allows CMLs to participate in a broader spectrum of calcium-mediated responses, which is particularly advantageous in adapting to fluctuating environmental stimuli.

CMLs also frequently contain distinctive N- or C-terminal extensions that add unique dynamics to their structure and influence how they interact with target proteins. These extensions enhance CML specificity and versatility, distinguishing them further from CaMs by providing tailored interactions with downstream CBPs. Such structural diversity across CaMs and CMLs is central to their role in managing calcium signalling responses to a range of physiological triggers, such as biotic stresses from pathogens and herbivores [68].

To fully understand the nuanced roles of CaMs and CMLs in plant immunity and stress adaptation, it is critical to study their specific conformational dynamics and interactions. Despite progress, there remain gaps in our knowledge about their downstream target proteins and the precise molecular pathways they activate. Ongoing research seeks to uncover these pathways, providing deeper insight into Ca^{2+} signalling and the potential for enhancing plant resilience through better understanding of these proteins' roles in stress responses.

11.6.1.1 Calmodulins and calmodulin-like proteins in biotic stress

CaMs and CMLs are key players in plant responses to biotic stress, such as pathogen invasion or herbivore attacks. These CBPs serve as vital sensors that decode calcium signals, activating defences to enhance plant resistance. Below is an expanded outline of their roles, interactions, and mechanisms in managing biotic stress [16, 69–71, 74–77].

11.6.1.2 Calcium signalling in response to biotic stress

During biotic stress, plants often experience a surge in intracellular Ca^{2+} levels. This influx acts as a rapid signalling trigger that initiates various defence responses. CaMs and CMLs bind to these Ca^{2+} ions, adopting new conformations that enable them to interact with essential target proteins, including transcription factors and kinases that activate defence-related pathways. This calcium-driven activation mechanism underpins the plant's early response to biotic stress.

11.6.1.3 Roles of CaMs and CMLs in defence

As part of the EF-hand family of CBPs, CaMs and CMLs contribute in several ways to plant defence mechanisms:

- Transcriptional regulation: CaMs and CMLs regulate transcriptional changes critical for effective plant defence. They interact with calmodulin-binding transcription factors (CBFs), like CBP60 and CAMTA, which modulate the expression of defence genes. For example, CBP60 enhances SA signalling, pivotal for SAR against pathogenic microbes.
- Activation of protein kinases: CaMs, upon binding Ca^{2+}, activate CDPKs. These kinases then phosphorylate target proteins, thereby activating various defence pathways crucial for plant resistance to biotic stress.
- Interplay with signalling molecules: CaMs and CMLs integrate signals from other molecules, such as ROS and plant hormones like JA and SA. This interaction allows plants to fine-tune their defences based on the type and intensity of the stress encountered, offering adaptability to a range of biotic challenges.

11.6.2 Mechanisms of CaM and CML-mediated biotic stress response

CaMs and CMLs manage plant defences through several complex processes:

- Decoding calcium oscillations: Biotic stress often leads to specific calcium oscillation patterns. CaMs and CMLs are thought to decode these patterns, enabling plants to mount appropriate responses. Although the precise decoding mechanisms remain unclear, this ability may allow plants to tailor their defences based on the specific nature of the threat.

- Target protein interactions: CaMs and CMLs interact with an array of downstream targets, including transcription factors and enzymes essential for defence responses. The adaptability and specificity of these interactions are crucial for orchestrating effective responses. Despite significant advances, many CML targets are still unknown, highlighting the need for further investigation.
- Coordination of calcium channels: The regulation of calcium transients during biotic stress involves coordinated activity among various calcium-permeable channels and pumps. CaMs and CMLs assist in regulating these channels, ensuring the generation of precise calcium signatures required to initiate defence mechanisms.

11.7 Pathogenesis-related proteins

PR proteins play a central role in plant defence against pathogens, with their presence marking an activated immune response. This chapter provides an in-depth exploration of PR proteins, covering their structural classes, defence mechanisms, regulatory pathways, and applications in crop resilience. Additionally, we address the complexity of PR protein signalling networks, highlighting recent advances in molecular plant pathology and genetic engineering applications aimed at enhancing plant resistance [78–81].

PR proteins were first identified in the context of SAR, a durable form of broad-spectrum immunity activated throughout the plant after initial infection. The discovery of PR proteins revolutionized plant pathology, leading to their classification based on sequence similarity, biochemical function, and immune role. Over 17 PR protein families have been characterized, with functions spanning cell wall fortification, signalling enhancement, and direct antimicrobial activity. The following are the different classes of plant PR proteins [82–89].

11.7.1 Classification and mechanisms of PR proteins

11.7.1.1 PR-1 proteins

PR-1 proteins are often the most abundant in response to infection and serve as key markers of the immune response. They are implicated in sterol binding, which disrupts pathogen cell membranes. PR-1a from tobacco (NtPR-1a), for example, has shown antifungal properties against *Phytophthora infestans*, targeting fungal sterols and inhibiting their growth. Additionally, *Arabidopsis* PR-1 proteins are upregulated by SA, which modulates the broader SAR response to biotrophic pathogens.

PR-1 proteins are among the best-studied PR proteins due to their consistent expression in response to pathogens and stress. These proteins are part of the SCP/TAPS (SCP: Scavenger receptor cysteine-rich; TAPS: Tpx-1/Ag5/PR-1/Sc7) superfamily, typically characterized by a conserved domain with six cysteine residues involved in disulfide bridge formation, stabilizing the protein's structure. PR-1 proteins are small, typically around 15–20 kDa, and are acidic, which may contribute to their stability in the extracellular space where they are commonly located.

PR-1 proteins are often regarded as markers of SAR due to their upregulation in response to the activation of this long-lasting, broad-spectrum immune response. Their precise mechanism remains largely unresolved, but it is suggested that they bind to sterols, interfering with pathogen development by disrupting cell membrane functions essential for the pathogen's growth and reproduction.

PR-1 proteins interact primarily within the extracellular matrix, where they may bind to membrane sterols that are essential for pathogen cell membrane integrity. One known interaction is with the receptor-like kinase FLS2 in *Arabidopsis*, which activates immune signalling upon perception of flagellin, a common bacterial PAMP. This interaction suggests that PR-1 proteins might work with PRR complexes at the cell membrane to enhance pathogen detection.

Upon binding, PR-1 triggers downstream signalling cascades, often involving MAPK pathways, which further amplify defence responses, including the production of ROS and transcriptional activation of additional defence genes. These responses lead to localized cell death, a phenomenon known as the hypersensitive response (HR), that limits pathogen spread.

11.7.1.2 PR-2 proteins: β-1,3-glucanases

β-1,3-glucanases target glucan in fungal cell walls, fragmenting pathogen structures and releasing oligoglucans that act as elicitors to amplify immune responses. The PR-2 family member in barley is effective against *Blumeria graminis*, the pathogen causing powdery mildew, where glucanase activity degrades fungal cell walls and aids in signal transduction for subsequent defence activation.

PR-2 proteins are β-1,3-glucanases, enzymes that break down β-1,3-glucans in pathogen cell walls, particularly in fungi. These enzymes are highly active against pathogens with glucan-rich cell walls, such as *Phytophthora* spp. and other oomycetes, facilitating the degradation of the pathogen's structural components.

Structurally, PR-2 proteins contain a catalytic domain that binds and hydrolyzes β-1,3-glucans, leading to the release of oligosaccharides. These oligosaccharides may act as DAMPs, further alerting the plant immune system to pathogen invasion.

PR-2 proteins are known to work in synergy with other PR proteins, such as PR-3 chitinases. For example, β-1,3-glucanase activity can expose chitin in fungal cell walls, allowing PR-3 chitinases to further degrade the cell wall, creating a stronger antimicrobial effect. Additionally, β-1,3-glucanases may release glucan oligosaccharides, which can serve as elicitors of immune responses.

Once β-1,3-glucan degradation products are recognized, they initiate downstream signalling pathways involving SA-dependent responses. SA is critical for the activation of SAR, making PR-2 proteins integral to long-term immunity. MAPK cascades are also activated in response to β-glucan degradation, leading to the production of antimicrobial compounds and additional PR proteins.

11.7.1.3 PR-3, PR-4, PR-8, and PR-11 proteins: chitinases

Chitinases break down chitin, a vital component of fungal cell walls. This disrupts fungal integrity and elicits immune responses. The chitinase CHIT42 from *Trichoderma* is frequently used in transgenic crops for its effectiveness against fungal pathogens like *Rhizoctonia solani*, known to cause root rot. The release of chitin fragments further activates additional defences, establishing an immune feedback loop critical for robust responses to fungal infections.

PR-3, PR-4, and PR-8 classes encompass various chitinases that degrade chitin, a major structural component of fungal cell walls. PR-3 proteins are typical chitinases, PR-4 proteins contain chitin-binding domains without hydrolytic activity, and PR-8 proteins are primarily lysozymes with chitinase-like activity.

PR-3 proteins, in particular, have catalytic domains that bind and cleave chitin polymers, disrupting fungal cell wall integrity and directly hindering pathogen growth. PR-4 proteins, while lacking enzymatic activity, still contribute by binding chitin and blocking pathogen growth through steric hindrance.

Chitinases often act synergistically with β-1,3-glucanases (PR-2) in pathogen cell wall degradation. This interaction potentiates immune responses as each enzyme targets distinct components of the cell wall, ensuring comprehensive pathogen degradation.

Recognition of chitin or its fragments by receptors like CERK1 initiates downstream signalling, resulting in JA-dependent responses, which are crucial in defence against necrotrophic pathogens. These pathways lead to the activation of defence genes, localized cell death, and the strengthening of cell walls with callose deposition.

11.7.1.4 PR-5 proteins: thaumatin-like proteins (TLPs)

TLPs disrupt fungal membranes by binding to specific glucans and permeabilizing cells. The PR-5 protein from barley (HvTLP) shows strong antifungal properties, particularly against root-infecting pathogens like *Fusarium* spp., where it disrupts cell membrane integrity and suppresses fungal growth. PR-5 proteins also aid in SAR signalling, highlighting their versatile defensive roles.

PR-5 proteins, also known as TLPs, have a characteristic structure stabilized by disulfide bonds that confer resistance to proteolysis. These proteins disrupt fungal cell membranes by binding to glucans and are thought to have antifungal and antiviral activities due to their ability to permeabilize pathogen cell walls.

PR-5 proteins are found in various plant tissues and are highly inducible by pathogens. Their accumulation results in enhanced permeability of pathogen membranes, effectively limiting pathogen growth and spread.

PR-5 proteins often interact with glucan-binding proteins on pathogen surfaces, causing membrane disruptions that compromise the pathogen's integrity. This interaction is closely linked to SA-dependent pathways, particularly in the activation of SAR. Additionally, TLPs can modulate oxidative burst, influencing downstream signalling through ROS production, which acts as a secondary messenger in immune responses.

11.7.1.5 PR-9 proteins: peroxidases

Peroxidases generate ROS in response to pathogens, resulting in an oxidative burst that cross-links cell wall polymers, fortifying barriers to pathogen ingress. The peroxidase gene OsPOX1 in rice promotes lignin formation in response to bacterial pathogens, enhancing resistance to leaf blight. This PR class exemplifies how ROS production not only creates a hostile environment for pathogens but also structurally strengthens plant tissue.

PR-9 proteins, or peroxidases, catalyze the reduction of hydrogen peroxide to water and oxygen, a process that produces ROS as a byproduct. These ROS play dual roles, serving as antimicrobial agents and as signalling molecules that activate additional immune responses.

Peroxidases are often associated with cell wall reinforcement, facilitating cross-linking of cell wall components that prevent pathogen entry.

ROS produced by PR-9 proteins act as signals that interact with hormone pathways, particularly those mediated by SA and ethylene. This signalling cascade leads to the activation of defence genes, further ROS production, and lignin synthesis, which contributes to physical barriers against pathogen invasion.

11.7.1.6 Other PR proteins

- PR-6 (protease inhibitors): Target proteases from herbivores or pathogens, deterring feeding or inhibiting microbial protein breakdown mechanisms. An example is the tomato proteinase inhibitor II, effective against insect herbivory by interfering with their digestive enzymes.
- PR-10 (ribonucleases and RNases): Interfere with pathogen RNA, thus limiting their replication. PR-10 proteins from *Pisum sativum* show antiviral properties, offering defence against RNA viruses.
- PR-12 to PR-15: Include antimicrobial peptides and defensins that destabilize pathogen membranes or regulate ion balance, such as PDF1.2 from *Arabidopsis*, a defensin protein effective against both fungal and bacterial infections.

11.8 Callose synthase

Callose, a β-1,3-glucan polysaccharide with occasional β-1,6 branching, is critical in plant structural integrity and defence. It is deposited in the cell wall and serves as a barrier against pathogen invasion and physical damage. In response to biotic stress, such as pathogen attacks, callose deposition rapidly increases in affected areas to reinforce cell walls, preventing pathogen entry and movement within plant tissues. Callose synthase enzymes, encoded by a family of *glucan synthase-like (GSL)* genes, catalyze the polymerization of glucose into callose at strategic locations, enhancing the plant's physical and biochemical defence. These enzymes are particularly responsive to pathogenic attacks, drought, and other stresses, underscoring their central role in plant defence mechanisms [90].

11.8.1 Structure of callose synthase

Callose synthase is a multi-domain enzyme embedded in the plasma membrane, where it polymerizes glucose into β-1,3-glucan chains to form callose. The enzyme structure includes an N-terminal regulatory domain, multiple transmembrane helices that secure it within the membrane, and a C-terminal catalytic domain responsible for the glucan polymerization activity.

The N-terminal domain is crucial for regulatory functions, including binding to signalling molecules like calcium ions and interacting with other proteins. Multiple transmembrane domains span the lipid bilayer, facilitating the transport of glucose substrates across the membrane. The C-terminal region contains catalytic residues and domains essential for β-1,3-glucan synthesis, including conserved glycosyltransferase motifs required for callose production. Additionally, calmodulin-binding regions and phosphorylation sites have been identified in callose synthase, making it sensitive to calcium levels and kinase activity. These structural features allow the enzyme to respond rapidly to calcium signalling cascades triggered by PAMPs, indicating that callose synthase plays a central role in early immune signalling [90, 91].

11.8.2 Role of callose synthase in plant defence mechanisms

The structural adaptability of callose synthase allows it to integrate into plant defence pathways, where it directly contributes to pathogen containment. When pathogens attack, callose synthase initiates localized callose deposition around infection sites, sealing off infected areas from surrounding tissues. This rapid deposition is facilitated by the enzyme's response to calcium signals, which modulate both its localization and activity.

Callose synthase also contributes to the HR by producing callose deposits that isolate cells undergoing programmed cell death, thereby containing the pathogen within a restricted area. This strategy is particularly effective against biotrophic pathogens that rely on living host cells to thrive. The enzyme's involvement in HR is supported by research in *Nicotiana tabacum*, where callose deposition closely coincides with HR-related signalling and subsequent cell death, establishing it as a physical and signalling reinforcement against pathogen spread [90–94]).

11.8.3 Examples of callose synthase roles

1. **Pathogen restriction and callose plug formation:** In *A. thaliana*, *GSL5* plays a crucial role in the pathogen-triggered callose synthesis. Mutants lacking functional *GSL5* exhibit increased vulnerability to fungal pathogens, suggesting that callose synthase activity is essential for forming callose plugs that physically block pathogen spread.
2. **Regulation of plasmodesmata and intercellular signalling:** In rice (*Oryza sativa*), callose synthase encoded by *GSL7* plays a regulatory role in plasmodesmatal callose accumulation. This regulation controls cellular communication under both normal and stress conditions, modulating intercellular signalling during infection or environmental stress.
3. **Response to abiotic stressors:** Callose synthase has also been linked to increased resilience under drought, salinity, and cold stress. In tomato (*Solanum lycopersicum*), studies found that drought-induced callose synthesis reinforced the cell wall, reducing the impact of water stress on cellular structures. This drought resilience is partly attributed to callose synthase's structural response to calcium signalling triggered by abiotic stress conditions.

11.9 Phytohormones

Phytohormones are naturally occurring organic compounds that regulate physiological processes in plants, acting as signalling molecules to mediate responses to environmental cues. These hormones play a crucial role in coordinating plant defence mechanisms against a variety of biotic and abiotic stresses, including pathogens, herbivores, drought, and salinity. The complexity of plant responses to stressors is underscored by the intricate interactions among different phytohormones, which can either act synergistically or antagonistically to optimize survival and fitness.

Phytohormones are classified into several categories, including SA, JA, ET, ABA, auxins, cytokinins, gibberellins (GAs), and brassinosteroids (BRs). Each of these hormones has distinct roles in mediating growth, development, and stress responses, yet they also interact extensively with one another to form a cohesive signalling network. This interplay is crucial for the establishment of effective defence strategies, as plants often face multiple simultaneous stresses in their natural environments.

Recent research has emphasized the concept of conserved hubs within hormone signalling networks, where certain proteins or pathways serve as critical points of integration for multiple signals. For example, NPR1 in the SA pathway acts as a master regulator, while JAZ proteins and MYC transcription factors are central to JA signalling. These hubs not only facilitate the execution of defence responses but also allow for flexibility and adaptability in the face of varying stress conditions.

The importance of phytohormones in plant defence has gained significant attention in the context of agricultural practices, particularly as global climate change increases the prevalence of stressors that affect crop yields. Understanding the mechanisms of hormone signalling and their interactions offers potential strategies for enhancing plant resilience through breeding or biotechnological approaches [95–97].

11.9.1 Salicylic acid

SA is a key phytohormone involved in plant defence, particularly against biotrophic pathogens—organisms that extract nutrients from living host tissues. When a plant is attacked, SA is rapidly synthesized and accumulated at the site of infection, leading to the activation of local and SAR. This systemic response is crucial for preparing uninfected parts of the plant for potential future attacks, effectively 'priming' them for a quicker and more robust defensive response.

Research has shown that the induction of SA synthesis is a critical response to various pathogens, including fungi and bacteria. For instance, upon infection with *Pseudomonas syringae*, a well-studied bacterial pathogen, *Arabidopsis* plants exhibit a rapid increase in SA levels. This increase is followed by the expression of PR genes, which encode proteins that have antifungal and antibacterial properties. This mechanism underlies the plant's ability to mount a systemic response to pathogens [98].

11.9.1.1 SA biosynthesis pathway and signal transduction

The biosynthesis of SA occurs primarily through the isochorismate pathway, which begins with the conversion of chorismate to isochorismate, followed by further modification to produce SA. This pathway is activated during pathogen attacks and is essential for the rapid accumulation of SA in response to stress.

Once synthesized, SA is transported throughout the plant, and its perception is mediated by receptors, particularly NPR1. NPR1 acts as a master regulator of the SA signalling pathway, facilitating the transcription of PR genes through its interaction with TGA transcription factors. When SA binds to NPR1, it undergoes a conformational change that allows it to translocate to the nucleus, where it activates the expression of defence-related genes [98, 99].

11.9.1.2 Mechanisms and pathways of SA-mediated defence

SA activates the expression of numerous PR genes, such as PR1, PR2, and PR5, which play vital roles in defending against pathogens. These proteins are involved in various defence mechanisms, including the production of antimicrobial compounds and the reinforcement of cell walls. Moreover, SA is linked to the generation of ROS, which serve as signalling molecules in the defence response. ROS not only plays a role in cell signalling but also contributes to the HR, characterized by localized cell death that limits the spread of pathogens.

In *Arabidopsis*, mutants deficient in SA biosynthesis, such as sid2 mutants, demonstrate significantly reduced resistance to *P. syringae*, underscoring the importance of SA in plant immunity. These findings illustrate how effective SA signalling is in conferring resistance to biotrophic pathogens [100].

11.9.1.3 Interactions with other hormones

SA exhibits an antagonistic relationship with JA, which is primarily involved in defence against necrotrophic pathogens and herbivores. The balance between SA and JA signalling pathways is critical for determining the plant's defensive strategy. For instance, in *Arabidopsis*, heightened SA signalling can inhibit JA-dependent defences, potentially increasing susceptibility to herbivores like *Spodoptera exigua*.

The cross-talk between SA and JA not only shapes the plant's response to pathogens but also reflects the complexity of hormonal regulation in plant defence strategies. This interaction is particularly relevant in the context of co-infection scenarios, where plants face simultaneous attacks from biotrophic and necrotrophic pathogens [96].

11.9.1.4 Conserved elements and SA-specific hubs

NPR1 is a pivotal hub within the SA signalling network. It not only regulates the expression of defence genes but also interacts with other signalling pathways, facilitating a coordinated response to multiple stresses. The integration of SA signals with those from other hormones highlights the complexity of plant defence strategies. Recent studies have shown that NPR1 can modulate the activity of various transcription factors, enhancing the plant's capacity to adapt to changing environmental conditions.

Additionally, the interaction of NPR1 with TGA transcription factors exemplifies how conserved hubs contribute to a robust defence response. These hubs are essential for integrating multiple signals and orchestrating the plant's defence mechanisms against a variety of stresses [101, 102].

11.9.2 Jasmonic acid

JA is a crucial phytohormone involved in mediating plant defence against necrotrophic pathogens and herbivores. Unlike SA, which primarily defends against biotrophic pathogens, JA is particularly effective in activating wound responses and fortifying plant tissues against physical damage caused by herbivores. JA's role in plant defence is multifaceted, encompassing the regulation of various physiological processes that enhance resistance.

When a plant experiences tissue damage due to herbivore feeding or pathogen invasion, JA levels rise significantly. This increase triggers a cascade of responses aimed at fortifying the plant's defences. For instance, the activation of protease inhibitors and secondary metabolites occurs, deterring herbivores and inhibiting pathogen growth [95–97].

11.9.2.1 JA biosynthesis and signal transduction

JA is synthesized from linolenic acid, which is released from chloroplast membranes during stress. The pathway leading to JA biosynthesis involves several enzymatic steps, ultimately producing the active form of JA, jasmonoyl isoleucine (JA-Ile). The perception of JA-Ile is mediated by the COI1 receptor, which is part of a complex that regulates the degradation of JAZ proteins. JAZ proteins are negative regulators of JA signalling, and their degradation upon JA-Ile binding releases MYC transcription factors, allowing for the expression of JA-responsive defence genes.

11.9.2.2 defence mechanisms and JA-mediated signalling pathways

JA signalling is pivotal in activating a suite of defence-related genes that encode proteins involved in producing protective compounds. For example, JA induces the expression of protease inhibitors, which inhibit the digestive enzymes of herbivores, thus reducing their ability to feed on plant tissues. Additionally, JA promotes the biosynthesis of secondary metabolites, such as flavonoids and alkaloids, that can deter herbivore feeding and provide antimicrobial properties.

Experimental studies have demonstrated that elevated JA levels enhance resistance in crops against various pests, including caterpillars and beetles. For example, research on *Brassica napus* has shown that JA treatment significantly increases the expression of genes involved in defence, leading to improved resistance against herbivores.

11.9.2.3 JA–ET synergy and antagonism with SA

JA works synergistically with ET, particularly in response to herbivory. The cooperation between JA and ET enhances the expression of defensive compounds, facilitating a stronger plant response to physical damage. For instance, during herbivore attacks, JA and ET jointly promote the expression of genes involved in synthesizing protective proteins, leading to a comprehensive defence mechanism.

However, the relationship between JA and SA is often antagonistic. High levels of SA can suppress JA signalling, potentially leaving plants more vulnerable to herbivore attacks. This antagonism illustrates the complex interplay between different hormonal pathways and emphasizes the need for plants to fine-tune their responses based on the type of threat they encounter.

11.9.2.4 Conserved components and JA hubs

Key components of the JA signalling pathway, such as JAZ proteins and MYC transcription factors, serve as crucial regulatory hubs. The degradation of JAZ proteins in response to JA-Ile binding is a critical step that activates defence gene expression. MYC transcription factors, once released from JAZ repression, can activate a wide array of JA-responsive genes, highlighting the importance of this regulatory mechanism in plant defence.

Research has shown that the JAZ–MYC complex plays a central role in mediating plant responses to both biotic and abiotic stresses. For example, JAZ proteins not only regulate JA signalling but also interact with other signalling pathways, demonstrating the interconnectedness of hormonal regulation in plant defence strategies.

11.9.3 Ethylene (ET)

11.9.3.1 ET's dual role in abiotic and biotic stress

ET is a gaseous phytohormone known for its multifaceted roles in plant growth and development, as well as in mediating responses to both abiotic and biotic stresses. ET is particularly important in regulating responses to mechanical stresses and in modulating other hormone pathways. Its involvement in plant defence mechanisms is most pronounced in response to necrotrophic pathogens and herbivory, where it enhances the plant's ability to mount effective defence responses.

In addition to its role in biotic stress responses, ET is also crucial for abiotic stress tolerance. Under conditions such as drought or salinity, ET signalling helps plants adjust their growth and development, allowing them to cope with unfavourable environmental conditions. This dual role of ET underscores its significance in maintaining plant health in the face of diverse challenges [95–97].

11.9.3.2 ET biosynthesis and perception mechanisms

ET biosynthesis begins with the amino acid methionine, which is converted to 1-aminocyclopropane-1-carboxylic acid (ACC) through a series of enzymatic reactions. ACC synthase is a key enzyme in this pathway, regulating the production of ET. Once synthesized, ET is perceived by a family of receptors that initiate downstream signalling pathways.

The signalling cascade involves EIN2 and EIN3/EIL transcription factors, which are activated upon ET perception. These transcription factors play a central role in mediating ET-responsive gene expression, contributing to the plant's overall defence capabilities [120].

11.9.3.3 defence mechanisms and pathway activation

ET signalling is particularly important in enhancing plant responses to necrotrophic pathogens and herbivores. When a plant is attacked, ET signalling pathways can activate the expression of various defence-related genes, including those involved in cell wall modification and the synthesis of secondary metabolites. For instance, ET enhances the expression of genes that contribute to lignin and suberin deposition in cell walls, strengthening plant tissues and making them less palatable to herbivores.

Research has demonstrated that ET signalling is integral to regulating cell wall modifications during pathogen invasion. For example, in response to infection by *Botrytis cinerea*, a necrotrophic fungus, ET promotes the synthesis of defensive compounds that inhibit fungal growth.

11.9.3.4 Conserved ET hubs and interaction with JA and ROS

EIN2 and EIN3 are central regulators in the ET signalling pathway. Their activation not only triggers ET-responsive gene expression but also integrates signals from other hormonal pathways, including JA and ROS signalling. The cooperation between ET and JA is particularly notable in enhancing plant defence during herbivory.

Additionally, ROS serve as secondary messengers that amplify ET signals, contributing to a robust immune response. The interplay between ET, JA, and ROS is essential for establishing a coordinated defence strategy, allowing plants to effectively respond to multiple stressors simultaneously.

11.9.4 Abscisic acid

ABA is a phytohormone primarily known for its role in regulating plant responses to abiotic stresses, particularly drought and salinity. One of the critical functions of ABA is its ability to induce stomatal closure, which helps to prevent water loss during periods of drought. This mechanism is vital for maintaining plant water status and overall health, especially in arid conditions.

In addition to its role in water regulation, ABA is increasingly recognized for its contributions to plant defence mechanisms. ABA signalling can enhance a plant's resilience against pathogen attacks by priming defence responses and modulating other hormonal pathways. For instance, ABA can influence the expression of PR genes and promote the synthesis of protective compounds [95–97].

11.9.4.1 Biosynthesis and perception of ABA

ABA is synthesized from carotenoids in response to environmental stresses. The biosynthesis pathway involves several enzymatic steps that convert carotenoids into ABA, with the final steps being regulated by specific enzymes such as 9-cis-epoxycarotenoid dioxygenase (NCED). The perception of ABA is mediated by PYR/PYL/RCAR receptors, which initiate downstream signalling cascades upon binding to ABA.

The signalling pathway typically involves SnRK2 kinases, which phosphorylate target proteins, leading to the activation of transcription factors that regulate stress-responsive genes. This signalling cascade is crucial for enabling plants to respond effectively to changing environmental conditions.

11.9.4.2 Cross-talk and trade-offs in growth and defence

While ABA is essential for enhancing stress tolerance, it also presents trade-offs between growth and defence. The antagonism between ABA and SA signalling can impact a plant's overall susceptibility to diseases. For example, under drought conditions, elevated ABA levels may suppress SA-mediated defence responses, leading to increased vulnerability to pathogens like *B. cinerea*.

Research has shown that the balance between ABA and other phytohormones is critical for optimizing growth under stress conditions. The ability of plants to manage this balance can significantly influence their fitness and survival in the face of environmental challenges.

11.9.4.3 Key ABA regulatory hubs and integrators

SnRK2 kinases and ABF transcription factors are key integrators of ABA signalling. These components play crucial roles in linking ABA responses with other hormonal pathways, ensuring a coordinated response to stress. Recent studies have highlighted the importance of ABF transcription factors in modulating the expression of genes involved in both ABA signalling and pathogen defence.

Moreover, the cross-talk between ABA and other phytohormones, particularly SA, emphasizes the complexity of hormonal interactions in plant defence strategies. Understanding these interactions is essential for developing strategies to enhance plant resilience to various stresses.

11.9.5 Auxins

Auxins are a class of phytohormones primarily associated with promoting cell elongation and growth. However, recent research has revealed their significant roles in plant defence mechanisms as well. Auxins can influence the plant's susceptibility to pathogens by modulating growth responses during stress conditions. This dual role highlights the complexity of auxin signalling in balancing growth and defence strategies.

In the context of plant–pathogen interactions, auxins can be manipulated by pathogens to promote their growth and survival. For instance, some pathogens produce auxin-like compounds that alter plant development, making the plant more susceptible to infection. Understanding these interactions is crucial for developing effective plant defence strategies [95–97].

11.9.5.1 Auxin biosynthesis and signalling pathways

Auxins are synthesized through two primary pathways: the tryptophan-dependent pathway and the tryptophan-independent pathway. The most well-known auxin, indole-3-acetic acid (IAA), is synthesized from tryptophan via several enzymatic steps. Once synthesized, auxins are transported throughout the plant and perceived by receptor proteins, primarily the TIR1 receptor.

The TIR1 receptor is part of the SCF (SKP1-CUL1-F-box) ubiquitin ligase complex, which regulates the degradation of Aux/IAAs, negative regulators of auxin signalling. This degradation allows the activation of ARF (auxin response factor) transcription factors, leading to the expression of auxin-responsive genes involved in various physiological processes.

11.9.5.2 Conserved pathways and manipulation by pathogens

Pathogens can hijack auxin signalling pathways to increase their susceptibility to infection. For instance, certain pathogens synthesize auxin-like compounds that mimic natural auxins, altering the host plant's growth and development. This manipulation can facilitate pathogen entry and spread, demonstrating the critical interplay between plant hormones and pathogen strategies.

Research has shown that auxin signalling pathways can be influenced by various stressors, including biotic attacks. The complex regulation of auxin levels in response to stress is essential for maintaining plant health and resilience.

11.9.5.3 Auxin-specific hubs and their interactions with defence hormones

The TIR1 receptor plays a pivotal role in integrating auxin signals with those from other phytohormones, particularly SA. The cross-talk between auxin and SA signalling pathways is critical for managing the plant's defence response. For example, studies have indicated that auxin can modulate SA-mediated defence responses, highlighting the need for a coordinated approach to plant defence strategies.

Understanding these interactions is essential for developing effective agricultural practices that enhance plant resilience to stressors while optimizing growth.

11.9.6 Cytokinins

Cytokinins are phytohormones primarily involved in regulating cell division and differentiation. They play crucial roles in plant growth and development, but they also contribute to plant defence mechanisms. Cytokinins are essential for nutrient allocation during pathogen attacks, helping to redirect resources to bolster immune responses.

Research has shown that cytokinins can enhance the expression of PR genes, contributing to increased resistance against various pathogens. For instance, cytokinin treatment has been shown to improve resistance to *P. syringae* in *Arabidopsis*, underscoring their significance in plant immunity [95–97].

11.9.6.1 Biosynthesis and signal transduction

Cytokinins are synthesized primarily in the roots and transported to the shoots. The biosynthesis of cytokinins involves several key enzymes, including isopentenyltransferase (IPT). Once synthesized, cytokinins signal through a two-component system involving receptor kinases, such as AHKs (*Arabidopsis* histidine kinases), which activate downstream signalling pathways.

The signalling cascade culminates in the activation of response regulators that mediate the expression of cytokinin-responsive genes, influencing growth and defence mechanisms.

11.9.6.2 Defence role and cross-talk with SA and JA pathways
Cytokinins have been shown to interact with other phytohormones, particularly SA and JA, in regulating defence responses. For example, cytokinin-induced PR gene expression has been linked to enhanced resistance against pathogens, indicating a cooperative role in defence signalling. This cross-talk between cytokinins and other hormones emphasizes the complexity of plant defence strategies and the need for integrated responses.

Recent studies have highlighted the importance of cytokinin signalling in mediating plant responses to simultaneous biotic and abiotic stresses, further illustrating the interconnectedness of hormonal regulation in plant defence.

11.9.6.3 Conserved hubs in cytokinin signalling
The type-A and type-B response regulators serve as conserved hubs in cytokinin signalling. These regulators play crucial roles in integrating cytokinin signals with those from other hormonal pathways, facilitating a coordinated defence response. Research has shown that the interplay between cytokinins and other hormones, particularly SA and JA, is essential for optimizing plant defence strategies.

Understanding the dynamics of cytokinin signalling in the context of plant defence offers potential avenues for improving crop resilience and performance in the face of environmental challenges.

11.9.7 Gibberellins

GAs are phytohormones primarily associated with promoting growth and developmental processes, such as seed germination and stem elongation. However, their role in mediating defence responses is complex, as GAs can influence the trade-off between growth and immunity. While GAs promote growth under favourable conditions, they can suppress defence mechanisms, particularly under low-stress conditions.

Research has shown that elevated GA levels can inhibit the expression of defence-related genes, making plants more susceptible to pathogens. For example, GA treatment has been shown to enhance susceptibility to *B. cinerea*, highlighting the growth-defence trade-off that plants must navigate [95–97].

11.9.7.1 GA biosynthesis and signal transduction
GAs is synthesized from geranylgeranyl diphosphate through a series of enzymatic reactions. The signalling pathway involves the GID1 receptor, which, upon binding to GAs, activates downstream responses. The interaction between GID1 and DELLA proteins is a critical component of GA signalling.

DELLA proteins act as negative regulators of GA signalling, and their degradation in response to GA binding allows for the activation of GA-responsive genes. This mechanism is essential for promoting growth while also highlighting the potential for GAs to suppress defence responses.

11.9.7.2 Mechanisms of GA-mediated immune modulation

The influence of GAs on plant immunity underscores the delicate balance between growth and defence. Under conditions of low stress, elevated GA levels can suppress the expression of defence genes, rendering plants more susceptible to pathogens. This modulation is particularly evident in crops, where optimal growth conditions may inadvertently increase vulnerability to diseases.

Understanding the mechanisms by which GAs influence immune responses is crucial for developing strategies to enhance plant resilience in agricultural systems.

11.9.7.3 Conserved hubs and GA cross-talk

DELLA proteins serve as key hubs in GA signalling and are integral to the cross-talk between GAs and other phytohormones, particularly SA and JA. The interactions between DELLA proteins and other signalling pathways highlight the complexity of hormonal regulation in plant defence.

Research has indicated that DELLA proteins can modulate the expression of defence-related genes, suggesting that manipulating GA signalling may provide avenues for enhancing resistance to pathogens while maintaining optimal growth.

11.9.8 Brassinosteroids

BRs are a class of phytohormones known for their role in promoting plant growth and development. However, they also contribute significantly to stress tolerance and immune activation. BRs enhance plant resilience to various abiotic stresses, such as drought and salinity, while also playing a role in defence against biotic stresses.

Research has demonstrated that BRs can improve plant tolerance to stress by modulating gene expression and physiological responses. For instance, BR treatment has been shown to enhance the expression of defence-related genes, contributing to increased resistance against pathogens [95–97].

11.9.8.1 BR biosynthesis and signalling pathways

BRs are synthesized from campesterol through a series of enzymatic conversions. The signalling pathway involves the BRI1 receptor, which, upon BR binding, initiates downstream signalling cascades that promote growth and defence responses.

The signalling cascade typically activates a series of transcription factors, including BZR proteins, which regulate the expression of BR-responsive genes. This regulation is crucial for integrating growth and defence responses, allowing plants to adapt to changing environmental conditions.

11.9.8.2 BR-mediated defence mechanisms and synergistic pathways

BRs cooperate with other phytohormones, particularly SA and JA, to enhance plant defence responses. This cooperation is evident in the regulation of defence-related genes, where BR signalling can amplify the effects of SA and JA in promoting resistance against pathogens.

Research has shown that BRs can enhance the expression of PR genes, leading to improved resistance against various pathogens. For example, studies on *Arabidopsis* have highlighted the synergistic effects of BRs with other hormones in modulating plant defence.

11.9.8.3 Conserved BR hubs and cross-talk with other hormones

BZR transcription factors serve as critical hubs in BR signalling and play a pivotal role in integrating signals from other hormonal pathways. The interactions between BRs and other hormones, particularly SA and JA, underscore the complexity of hormonal regulation in plant defence strategies.

Understanding the dynamics of BR signalling in the context of plant defence offers potential avenues for enhancing crop resilience and performance in the face of environmental challenges.

11.10 Heat shock proteins

Heat shock proteins (HSPs) and molecular chaperones are pivotal to the stress response in plants, helping maintain cellular integrity and function under a wide range of stress conditions. Plants, being sessile organisms, are continuously exposed to fluctuating environmental conditions, including temperature extremes, drought, salinity, and pathogen attacks. The ability to rapidly and effectively respond to these stressors is crucial for survival and fitness. HSPs are a diverse family of proteins that assist in protein folding, refolding, and degradation, ensuring that other proteins maintain their functional conformations during stress. This review aims to explore the different classes of HSPs, their structures and functions, and their roles in plant defence mechanisms, drawing insights from key studies in the field [45, 103, 104].

11.10.1 Classification of HSPs

HSPs are classified into families based on their molecular weight and specific functions. The major classes include:

- **HSP100:** This class is characterized by its ability to disaggregate proteins and facilitate their proteolysis. HSP100 proteins are crucial for the recovery of cells from stress conditions and are involved in the clearance of damaged proteins. HSP100 typically forms hexameric structures that use ATP hydrolysis to drive protein disaggregation. The AAA+ (ATPases Associated with various cellular Activities) domain is critical for their function, allowing these proteins to unfold and refold substrate proteins.
- **HSP90:** HSP90 proteins are essential for folding and stabilizing many client proteins, including those involved in signal transduction and protein trafficking. They play critical roles in the regulation of various signalling pathways. HSP90 is composed of two domains: the N-terminal domain, which binds ATP, and the C-terminal domain, which is involved in client protein interactions. HSP90 undergoes significant conformational changes upon ATP binding and hydrolysis, enabling it to stabilize client proteins.

- **HSP70:** This family is vital for the folding of nascent polypeptides and the refolding of denatured proteins. HSP70 proteins also participate in transporting proteins across membranes and in targeting misfolded proteins for degradation. It contains a highly conserved ATPase domain that is responsible for binding and releasing substrate proteins. The substrate-binding domain recognizes and binds to exposed hydrophobic regions of unfolded proteins, facilitating their proper folding.
- **HSP60:** Often localized in mitochondria and chloroplasts, HSP60 proteins assist in the proper folding of mitochondrial and plastid proteins, ensuring their functionality. Structurally, HSP60s form barrel-shaped oligomers that encapsulate substrate proteins, providing a protected environment for proper folding. This compartmentalization is crucial for the folding of larger proteins that may require additional time and conditions to achieve their native state.
- **Small HSPs (sHSPs):** These proteins are crucial for preventing protein aggregation and protecting cells under stress conditions. They typically function as holdases, binding to unfolded or misfolded proteins. These typically exist as oligomers and have a conserved alpha-crystallin domain that allows them to bind to denatured proteins. Their flexible structure allows them to interact with a variety of substrate proteins, preventing aggregation.

11.10.2 HSPs as molecular chaperones

HSPs are integral to the molecular chaperone network within plant cells, particularly under stress conditions. They assist in the proper folding of proteins, preventing misfolding and aggregation, which can lead to cellular dysfunction. Molecular chaperones function through several mechanisms, as follows.

11.10.2.1 Protein folding

HSPs promote the correct folding of nascent polypeptides as they emerge from ribosomes. This is particularly important for multi-domain proteins that require specific folding pathways.

11.10.2.2 Refolding of denatured proteins

Under stress, proteins can become denatured due to heat or oxidative damage. HSPs bind to these unfolded proteins, preventing aggregation and facilitating their refolding into functional conformations.

11.10.2.3 Targeting for degradation

When proteins are irreparably damaged, HSPs can target them for proteolytic degradation through the ubiquitin–proteasome system, ensuring that damaged proteins do not accumulate and interfere with cellular functions.

In a study on *Arabidopsis*, HSP70 was shown to enhance the recovery of the photosynthetic enzyme Rubisco during heat stress, ensuring continued photosynthetic efficiency. This underscores the importance of HSPs in maintaining critical metabolic processes under stress conditions.

11.10.2.4 HSPs in response to biotic stress

Plants encounter a variety of biotic stressors, including pathogens and herbivores. The expression of HSPs is often upregulated in response to these attacks, serving as part of the plant's immune response.

HSPs are not only involved in protein folding but also play significant roles in the signalling pathways that regulate plant defence mechanisms:

1. Interaction with pathogen effectors: Some HSPs interact with effector proteins from pathogens, modulating the host's immune responses. This interaction can either facilitate pathogen invasion or enhance the plant's defensive capacity.
2. Regulation of defence-related genes: HSPs can influence the expression of genes involved in the synthesis of defensive compounds, such as phytoalexins and pathogenesis-related (pr) proteins.

Example: HSP90 is crucial in the signalling pathways of plant hormones like SA and JA. These hormones are integral to the activation of defence responses against biotic stresses. HSP90 stabilizes key signalling proteins, enabling the plant to mount an effective defence against pathogens.

11.11 Lipid transfer proteins

Lipid transfer proteins (LTPs) are ubiquitous small, soluble proteins found primarily in plants. These proteins bind hydrophobic molecules, such as lipids, and transport them across membranes, which is crucial for cell growth, cuticle formation, and immune responses. The ability of LTPs to function in harsh environmental conditions, such as drought and pathogen attacks, reflects their structural robustness.

Studies emphasize the role of LTPs in diverse processes, highlighting both the functional and structural diversification among species and plant tissues. LTPs are typically found in seeds, leaves, roots, and floral organs, indicating their involvement in key biological activities, from development to stress response [105–110].

11.11.1 Structural characteristics of LTPs

The molecular structure of LTPs contributes directly to their stability and functionality. All LTPs share a common core of 4–5 tightly packed alpha-helices, which create a hydrophobic cavity that binds lipid molecules. Additionally, disulfide bridges stabilize the alpha-helical structure, rendering these proteins resistant to heat and proteolysis.

11.11.2 Structural classification

11.11.2.1 lTP1 family

- **Molecular weight:** ~9 kDa.
- **Structure:** Comprises five alpha-helices connected by loops, creating a tunnel-like hydrophobic core.
- **Role:** Involved in the transfer of phospholipids and cutin monomers.
- **Example:** *A. thaliana* AtLTP1, which binds fatty acids and contributes to cuticle formation.

11.11.2.2 LTP2 family

- **Molecular weight:** ~7 kDa (smaller than LTP1)
- **Structure:** Shares a similar helical structure with LTP1 but with a slightly different arrangement of helices, allowing for binding specificity to smaller lipids.
- **Role:** Facilitates lipid exchange between cellular membranes.
- **Example:***Oryza sativa* OsLTP2, which is involved in lipid transport under stress conditions.

11.11.2.3 Atypical LTPs

- These include subclasses like LTPc, LTPd, and LTPg. These proteins have evolved to bind specialized lipid molecules or perform functions beyond simple lipid transport.
- **Example:** LTPg proteins are often linked to stress responses and pathogen defence, while LTPc contributes to the formation of the plant cuticle.

11.11.3 defence response against pathogens

LTPs have been implicated in plant immune responses, especially during pathogen attack. These proteins are often upregulated when plants are under attack by fungi or bacteria. They transport lipids needed for reinforcing the plant cell wall or signalling molecules involved in activating immune responses.

AtDIR1 in *A. thaliana* plays a key role in SAR. During an infection, it facilitates the long-distance movement of lipid-based signals, triggering immune responses in distal tissues. *B. napus* BnLTP2 contributes to pathogen resistance by transporting lipid molecules that strengthen the cell wall, preventing pathogen invasion.

LTPs represent a versatile and essential class of proteins in plants, performing a wide range of physiological functions. From ensuring proper cuticle formation to responding to environmental stresses and pathogens, LTPs are integral to plant health and development. Their diverse structural arrangements allow them to bind various lipids and function across different cellular environments.

As research continues to explore the roles of LTPs, new insights into their mechanisms and potential applications emerge. For example, engineering crops to express specific LTPs could enhance drought tolerance or pathogen resistance, making agriculture more sustainable. Future studies may also focus on how LTPs interact with other proteins and signalling molecules to coordinate complex biological processes.

11.12 Antimicrobial peptides

Plant antimicrobial peptides (AMPs) are crucial defence molecules that provide the first line of protection against pathogenic microbes. These peptides play an essential role in the innate immune system by disrupting pathogen membranes, inhibiting microbial enzymes, or acting as signalling molecules to trigger immune responses. This review focuses on several classes of plant AMPs, including defensins, thionins,

LTPs, and hevein-like peptides, exploring their structural features, antimicrobial functions, and their role in plant innate immunity [111–115].

11.12.1 Classes of plant AMPs

Plant AMPs exhibit a high degree of structural diversity, contributing to their ability to combat a wide range of pathogens. Below is an exploration of key classes of plant AMPs and their structural features, as follows.

11.12.1.1 Defensins

Defensins are small, cysteine-rich peptides widely distributed in plants, animals, and fungi. They are known for their ability to disrupt microbial membranes by binding to specific phospholipids, such as phosphatidylinositol phosphates.

Plant defensins typically consist of 45–54 amino acids stabilized by four to five disulfide bridges, which create a compact and stable three-dimensional structure resistant to proteolysis and environmental stress. Defensins adopt a β-sheet-rich structure with a characteristic fold stabilized by cysteine residues. Plant defensins interact with microbial membranes by binding to negatively charged phospholipids. This interaction leads to pore formation and membrane destabilization, causing leakage of ions and cell death. Some defensins also interfere with intracellular targets, such as enzymes or signalling proteins.

MsDef1: A defensin from *Medicago sativa* with antifungal activity against *Fusarium* species.

Psd1: A defensin from *P. sativum* that targets fungal cell walls.

NaD1: A defensin from *Nicotiana alata* that forms dimers and exhibits strong antifungal activity.

11.12.1.2 Thionins

Thionins are small peptides, usually 45–48 amino acids long, rich in basic residues and cysteine. They are primarily involved in plant defence against fungal and bacterial pathogens.

Thionins are characterized by a compact α-helical and β-sheet arrangement stabilized by disulfide bonds. This rigid structure allows them to penetrate microbial membranes effectively. α-thionins in barley disrupt fungal membranes and prevent hyphal growth. Crambin, a well-studied thionin, demonstrates how these peptides fold into compact, membrane-active structures.

Thionins target microbial membranes, forming ion-permeable channels that cause leakage of cellular contents. Their high net positive charge facilitates electrostatic interactions with negatively charged microbial membranes, making them highly effective against pathogens.

11.12.1.3 Hevein-like peptides

Hevein-like peptides contain chitin-binding domains and are primarily active against fungal pathogens by targeting fungal cell walls.

Hevein-like peptides are characterized by a conserved cysteine-rich motif and a chitin-binding domain. This domain allows them to bind to the chitin in fungal cell walls, disrupting fungal integrity. Hevein from rubber trees exhibits antifungal activity by binding fungal chitin. WGA (wheat germ agglutinin) binds to N-acetylglucosamine residues in fungal and bacterial pathogens.

These peptides interfere with fungal cell wall synthesis by binding to chitin, a crucial component of fungal cell walls. This binding prevents fungal growth and hyphal extension.

11.12.1.4 Role of AMPs in plant innate immunity

Plant AMPs not only act as antimicrobial agents but also play critical roles in the innate immune system by triggering immune responses, signalling defence pathways, and acting as molecular markers of infection.

The primary role of AMPs in plant immunity is to directly kill invading microbes. Peptides like defensins and thionins bind to microbial membranes, forming pores or disrupting lipid bilayers, leading to ion leakage and cell death. This rapid response provides plants with an immediate defence mechanism to control infections.

In addition to their antimicrobial activity, many AMPs function as signalling molecules that activate defence-related pathways in plants. For example, defensins can trigger the production of ROS and activate MAPK cascades, leading to the expression of defence genes.

Some AMPs contribute to the establishment of SAR, a long-lasting immune response that provides broad-spectrum protection against a variety of pathogens. Peptides like LTPs are involved in signalling between local infection sites and distal tissues, priming the entire plant for future attacks.

11.13 Transcription factors

Transcription factors (TFs) are essential regulators in the complex signalling networks that govern plant immune responses. These TFs help plants detect pathogens and rapidly alter gene expression to defend against microbial invasion [116–119].

11.13.1 WRKY transcription factors in plant immunity

WRKY transcription factors play a pivotal role in both basal and inducible plant defences, regulating immune responses through transcriptional control of defence-related genes. WRKY proteins contain a conserved WRKYGQK domain and typically bind to W-box elements in the promoters of their target genes. WRKY TFs form a complex regulatory network, integrating signals from biotic and abiotic stress pathways.

PTI is the first line of defence, where PRRs on the plant cell surface detect PAMPs. WRKY TFs like WRKY33 and WRKY70 are rapidly induced after PRR activation, amplifying the defence response. During ETI, WRKY TFs are involved in HR cell death, which confines the spread of pathogens.

WRKY33 plays a significant role in mediating resistance to necrotrophic fungi such as *B. cinerea* by regulating jasmonate (JA) signalling and activating the biosynthesis of camalexin, a phytoalexin.

WRKY transcription factors are heavily involved in hormonal cross-talk between SA, JA, and ET pathways. WRKY70, for example, acts as a positive regulator of SA-dependent immunity while repressing JA signalling, fine-tuning the trade-off between biotic stress responses and growth regulation.

WRKYs are part of self-regulating loops to prevent runaway immune responses. For example, WRKY28 activates genes that enhance defence, but it also triggers negative regulators that mitigate the immune response to avoid excessive energy costs. WRKY6 negatively regulates cell death, ensuring localized immune responses do not compromise the plant's growth.

11.13.2 NAC transcription factors: coordinators of stress and immunity

NAC TFs, named after the NAM, ATAF, and CUC gene families, are multifunctional and play significant roles in plant immunity and abiotic stress responses. NAC proteins bind to cis-elements in the promoters of defence genes, modulating immune responses.

Several NAC TFs regulate HR, a programmed cell death mechanism that restricts pathogen growth. For instance, NAC019 induces the expression of ROS-producing genes, driving localized cell death at infection sites.

OsNAC4 in rice regulates HR during bacterial infections, balancing between immune activation and limiting unnecessary cell death.

NAC TFs modulate cross-talk between SA and JA pathways to ensure appropriate immune responses. They are also involved in abiotic stress responses, such as drought and salinity, which affect plant immunity. By integrating signals from various stresses, NAC proteins optimize the balance between growth and defence.

Some NAC TFs recruit chromatin remodelling factors to modify the expression of defence genes. For example, stress-induced acetylation of NAC promoters leads to increased expression under pathogen attack, enhancing immunity.

11.13.3 MYB transcription factors: secondary metabolism and structural defence

MYB TFs are key regulators of secondary metabolites like phenolics and flavonoids, which have antimicrobial properties. MYB TFs also play a role in reinforcing the plant cell wall by regulating lignin biosynthesis. This structural fortification serves as a physical barrier to pathogen entry.

Secondary metabolites, such as flavonoids and alkaloids, accumulate during infections to inhibit pathogen growth. MYB TFs activate the biosynthesis of these compounds in response to microbial invasion.

AtMYB44 in *Arabidopsis* modulates SA-dependent signalling pathways and enhances the production of flavonoids, which have both direct antimicrobial effects and signalling roles in defence.

Increased lignin deposition, regulated by MYB TFs, strengthens plant cell walls and prevents pathogen penetration. MYB85, for instance, enhances lignin production under pathogen stress, providing resistance to *Fusarium* infections.

11.13.4 bZIP transcription factors: key modulators of hormonal pathways

Basic leucine zipper (bZIP) transcription factors are critical regulators of hormone signalling, particularly ABA, SA, and JA pathways. bZIP proteins modulate responses to both biotic and abiotic stresses, coordinating immune responses with growth and environmental adaptation. bZIP TFs help plants balance immunity with environmental stress responses, such as drought and salt stress. By fine-tuning hormone levels, these TFs ensure that plants can manage multiple challenges simultaneously without compromising survival.

- **bZIP60:** Involved in the unfolded protein response (UPR), bZIP60 enhances resistance against viral infections by managing protein folding in the endoplasmic reticulum.
- **bZIP10:** Activates SA-responsive defence genes, contributing to SAR.

11.13.5 Mechanisms of transcriptional regulation in plant immunity

Transcription factors serve as hubs for hormonal interactions, integrating signals from SA, JA, and ABA pathways. SA is typically associated with resistance to biotrophic pathogens, while JA and ET responses are effective against necrotrophs. WRKY, NAC, and bZIP TFs coordinate these responses to ensure efficient immunity.

WRKY70 positively regulates SA responses while repressing JA signalling, ensuring that immune responses are appropriate for the type of pathogen encountered.

Many TFs recruit chromatin modifiers to activate or repress defence-related genes. Histone acetylation, methylation, and DNA methylation are epigenetic marks that determine the accessibility of transcriptional machinery to immune genes. These dynamic changes allow plants to respond rapidly to pathogens and adapt to changing environmental conditions.

11.14 Conclusions

In conclusion we expect that in the next few years we will see a substantial increase in our understanding of the processes of MAMPs perception and signal transduction in plants through the deployment of cross disciplinary approaches and ever-expanding ranges of molecular experimental tools. The diversity and extent of MAMPs/PAMPS are relatively poorly understood, despite their crucial function in immunity. A few numbers of MAMPs/PAMPs as described have been the subject of the majority of studies. The discovery of novel MAMPs/PAMPs will shed light on the molecular and evolutionary processes that underlie host–pathogen interactions, and a better knowledge of the ways in which these molecules trigger defence responses could significantly enhance plant health and disease resistance.

Increased PRR synthesis and R protein integration in plants through engineering can improve plants' capacity to recognize microbes. When the integrated proteins from several NLRs—which identify distinct effectors—are combined into a single NLR, many disease-resistant plant kinds can be produced. Other commonly used tactics that can cause resistance by acting against the pathogen include the overexpression of inhibitor proteins, intermediate elements of defence signalling pathways, hydrolytic enzymes that target the microbial cell wall, and secondary metabolites that function as antimicrobial components.

References

[1] Ngou B P M, Ding P and Jones J D G 2022 Thirty years of resistance: zig-zag through the plant immune system *Plant Cell* **34** 1447–78

[2] Jones J D G and Dangl J L 2006 The plant immune system *Nature* **444** 323–9

[3] Zipfel C 2014 Plant pattern-recognition receptors *Trends Immunol.* **35** 345–51

[4] Zipfel C, Kunze G, Chinchilla D, Caniard A, Jones J D G, Boller T *et al* 2006 Perception of the bacterial PAMP EF-Tu by the receptor EFR restricts agrobacterium-mediated transformation *Cell* **125** 749–60

[5] Boutrot F and Zipfel C 2017 Function, discovery, and exploitation of plant pattern recognition receptors for broad-spectrum disease resistance *Annu. Rev. Phytopathol.* **55** 257–86

[6] Restrepo-Montoya D, Brueggeman R, McClean P E and Osorno J M 2020 Computational identification of receptor-like kinases 'RLK' and receptor-like proteins 'RLP' in legumes *BMC Genomics* **21** 459

[7] Macho A P and Zipfel C 2014 Plant PRRs and the activation of innate immune signalling *Mol. Cell* **54** 263–72

[8] Fischer I, Diévart A, Droc G, Dufayard J F and Chantret N 2016 Evolutionary dynamics of the leucine-rich repeat receptor-like kinase (LRR-RLK) subfamily in Angiosperms *Plant Physiol.* **170** 1595–610

[9] Schweizer P, Jeanguenat A, Mösinger E and Métraux J P 1994 Plant protection by free cutin monomers in two cereal pathosystems *Current Plant Science and Biotechnology in Agriculture Advances in Molecular Genetics of Plant–Microbe Interactions: Vol 3 Proc. of the 7th Int. Symp. on Molecular Plant–Microbe Interactions (Edinburgh)* M J Daniels, J A Downie and A E Osbourn (Dordrecht: Springer) pp 371–4

[10] Walker J C 1994 Structure and function of the receptor-like protein kinases of higher plants *Plant Mol Biol.* **26** 1599–609

[11] Yu X, Feng B, He P and Shan L 2017 From chaos to harmony: responses and signalling upon microbial pattern recognition *Annu. Rev. Phytopathol.* **55** 109–37

[12] Yu T Y, Sun M K and Liang L K 2021 Receptors in the induction of the plant innate immunity *Mol. Plant Microbe Interact.* **34** 587–601

[13] Yuan M, Jiang Z, Bi G, Nomura K, Liu M, Wang Y *et al* 2021 Pattern-recognition receptors are required for NLR-mediated plant immunity *Nature* **592** 105–9

[14] Gubaeva E, Gubaev A, Melcher R L J, Cord-Landwehr S, Singh R, El Gueddari N E *et al* 2018 Slipped sandwich' model for chitin and chitosan perception in *Arabidopsis Mol. Plant Microbe Interact.* **31** 1145–53

[15] Cao Y, Liang Y, Tanaka K, Nguyen C T, Jedrzejczak R P, Joachimiak A *et al* 2014 The kinase LYK5 is a major chitin receptor in Arabidopsis and forms a chitin-induced complex with related kinase CERK1 *eLife* **3** e03766

[16] Cheval C, Samwald S, Johnston M G, de Keijzer J, Breakspear A, Liu X *et al* 2020 Chitin perception in plasmodesmata characterizes submembrane immune-signalling specificity in plants *Proc. Natl Acad. Sci.* **117** 9621–9

[17] Wan J, Zhang X C, Neece D, Ramonell K M, Clough S, Kim S yong *et al* 2008 A LysM receptor-like kinase plays a critical role in chitin signalling and fungal resistance in *Arabidopsis Plant Cell* **20** 471–81

[18] Zhang B, Ramonell K, Somerville S and Stacey G 2002 Characterization of early, chitin-induced gene expression in rabidopsis *Mol Plant-Microbe Interact.* **15** 963–70

[19] Danna C H, Millet Y A, Koller T, Han S W, Bent A F, Ronald P C *et al* 2011 The *Arabidopsis* flagellin receptor FLS2 mediates the perception of Xanthomonas Ax21 secreted peptides *PNAS* **108** 9286

[20] Chinchilla D, Bauer Z, Regenass M, Boller T and Felix G 2006 The *Arabidopsis* receptor kinase FLS2 Binds flg22 and determines the specificity of flagellin perception *Plant Cell* **18** 465–76

[21] Chinchilla D, Zipfel C, Robatzek S, Kemmerling B, Nürnberger T, Jones J D G *et al* 2007 A flagellin-induced complex of the receptor FLS2 and BAK1 initiates plant defence *Nature* **448** 497–500

[22] Zipfel C, Robatzek S, Navarro L, Oakeley E J, Jones J D G, Felix G *et al* 2004 Bacterial disease resistance in *Arabidopsis* through flagellin perception *Nature* **428** 764–7

[23] Vetter M M, Kronholm I, He F, Häweker H, Reymond M, Bergelson J *et al* 2012 Flagellin perception varies quantitatively in *Arabidopsis thaliana* and its relatives *Mol. Biol. Evol.* **29** 1655–67

[24] Eckardt N A 2008 Chitin signalling in plants: insights into the perception of fungal pathogens and rhizobacterial symbionts *Plant Cell* **20** 241

[25] Hayafune M, Berisio R, Marchetti R, Silipo A, Kayama M, Desaki Y *et al* 2014 Chitin-induced activation of immune signalling by the rice receptor CEBiP relies on a unique sandwich-type dimerization *Proc. Natl Acad. Sci.* **111** E404–413

[26] Li G B, Liu J, He J X, Li G M, Zhao Y D, Liu X L *et al* 2024 Rice false smut virulence protein subverts host chitin perception and signalling at lemma and palea for floral infection *Plant Cell* **36** 2000–20

[27] Sánchez-Vallet A, Mesters J R and Thomma B P H J 2015 The battle for chitin recognition in plant–microbe interactions *FEMS Microbiol. Rev.* **39** 171–83

[28] Yamada K, Yamaguchi K, Yoshimura S, Terauchi A and Kawasaki T 2017 Conservation of chitin-induced MAPK signalling pathways in rice and Arabidopsis *Plant Cell Physiol.* **58** 993–1002

[29] Martin G B, Bogdanove A J and Sessa G 2003 Understanding the functions of plant disease resistance proteins *Annu. Rev. Plant Biol.* **54** 23–61

[30] Yamaguchi T, Yamada A, Hong N, Ogawa T, Ishii T and Shibuya N 2000 Differences in the recognition of glucan elicitor signals between rice and soybean: β-glucan fragments from the rice blast disease fungus *Pyricularia oryzae* that elicit phytoalexin biosynthesis in suspension-cultured rice cells *Plant Cell* **12** 817–26

[31] Klee H J 2002 Control of ethylene-mediated processes in tomato at the level of receptors *J. Exp. Bot.* **53** 2057–63

[32] Umsza-Guez M A, Díaz A B, Ory I de, Blandino A, Gomes E and Caro I 2011 Xylanase production by *Aspergillus awamori* under solid state fermentation conditions on tomato pomace *Braz. J. Microbiol.* **42** 1585

[33] Boller T and Felix G 2009 A renaissance of elicitors: perception of microbe-associated molecular patterns and danger signals by pattern-recognition receptors *Annu. Rev. Plant Biol.* **60** 379–406

[34] Boller T and He S Y 2009 Innate immunity in plants: an arms race between pattern recognition receptors in plants and effectors in microbial pathogens *Science* **324** 742–4

[35] Tanaka K and Heil M 2021 Damage-associated molecular patterns (DAMPs) in plant innate immunity: applying the danger model and evolutionary perspectives *Annu. Rev. Phytopathol.* **59** 53–75

[36] Kunze G, Zipfel C, Robatzek S, Niehaus K, Boller T and Felix G 2004 The N terminus of bacterial elongation factor Tu elicits innate immunity in Arabidopsis plants *Plant Cell* **16** 3496–507

[37] Jamieson P A, Shan L and He P 2018 Plant cell surface molecular cypher: receptor-like proteins and their roles in immunity and development *Plant Sci.* **274** 242–51

[38] Lolle S, Stevens D and Coaker G 2020 Plant NLR-triggered immunity: from receptor activation to downstream signalling *Curr. Opin. Immunol.* **62** 99–105

[39] Monteiro F and Nishimura M T 2018 Structural, functional, and genomic diversity of plant NLR proteins: an evolved resource for rational engineering of plant immunity *Annu. Rev. Phytopathol.* **56** 243–67

[40] Tamborski J and Krasileva K V 2020 Evolution of plant NLRs: from natural history to precise modifications *Annu. Rev. Plant Biol.* **71** 355–78

[41] Maekawa T, Kufer T A and Schulze-Lefert P 2011 NLR functions in plant and animal immune systems: so far and yet so close *Nat. Immunol.* **12** 817–26

[42] Duxbury Z, Ma Y, Furzer O J, Huh S U, Cevik V, Jones J D G *et al* 2016 Pathogen perception by NLRs in plants and animals: parallel worlds *BioEssays* **38** 769–81

[43] van Wersch S, Tian L, Hoy R and Li X 2020 Plant NLRs: the whistleblowers of plant immunity *Plant Commun.* **1** 100016

[44] Cadiou L, Brunisholz F, Cesari S and Kroj T 2023 Molecular engineering of plant immune receptors for tailored crop disease resistance *Curr. Opin. Plant Biol.* **74** 102381

[45] Jacob P, Hirt H and Bendahmane A 2017 The heat-shock protein/chaperone network and multiple stress resistance *Plant Biotechnol. J.* **15** 405–14

[46] Wu C H, Abd-El-Haliem A, Bozkurt T O, Belhaj K, Terauchi R, Vossen J H *et al* 2017 NLR network mediates immunity to diverse plant pathogens *Proc. Natl Acad. Sci.* **114** 8113–8

[47] Ngou B P M, Jones J D G and Ding P 2022 Plant immune networks *Trends Plant Sci.* **27** 255–73

[48] Bent A F and Mackey D 2007 Elicitors, effectors, and R genes: the new paradigm and a lifetime supply of questions *Annu. Rev. Phytopathol.* **45** 399–436

[49] Chisholm S T, Coaker G, Day B and Staskawicz B J 2006 Host–microbe interactions: shaping the evolution of the plant immune response *Cell* **124** 803–14

[50] Dinesh-Kumar S P, Tham W H and Baker B J 2000 Structure–function analysis of the tobacco mosaic virus resistance gene N *Proc. Natl Acad. Sci.* **97** 14789–94

[51] Symons A, Beinke S and Ley S C 2006 MAP kinase kinase kinases and innate immunity *Trends Immunol.* **27** 40–8

[52] Asai T, Tena G, Plotnikova J, Willmann M R, Chiu W L, Gomez-Gomez L *et al* 2002 MAP kinase signalling cascade in Arabidopsis innate immunity *Nature* **415** 977–83
[53] Cristina M, Petersen M and Mundy J 2010 Mitogen-activated protein kinase signalling in plants *Annu. Rev. Plant Biol.* **61** 621–49
[54] Bosamia T C, Agarwal P, Gangapur D R, Mathew P N, Patel H K and Agarwal P K 2024 Transcriptome sequencing of rectretohalophyte aeluropus lagopoides revealed molecular insight of salt stress adaptation *J. Plant Growth Regul.* https://doi.org/10.1007/s00344-023-11222-6
[55] Meng X and Zhang S 2013 MAPK cascades in plant disease resistance signalling *Annu. Rev. Phytopathol.* **51** 245–66
[56] Li L, Nelson C J, Trösch J, Castleden I, Huang S and Millar A H 2017 Protein degradation rate in *Arabidopsis thaliana* leaf growth and development *Plant Cell* **29** 207–28
[57] Pitzschke A, Schikora A and Hirt H 2009 MAPK cascade signalling networks in plant defence *Curr. Opin. Plant Biol.* **12** 421–6
[58] Dodds P N and Schwechheimer C 2002 A breakdown in defence signalling *Plant Cell* **14** S5–8
[59] Pan L and Smet I D 2020 Expanding the mitogen-activated protein kinase (MAPK) universe: an update on MAP4Ks *Front. Plant Sci.* **11** 1220
[60] Taj G, Agarwal P, Grant M and Kumar A 2010 MAPK machinery in plants: recognition and response to different stresses through multiple signal transduction pathways *Plant Signal. Behav.* **5** 1370
[61] Van Gerrewey T and Chung H S 2024 MAPK cascades in plant microbiota structure and functioning *J. Microbiol.* **62** 231–48
[62] Bigeard J and Hirt H 2018 Nuclear signalling of plant MAPKs *Front. Plant Sci.* **9** 469
[63] Tena G, Boudsocq M and Sheen J 2011 Protein kinase signalling networks in plant innate immunity *Curr. Opin. Plant Biol.* **14** 519–29
[64] Hardie D G 1999 Plant protein serine/threonine kinases: classification and functions *Annu. Rev. Plant Biol.* **50** 97–131
[65] Zhou J M and Zhang Y 2020 Plant immunity: danger perception and signalling *Cell* **181** 978–89
[66] Thulasi Devendrakumar K, Li X and Zhang Y 2018 MAP kinase signalling: interplays between plant PAMP- and effector-triggered immunity *Cell. Mol. Life Sci.* **75** 2981–9
[67] Zeng H, Zhu Q, Yuan P, Yan Y, Yi K and Du L 2023 Calmodulin and calmodulin-like protein-mediated plant responses to biotic stresses *Plant Cell Environ.* **46** 3680–703
[68] Perochon A, Aldon D, Galaud J P and Ranty B 2011 Calmodulin and calmodulin-like proteins in plant calcium signalling *Biochimie* **93** 2048–53
[69] Yang T and Poovaiah B W 2002 A Calmodulin-binding/CGCG Box DNA-binding protein family involved in multiple signalling pathways in plants *J. Biol. Chem.* **277** 45049–58
[70] Kang C H, Jung W Y, Kang Y H, Kim J Y, Kim D G, Jeong J C *et al* 2006 AtBAG6, a novel calmodulin-binding protein, induces programmed cell death in yeast and plants *Cell Death Differ.* **13** 84–95
[71] Zvereva A S, Klingenbrunner M and Teige M 2024 Calcium signalling: an emerging player in plant antiviral defence *J. Exp. Bot.* **75** 1265–73
[72] Ryder L S, Dagdas Y F, Kershaw M J, Venkataraman C, Madzvamuse A, Yan X *et al* 2019 A sensor kinase controls turgor-driven plant infection by the rice blast fungus *Nature* **574** 423–7

[73] Jun T, Shupin W, Juan B and Daye S 1996 Extracellular calmodulin-binding proteins in plants: purification of a 21-kDa calmodulin-binding protein *Planta* **198** 510–6
[74] Yang T and Poovaiah B W 2003 Calcium/calmodulin-mediated signal network in plants *Trends Plant Sci.* **8** 505–12
[75] Zielinski R E 1998 Calmodulin and calmodulin-binding proteins in plants *Annu. Rev. Plant Physiol. Plant Mol. Biol.* **49** 697–725
[76] Snedden W A and Fromm H 1998 Calmodulin, calmodulin-related proteins and plant responses to the environment *Trends Plant Sci.* **3** 299–304
[77] Bouché N, Yellin A, Snedden W A and Fromm H 2005 Plant-specific calmodulin-binding proteins *Annu. Rev. Plant Biol.* **56** 435–66
[78] dos Santos C and Franco O L 2023 Pathogenesis-related proteins (PRs) with enzyme activity activating plant defence responses *Plants* **12** 2226
[79] Kaur A, Kaur S, Kaur A, Sarao N K, Sharma D, Kaur A *et al* 2022 Pathogenesis-related proteins and their transgenic expression for developing disease-resistant crops: strategies progress and challenges *Case Studies of Breeding Strategies in Major Plant Species* (IntechOpen)
[80] Bol J F, Linthorst H J M and Cornelissen B J C 1990 Plant pathogenesis-related proteins induced by virus infection *Annu. Rev. Phytopathol.* **28** 113–38
[81] Loon L C, van, Rep M and Pieterse C M J 2006 Significance of inducible defence-related proteins in infected plants *Annu. Rev. Phytopathol.* **44** 135–62
[82] Ebrahim S, Usha K and Singh B 2011 Pathogenesis related (PR) proteins in plant defence mechanism *Plants (Basel)* **12** 2226
[83] Edreva A 2005 Pathogenesis-related proteins: research progress in the last 15 years *Gen. Appl. Plant Physiol.* **31** 105–24
[84] Linthorst H J M and Van Loon L C 1991 Pathogenesis-related proteins of plants *Crit. Rev. Plant Sci.* **10** 123–50
[85] Sels J, Mathys J, Dc Coninck B M A, Cammue B P A and De Bolle M F C 2008 Plant pathogenesis-related (PR) proteins: a focus on PR peptides *Plant Physiol. Biochem.* **46** 941–50
[86] Kitajima S and Sato F 1999 Plant pathogenesis-related oroteins: molecular mechanisms of gene expression and protein function *J. Biochem.* **125** 1–8
[87] Sharma A, Sharma A, Kumar R, Sharma I and Vats A K 2021 PR Proteins: key genes for engineering disease resistance in plants In *Crop Improvement* (Boca Raton, FL: CRC Press)
[88] Jain D and Khurana J P 2018 Role of pathogenesis-related (PR) proteins in plant defence mechanism *Molecular Aspects of Plant–Pathogen Interaction* ed A Singh and I K Singh (Singapore: Springer) pp 265–81
[89] Vanloon L C and Vanstrien E A 1999 The families of pathogenesis-related proteins, their activities, and comparative analysis of PR-1 type proteins *Physiol. Mol. Plant Pathol.* **55** 85–97
[90] Verma D P S and Hong Z 2001 Plant callose synthase complexes *Plant Mol. Biol.* **47** 693–701
[91] Wang B, Andargie M and Fang R 2022 The function and biosynthesis of callose in high plants *Heliyon* **8** e09248
[92] Flors V, Ton J, Jakab G and Mauch-Mani B 2005 Abscisic acid and callose: team players in defence against pathogens? *J Phytopathol.* **153** 377–83

[93] Ellinger D and Voigt C A 2014 Callose biosynthesis in arabidopsis with a focus on pathogen response: what we have learned within the last decade *Ann. Bot.* **114** 1349–58

[94] Li N, Lin Z, Yu P, Zeng Y, Du S and Huang L J 2023 The multifarious role of callose and callose synthase in plant development and environment interactions *Front. Plant Sci.* **14** 1183402

[95] Berens M L, Berry H M, Mine A, Argueso C T and Tsuda K 2017 Evolution of hormone signalling networks in plant defence *Annu. Rev. Phytopathol.* **55** 401–25

[96] Blázquez M A, Nelson D C and Weijers D 2020 Evolution of plant hormone response pathways *Annu Rev. Plant Biol.* **71** 327–53

[97] Pieterse C M J, Does D V D, Zamioudis C, Leon-Reyes A and Wees S C M V 2012 Hormonal modulation of plant immunity *Annu. Rev. Cell Dev. Biol.* **28** 489–521

[98] Ding P and Ding Y 2020 Stories of salicylic acid: a plant defence hormone *Trends Plant Sci.* **25** 549–65

[99] Chen Z, Malamy J, Henning J, Conrath U, Sánchez-Casas P, Silva H *et al* 1995 Induction, modification, and transduction of the salicylic acid signal in plant defence responses *Proc. Natl Acad. Sci.* **92** 4134–7

[100] Kumar V and Almomin S 2018 Plant defence against pathogens: the role of salicylic acid *Res. J. Biotechnol.* **13** 97–103

[101] Klessig D F, Durner J, Noad R, Navarre D A, Wendehenne D, Kumar D *et al* 2000 Nitric oxide and salicylic acid signalling in plant defence *Proc. Natl Acad. Sci.* **97** 8849–55

[102] Shah J 2003 The salicylic acid loop in plant defence *Curr. Opin. Plant Biol.* **6** 365–71

[103] Park C J and Seo Y S 2015 Heat shock proteins: a review of the molecular chaperones for plant immunity *Plant Pathol. J.* **31** 323

[104] Al-Whaibi M H 2011 Plant heat-shock proteins: a mini review *J King Saud Univ. Sci.* **23** 139–50

[105] Blein J P, Coutos-Thévenot P, Marion D and Ponchet M 2002 From elicitins to lipid-transfer proteins: a new insight in cell signalling involved in plant defence mechanisms *Trends Plant Sci.* **7** 293–6

[106] Salminen T A, Blomqvist K and Edqvist J 2016 Lipid transfer proteins: classification, nomenclature, structure, and function *Planta* **244** 971–97

[107] Ng T B, Cheung R C F, Wong J H and Ye X 2012 Lipid-transfer proteins *Pept. Sci.* **98** 268–79

[108] Kader J C 1997 Lipid-transfer proteins: a puzzling family of plant proteins *Trends Plant Sci.* **2** 66–70

[109] Kader J C 1996 Lipid-transfer proteins in plants *Annu. Rev. Plant Physiol. Plant Mol. Biol.* **47** 627–54

[110] Yeats T H and Rose J K C 2008 The biochemistry and biology of extracellular plant lipid-transfer proteins (LTPs) *Protein Sci.* **17** 191–8

[111] Montesinos E 2023 Functional peptides for plant disease control *Annu. Rev. Phytopathol.* **61** 301–24

[112] Marcos J F, Muñoz A, Pérez-Payá E, Misra S and López-García B 2008 Identification and rational design of novel antimicrobial peptides for plant protection *Annu. Rev. Phytopathol.* **46** 273–301

[113] Azmi S and Hussain M K 2021 Analysis of structures, functions, and transgenicity of phytopeptides defensin and thionin: a review *Beni-Suef Univ. J. Basic Appl Sci.* **10** 5

[114] Contreras G, Shirdel I, Braun M S and Wink M 2020 Defensins: transcriptional regulation and function beyond antimicrobial activity *Dev.Comp Immunol.* **104** 103556
[115] Li J, Hu S, Jian W, Xie C and Yang X 2021 Plant antimicrobial peptides: structures, functions, and applications *Bot. Stud.* **62** 5
[116] Seo E, Choi D and Choi 2015 Functional studies of transcription factors involved in plant defences in the genomics era *Brief. Funct. Genom.* **14** 260–7
[117] Ng D W K, Abeysinghe J K and Kamali M 2018 Regulating the regulators: the control of transcription factors in plant defence signalling *Int. J. Mol. Sci.* **19** 3737
[118] Chacón-Cerdas R, Barboza-Barquero L, Albertazzi F J and Rivera-Méndez W 2020 Transcription factors controlling biotic stress response in potato plants *Physiol. Mol. Plant Pathol.* **112** 101527
[119] Wani S H, Anand S, Singh B, Bohra A and Joshi R 2021 WRKY transcription factors and plant defence responses: latest discoveries and future prospects *Plant Cell Rep.* **40** 1071–85
[120] Yang T, Lv R, Li J, Lin H and Xi D 2018 Phytochrome A and B negatively regulate salt stress tolerance of Nicotiana tobacum via ABA–jasmonic acid synergistic cross-talk *Plant Cell Phys.* **59** 2381–93

www.ingramcontent.com/pod-product-compliance
Ingram Content Group UK Ltd.
Pitfield, Milton Keynes, MK11 3LW, UK
UKHW051905100926
8556IPUK00005B/767

9 780750 356718